Swarm Robotics

Heiko Hamann

Swarm Robotics

A Formal Approach

Second Edition

Heiko Hamann
Department of Computer and Information Science
University of Konstanz
Konstanz, Germany

ISBN 978-3-032-10583-7 ISBN 978-3-032-10584-4 (eBook)
https://doi.org/10.1007/978-3-032-10584-4

This Springer imprint is published by the registered company Springer Nature Switzerland AG
The registered company address is: Gewerbestrasse 11, 6330 Cham, Switzerland

Foreword

Swarm robotics is an emerging field of robotics that focuses on the coordination of multiple robots to perform tasks collectively, drawing inspiration from the behavior of social insects such as ants, bees, and termites. This interdisciplinary research field integrates concepts from biology, artificial intelligence, computer science, and engineering to create systems capable of solving challenging problems and performing complex tasks through decentralized decision-making and collaboration. The appeal of swarm robotics lies in its potential to harness the power of many simple robots to achieve sophisticated collective behaviors, thereby addressing challenges that are difficult or impossible for single robots to tackle.

The foundational principles of swarm robotics are rooted in the study of swarm intelligence, which emphasizes the ability of large groups of agents to effectively solve problems through local interactions, decentralized control, and self-organized behavior. These principles imply that individual robots, operating with simple rules and capabilities, can collaborate successfully without the need for a central control system.

In the context of robotics, swarm systems, usually addressed to as robot swarms, typically consist of numerous homogeneous or heterogeneous robots that can locally communicate and interact with one another and their environment. One of the defining characteristics of robot swarms is their decentralized nature, which contrasts sharply with traditional robotic systems that rely on centralized control and hierarchical decision-making. This decentralization leads to several advantages: (i) scalability, as robot swarms can easily scale up or down by adding or removing robots without the need for reprogramming—this scalability is particularly beneficial for tasks requiring flexibility and adaptability; (ii) robustness, thanks to the redundancy inherent in robot swarms: if one or more robots fail, the remaining ones can continue to operate and complete the assigned tasks; (iii) efficiency, as the collective behaviors often lead to increased efficiency in task execution; for example, when searching for an object, robots in a robot swarm can cover a larger area in parallel and find the object faster compared to a single robot working alone; and (iv) flexibility, as a robot swarm, especially if heterogeneous and therefore comprising robots with many different capabilities, can easily adapt to different environments and task requirements.

Despite its advantages, swarm robotics also faces significant challenges. One of the primary difficulties is achieving effective coordination among agents. While simple local interactions can lead to emergent behaviors, ensuring that these behaviors align with the intended objectives of the swarm is complex. Indeed, the design of algorithms that facilitate efficient collaboration and task allocation among robots in the swarm is a critical area of ongoing research, to which *Swarm Robotics: A Formal Approach* by Heiko Hamann gives an important contribution.

After a few introductory chapters that give the reader the necessary conceptual tools at the basis of the swarm robotics research field, the book rigorously explores the principles underlying swarm robotics through the lens of formal methodologies: in fact, the key contribution of this book is its emphasis on formal models. Hamann illustrates how formal models and methods can be employed to design robot swarms that operate effectively within their environments. By leveraging such formal models and methods, researchers and engineers can predict the outcomes of swarm interactions, enhancing the reliability of these systems, which is particularly important for future real-world applications.

The book, that is intended for master or graduate students interested in distributed robotic systems, as well as for researchers that want to acquire knowledge in the field of swarm robotics, is an exciting reading that I strongly recommend.

Marco Dorigo
Institut de Recherches
Interdisciplinaires et de
Développements en Intelligence
Artificielle (IRIDIA)
Université Libre de Bruxelles (ULB)
Brussels, Belgium

Preface to the Second Edition

I chose to write this Preface as a dialogue. Part of the inspiration came from the recent trend of auto-generated podcast interviews, which have had mixed reviews. But even more, I was influenced by the playful and thoughtful conversations in Douglas Hofstadter's book *Gödel, Escher, Bach*, a work I truly admire.

Naturally, the conversation that follows is fictional. I made up the questions and wrote the answers, sometimes using too much irony. Still, everything is based on real motivations and experiences. I wanted to share the story behind this second edition in a way that feels open and thoughtful. Please don't take every word literally, but I hope you'll think about the ideas I share.

Host: Heiko, welcome, and congratulations on the second edition of your book! It has been more than seven years since the first edition was released. What sparked the idea to write a new one?

Heiko: Thanks! Honestly, it's something I'd been thinking about for a while. The field didn't stand still, far from it. I continued to monitor developments, not just in swarm robotics, but also more broadly in AI and robotics. I was taking notes year by year; this is missing from the book, and this is missing from the book...Eventually, it became clear that the first edition was starting to show its age.

Host: What kind of developments pushed you in that direction?

Heiko: The AI landscape has changed dramatically since 2018. We've seen deep learning evolve, reinforcement learning mature, and then this major leap with large language models (LLMs).

Host: Interesting, so developments in centralized AI, like LLMs, actually influence how we think about decentralized systems like swarms?

Heiko: Well, LLMs are mostly driven by natural language processing and profit from these gigantic corpora. In robotics, we often lack access to these beautiful, large datasets. These extreme successes in AI and machine learning are gradually making their way into mobile robotics and, ultimately, swarm robotics. However, that kind of progress inevitably raises questions about how we think of intelligence, centralized versus distributed, designed versus emergent, and those are questions at the heart of swarm robotics.

Host: So even though LLMs aren't swarms, they still affect how we think about swarms?

Heiko: I think so, yes. LLMs represent a different end of the spectrum. Still, they shift the baseline for what we consider 'smart.' And that makes it all the more important to sharpen our understanding of decentralized intelligence and how to design for it.

Host: And that's where this new edition comes in?

Heiko: Right. I wanted to bring the book up to speed, not just by adding more material, but by rethinking parts of it, expanding the theoretical side, and anchoring it more firmly in what we now know.

Host: What kinds of new material are we talking about here? Are there completely new topics in this edition?

Heiko: Yes, quite a few. I've added new sections on machine learning, including dedicated sections on reinforcement learning, inverse reinforcement learning, and evolutionary swarm robotics.

Host: That's interesting. How does that connect to swarms? Learning usually sounds like something an individual does.

Heiko: True, that's part of what makes it interesting. In swarms, we often rely on emergent effects through hand-coded reactive behaviors; however, we can also attempt to have the robots learn these behaviors. We need to think carefully about what 'reward' means in a decentralized system, or we will learn from centralized approaches and later execute them in a decentralized manner.

Host: So you're asking: how do we define the reward function for a swarm?

Heiko: And how local observations, that is, what a single robot sees, can lead to global behaviors, through learning. With inverse reinforcement learning, we can even try to infer collective objectives from observed behaviors.

Host: So instead of writing rules, you let learning or evolution do the work?

Heiko: Yes. The book focuses on the formal approach, but learning and search-based methods can also be powerful. We can't hand-design everything. Evolution and learning allow us to search the space of possibilities.

Host: And how do modern approaches to world models fit into that picture? You mention them in the new edition.

Heiko: World models are becoming increasingly relevant. The idea is that agents, whether robots or learning algorithms, build internal models of their environment and use them to plan or generalize.

Host: So, swarms will soon be able to imagine futures?

Heiko: You could say that. At least swarms could predict consequences, which opens up new avenues for coordinated planning and adaptive behavior. It also connects our field more closely with developments in cognitive robotics and AI.

Host: What are some of the other biggest changes in this new edition?

Heiko: One major addition is a whole new chapter on scalability.

Host: Scalability has always been part of swarm robotics. What's new about your take?

Heiko: True, we always cared about it. However, we have now formalized it much more clearly: we examine the impact of larger swarm sizes, swarm density, and how algorithms and behaviors scale.

Host: And you're talking here about designing the swarm system beforehand. Like calculating optimal swarm densities and predicting communication bottlenecks?

Heiko: Yes, but not only that. I also address the concept of 'online' scalability, meaning robots are aware of swarm density after deployment and can react accordingly at runtime. That's a subtle but important shift.

Host: That sounds quite technical. Why does scalability matter so much?

Heiko: Because it's fundamental. Our swarm models may also impact other systems.

Host: You mean the implications go beyond robotics? Into other fields?

Heiko: I believe the way we now think about scalability in swarm robotics has potential far beyond our field. These ideas could be useful for distributed computing, physics and chemistry, and even future large-scale AI infrastructures. It's a general design challenge and system property, not just a robotics problem.

Host: That's a bold claim. However, it makes sense; swarms are essentially testbeds for scalable systems.

Heiko: Exactly. They let us explore how simple agents can achieve robust global behavior, and how that breaks when you scale things up.

Host: That sounds like a pretty fundamental rethink. Where did that come from?

Heiko: Actually, it started with the first edition of this book. As the author, I was compelled to revisit the basics. And when you do that seriously, you suddenly notice some gaps. Even in things you've worked on for years. That reflection pushed me to develop a more principled approach to scalability.

Host: You also mention more of the fundamental flocking models now. These are a pretty central piece in collective behavior. Was there really something missing from the first edition?

Heiko: Some of these foundational models—like Couzin's and Vicsek's for flocking, hadn't found their way into the first edition.

Host: So now the aim is to be more comprehensive? Cover both biological and robotic models?

Heiko: Exactly. Now that we're expanding the book, I wanted to provide readers with a more comprehensive picture. That means including the fundamental swarm models, flocking dynamics, and the key biological inspirations. I hope that readers walk away with a broader understanding, not just of what works in robotics, but also of where the ideas originated.

Host: So we've covered the classics in modeling collective motion. But you hint that we shouldn't take them as the final word?

Heiko: Yes, exactly. These models have been very influential, shaping both swarm robotics and theoretical biology. But we shouldn't treat them as sacred. In fact, some researchers would consider these models to be incorrect today.

Host: That's quite a statement! Why this harsh verdict? What's changed?

Heiko: A lot of recent findings in neuroscience and biology suggest that animals may not follow simple, egocentric alignment rules. Instead, they might use more complex, world-centered (allocentric) processing. For example, models based on ring-attractor networks show that coherent group behavior can emerge without explicit alignment. These ideas encompass different dynamics and could better explain phenomena such as criticality and selective attention in swarms.

Host: I see, so you also put in more of these biological fundamentals to make researchers in swarm robotics aware. Let's talk tools. Robotics evolves fast. What changed on the technical side?

Heiko: A lot. ROS 2 is a big one. At the time of the first edition, there was only ROS 1, which was never really designed for large multi-robot systems. ROS 2 introduces native support for decentralized, asynchronous communication, making it a viable option for swarms. The only problem is that it is still struggling a lot with scalability.

Host: So the tech finally caught up?

Heiko: In part, yes. We also now have more options for simulation environments and standardized software stacks.

Host: Besides theory, did you also add new examples or case studies?

Heiko: Yes! I am now placing more emphasis on field robotics. One example from my own lab I'm especially excited about involves autonomous sailboats. We now have a scenario on Lake Constance, where small, fully autonomous robotic sailboats operate as a team on open water.

Host: Wait, sailboats? That's not something you see every day in robotics books.

Heiko: True. But that's the point. It's a new challenge space, with difficult conditions, long horizons, and natural disturbances. Swarm robotics in its rawest form. And it's beautiful. There's something about sailboats autonomously coordinating in the middle of a lake that captures the imagination.

Host: While updating the book, the world outside the lab also changed. How did broader geopolitical events affect your thinking, especially the war in Ukraine?

Heiko: That war has been a turning point for robotics. We've seen drones, often small, cheap, and remotely piloted, play a central role. It's deeply unsettling.

Host: Do you feel that swarm robotics is now closer to deployment in conflict scenarios?

Heiko: Russia's invasion of Ukraine has shown how loosely coordinated drone swarms can have a strategic impact. Every researcher in swarm robotics feels this pressure more than ever before. There's dual-use potential at nearly every corner, and one needs to make a decision. By the way, did you know that today is Ukraine's Independence Day?

Host: No, I didn't know that. That adds gravity to the discussion. But not only did the world outside the lab change; you also changed labs temporarily. I heard that you did most of the writing during your sabbatical. Tell us about that journey.

Heiko: Yes, that was the best part! I worked on the book in four amazing places: Boston, Granada, Bogotá, and Cape Town. Each place gave me something different.

Host: Let's hear it, city by city?

Heiko: Sure. First stop: Boston, USA, where I was kindly hosted by Carlo Pinciroli at Worcester Polytechnic Institute. I've visited MIT, Harvard, and Amazon Robotics. The entire region is home to brilliant roboticists who are accomplishing remarkable feats. Being around that kind of energy really helped shape my thinking. Many thanks to Nancy Lynch, Luca Carlone, Andreas Kolling, and Justin Werfel, and special thanks to Carlo Pinciroli for hosting me!

Host: Next Granada, southern Spain?

Heiko: A beautiful, calm city. I had a workplace near the Alhambra. Perfect for rewriting, reorganizing, and tightening up chapters. I was kindly hosted by Hector Garcia de Marina at Universidad de Granada. His team showed me their fixed-wing drones, which were very inspiring, and many thanks!

Host: Next was Cape Town, South Africa, right?

Heiko: In Cape Town, I was kindly hosted by Geoff Nitschke at the University of Cape Town. It was an outstanding balance to work very focused during the week and then go on hiking trips with Geoff on weekends. Many thanks to Geoff for that experience!

Host: And finally Bogotá?

Heiko: I stayed in a lovely apartment with a view over Bogotá. As I typed the final sentence of the book, a hummingbird hovered outside the window, just for a moment, then vanished. Whether it came to mark the occasion, I cannot say. But I took it as a sign that the manuscript was finally done.

Host: Wow, that sounds almost too perfect.

Heiko: Well, to be honest, that was a Herzog-inspired exaggeration, borrowed from his tale about finishing *Of Walking in Ice*, I think it was. I had a wonderful time in Bogota, and I'm very grateful to my host, Felix Mohr. In reality, I worked hard on the book well into the following semester, while teaching, and continued through the summer. As Thomas Mann put it, "A writer is someone for whom writing is more difficult than it is for other people." That line resonated more than once.

Host: Obviously, a long journey across the globe and in writing. You want to name names?

Heiko: O yeah, definitely. I am grateful to my colleagues and collaborators who provided valuable feedback on earlier drafts of several chapters. Naturally, any remaining errors or shortcomings are entirely my own. Special thanks to Marco Dorigo for reviewing the new scalability chapter and for contributing the foreword despite his busy schedule, Matan Yah Ben Zion, who helped me with new material on active matter, Andrew Vardy, who visited me in Konstanz and also reviewed the scalability chapter, Andreagiovanni Reina, who did some last-minute proofreading and discussed the structure of the book, Tanja Kaiser for pointing me to relevant literature on LLMs, my Ph.D. student Julian Petzold, who designed an earlier version of the Millennium Bridge exercise and one of the exercises on collective decision-making, my Ph.D. student Pranav Kedia, for proofreading two chapters, my Ph.D. student Julian Kaduk, for his help with the literature on human-swarm interaction. Many thanks also for granting permission to print new photos in this second edition: Elisabeth Böker, Andreagiovanni Reina, and Mohsen Raoufi. Many thanks to my colleagues in Konstanz who are essential elements in creating a lively and stimulating environment for ideas: Andreagiovanni Reina, Liang Li, Jonas Kuckling, Clemens Bechinger, visiting Bernd Meyer, and everybody at the CASCB research cluster, but especially Iain Couzin and Oliver Deussen. Many thanks to my colleagues and friends in Berlin, especially Pawel Romanczuk, Mohsen Raoufi, David Mezey, Yating Zheng, Palina Bartashevich, and everybody at the SCIoI research cluster, but especially Jens Krause and Oliver Brock. Many thanks to my other colleagues based in Germany working on swarm robotics: Sanaz Mostaghim and Roderich Groß. Thank you all for your continued support in sparking ideas and helping in so many ways. Also, many thanks to all other colleagues and friends who offered feedback or inspiration during the last years. My sincere thanks go to the great team at Springer: Mary E. James, Brian Halm, Pradheepa Vijay, and Kaitlyn Bonomo. And heartfelt thanks to my fantastic team back in Konstanz, who always manage to cheer me up on hard work days and keep everything running while I occasionally disappear into book mode.

Host: Final question: What do you hope readers take away from this second edition?

Heiko: That swarm robotics is evolving faster than ever, and it is becoming one of the most globally relevant technologies of our time. Whether you're a student, a researcher, or simply curious about how coordination works at scale, I hope this book offers not just answers, but also questions worth exploring.

Konstanz, Germany — Heiko Hamann
August 2025

Preface to the First Edition

My major motivation to think about swarm robotics is the question of how probabilistic local actions of small robots sum up to rational global patterns shown by the swarm. For the engineer, it has potential to let a dream come true because complex problems may be solved by designing simple collaborating components. For the scientist, it has potential to help explain grand questions about consciousness, human societies, and the emergence of complexity.

Swarm robotics is a maturing field that deserves a book being dedicated to it completely. Until now there were many interesting books also treating swarm robotics besides other subjects, but there is no fully dedicated book yet. As a resource, especially for young researchers and students, a book seems important. I am trying to fill this gap. This book tells the story of designing maximally scalable and robust robot systems, which turns out to be challenging with today's methodology. Therefore, join me on our journey to find better and novel approaches to designing decentralized robot systems.

Like many other books, this one has a quite long history of origins. In 2013, I started to teach a quite special master's course for computer scientists called "swarm robotics" at the University of Paderborn, Germany. When I was creating the course, it was particularly hurtful not to have a book that teaches swarm robotics in a complete piece. Of course, there are books that contain material relevant to swarm robotics, such as the books on swarm intelligence by Bonabeau et al. [91] and Kennedy and Eberhart [419] and the great book on bioinspired AI by Floreano and Mattiussi [247]. However, they do not define a complete curriculum. So I had to go through what any teacher has to do when creating a course from scratch: define a canon of the most relevant subjects that each student needs to know and create all the teaching material. Probably back then I already had in mind that writing a book may help the field and be a reasonable option. This was certainly confirmed when students were desperately asking for accompanying reading material and I had to keep telling them: "Yes, that would be great but nobody has yet written that book on swarm robotics." So I started the long writing process slowly back in 2013 and students kept asking me for that book over the next years. In the years between 2013 and 2016, I taught that course four times and was able to improve the material. During the main writing period in

2017, I had moved on to another professorship in Lübeck, Germany; meanwhile, I was happy that I often could just write down what was already in my head from teaching the course. However, for a more complete picture of swarm robotics, it was necessary to add much more material going clearly beyond the content of a course. One can tell from the references in this book that I tried hard not to leave out relevant papers. It is still possible, however, or rather quite likely, that I have overlooked something or somebody maybe very relevant to swarm robotics. In that case, please accept my apologies and please let me know about it.

I hope that this book will help at least a few lecturers out there, in the case that they teach either a course fully dedicated to swarm robotics or a course that partially contains material on swarm robotics. You should not go through the same pains that I had when preparing a course out of nothing. Hopefully, also students find it useful, easy to follow, and maybe even a bit entertaining. Young researchers, who want to study the field and to do research on swarm robotics, will hopefully also find some useful information in here as a starting point. At least the problem of overlooking an important paper may be less likely. Finally, I also hope for enthusiastic swarm robotics hobbyists, who dare to look into such a rather scientific book with all that hard-to-follow terminology. Again, I have tried hard to make this text accessible pretty much to everybody or at least to those who know a bit about computer science or (computational) biology.

This book, of course, is only possible because there were many people who either helped me directly with this book or indirectly by discussing swarm robotics with me over the last decade or by letting me know about relevant new or old papers. I want to thank Marco Dorigo, Thomas Schmickl, Payam Zahadat, Gabriele Valentini, Yara Khaluf, Karl Crailsheim, Ronald Thenius, Ralf Mayet, Jürgen Stradner, Sebastian von Mammen, Michael Allwright, Mostafa Wahby, Mohammad Divband Soorati, Tanja Kaiser, Eliseo Ferrante, Nicolas Bredeche, Sanaz Mostaghim, Jon Timmis, and Kasper Støy. In addition, I want to thank my students attending the swarm robotics course in Paderborn between 2013 and 2016 for asking so many interesting questions, for sharing their solutions to problems of swarm robotics, for motivating me with their enthusiasm, and for challenging me with skeptical questions.

My sincere thanks go to my contacts at Springer Science+Business Media for their patience and for pushing me in the right moment: Mary E. James, Rebecca R. Hytowitz, Murugesan Tamilsevan, and Brian Halm.

Many thanks also for granting permissions to print photos: Thomas Schmickl, Francesco Mondada, Farshad Arvin, José Halloy, Ralf Mayet, Katie Swanson, and Martin Ladstätter.

Lübeck, Germany Heiko Hamann

Competing Interests The author has no competing interests to declare that are relevant to the content of this manuscript.

Contents

Chapter 1
Introduction to Swarm Robotics

But it was one thing to release a population of virtual agents inside a computer's memory to solve a problem. It was another thing to set real agents free in the real world.
—Michael Crichton, Prey

The flying swarm is immediately sent into the "cloud-brain" formation, and its collective memory reawakens.
—Stanisław Lem, The Invincible

In fact, the colony is the real organism, not the individual.
—Daniel Suarez, Kill Decision

Abstract We introduce fundamental concepts of swarm robotics and provide a little overview. Swarm robotics is a complex approach that requires an understanding of how to define swarm behavior, whether there is a minimum size of swarms, and what the requirements and properties of swarm systems are. We define self-organization and develop an understanding of feedback systems. Swarms do not necessarily need to be homogeneous but can consist of different types of robots, making them heterogeneous. We also discussed the interaction of robot swarms with human beings as a factor.

Swarm Robotics

Swarm robotics is the study of how to make robots collaborate and collectively solve a task that would otherwise be impossible to solve by a single one of these robots.

This true collaboration is the beauty of swarm robotics; it is the ideal of teamwork. Every swarm member contributes equally, and everyone shares the same higher-level objectives. Public perception of swarm robotics seems to overemphasize the uncanny aspects, though. Maybe you have read Stanisław Lem's *The Invincible*, Michael Crichton's *Prey*, Frank Schätzing's *The Swarm*, or Daniel Suarez' *Kill Decision*. Maybe you have wondered why most of these robot swarms or natural swarms are evil or at least threatening. Unfortunately, these writers imagine our swarm future as

H. Hamann, *Swarm Robotics*,
https://doi.org/10.1007/978-3-032-10584-4_1

a Dystopia (i.e., a frightening anti-utopia). An aspect that should not be left undiscussed, but we leave it for later when we conclude the chapter with a discussion of future technology. Maybe you have also wondered whether the assumptions of Lem, Crichton, Schätzing, or Suarez are actually realistic. The main question is how robot swarms can be implemented, and that is what we will study in this book.

While swarm robotics, with its immediate application of collaborating robots, gives you many robots that are at your service, it also helps to develop possibly even more critical methods for the future of engineering. Today's technology has reached high levels. Among the most complex systems human beings have ever created are CERN's Large Hadron Collider and NASA's space shuttle. Other complex systems that have been engineered are CPUs (exponential increase of transistors per chip over time) and state-of-the-art (monolithic) robots [67, 506]. Common to these systems is that they are almost impossible to understand by a single person because they contain many interacting components, and many of these components are complex within themselves. To sustain command over the complexity of engineered products. Also in the future, we may reach a degree of complexity, when a paradigm shift in our methods is necessary [257]. Is there a method to generate and organize complex systems that does not rely on conquering a vast heterogeneous set of components and sub-components? Is it possible to create complex systems based on simple components? Fortunately, the answer is yes, and we find examples of such systems in Nature. Examples are swarms, flocks, and herds (see Fig. 1.1). Thousands of starlings self-organize in flocks and show collective motion. Shoals of fish move in circles, forming significant, confusing clusters that make it difficult for predators to perceive individual fish. Locusts March in long aggregates through the desert, and ants show the diversity of behaviors that is known to most of us. All these systems are composed of a high number of similar units. Each of these units is relatively simple and has rather few capabilities relative to the complexity of the whole system that they form. These natural swarms embody the extraordinary concept of a complex system that is composed of simple components that interact based on simple rules. From an engineering point of view, it would be appealing to know how to engineer such systems, and from the scientific point of view, it would be appealing to see how these systems operate. These two assignments are the two threads leading through this book, and luckily, they are directly interconnected. If you can build something, then you have really understood it; and if you have understood it, then you can build it.

Getting back to the quotations from Crichton and Lem at the top of this Chapter, it will be an additional challenge to implement the swarm behaviors not in simulated agents but in embodied agents, that is, robots. It will also be challenging to distinguish whether a certain observed behavior is a collective effect (e.g., a collective memory) or a mere individual reaction by a robot itself.

Besides the engineering challenge, there is also a much deeper challenge of uniting the two perspectives that each swarm necessarily has. Some features are only present on the swarm level, and some features are only present on the level of an individual swarm member. As observers, we see the overall shape of the swarm, its dynamics, and we can often understand the purpose of each motion or the architectures that these swarms construct. For the individual robot, ant, or fish, most of these features

Fig. 1.1 Swarms of honeybees (license Creative Commons, CC0), termites (photo by Bernard Dupont, CC BY-SA 2.0), a flock of geese (CC0), a flock of starlings (photo by Nir Smilga, CC BY 2.5), a school of fish (CC0), and a herd of sheep (CC0)

are hidden, and yet it chooses the correct actions most of the time. To understand how the summation of simple behaviors results in qualitatively new features on the swarm level, which cannot possibly be understood by looking at its parts, is the most intriguing problem of swarm intelligence and swarm robotics. Solving this riddle will have far-reaching consequences because it is undoubtedly related to neuroscience, sociology, and physics.

1.1 Initial Approach to Swarm Robotics

1.1.1 What is a Swarm?

It is interesting to notice that there seem to be no explicit definitions of swarms in the literature. Common contributions, for example, from the field of swarm intelligence dare to keep a gap in this concern [91, 419]. Instead, one finds definitions of swarming behavior.

Swarming Behavior

In biology, a common definition of swarming behavior is "aggregation often combined with collective motion."

This largely aligns with our previous examples. A swarm can be seen as an ensemble of individual units that achieves improved system-level performance via information sharing, coordination, and/or cooperation; this is often in pursuit of a public good that would be inaccessible to individuals alone [476]. The importance of swarms as a concept is also communicated in the language itself. For example, in English, there are words for several swarm behaviors, such as flocking in the case of birds, shoaling or schooling in the case of fish, and herding in the case of quadrupeds. Consequently, we owe a definition of a swarm and conclude for now that a swarm is defined by its behavior.

1.1.2 How Big is a Swarm?

A question in swarm robotics that often raises a smile is the question of the appropriate size of a swarm. What seems to be clear is that it has to be many. But how many exactly? Fortunately, in this case, a definition by Beni [72] is found in the literature, although he defines swarm size by what it is not: "not as large as to be dealt with statistical averages" and "not as small as to be dealt with as a few-body problem." Hence, for a proper swarm size N after Beni [72] we get $10^2 < N \ll 10^{23}$ which should be interpreted in the way that a swarm should not be "Avogadro-large." The Avogadro constant is $N_A \approx 6.02 \times 10^{23}$ 1/mol, whereas "mol" is the amount of substance containing as many elementary entities as in 12 g of pure carbon-12. The size of this upper bound might seem unreasonably high at first glance, but it is helpful because it has a qualitative meaning. Avogadro-large systems, such as gases, liquids, or solids, are typically treated statistically.

The lower bound is also defined method-relative so as not to be dealt with as a "few-body" problem. If we investigate the complexity of few-body problems starting from small group sizes, then we can start from just one. A system with only one individual is certainly missing all interesting features, such as communication and

Fig. 1.2 Example trajectory from the three-body problem (Newtonian mechanics, two heavy stationary masses and a satellite in motion) as an example of complexity in "few-body" problems

collaboration. So we should go to two, which is the smallest possible social group, also called a dyad. The transition from one to two individuals changes the system fundamentally [74]. An example of a few-body problem is the two-body problem from classical mechanics. The problem is to determine the motion of two interacting bodies, such as a planet and its moon, based on Newtonian mechanics. Remarkably, the three-body problem shows already a qualitative difference (see Fig. 1.2) because it turns out to be a difficult mathematical problem [57]. Therefore, one might argue that the three-body problem belongs not to the class of simple few-body problems. Consequently, the chosen lower bound of 10^2 might arguably be already too high, while a lower bound of three might also be too extreme.

Interestingly, this problem is related to the sorites paradox (sorites from soros, Greek for heap). It is the question of how many grains a heap must contain to be a heap. Sounds familiar? Although the question sounds silly, it is actually much deeper than one may think. One is tempted to solve the problem with a threshold (e.g., 10,000 grains), only noticing then that it is unfair to deny the 9,999-grain aggregation being a heap. In the end, it is a problem of vagueness in our natural languages that was discussed by such famous players as Russell [688].

The way out of this dilemma is to prevent setting thresholds and to stay with a method-relative definition. A swarm is not necessarily defined by its size but rather by its behavior.

Swarm Size

If it implements a swarm behavior, then it should be considered a swarm and a swarm of $N = 3$ swarm members might show a verifiable swarm behavior.

The properties and requirements of swarm behaviors are discussed in the following section.

1.1.3 *What is Swarm Robotics?*

First ideas on distributed problem solving in a software approach were published, for example, by Smith [737]. These are initial ideas for protocols in multi-agent systems, similar to later auction protocols. Initial concepts for coordinating multiple robots were developed at the Jet Propulsion Laboratory in 1977, later known as *Purple Pigeons* [1]. This was a multi-rover mission concept for Mars, featuring two landers that each deploy two rovers to explore the terrain collaboratively. A mission that never flew.

Early first steps toward controlling multi-robot systems were taken, for example, by Freund [258]. This marks the transition from single-robot control toward coordinated robotic behavior of multiple robots in shared workspaces. The shared working space has a collision risk and requires careful coordination between robot controllers. Multi-robot path finding (MAPF) was immediately identified as a key challenge by Freund and Hoyer [259]. They proposed a cooperative strategy where each robot plans locally but considers global conflict resolution. This approach enables safe navigation in shared environments by hierarchical coordination.

First implementations of mobile swarm robots go back to Maja J. Matarić and her *Nerd Herd*, see Matarić [521–523]. Matarić [522] studied collective robot behavior using simple rules, local sensing, and limited communication, achieving scalability and robustness without central control. Matarić [521] studied basic interaction primitives, such as collision avoidance, following, dispersion, aggregation, homing, and flocking. Experiments involving up to five robots were reported.

Another early work is that of Kube and Zhang [448]. They explore how multiple behavior-based robots can collectively achieve tasks. They define five mechanisms inspired by social insects (e.g., following behavior, external stimuli trigger collective behaviors, detection of close-by robots). They tested their approach in simulations and with a swarm of five mobile robots. Dudek et al. [213] propose a classification system for swarm robot systems based on key dimensions: swarm size, communication range, communication topology, communication bandwidth, swarm reconfigurability, swarm unit processing ability, and swarm composition (homogeneous or heterogeneous). The taxonomy aims to facilitate the comparison of different swarm architectures and highlight their strengths, constraints, and tradeoffs.

Swarm Robotics by Dorigo and Şahin

A definition following Dorigo and Şahin [197] is: "Swarm robotics is the study of how a large number of relatively simple, physically embodied agents [The concept of "agents" is well-defined and complex, see Russell and Norvig [690] and Sect. 3.5.2.] can be designed such that a desired collective behavior emerges from the local interactions among agents and between the agents and the environment."

The term "relatively simple" agents can furthermore be specified task-relatively as relatively incapable or inefficient robots relative to the considered task. The fact that local interactions between agents and the environment should be possible requires robots to have local sensing, and probably also communication capabilities. In fact, (local) communication is often considered a key feature of swarms [102]. Communication is also a requirement to allow for collaboration and cooperation between swarm members. Collaboration, in turn, is required to go beyond a mere parallelization in a swarm system. In the simplest form of parallelization, for example, each processor receives a work package that is then processed without any other required communication between processors until the work package is completed. Although this simple form of parallelization is an option for robot groups (imagine a cleaning task with each robot individually cleaning a small, assigned area), in swarm robotics, we want to clearly go beyond that concept and generate an additional performance boost by collaboration between robots. "Collective behavior" is a technical term common in sociology [282], which refers to spontaneously emerging behavior. The studies on collective animal behavior started later [106]. The term collective behavior is used in swarm robotics research rather loosely. It seems sufficient to read it with the simple meaning "behavior of the whole swarm," that is, the resulting overall behavior of all swarm members. Notably, the terms "cooperation" and "collaboration" are often used interchangeably in the swarm robotics literature. We want to make an effort here to be more precise. We define coordination, cooperation, and collaboration:

Coordination, Cooperation, Collaboration

- **Coordination**: In swarm robotics, coordination is about organizing and synchronizing activities to ensure smooth functioning and to avoid conflicts. So this is the weakest form of interaction among these three terms. The term is also different from the other two as it describes what the robots do instead of how they do it.
- **Cooperation**: When robots cooperate, it is about working side by side toward a shared goal. The communication of robots can be minimal as they tend to work independently. Tasks are divided. Robots make decisions individually about their assigned sub-task. A cooperative task would inherently require a coordinated approach by the robot swarm. Cooperation is suitable for tasks that can be easily divided and assigned. For example, robots divide a search area and each explores independently, with minimal communication.
- **Collaboration**: Robots collaborating are working together interactively to achieve a shared goal. Both tasks and decision-making are shared. Frequent interaction, communication, and information sharing are essential. For example, a swarm of robots collaborates to carry a large object that one robot cannot lift alone. They must synchronize their movements and adjust in real-time.

Beni [72] defines an intelligent swarm as "a group of non-intelligent robots forming, as a group, an intelligent robot. In other words, a group of "machines" capable of forming "ordered" material patterns "unpredictably." This definition requires, however, many other involved concepts such as intelligence, order, and unpredictability, and is not pursued here. Instead, we note his definition of properties of a robot swarm: decentralized control, lack of synchronicity, simple (quasi) identical members or quasi-homogeneous members which are mass produced [72]. Decentralized control is the opposite of centralized control, which is the control of remote units by a central unit.

Decentralized Control

Decentralized control is a paradigm in which each robot controls itself based only on local information and interactions, without being directly commanded by other robots or a central authority.

While a swarm of robots can be described as a distributed system, decentralized control refers explicitly to the absence of a central authority, with each robot acting autonomously based on local information and interactions. The lack of synchronicity means that the swarm is asynchronous, which is the lack of a global clock to which each swarm member could have access. In turn, a swarm has to synchronize if necessary explicitly.

Sharkey [729] discusses different types of swarm robotics. She argues that early approaches to swarm robotics were nature-inspired and "minimalist approaches" that limited the capabilities primarily of the hardware, which seems not to be a consensus in the community anymore. Hence, she identifies *scalable swarm robotics* which is not minimalist and not directly nature-inspired, *practical minimalist swarm robotics*, which is not directly nature-inspired, and *nature-inspired minimalist swarm robotics* which is, if you like, the original.

Zakiev et al. [892] review the evolving terminology in swarm robotics and distinguish it from related fields, such as multi-robot systems, multi-agent systems, and sensor networks. They examine various definitions and identify key properties that define a robotic swarm, including scalability, local sensing and communication, distributed control, homogeneous agents, and individual simplicity or incapability. These criteria seem essential for differentiating swarm robotics from other concepts and need to be respected for consistent terminology.

1.1.4 Why Swarm Robotics?

When the benefits of swarm robotics systems are discussed, typically three main advantages are mentioned [197]. The first advantage is that these systems are robust. Robustness is defined as fault tolerance and fail-safety, and it is achieved by massive redundancy and the avoidance of single points of failure. A swarm is

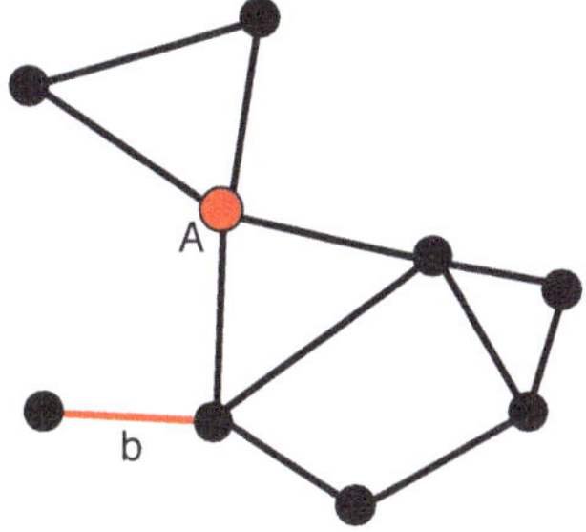

Fig. 1.3 In this graph, node A and edge b represent a single point of failures because if they fail, the graph is separated into several components

quasi-homogeneous and hence each robot can be substituted by any other. Note that redundancy also requires a specific minimal swarm size. Above, we have mentioned that a group of size three could be considered a swarm depending on the methodology. That is still a fair assessment, but it is also clear that a swarm of size three does not have the same level of fail-safety as a swarm of size 100.

The control of the swarm is decentralized and therefore a single-point-of-failure does not exist. Every robot interacts only with direct neighbors, and it also stores only locally obtained data (i.e., local information). Losing a robot due to any failure consequently has only a local effect, which is easily overcome. In an appropriately organized robot swarm, even a certain percentage of robots may be lost without a significant impact on its effectiveness. The efficiency might drop, but the given task is still solved completely.

Robustness and Resilience

> **Robustness**
>
> Robustness is defined as a system's ability to maintain performance despite known, bounded disturbances.

Robustness typically relies on redundancy and fault tolerance. We give an example for a more detailed understanding of robustness. A network of swarm robots can be interpreted graph-theoretically. Say, for the given task, the swarm must stay connected in one large connected component. A situation that is to be avoided to guarantee robustness is that the removal of one edge results in a separation of the graph into two components. The explicit determination of whether a particular link is crucial is non-trivial and complex in the given decentralized setting. An easy solution would be to guarantee a specific minimal robot density (i.e., number of robots per area) that implies a certain average degree (i.e., number of neighboring robots in communication range) of each robot. This is only a probabilistic guarantee, which is, however, typical for swarm robotics systems. Robustness can be considered in graphs. In Fig. 1.3 node A and edge b represent a single point of failure. If A or b

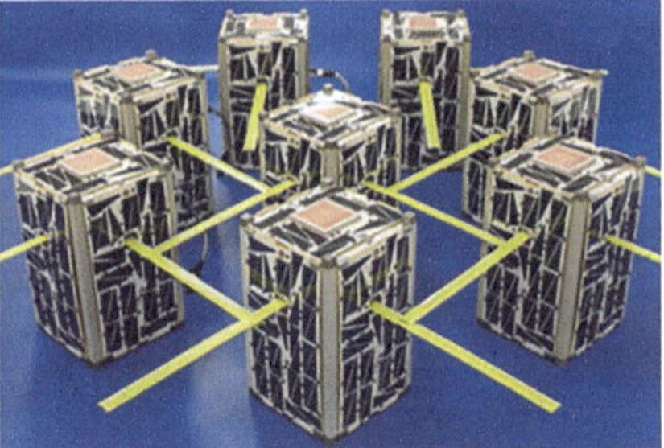

Fig. 1.4 Robustness by redundancy: instead of launching a single space probe (left), one can launch a group of probes or satellites (right, project EDSN—Edison Demonstration of Smallsat Networks) (photos by NASA, public domain)

would fail, then the graph would be separated into several connected components (cf. vertex separator in graph theory). The graph can represent a communication network formed by swarm robots and their communication range. Initially, each robot can reach each other robot either explicitly via multi-hop communication or indirectly by diffusion of information through the network (e.g., sequences of local broadcasts). Once robot A fails or communication along edge b is not possible anymore, the robot swarm collapses into two subpopulations that cannot communicate with each other anymore.

A good example for robustness by redundancy is the following. In today's space flight, typically, single space probes are launched that constitute a central approach (see Fig. 1.4). Whenever a fatal failure occurs in this probe, the whole mission fails. Hence, no effort is spared to perfect these probes and to test them in any possible way. An alternative would be to produce a high number of small and possibly simple probes, in the ideal case, for the same amount of money. The limited capabilities of each swarm probe (e.g., radio signal strength, resolution of the camera) are compensated for by cooperation. Losing a certain percentage of swarm probes would not endanger the whole project because the probe swarm stays effective. A robust system is a reliable system, and a reliable system can be created based on unreliable parts via redundancy [54].

Looking back at the history of swarm robotics, it seems fair to say that the property of robustness was over-promised. Meanwhile, we know better that individual failing robots can have a global impact on the swarm [80, 101]. For example, in a collective motion scenario, a robot with broken locomotion may cause the whole swarm to stop if there is too strong cohesion and no other mechanism that pushes the swarm to leave this one robot behind. Also, adversarial attacks should be considered, where a single robot injects wrong information through communication. Many swarm robotics control algorithms are not robust to such an attack.

A related term that did not appear in early literature on swarm robotics is resilience.

Resilience

Resilience is a system's ability to withstand or recover from unknown and unmodeled disruptions, often through adaptive reconfiguration. It includes the capacity for system-wide transformation to preserve or restore functionality.

Prorok et al. [654] introduce a formal framework and taxonomy for designing resilient multi-robot systems. Resilience describes the ability to adapt to unexpected or unmodeled disruptions, including recovery and reconfiguration at the system level. The authors classify disturbances, identify three operational approaches (pre-, intra-, and post-operative), and apply the resilience framework to perception, planning, and control. By distinguishing resilience from robustness, Prorok et al. [654] highlight the importance of adaptive and system-wide transformations as key to achieving long-term functionality under real-world uncertainty.

New approaches in swarm robotics try to implement actual robustness and resilience, see Sects. 6.6 and 6.6.3.

Flexibility

The second advantage is flexibility. Again, due to quasi-homogeneity, there is no specialization in terms of hardware. Each robot is capable of completing any task that another robot can. The swarm can adapt to a wide range of tasks because often specialization in terms of hardware is not necessary. The robots overcome their limitations in capabilities through cooperation. For example, too weak actuators are compensated for by the collective transport of objects, too low signal intensities are compensated for by forming multi-hop communication lines of robots, and too small dimensions are compensated by self-assembling robots that collectively cross holes or other difficult obstacles.

Scalability

The third advantage is scalability that we will discuss at length in Chap. 2. The applied methods, such as the control algorithms of the robots, scale to any size of the swarm.

Scalability

Scalability enables a robot swarm to "maintain its function while increasing its size without the need to redefine the way its parts interact" [203].

This is because each robot interacts only with its local neighborhood. Methods that do not scale, such as broadcasts to the whole swarm or processes that depend on surveying large fractions of the swarm, are prohibited. For scalability, the guarantee of a constant robot density might be necessary. Still, the overall swarm size is allowed to be chosen freely when the area in which the swarm operates is scaled up accordingly.

An example of the limits of scalability in classical computer systems is Internet services. An increasing number of servers answer a growing number of users. The difficulty is that, typically, some bottleneck exists. For example, incoming service requests might need to be registered at a central machine. As a consequence, response times do not stay constant with increasing number of users, and they also do not scale linearly but exponentially.

1.1.5 What is Not Swarm Robotics?

Sometimes, multi-agent or multi-robot systems can be difficult to distinguish conceptually from swarm robotics systems. Multi-agent systems, however, often rely on the distribution of or access to global information, broadcasts, sophisticated communication protocols (e.g., negotiations) that require a reliable robot–robot communication, explicit assignments of roles that require robots to identify individual robots, or to know the total swarm size [240, 604]. These systems typically do not scale, also because of the applied, non-scalable technologies, such as Bluetooth or WLAN (wireless local area network, aka WiFi). Systems that are not fully distributed and have central elements are typically not considered part of swarm robotics.

1.2 Early Investigations and Insights

Next, we discuss two essential and fundamental features of swarm robotics systems. As the first subject, we discuss communication because swarms sometimes communicate in unexpected ways. They can communicate through the environment, and we distinguish between explicit and implicit communication. As the second subject, we specify the concepts of the swarm level (macroscopic) and the level of the individual swarm member (microscopic). From observing the macroscopic behavior, we should be able to infer the microscopic behavior and vice versa. However, it turns out that this is a fundamental challenge of swarm robotics.

1.2.1 Communication

As discussed above, communication is essential to allow for collaboration in a swarm. In swarm robotics, we distinguish between two kinds of communication: explicit and implicit communication. Explicit communication is the classical concept of communication, that is, there is a communication channel between sender and receiver, and an explicit message is sent. Implicit communication does not have an explicit communication channel or an explicit message. It is also called cue-based communication. For example, by merely perceiving the presence of another robot, a robot might adapt its behavior accordingly. Hence, one can say that a robot communicated implicitly with the other one. In biology, there are two classes of interaction between animals at a distance. First, a cue is an unintentional index. For example, a trail in the snow is left as a cue by one animal, which another animal can then exploit. For instance, it can choose to follow the trail if it is hunting for prey, or it decides to avoid the trail if it wants to avoid approaching it. Second, a signal is an intentional index. For example, an alarm cry of a bird is an intentional warning to its peers.

Another relevant communication concept is that of stigmergy, which is a form of cue-based communication [295].

Stigmergy

Stigmergy is cue-based communication by way of the environment. It is a mechanism of indirect coordination, where agents influence each other by modifying and responding to a shared environment.

An example is pheromone trails in ants. An ant drops pheromone on the ground, which is information for ants that pass this position later. Hence, it can be seen as a form of delayed communication. It is like using the environment as a blackboard where you can leave and edit messages for whoever may pass by later. Other examples of stigmergy operate on other materials that are not exclusively used for communication. For example, building material in nest construction serves as an indicator of where to continue to add material, or in other situations, aggregations of objects serve as messages.

1.2.2 Two Levels: Micro and Macro

Naturally, swarm systems consist of two levels. First, there is a microscopic level (also called local level), which is the level of the individual robot, the level, which is the local view dominated by limited knowledge, limited sight distance, uncertainty, and primitive actions. Second, there is a macroscopic level, which is the level of the whole swarm, the level which is the global view dominated by full knowledge, full overview, at which the overall task is defined. Often, the terms "local" for microscopic and "global" for macroscopic are used. As shown in Fig. 1.5, the micro and macro level figuratively correspond to the microscope used to explore the microscopic worlds (e.g., intracellular communication) and the telescope used to explore the macroscopic worlds (e.g., dynamics of galaxies).

The existence of these two levels implies a grand challenge for the design of swarm robotics systems. The task of the swarm is defined on the macro-level (swarm

Fig. 1.5 Two levels, micro and macro, figuratively microscope (Binocular compound microscope, Carl Zeiss Jena, 1914) and telescope (photo by Azmie Kasmy, CC BY-SA 3.0)

behavior), while the implementation of the robot controller is to be done on the micro-level (executable code for the robot's microcontroller).

Furthermore, there is also a classification of swarm models based on these two terms. Macroscopic models abstract away microscopic details; that is, they do not model all properties of individual robots (e.g., their position, internal state). Microscopic models, in turn, represent each robot explicitly with all necessary details (e.g., complete trajectories of all robots, a robot's state of its internal memory). Sometimes researchers distinguish modeling approaches in Lagrangian and Eulerian [618]. Lagragian refers to microscopic models that represent properties of each individual explicitly (e.g., velocities and internal states). Eulerian refers to macroscopic models that represent only group properties in continuum equations (e.g., density of the swarm with partial differential equations).

Sometimes it is difficult to understand and accept that there really is a macroscopic level in the system. Often, one can break down seemingly macroscopic features into smaller pieces in a reductionist approach, and hence, macro-effects vanish. However, some studies address macroscopic concepts, such as collective perception [711, 804] and collective memory [158].

Collective Perception

In collective perception, objects or environmental features are not recognized by individual robots alone, but emerge from the distributed sensing and interactions of the swarm as a whole.

Each individual robot has only access to a limited amount of information (local information) that is not enough to infer the overall property of the investigated object or the global feature of the environment. The information is strictly distributed within the swarm, and only the whole swarm can perceive and create awareness.

Collective Memory

A collective memory can emerge "where the previous history of group structure influences the collective behaviour exhibited as individual interactions change" [158].

Again, the respective information cannot be localized within the swarm but instead is distributed over all swarm members on the macroscopic level. The "cloud-brain" mentioned in the quotation of Lem at the beginning of this chapter would be another example of a macroscopic feature in the form of collective memory. Still, any particular operation within this memory could be broken down into operations within an individual robot, just as one can do for a regular computer memory. However, the system is only fully grasped when we accept macro-features and when we look at the big picture. This way, the macroscopic level would be only epistemological (i.e., a model assumption that helps to understand the system but is not necessarily physically existent), such that one could still argue that it does not really exist. Yet, that is a philosophical discussion that would go beyond the scope of this book.

1.3 Self-organization, Feedbacks, and Emergence

An essential concept in swarm robotics is self-organization. All robots act based on local perception, while the swarm is supposed to complete a task that is defined on the macroscopic level. The idea of self-organization explains exactly this relation between micro and macro and answers the question of how macro-structures can be generated based on micro-interactions only. In a populist but concise statement, self-organization can be described as a generator of "order from noise." It describes how spatial, temporal, and spatiotemporal patterns emerge in open systems driven away from thermal equilibrium. Open systems permanently "consume" energy and therefore they can generate structures. We cannot create order without wasting energy as described by the second law of thermodynamics. Animals have a metabolism and digest food that contains chemically bound energy. Robots have batteries that provide power to run the motors, CPU, sensors, and actuators.

Self-organization

Self-organization is based on four components: positive feedback, negative feedback, multiple interactions, and balance of exploitation and exploration [91, 580].

Feedbacks introduce dynamical non-linearities to the system, which are necessary to generate complex behaviors. Multiple interactions occur because self-organizing systems consist of many entities that interact by physical contact or by means of direct or indirect communication. The historic origin of self-organization is systems known from the field of physics, that is, non-living systems [315]. For example, the formation of a crystal transforms a gas or a liquid into a solid form. An initial fluctuation (random event) forms a crystal nucleus, which is then reinforced by the crystal growth. Hence, the above statement "order from noise" applies here. The form of the crystal is to a large extent determined by the underlying atomic and molecular characteristics of the material. Still, there is room for variety, as known from the phenomenon of snowflakes. The balance between exploitation and exploration in a self-organizing system makes it adaptive. Even if repeating a certain action seems the only profitable strategy for now (i.e., exploitation), it makes sense to keep exploring, for example, the environment and keep checking for changes. Only if the swarm learns about these changes can it adapt to them and switch its behavior. An example is an ant colony that found a great food source but still keeps exploring alternative food sources. The currently exploited food source may deplete, or a predator may increase the cost of getting there.

The two most important of the four components of self-organization are the feedbacks. They provide a concrete and rich schema to identify and interpret self-organizing systems. Next, we study positive and negative feedback in more detail.

1.3.1 Feedbacks

Feedback is a seemingly simple concept, but not always easy to identify, and sometimes it is challenging to distinguish positive from negative feedback in a self-organizing system. In the following, we restrict ourselves to a simple definition, and later in the book, we try to identify feedback in systems of swarm robotics.

The effect of positive feedback is a force that increases, inflates, and escalates quantities. In self-organization, positive feedback is important because it breaks the initial randomly uniform distribution of system properties (homogeneity, without irregularities). A fluctuation creates initially small deviations that are then reinforced by positive feedback. Examples are the snowball effect, path formation, and a bubble in a financial market. They require only a small initial trigger, a snowflake starting to roll downhill, a human being walking across an unmown hayfield, or rumors about the future profitability of Internet companies, which can then potentially escalate to a process that makes a difference. The economic view[1] is that an individual

> may think there is an opportunity for speculation: to buy the stock now and sell to other buyers at a profit later. [...] the initial increase in demand creates a positive feedback, leading to further increases in demand.

Another example is the insane requirement of permanent growth in capitalism to allow for the payment of interest, which requires economies to grow exponentially. Unfortunately, our earthly resources are limited, which means that even processes

[1] Open-access book "The Economy," http://www.core-econ.org/the-economy/book/text/11.html#118-modelling-bubbles-and-crashes.

that are driven by positive feedback need to stop at some time. Even the biggest snowball rolls to a stop once it has reached the bottom. A path across a hayfield will not attract an infinite number of people. Every bubble ends in a stock market crash. Another positive feedback process is the so-called Matthew effect from sociology. In reference to the Gospel of Matthew, it is the effect of the increasing gap between rich and poor: the rich get richer and the poor get poorer. This effect is, for example, observed in the creative industries, where small initial advantages in the sales figures of a medium generate attention and hence promote further sales. In swarm robotics, we can make use of positive feedback, for example, in collective decision-making, where we want to see a decision on the global level in the form of a consensus (all swarm members agree on one option).

A system that has only positive feedback will explode because there is no limit; that is, it will generate snowballs of infinite size or infinite stock prices. Every system operating on finite resources also has negative feedback components. Negative feedbacks stop positive feedback processes due to limited resources. For example, a positive feedback process that reinforces the majority in opinion dynamics is finally stopped because we are running out of voters who are of the opposite opinion. The effect of negative feedback is a force that reduces deviations, diminishes majorities, and damps overshoots. Examples are control systems such as the centrifugal governor or the baroreflex, which keep the blood pressure constant. An increased blood pressure results in a reduction of the heart frequency, which then decreases the blood pressure. Too low blood pressure, in turn, inhibits the baroreflex, which then increases the blood pressure. The centrifugal governor is a nice piece of engineering that can be used to regulate the rotational speed of a steam engine (see Fig. 1.6). The centrifugal force lifts masses that, in turn, regulate the intake of steam via a valve. In swarm robotics, we can make use of negative feedback, for example, to keep the swarm in a specific equilibrium state or to keep the swarm away from too extreme system states.

1.3.2 Examples of Self-organizing Systems

A dynamic example is Rayleigh-Benard convection. When a relatively flat layer of a fluid (e.g., oil or water) is heated homogeneously from below, flows of warm and cool fluid are generated. Based on initial fluctuations, these flows self-organize into convection cells with an upflow of warm fluid in one half of the cell and a downflow of cold fluid in the other half.

Another example is that of the laser (acronym for light amplification by stimulated emission of radiation), which is maybe counterintuitive. Laser light is a coherent light wave that is created by "pumping" energy into a gain medium (e.g., a gas). From the point of view of self-organization, the atoms in the medium synchronize and emit light in a coordinated way. Hermann Haken contributed a theory of lasers, which led to the development of a theory called "synergetics" [314, 315]. Synergetics is basically a clever method to reduce systems described by multiple differential equations down to

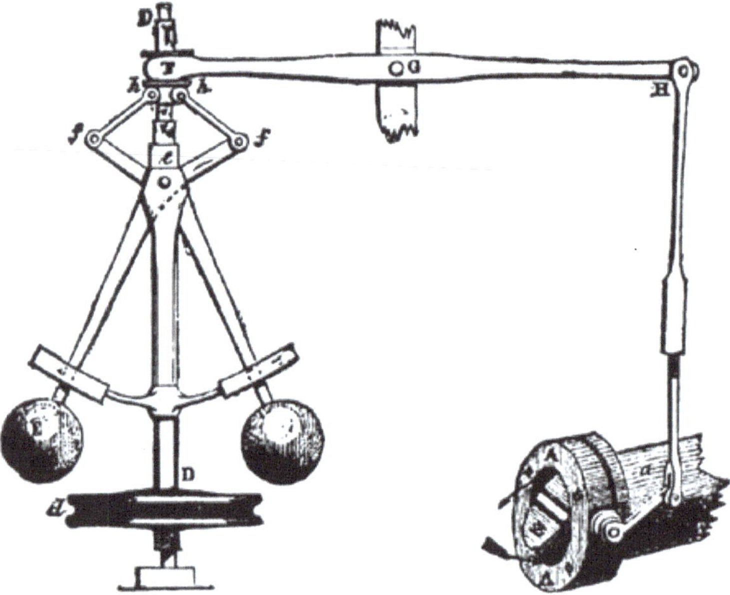

Fig. 1.6 Centrifugal governor as example of negative feedback (public domain)

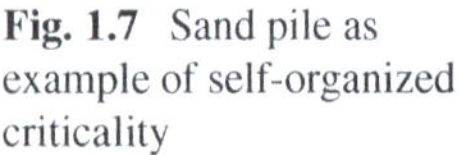

Fig. 1.7 Sand pile as example of self-organized criticality

a few differential equations while still representing the essential system components. Synergetics can also be seen as a theory of self-organization, although it seems not to have high potential for generalization.

A third phenomenon of interest is self-organized criticality, which is, for example, shown in sand piles of increasing size [48]. When a constant flow of sand is added to a pile, the pile increases in its height and also the sides are getting steeper (see Fig. 1.7). The friction between grains functions as a stabilizing force, which is counteracted by gravity, which implies an increasing force due to the growing weight of the pile. In the moment when the force of gravity equals the force of friction, the system is critical. When even more sand is added, this criticality is released by an avalanche

Fig. 1.8 Examples of natural pattern formation: pigmentation of fur and sea shells (photos CC0)

that decreases the steepness of the pile. An interesting characteristic of this system is the distribution of the sizes of avalanches, which is a power-law distribution. Small avalanches are exponentially more likely than big avalanches, which have, however, non-zero probabilities. That is why one speaks of a long-tail distribution. To summarize, we note that energy is added to the system by placing sand grains on top (potential energy), multiple interactions are the physical contacts between neighboring grains, and order is shown in the power-law distributed avalanches, that is, temporally separated events in contrast to the constant flow of added sand.

The behavior of a self-organizing system can also be as simple as seen in systems showing the Brazil nut effect [301, 567]. This is a simple experiment that might entertain the reader while having breakfast if cereals are preferred. Say a box contains a granular medium such as nuts or balls of different sizes. Energy is added to the box by shaking it. Observed is a sorting effect: big grains (or nuts, balls, etc.) end up on top of the box, and small grains are found at the bottom. This effect is generated by small grains filling the gaps beneath bigger grains while these are jumping up due to the shaking process.

Examples of self-organization in living systems are pattern formation and behaviors in insects. Pattern formation is found in developmental processes in animals such as embryology [283, 303, 379, 402, 874], the pigmentation of fur [118, 583], and the pigmentation of sea shells [538, 540, 541] (see Fig. 1.8). The multitude of interactions is implemented by communicating cells, and energy is added via the cells' metabolism. Social insects often show complex behaviors, for example, in foraging or nest building. It seems possible that some simple behaviors in these animals are genetically programmed, that is, they might be hardwired in their brains. It seems, however, unlikely that this is true for many of their behaviors. A more plausible explanation seems that evolution selects behavioral rules that capitalize on principles of self-organization [247]. These simple rules generate the necessary behavior not directly but via interactions between individuals that implement a self-organizing process. This way, the required behavioral complexity is outsourced to some extent into the multiple interactions between many individuals sharing similar genetic material.

Self-organization is, by some authors, even linked to such prominent research questions as that for the origin of life. The chemical origin of life might be based on hypercycles and autocatalytic systems [224, 225]. The evolution of non-cellular life may have been based on adsorption on grains. The evolution of cellular life, in turn, may have been based on the self-assembly of viruses. Other examples of self-organizing systems are percolation [298] and diffusion-limited aggregation [870]. Percolation is a phenomenon observed in various scenarios, including forest fires and the filtering of fluids through porous materials. In a forest, the density of trees influences whether the fire can spread through the entire forest or be stopped. Similarly, the density of holes in a porous material determines its permeability. Diffusion-limited aggregation is basically an aggregation process of particles approaching a seed from many different directions in random walks. Interesting tree structures are formed this way.

1.3.3 Emergence

Emergence is a difficult-to-define and vague philosophical concept. Still, it is of interest for swarm robotics because it might be a promising approach for designing robust and complex systems. It is the idea that totally new properties might emerge on a higher level that concepts of the lower level cannot describe (cf. also holism and reductionism). Many definitions of emergence use concepts in dependence on Aristotle's "the whole is greater than the sum of its parts," surprise of an observer, fundamental novelty, or unpredictability [2, 90, 172, 179, 449, 450, 854, 873]. Holland [362] just throws up his hands in defeat: "It is unlikely that a topic as complicated as emergence will submit meekly to a concise definition, and I have no such definition to offer." Johnson [390] discusses the striking observation of maturing in ant colonies that takes longer than the life of an individual ant: "How does the whole develop a life cycle when the parts are so short-lived? It would not be wrong to say that understanding emergence begins with unraveling this puzzle." Concerning the prediction of emergent systems Johnson [390] refers to Gerald Edelman's experience: "You never really know what lies on the other end of a phase transition until you press play and find out. That is the lesson of Gerald Edelman's recipe for simulating a flesh-and-blood organism: you set up a system of various pattern-recognition devices and feedback loops, connecting the virtual organism to a simulated environment. And then you see what happens." It is that potential unpredictability of such systems that complicates the design of control algorithms for swarm robots.

Researchers from the field of swarm robotics have addressed the problem of defining emergence as well.

Emergence

Bayindir and Şahin [63] state: "Emergence is a key property in complex systems, which means the behavior of the complex system cannot be understood by examining only the components of the system. Although the components of the complex system can be simple, the resultant system may be complex because of the interactions of the system components."

Following Beni [72] and Şahin [167], emergence can also serve as an indicator that a particular control algorithm is consistent with the principles of swarm robotics. Bjerknes et al. [81] claim: "The algorithm meets the criteria for swarm robotics [ ] We have a highly robust and scalable swarm of homogeneous and relatively incapable robots with only local sensing and communication capabilities, in which the required swarm behaviors are truly emergent." Dorigo et al. [200] have a simple definition for ant systems to offer: "Solutions to problems faced by a colony are emergent rather than predefined."

The philosophical concept of emergence goes back to John Stuart Mill [556, 746] and is described as the "concept of genuinely new kinds of properties produced by Nature that cannot be reduced" [721]. An underlying assumption is that there are irreducible properties, for example, in a way that "psychology is not applied biology, nor is biology applied chemistry" [25]. The importance of interactions among many entities was stressed by Prigogine [650]: "As long as we consider merely a few particles, we cannot say if they form a liquid or gas." But once there are many of these particles, we can observe them in different phases (e.g., liquid or gaseous): "Phase transitions correspond to emerging properties" [650]. The question of whether the true novelty of these emerging properties can really exist is again rather philosophical [164, 165]. Prigogine [650] is positive and summarizes his position in poetic language: "Nature is indeed related to the creation of unpredictable novelty, where the possible is richer than the real."

From the programmer's point of view in swarm robotics, emergence can be understood as "programming between the lines." The programmer defines the local behavior of an individual robot via the robot's control program without explicitly defining the emergent, global behavior, which an individual robot cannot perceive. Emergence is an intriguing concept because it constitutes the engineer's dream—the construction of complex systems by an inproportionate small effort.

1.4 Other Sources of Inspiration

Following Şahin [167] there are additional sources of inspiration in addition to the above examples of self-organization. Also Baluška and Levin [53] remind us in their article "On having no head" that cognition does not necessarily require a brain (see Fig. 1.9). Even tissue cells, the cytoskeleton (filaments in the cell that can contract and

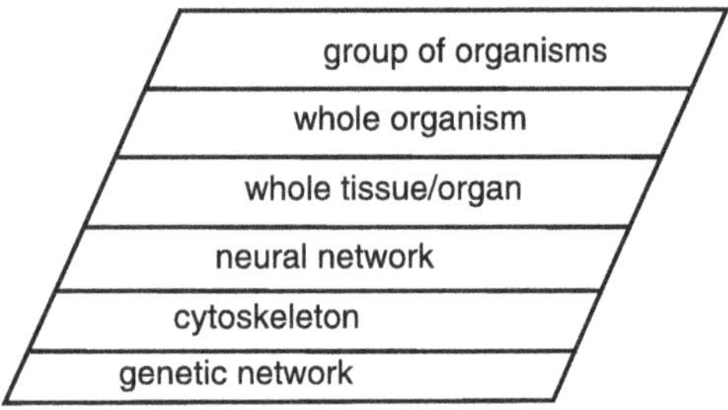

Fig. 1.9 Cognition on different levels in biological systems [53]

influence cell migration), and the underlying genetic network can sense environmental changes and react to them. An important source of inspiration is, for example, the slime mold [4]. The slime mold is an amoeba that can operate in three modes of operation: a single amoeba, a swarm of amoebas in collective motion, and an aggregate of amoebas forming a pseudo-organism. A slime mold amoeba excretes a chemical substance called cAMP in small portions periodically. If the local environment of the amoeba exceeds a given threshold, the amoeba excretes a much higher amount of cAMP molecules (positive feedback). Amoebas walk uphill in their local cAMP field (i.e., toward higher concentrations of cAMP). A trail-like formation toward the center of an aggregation forms. Finally a snail-like pseudo-organism is formed with up to 10^5 cells, which is called the "slug" state of a slime mold.

That way, the slime mold shows modes of operation with different degrees of coupling between amoebae. With a single, isolated amoeba, there is no coupling. The collective motion of amoebas is a typical swarm mode with moderate coupling. The pseudo-organism mode is characterized by an intensive, physical coupling that goes beyond the typical cooperation in swarms. Temporarily, such an intensive coupling is, however, also found in swarms. For example, ants are known to self-assemble themselves into chains of physically connected ants to form bridges between leaves or over small ditches [271].

Another source of inspiration is quorum sensing in bacteria. Similarly to the slime molds, bacteria also show intelligent behavior based on interactions. It is important to note that these unicellular organisms control themselves by different means compared to vertebrates that utilize nerve cells. The behavior of unicellular organisms is generated by intracellular communication that relies on intracellular signals, diffusion processes, and sophisticated processes that implement the interaction between the cell and its environment through its membrane. Bacteria can detect the local density of bacteria. This information is used to coordinate processes that are only efficient if many of the population do them at the same time.

Another potential source of inspiration is "amorphous computing," a general concept was introduced by Abelson et al. [3], Bachrach and Beal [43]. The investigations reported under the name of amorphous computing are focused on finding programming paradigms for a continuous "smart" medium, or rather for a discrete approximation of such a medium. It is composed of immobile particles and can be thought of as being similar to a sensor array (cf. sensor/actuator networks). The approach is biologically inspired by the cooperation of cells in natural multicellular organisms, the hardware is assumed to be unreliable, and the network topology is assumed to be unknown. Essentially, it is research on programming languages that are defined on the continuous medium abstraction, and the challenge is to find compilers that translate these abstract descriptions into executable code for the sensor nodes [881–883].

1.5 Homogeneous and Heterogeneous Swarms

As discussed, the standard robot swarm is supposed to be homogeneous to allow for mass production, high redundancy, and robustness. However, there are also concepts of heterogeneous swarms [196, 212, 417, 652]. In a heterogeneous swarm, the complexity of an individual robot can be further decreased, even though in the homogeneous approach, a robot is already supposed to be simple. Following the heterogeneous approach, the robots can be designed as a small set of different robot types that can then be put together in appropriate sub-sets to enable the group to complete the respective task. Heterogeneity lowers the degree of redundancy, which may reduce robustness if not considered carefully.

The Swarmanoid project [196] presented one of the first large-scale heterogeneous swarm robotics systems. It combined three robot types with complementary abilities: foot-bots for ground mobility and transport, hand-bots for climbing and manipulation, and eye-bots for aerial exploration and sensing. The system employed a shared hardware and software architecture. A dedicated simulator (ARGoS) was extended to support scalable design and testing.

The central demonstration involved a search-and-retrieval task in a human-made environment: eye-bots explored and built a communication backbone, foot-bots transported hand-bots, and hand-bots retrieved objects from shelves. This experiment showed that heterogeneous swarms can operate in complex three-dimensional environments and achieve levels of flexibility and robustness that homogeneous systems cannot provide.

1.6 The Human Factor

Although many applications of swarm robotics focus on ideas such as automation and operating in environments that are dangerous for human beings, the human factor should not be ignored. Robot swarms might operate in environments shared with human beings, or individual robots within the swarm might be controlled by a human remotely. An example study of human–swarm interaction is reported by Kapellmann-Zafra et al. [404]. They investigate how a human operator can steer the robots based on their local view only. A situation that one would face in reality is that no system fuses the sensor data from all robots and creates at least a semi-global view for the human operator.

Given that the public perception of swarm robotics is rather negative because of biased science-fiction stories, it is surprising that there are not many studies investigating the psychological aspect of robot swarms. A positive exception is the study by Podevijn et al. [642] who investigated the influence of robot swarms on the psychophysiological state of humans. They have found that "the number of robots to which a human is exposed has a significant impact on the psychophysiological state of the human and that higher numbers of robots provoke a stronger response." That may be the expected result, and most of us can understand that having many agents (robots, animals, human beings) around ourselves and close to ourselves can be at least unpleasant depending on the context. Hence, we should investigate options for how robot swarms can be designed such that they are not perceived as mischiefs coming in a high number of mischievous entities.

Another aspect of the human factor is the software engineer of robot swarm controllers and required or helpful interfaces between the human engineer and the robot swarms. McLurkin et al. [536] address that issue. Given that one works with possibly hundreds or thousands of robots, it is essential to allow for simple debugging that does not require handling robots individually. The challenge is that, while an individual robot is already complex in itself, the emergent robot group dynamics may be even more complex. A structured, sophisticated design approach is hence even more important.

Finally, there is also the aspect of interfacing between a human supervisor and a robot swarm. Research on Human–Swarm Interaction (HSI) is discussed in Sect. 6.2. HSI refers to the study and design of how humans interact with and influence robot swarms. Unlike traditional human-robot interaction, which deals with single robots, HSI studies the complexities of interacting with large numbers of autonomous robots. A central challenge is that one needs to combine autonomy with human oversight while swarm systems are designed for self-organization. Real-world applications often require human input, especially during deployment, recovery, or in safety-critical situations. HSI enables shared control, where human operators intervene selectively, guiding or correcting swarm behavior while preserving the benefits of distributed autonomy.

Effective HSI depends on the interface through which humans communicate with the swarm. These interfaces can range from visual displays and auditory feedback to tactile systems and more immersive technologies such as virtual and augmented reality. Granularity of control describes the level at which a human operator can intervene in the swarm's behavior. A simple form can be direct control of individual robots (micro-level) but it could also be high-level commands affecting the entire swarm (macro-level). This introduces a tradeoff: fine-grained control can be cognitively demanding and challenging to scale, while high-level control may lead to disengagement or reduced situational awareness.

Due to the complexity and volume of data in a robot swarm, filtering and summarizing information for the human operator is essential. Instead of overwhelming the user with raw sensor data or the complete swarm state, interfaces can aggregate information through abstraction. Event-driven updates and saliency filters help prioritize what the operator sees, focusing attention on anomalies or mission-relevant events.

1.7 Implementations in Hardware and Software

Later in this book, we go through several example scenarios of swarm robotics (see Chap. 5), and we also discuss hardware aspects of swarm robots in some detail (see Chap. 3). For now, we only glance at some examples of scenarios and a few swarm robotics projects. Finally, we shortly discuss the use of robot simulators and potential future applications of swarm robotics.

1.7.1 Example Tasks and Swarm Robotic Projects

Many tasks that robot swarms have completed are based on the motion of robots and the detection of swarm mates. Examples for such tasks are aggregation, self-organization into a lattice formation or dispersion, flocking, mere covering of an area, and exploration tasks. Tasks of higher complexity include the application of additional, special-purpose sensors or necessitate saving acquired information, such as in the deployment of distributed antennas or arrays of mobile sensors, mapping of the environment, and creation of gradient fields (i.e., robots operate as guideposts). In manipulation tasks, special actuators are needed, for example, in foraging behaviors, mining, and containment of oil spills. A particularly interesting kind of tasks includes the interaction with a second group of agents, for example, shepherding, that is, a robot swarm herds a group of animals. In summary, the list of potential tasks is long and not necessarily different from the list of applications that a single robot can solve (see Chap. 5). However, a swarm of robots heavily parallelizes the execution of the task and adds further qualities, such as robustness and scalability.

There have already been many research projects with a focus on swarm robotics during the last 15 years. Here, we shortly summarize the ideas and objectives of three of them. One of the first projects on swarm robotics was the project *Swarm-bots: Swarms of self-assembling artifacts*[2] [201]. It lasted three and a half years from 2001 until 2005. The main objective was to design and control robots that self-organize and self-assemble inspired by swarm intelligence and the behavior of social insects. The *Swarm-bots* were designed as small mobile robots with the special capability of being able to connect physically. They have a small gripper directly attached to the main robot body that can grab a circumferential ring of a nearby robot. This way, the robots can aggregate, physically connect, and still move in the assembled state. One can say that the assembled robots form a bigger robot consisting of autonomous robot modules. An advantage of the assembled, bigger robot is that it can pass steeper

[2] http://www.swarm-bots.org/, funded by the European commission, grant FET IST-2000-31010.

slopes and bigger holes that would be impassable for a single robot (so somehow it implements the vision of the above quote from Lem's *The Invincible*). Another notable showcase was that of 18 swarm-bots aggregating in four lines and pulling a little girl lying on the floor. Here, the idea is that robots can work together and apply bigger physical forces together.

Another interesting mention is the project *I-SWARM: Intelligent Small World Autonomous Robots for Micro-manipulation*[3] [728]. It lasted four years from 2004 until 2008. The very ambitious objective was to build 100 extremely small robots of dimensions $1 \times 1 \times 1\,\text{mm}^3$. These extreme requirements were later released a bit to dimensions $3 \times 3 \times 3\,\text{mm}^3$. The engineering challenge in *I-SWARM* was to integrate novel actuators for locomotion, miniaturized powering with micro solar cells, miniaturized wireless communication, and integrated circuits for an on-board intelligence. Also, here the idea was to control the swarm of tiny robots by methods from swarm intelligence. In the end, the project did not deliver fully functional robots due to limited resources, but the developed technologies, the methodology, and the designs are ready to use.

The third project to mention here is the project *TERMES: Collective construction of 3D structures by termite-inspired climbing robots*[4] [627]. The project was active between 2011 and 2014. The objective of *TERMES* was to develop a swarm construction system (collective construction) based on collaborating robots that pick up building material, transport it, and release it on the construction site to build desired structures. The approach is roughly inspired by the behavior of social termites that build amazingly complex mounds. We will discuss the *TERMES* project in more detail in Chap. 5.

1.7.2 Future Civil Applications

As many other research topics connected to mobile robotics, swarm robotics also has not yet many real-world applications that are already on the market. One of the few exceptions is the application of underwater exploration and monitoring, which is in use in research applications [210, 552]. The breakthrough of mobile robotics applications is expected by many in the following years. This is not the first time in history that a breakthrough is predicted for robotics [272], hence one should carefully observe what is going to happen.

Besides the often recited standard applications of cleaning, monitoring, exploration, and guarding, there is a vast variety of ongoing research projects in swarm robotics, and many of them have high potential for future applications. An example is *Zooids: Building Blocks for Swarm User Interfaces* [433, 465]. The fascinating

---

[3] Funded by the European commission, grant IST FET-open 507006.

[4] http://www.eecs.harvard.edu/ssr/projects/cons/termes.html, funded by the Wyss Institute for Biologically Inspired Engineering, Harvard.

idea is to use small mobile robots on a desk as a user interface. Also, high potential has the idea of self-organization construction with swarm robots [348, 686].

Schranz et al. [718] provide a systematic survey of ongoing research platforms and industrial projects. They show that while no industrial product yet uses fully fledged swarm algorithms, several domains already employ basic swarm behaviors. Example applications that mostly use centralized control include agricultural robots, such as SwarmBot3.0[5] and Fendt Xaver,[6] warehouse logistics systems like Amazon's fleet of 10^6 robots,[7] and Ocado's fleet,[8] environmental monitoring platforms, such as Platypus [387] and subCULTron [780], as well as entertainment applications, such as drone light shows [841] with up to 11,787 drones.[9]

An intriguing set of long-term visions for swarm robotics is the following challenges. In the future, we may face the end of using antibiotics as we know them due to antimicrobial resistance. We may face the end of pesticides due to resistance. Our oceans are littered with millions of tons of plastic. The future of space flight may be the development of robust and scalable swarms of probes.

Nanorobotics [131, 360, 475] is based on the idea of building many micro- or nano-scale robots that operate, possibly even within the human body. They could be used to monitor, to deliver drugs carefully targeted, and could even fight bacteria. Hence, they could be an option once bacterial resistance against antibiotics will be a major problem and we will run out of effective antibiotics. A similar concept, but on bigger length-scales and hence more realistic, would be to use swarm robots in agriculture. Similarly to bacterial resistance, there are so-called super weeds that are created by large-scale applications of pesticides and that are resistant to most pesticides. The application of swarm robotics to agriculture would mean fighting pests mechanically instead of chemically. So we would go back to previous, possibly safer methods; only this time it would not be humans picking the plants and bugs, but many small robots.

Besides everyday cleaning tasks, such as facade cleaning, we also face the massive task of cleaning our oceans, which are littered with massive loads of plastic. Similarly, we will need to clean the space around our planet because even tiny pieces of garbage endanger satellites. Both of these cleaning tasks most likely need to be done by autonomous systems, and both need to be heavily parallelized. There seems to be no chance of doing it in a centralized way, just because of the sheer magnitude of the task. Similar techniques may prove to be useful in space flight, in general, because the robustness and scalability of swarm robotics will make space projects much more reliable and less of a gamble.

[5] https://www.swarmfarm.com.

[6] https://www.fendt.com/at/projekt-xaver.

[7] https://www.aboutamazon.com/news/operations/amazon-million-robots-ai-foundation-model.

[8] https://hbr.org/sponsored/2018/05/how-online-grocer-ocado-is-automating-warehouses-using-swarms-of-robots.

[9] https://www.guinnessworldrecords.com/news/2025/7/spectacular-light-show-featuring-11000-drones-creates-record-breaking-image-in-the-sky.

1.7.3 Future Defense Applications

Одночасно ми вдарили 400 маленькими дронами
We struck simultaneously with 400 small drones.

—*Volodymyr Oleksandrovych Zelenskyy*[10]

What has once been thought can never be taken back.

—*Friedrich Dürrenmatt, Die Physiker*

Besides these promising future applications, there are also unpleasant developments in weaponry. It is clear that a robust and scalable technology is also attractive for the defense sector, and ongoing work focuses on the automation of war [290]. This obvious thought is also reflected in fiction, where swarms of robots appear typically as threats. Unfortunately, there are already many projects and first military applications of robot swarms. This is, however, the curse of technology that there is dual use of it. The Swiss author Friedrich Dürrenmatt has summarized this dilemma in his play *The Physicists*: "Nothing that has been thought, can ever be taken back." One feels tempted to add also: what can be thought, will be thought. Many researchers have taken actions to create awareness and also to research the subject of *ethical robots* [815] and *machine ethics* [184]. Combining swarm robotics with artificial intelligence methods to build fully autonomous robot swarms is a future we must prepare for.

Swarm robotics raises several ethical challenges that go beyond technical design. Following Müller [585], accountability is a central concern: when a swarm causes harm through emergent behavior, it is unclear whether responsibility lies with the designers, operators, or the swarm system itself. Another difficulty is predictability and control, as emergent collective behavior is inherently problematic to certify for safety and reliability.

In the military domain, swarm robotics has attracted significant attention. DARPA's OFFSET program (Offensive Swarm-Enabled Tactics) explores the use of swarms of up to 250 heterogeneous robots (UGVs and UAVs) to support complex urban missions [144]. The focus lies on distributed exploration, coordinated motion, and dynamic task allocation under adversarial conditions. Another prominent example is the Perdix drone swarm, developed by the U.S. Department of Defense, which demonstrated the launch of 103 micro-drones from fighter jets.[11] These drones operate with decentralized consensus mechanisms, enabling aggregation and pattern formation behaviors. Such projects illustrate the strategic interest in scalable, resilient swarm systems, while also clearly indicating the dual-use nature of swarm robotics technologies.

[10] https://interfax.com.ua/news/general/1075341.html, *May 28, 2025.*

[11] https://www.cbsnews.com/news/60-minutes-autonomous-drones-set-to-revolutionize-military-technology/.

Altmann et al. [20] analyzed in the project *Preventive Arms Control for Small and Very Small Armed Aircraft and Missiles* the technological developments, risks, and arms control options for miniature armed uncrewed aircraft and missiles. It documents the rapid progress in this field [18, 19, 631]. These studies highlight how even systems with very low payloads can destabilize international security—mainly when used in swarms—and argue for the implementation of early preventive regulation and verification measures.

Since Russia's full-scale invasion of Ukraine in 2022, the conflict has seen frequent use of coordinated drone attacks resembling rudimentary swarm tactics, waves of inexpensive quadcopters and loitering munitions, such as the *Shahed-136*, have been launched in mass salvos to overwhelm defenses [453, 454]. These operations, however, remain human-controlled attacks rather than true autonomous swarms.[12]

Zelenskyy's above quote refers to an attack of Ukraine's Security Service on June 1, 2025. They carried out a large-scale coordinated drone strike against four Russian airbases located deep inside Russian territory: Belaya (Irkutsk Oblast), Olenya (Murmansk Oblast), Dyagilevo (Ryazan Oblast), and Ivanovo (Ivanovo Oblast). Using 117 first-person view drones (FPVs), the operation targeted Russia's strategic bomber fleet and reportedly destroyed or damaged 41 aircraft, inflicting an estimated 7 billion US dollars in losses.[13] The drones were flown remotely by Ukrainian operators. Each FPV was linked by short-range radio to the launch trucks, which served as forward control stations. The trucks then either relayed signals via cellular or satellite networks to operators outside Russia, or the drones followed pre-programmed/semi-autonomous routes until human pilots took over for the attack. To achieve surprise, they smuggled drones into Russia in trucks with retractable roofs, parked them near the air bases, and launched them remotely from Russian territory, thereby bypassing Russian air defenses. This marked the first Ukrainian drone strike to reach Siberia, demonstrating the reach of FPV tactics. The innovative use of trucked launch platforms and massed FPVs allowed Ukraine to evade electronic warfare, overwhelm defenses, and strike targets far from the front lines.

The war illustrates how rapidly swarm-like approaches can transition from laboratory demonstrations to large-scale armed conflict, underscoring the importance of preventive arms control discussions.

As researchers and engineers, we have to take the responsibility not by trying to prevent progress but by trying to avoid evil applications. Swarm intelligence and its application in the form of swarm robotics have a high potential to be of good use for humankind. We should use the novel technology of swarm robotics to solve a few of our many problems instead of using it to create even new problems.

[12] https://www.csis.org/analysis/russia-ukraine-drone-war-innovation-frontlines-and-beyond.

[13] https://www.understandingwar.org/backgrounder/russian-offensive-campaign-assessment-june-1-2025.

1.8 Further Reading

To learn more about applications of swarm robotics, read Schranz et al. [718] and Schranz et al. [719]. For a discussion of future aspects, read Dorigo et al. [204]. Summaries of swarm robotics research are the review paper by Brambilla et al. [102], a Scholarpedia article by Dorigo et al. [203], and two review papers by Bayindir and Şahin [63] and Bayindir [62]. A good idea to get a quick overview of some state of the art, may also be to read the special issue in *Science Robotics* from 2021 on swarm robotics edited by Floreano and Lipson [246]. It contains, for example, research on self-organizing robotic societies, robots that form superstructures, and the idea of reducing communication actually to enhance robustness in the swarm.

1.9 Exercises

Exercise 1.1 (*Define and Differentiate*) Define swarm, swarming behavior, and swarm robotics. Compare these definitions with those from biology and multi-agent systems.

Exercise 1.2 (*Micro vs. Macro Reasoning*) Give one example of a microscopic (individual-level) and one of a macroscopic (swarm-level) feature in swarm robotics. Do you think the macroscopic level truly exists in the system, or is it just a construct used by observers to explain what they see?

Exercise 1.3 (*Identify Feedback Loops*) Pick one example of swarm behavior (e.g., aggregation, flocking) and identify plausible positive and negative feedback.

Exercise 1.4 (*Emergence Exploration*) Describe an example of emergent behavior (e.g., collective memory or perception). Why can't it be explained by individual behavior alone?

Exercise 1.5 (*Swarm Robustness Evaluation*) Analyze the graph-theoretic view of robustness. What kind of graph topologies best preserve swarm connectivity under failure?

Exercise 1.6 (*Literature Contrast*) Read an excerpt from *The Invincible* or *Prey*. Discuss how fictional representations differ from real swarm robotics.

Exercise 1.7 (*Ethics in Application*) Choose a future swarm robotics application (e.g., nanomedicine, ocean cleaning). Identify three ethical risks and three mitigation strategies.

Chapter 2
Scalability

"Too many robots. Too many robots make too much of everything."

—Frederick Pohl, The Midas Plague

"Adding manpower to a late software project makes it later."

—Fred Brooks, The Mythical Man-Month

"In general it's probably better for them to live separately. They exist in loose swarms...However, they will unite in moments of danger"

—Stanislaw Lem, The Invincible

Abstract We explore swarm robotics scalability, explaining how cooperation–interference tradeoffs and coordination overhead create a concave performance function with an optimal density. Scalability in swarm robotics, defined as maintaining or improving performance with increasing swarm size, is a crucial system capability often ideally pursued through local communication and decentralized control. However, achieving perfect scalability is often hindered by coordination overhead and swarm density, as increasing the number of robots in a fixed operational area generally leads to interference and efficiency losses. Empirical observations frequently show a concave swarm performance function, peaking at an optimal swarm density where cooperation is balanced with resource contention, resulting in various scaling behaviors from linear to retrograde. Computational models from parallel computing, such as Amdahl's Law, Gustafson's Law, and Gunther's Universal Scalability Law (USL), offer frameworks for analyzing these scaling patterns. The SGF (solo, grupo, fermo) model mechanistically links individual robot states to macroscopic swarm performance and can even derive these classical scalability laws as special cases. A critical insight is the prevalence of two-phase performance or bimodal distributions, revealing that operating near optimal density can place a system precariously close to sudden, catastrophic breakdowns due to cusp bifurcations and hysteresis effects, highlighting the importance of designing for robust and online scalability.

H. Hamann, *Swarm Robotics*,
https://doi.org/10.1007/978-3-032-10584-4_2

2.1 Introduction

2.1.1 *Learning Objectives*

After reading this chapter, you should be able to:

- Understand the role of scalability in determining swarm performance across application scenarios.
- Define and distinguish between system size scalability and swarm density, and explain their effects on swarm efficiency.
- Identify the core principles of swarm robotics (local communication, local information, robustness) and their link to scalability.
- Explain the relationship between individual and swarm performance with respect to size and density.
- Recognize different scaling behaviors (linear, sublinear, superlinear, diminishing returns, retrograde) and their causes.
- Analyze the cooperation–interference tradeoff in swarms, where collaboration competes with resource contention.
- Apply computational models of scalability (Amdahl's Law, Gustafson's Law, Universal Scalability Law) and interpret their assumptions.
- Understand the basics of queuing theory (arrival rates, service rates, Markov chains) and its use in swarm analysis.
- Describe the SGF model as a mechanistic view of swarm performance.
- Recognize two-phase performance and bimodal distributions, and their relation to critical density thresholds.
- Draw analogies to traffic flow, biology, and chemistry to contextualize scalability challenges.
- Consider practical implications for swarm design, including online, robust, and deployment scalability, congestion control, and communication technologies.
- Conduct basic analyses and simulations of swarm scaling, such as queuing models or superlinear speedups.

2.1.2 *Guiding Questions*

- What is scalability in swarm robotics, and why is it a critical capability for swarm performance across application scenarios?
- How do system size (N) and swarm density (ρ) influence performance, and what are the key density states (sparse, medium, dense)?
- What is the cooperation–interference tradeoff, and how does it shape the relationship between collaboration opportunities and resource contention?
- Under what conditions can swarms achieve superlinear scalability, and what mechanisms or tasks enable such synergistic effects?

- How do classical models of parallelism, such as Amdahl's and Gustafson's Laws, provide insight into the limits and potential of swarm scalability?
- What does the Universal Scalability Law add to our understanding of swarm performance, especially under contention and coherence delays?
- How does the SGF model deepen our understanding of swarm behavior, and what does it reveal about critical density thresholds?
- What practical implications does scalability research have for swarm design, including online scalability, congestion control, and deployment logistics?
- How do communication technologies and protocols constrain or enable scalability in real-world swarm deployments?
- What broader insights into scalability can be gained by drawing analogies from traffic flow, biology, or chemistry?

2.2 Conceptual Foundations

The most frequently mentioned advantages of swarm robotics are robustness, flexibility, and scalability [72]. Scalability as a system capability is important because it significantly determines the swarm system's performance across many application scenarios. The dogma of swarm robotics binds us to local communication, local information, and to avoid any single point of failure. An approach that follows these ideals should, as a reward, result in a maximally scalable system.

2.2.1 *What Is Scalability?*

Scalability here refers to system size scalability, that is, improving performance or maintaining functionality of the swarm as the swarm size increases. We want to be able to deal with tasks of any size and demand. If there is more to be done, we want to be able to add more robots to complete the task on time. We define scalability as the capability of our robot swarm to stay efficient for any swarm size N (see Table 2.1 for an overview of variables and concepts). Scalability in robotic systems implies that the system can grow in size without incurring substantial efficiency losses or excessive overhead. The same applies to a shrinking swarm whose swarm performance should also not decrease disproportionately. We define scalability on swarm performance Π as $\Pi(N') \geq \eta c \Pi(N)$ for $N' = cN$ with swarm size scaling factor $c > 1$ and accepted overhead $\eta \in [1/c, 1]$. One could argue that the best we can get would be linear scaling: we double the swarm size from N to $N' = 2N$ (e.g., $c = 2$) and the swarm performance also doubles from Π to 2Π ($\eta = 1$). If we allow for some overhead to administrate the additional robots, we choose $\eta < 1$. We don't further specify the concept of swarm performance Π here; for now, we assume that an intuitive understanding is enough (e.g., think of the number of items retrieved by the

Table 2.1 Overview of swarm robotics and computing performance metrics with connections and interpretations

	Concept	Notation	Connections	Description
Swarm	Swarm size	N	–	Number of robots
	Swarm density	ρ	$\rho = N/A$	Robots per unit area
	Operation area	A		Bounded space for swarm operation
	Swarm performance	$\Pi(N)$, $\Pi(\rho)$	Throughput	Global performance of the swarm (e.g., tasks/time)
	Individual performance	π	$\Pi(N)/N$, efficiency	Average effort per robot
	k robots meeting prob.	Γ_k	–	Probability of finding at least k robots in a defined space
Computing	Throughput	X_N	$\Pi(N)$, $\Pi(\rho)$	Tasks completed per time unit
	Speedup (time)	$S_{\text{time}}(N)$	T_1/T_N	Latency reduction compared to serial case
	Speedup (throughput)	$S_{\text{thruput}}(N)$	$X(N)/X(1) = C(N)$	Relative throughput increase also known as relative capacity C
	Efficiency	$E(N)$	$S(N)/N = T_1/(NT_N)$	Normalized speedup or utilization
Overhead	General overhead	γ	–	Coordination and communication overhead
	Contention	α	USL parameter	Shared resource bottlenecks (e.g., space, communication)
	Coherence	β	USL parameter	Synchronization or coordination costs
Special scaling cases	Threshold swarm size	N_m	$\Pi(N_m) > 0$	Minimum swarm size needed for task completion
	Peak performance swarm size	N^*	$\arg\max \Pi(N)$	Swarm size with maximal swarm performance Π^*
	Linear scaling	$\Pi(N) = N\Pi(1)$	$C(N) = N$	Each added robot contributes equally
	Sublinear scaling	$\Pi(N) < N\Pi(1)$	$C(N) < N$	Positive marginal gains, realistic scenario
	Superlinear scaling	$\Pi(N) > N\Pi(1)$	$C(N) > N$	Collaborative amplification effects
	Marginal performance	$\Delta\Pi(N)$	$\frac{d\Pi}{dN}$ or $\Pi(N+1) - \Pi(N)$	Performance gained by adding one more robot
	Diminishing returns	$\Delta^2\Pi(N) < 0$	$\frac{d^2\Pi}{dN^2} < 0$	Saturation effect, second derivative negative, additional robots add less and less but positive marginal value
	Retrograde swarm performance	$\Delta\Pi(N) < 0$	$N > N^*$, $\frac{d\Pi}{dN} < 0$	First derivative negative, additional robots have negative marginal value, swarm performance decreases

swarm in a defined time interval in a foraging task). Also consider down-scaling, that is, studying the effect on swarm performance of reducing the swarm size N to $N' = N/2$. We define downward scalability as $\eta\Pi(N') \geq c\Pi(N)$ for $N' = cN$ $(c = 0.5)$ with swarm size scaling factor $0 < c < 1$ and accepted overhead $\eta \in [1, 1/c]$.

2.2.2 *Swarm Density*

An example for scaling is a foraging task where N robots obtain Π items in a given time period, while $2N$ robots obtain 2Π items in the same time. That would assume zero overhead, that is, to organize and coordinate the N additional robots we don't require any additional resource. For example, if robots share the same space, then we would expect that having double the number of robots would make them lose more time in collision avoidance actions. That means we should speak of swarm density, that is the number of robots per area unit (e.g., number of robots per square meter). A similar concept is known as robot density.

Swarm Density

Swarm density is

$$\rho = \frac{N}{A} \tag{2.1}$$

for a provided area size A of the area where the robots operate and swarm size N.

Notice that in many reported swarm studies, swarm performance is measured based on swarm size often without explicit mention of swarm density. However, in most cases area A is kept constant, such that in effect the swarm size N directly controls the swarm density ρ. An early mention of swarm density in literature is that by Matarić [522].

We should expect that there cannot be perfect scalability with increasing swarm density. In the extreme case, once the whole area is filled with robots they become immobile and cannot complete any reasonable task anymore. Instead, we could require that adding a robot (i.e., $N' = N + 1$) is complemented by adding space (i.e., $A' = A + \Delta A$). This way we could keep the swarm density ρ constant and ensure a positive marginal swarm performance gain $\Delta\Pi = \Pi(N') - \Pi(N) > 0$.

When speaking of swarm density in general, also notice that we can argue that there are three fundamental swarm density states: sparse swarms ρ_{sparse}, medium dense swarms ρ_{med}, and dense swarms ρ_{dense} with $\rho_{\text{sparse}} < \rho_{\text{med}} < \rho_{\text{dense}}$. In a sparse swarm, there is little contention about shared space but also few opportunities to collaborate as robots need to bridge long distances between themselves [429, 774]. In a dense swarm, there is a lot of contention about shared space, that is, robots most of the time try to avoid collisions and block each other [215, 492]. In turn, there is maximal opportunity for collaboration, however, robots cannot make much use of it as they can't move freely. We continue that thought later when introducing the SGF model in Sect. 2.5.2. A medium dense swarm ρ_{med} may balance freedom of movement and opportunities for collaboration best. Hence, there is a maximal swarm performance $\Pi^*(N^*)$ and an optimal (or critical) density $\rho^* = \frac{N^*}{A}$ in most scenarios of swarm robotics.

2.2.3 *Coordination Overhead*

When speaking of a robot swarm of size N showing swarm performance Π, it implies that there is an effort by the swarm that contributes to completing the actual task and an overhead γ that is required to coordinate the N robots to avoid interference in a self-organized way. We may envision an idealized swarm performance Π^{ideal} by the swarm that it would achieve in a perfect, overhead-free scenario. However, since we operate in the real world, we must account for an overhead $\gamma > 0$ and subtract it accordingly. We get for the actual swarm performance

$$\Pi = \Pi^{\text{ideal}} - \gamma. \tag{2.2}$$

This serves rather as a mental model of swarm performance as the idealized swarm performance Π^{ideal} cannot be measured but possibly calculated using an appropriate swarm model or simulation.

2.2.4 *Individual Performance*

Let us also introduce an average individual performance π, which is the average contribution from each robot.[1] We define the individual performance of a swarm robot as $\pi = \frac{\Pi}{N}$. So it includes a fraction γ/N of the overhead as in our distributed system the coordination effort is executed and shared across the swarm. Note two important properties of the individual performance π:

1. The individual performance $\pi(N)$ is a function of the swarm size N. As the swarm size changes, both performance and overhead are affected, which in turn influences π.
2. The maximal individual performance $\pi^*(N^*_{\text{ind}})$ typically occurs at a smaller swarm size than the maximum swarm performance $\Pi^*(N^*)$. That is, we often observe: $N^*_{\text{ind}} < N^*$. This indicates a tradeoff between individual efficiency and total system performance.

[1] Excuse the clash of notation here, as π is usually reserved for an agent's policy in machine learning.

2.2.5 *Key Question and Cooperation–Interference Tradeoff*

The key question of scalability is: If N robots achieve $\Pi(N) = \Pi^{\text{ideal}} - \gamma$, what can cN robots achieve, that is, the performance $\Pi(cN)$ of cN robots ($c > 0$, $c \neq 1$)? For example, if $c = 2$, would we get $\Pi(2N) = 2\Pi^{\text{ideal}} - \gamma$ or rather $\Pi(2N) = 2\Pi^{\text{ideal}} - 2\gamma$? Or even $\Pi(2N) = 2\Pi^{\text{ideal}} - 3\gamma$ or worse? So we are interested in knowing how these overheads scale. One illustrative way to think about this is to start from $\Pi(1) = 1 - 0$, assuming no overhead γ because there is nothing to coordinate. Then, as we increase the number of robots, coordination overhead may accumulate: $\Pi(2) = 2 - \gamma$, $\Pi(3) = 3 - 2\gamma$, and so on. This raises the question: does the overhead grow linearly or superlinearly with N?

The picture painted so far is a bit pessimistic. Having more robots always generates overhead. Is that true? Is there nothing to win by being part of a bigger team? There is luckily also an effect of cooperation on swarm performance. By adding more robots to the system we create also more opportunities to cooperate. Depending on the task, we may require cooperation and collaboration between robots (see Sect. 1.1.3 for a definition). We face a tradeoff between not oversharing resources (e.g., limited room and robot–robot interference) while still providing enough opportunities to cooperate. We discuss this in more detail in Sect. 2.3.3.

2.3 Empirical Observations

2.3.1 *Optimality and a Concave Function*

We expect that there is an optimal swarm density ρ^* which is well-balanced between potential for cooperation and interference and hence shows good swarm performance. A robot system that doesn't allow for enough cooperation may be hampered as it is too competitive; but it could also be too collaborative [320]. For example, if too many robots get recruited for a collective transport task, the system may perform worse than a smaller group of robots. Similarly, information can be undershared or overshared. For example, if information about early success is shared too much, the system may end up lacking explorative power. So we argue that in swarm robotics we are facing tradeoffs in three categories as shown in Fig. 2.1a: (1) utilization of shared resources, (2) degree of cooperation/collaboration, and (3) information flow. A typical system has to deal with several instances from one or more of these categories (e.g., two shared resources and one type of collaboration). They are usually mutually dependent, which causes complex design decisions. Across swarm densities that means that we have too sparse swarms (low swarm densities ρ) where we underuse shared resources, an optimal swarm density where all tradeoffs are well-balanced, and too dense swarms (high swarm densities ρ) where we deplete shared resources.

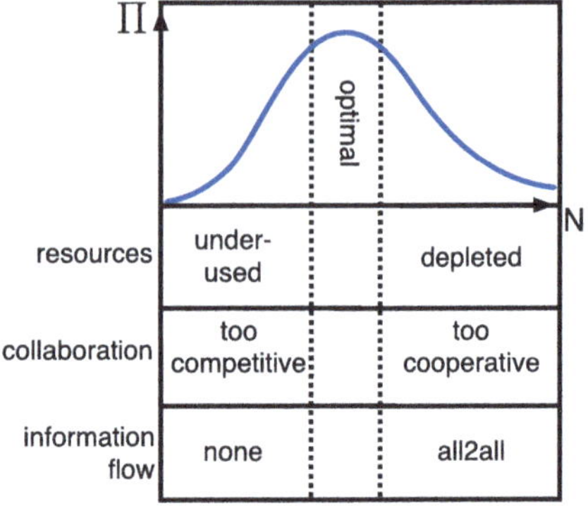

(a) Scalability challenges across swarm sizes for fixed area A (source: [320]).

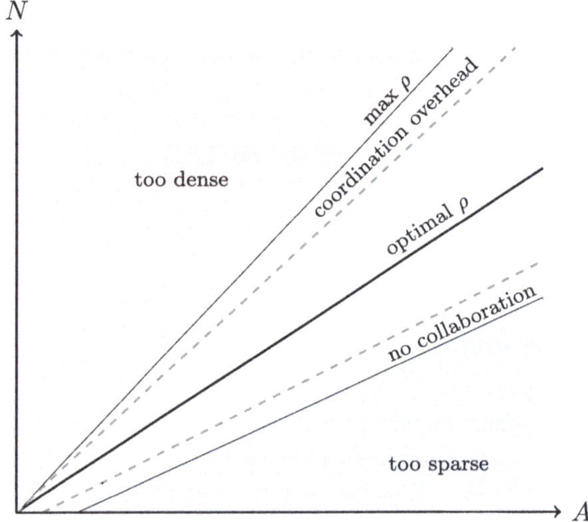

(b) Too sparse swarm, balanced swarm densities, and too dense swarm over swarm size N and operation area A.

Fig. 2.1 **a** Schema of how to interpret scalability challenges for different swarm densities (here as swarm size N assuming fixed area A). **b** Plot of several regions of swarm densities $\rho = N/A$

This is also visualized in Fig. 2.1b. We have at least three regions of swarm densities: too sparse with limited opportunities for collaboration (robots meet rarely), rather balanced or potentially optimal density, and too dense with a lot of coordination overhead (limited by a physically maximally possible robot density). Adding robots to sparse swarms (or reducing operation area) increases swarm performance (positive marginal gain). Adding robots to dense swarms (or reducing operation area) decreases swarm performance (negative marginal gain).

As a result, qualitatively we expect a swarm performance function as given in Fig. 2.1a: increase of swarm performance, optimality, and decrease of performance. That is a concave function with a defined maximum performance Π^* for optimal swarm density ρ^*. Note that we and also most examples from literature often plot these diagrams over swarm size N instead of swarm density ρ. This is almost always done under the assumption of constant operational area A. Plots of swarm performance Π as a function of swarm size N assuming constant swarm density (i.e., non-constant area A) are rare. Also, if the swarm system works properly, the swarm performance should scale well (e.g., linearly). We are more interested in the system's performance when it is challenged, that is, when swarm density is not constant.

In a more refined way, we can divide the robot swarm's performance and swarm density qualitatively into four regions (see Fig. 2.2d): sublinear increase, superlinear increase characterized by collaborations of robots, optimal swarm density (could also be called operating point), and decrease due to depleted shared resources (e.g., interference). The shape of the function shown in Fig. 2.2d occurs in many different swarm systems [65, 317, 330, 345, 427, 429, 464, 479, 682] such that it seems justified to define it as a general scheme for swarm performance [319, 322]. The existence of such a general shape is also reported by Østergaard et al. [607] in the context of multi-robot systems:

> We know that by varying the implementation of a given task, we can move the point of "maximum performance" and we can change the shapes of the curve on either side of it, but we cannot change the general shape of the graph.

In the following we revisit these thoughts by going through the example of a bucket brigade scenario.

2.3.2 *Example: Bucket Brigades*

The average performance of a robot swarm depends on the swarm density or the swarm size if the area in which the swarm operates is kept constant. This is readily visualized by the example of bucket brigades. A bucket brigade is a multistage transport scheme which can be seen as a form of task partitioning [23]. Forming a bucket brigade is useful once the swarm can cover most of the distance between source and sink and bucket brigades help to reduce interference between robots [473, 636]. Bucket brigades also occur in a number of ant species [339, 366].

We design our toy scenario in the following way. When there is only one robot ($N = 1$), then it has to pick up water at the one end and transport it to the other end by driving there. Similarly, for low numbers of robots, they have to go back and forth, whereas two generate a water flow double that of a single robot ($\Pi(2) = 2\Pi(1)$), four double that of two and so on (see Fig. 2.2a).

Say, once there are $N = 6$ robots, they are able to form a bucket brigade which means they stay fixed in their position and transport water only by handing it over to the next robot in the brigade. We assume that by this simple form of cooperation

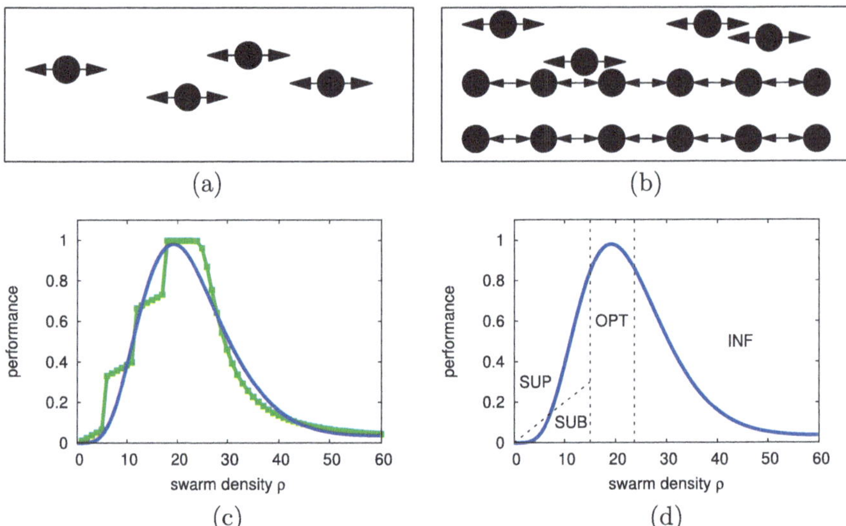

Fig. 2.2 Bucket brigade example for swarm performance (robots have to transport objects back and forth between the left and right side of the robot arena) and typical swarm performance function over swarm density $\rho = N/A$ for a fixed area $A = 1$ (without units). **a** Bucket brigade, $N = 4$ robots (**b**) Bucket brigade, $N = 16$ robots (**c**) Bucket brigade, performance. **d** Swarm performance showing four regions, SUP: superlinear, SUB: sublinear, OPT: optimal, INF: interference

these six robots are able to transport even more than double the amount of water compared to what three robots can do ($\Pi(6) > 2\Pi(3)$, see Fig. 2.2b). This constitutes a superlinear performance increase which is indicated by a big jump in Fig. 2.2c. A superlinear performance increase means that not only the overall performance increases with increasing team size ($\Pi(c_{\text{swarm}}N) > c_{\text{swarm}}\Pi(N)$, $c_{\text{swarm}} > 1$), but also the individual performance increases with increasing team size ($\pi(c_{\text{ind}}N) > \pi(N)$, $1 < c_{\text{ind}} < c_{\text{swarm}}$, see Sect. 2.4.5). This does not correspond to our everyday experience. Imagine you are studying in a group of students and by increasing the group size everyone of you gets more efficient. That is a great and very desirable effect but unfortunately only rarely observed in teams.

When more and more robots are added to the system, the robot density continues to increase and for a certain robot density ρ^* an optimal swarm performance Π^* is reached. When the robot density is further increased then the swarm performance actually decreases because robots interfere with each other and slow themselves down (negative marginal gain, retrograde swarm performance). When the robot density is increased even further then the performance continues to drop possibly down close to zero when almost all motion is stopped due to interference [479].

2.3.3 *Opportunities to Collaborate: When Bigger Is Better*

Swarm robotics demonstrates remarkable capabilities when robots physically collaborate to overcome challenges. By attaching physically to one another (e.g., Swarm-bots project by Dorigo et al. [201, 300]) or working in teams, they can solve problems that are beyond the ability of individual robots. When faced with a slope too steep for a single robot, they form a chain [601]. Each robot attaches to the next, redistributing weight and providing mutual support to climb together. Similarly, they can cross wide gaps as the robots link together to form a moving bridge [790].

When tasked with pulling sticks partially buried in the ground, robots pair up to combine their leverage [373]. One grips the stick at the top while the other pulls from below.

Kwa et al. [456] find in collective tracking that swarm adaptivity is achieved through modulating the network connectivity (i.e., the number of neighbors each agent communicates with). The swarm's tracking performance increases with connectivity until an optimum is reached and for higher connectivity the performance decreases (i.e., retrograde scaling).

These examples illustrate how swarms overcome physical limitations through teamwork. By attaching, redistributing forces, and synchronizing actions, they showcase the power of collective strength and adaptability. Note that also less obvious forms of teamwork exist as they are not directly physically collaborating. Examples for that would be foraging (see Sect. 5.3.8), collective-decision-making (see Chap. 8), or target tracking (see Sect. 5.3.9).

Clearly, we want to enable our robot swarm to solve problems collectively. Providing more robots results in a bigger and denser swarm that creates opportunities to collaborate. We face a tradeoff between sharing resources while providing enough opportunities to cooperate as mentioned above.

Next, we look into how measurements of swarm performance $\Pi(N)$ can be read to quickly grasp some of the likely underlying qualitative features in the swarm behavior concerning the level of collaboration in the system. We assume that we can measure the swarm performance over an interval of swarm densities ρ or as mentioned above on an interval of swarm sizes N (e.g., $N \in [0, 50]$) for fixed operational area A. We also assume that we can do repeated measurements in independent runs of the swarm system for each measured $\Pi(N)$ and that we average them (we discuss whether averaging over swarm performances is actually a clever idea in Sect. 2.5.3).

By looking at the initial phase of the performance curve (❸ in Fig. 2.3, for $N < 15$), we can obtain indications of how much robot–robot collaboration is done to complete the task (cf. other, more sophisticated efforts to derive group sizes from macroscopic measurements by Hamann et al. [332]). A fast increase of $\Pi(N)$ for small values of $N \in \{1, 2, 3\}$, shows that a small swarm is already sufficient to complete at least parts of the task (e.g., green curve of Fig. 2.3). Instead, if the curve has a slow start and $\Pi(N)$ shows a noticeable increase only for larger sizes N, it could indicate that robot–robot collaborations in larger groups are necessary. In most published swarm performance measurements, the initial increase of performance is

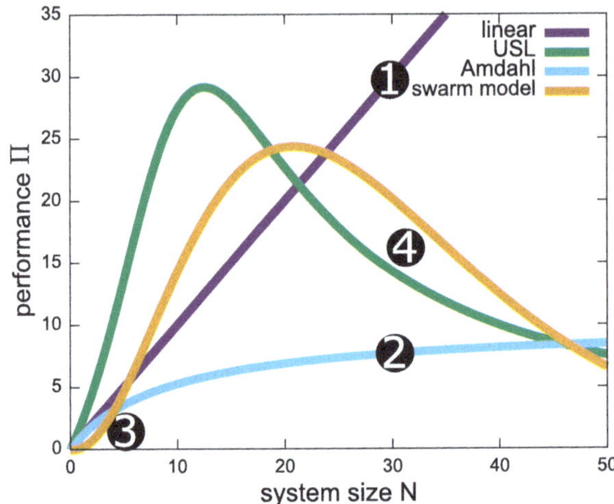

Fig. 2.3 Schematic plot of qualitatively different swarm performance $\Pi(N)$ curves as a function of the system size N. For large swarm sizes, the performance can (i) increase linearly (purple line, zone ❶), (ii) saturate to a constant value (blue line, zone ❷), or (iii) initially increase (zone ❸ of green and orange lines) and subsequently decrease (zone ❹ of green and orange lines). In scalability analysis, constant increase corresponds to Gustafson's law (see Sect. 2.4.2, [309]), saturation corresponds to Amdahl's law (see Sect. 2.4.2, [21]), and increase/decrease to Gunther's universal scalability law (USL, see Sect. 2.4.4, [305]). In comparison to a linear and a saturating function, we show two curves indicating a decrease in performance for too high densities to highlight important differences that can instruct us on qualitatively different behaviors. The initial phase (zone ❸) can show a slow (orange) or quick (green) increase of performance. The final phase, zone ❹, can show high- order (green) or almost linear (orange) decrease. These visual cues can give important insights about into swarm behavior and allow for its efficient design. *Source* Hamann et al. [335]

approximately linear (fast increase, large marginal gains). However, there are rare cases of published datasets showing a nonlinear (curved, initially small marginal gains) and slow increase [526]. Note that we do not focus on distinguishing between super- and sub-linear performance increases, instead we try to understand when to expect linear increases and when to expect nonlinear increases.

Several models that we describe in Secs. 2.4 and 2.5 can represent both linear (fast) and nonlinear (slow) increase despite their simplicity. Interestingly, similar nonlinear system behaviors can be observed in models from not directly related fields, such as PT2 lag elements in control theory, or residence times in cascades of stirred-tank reactors (tanks in series) [483]. In both of these examples, sequences of events or higher order time-delays introduce the observed nonlinearity. Comparable effects emerge in robot swarms when several robots need to collaborate in order to perform a given task.

To support our above claims, we show two minimal examples in which observing the system performance curve for small system sizes (❸, $N < 10$, in Fig. 2.3) allows us to estimate the necessary amount of robot–robot interactions to complete the task. If robots can perform the task without any/much help from other robots, then

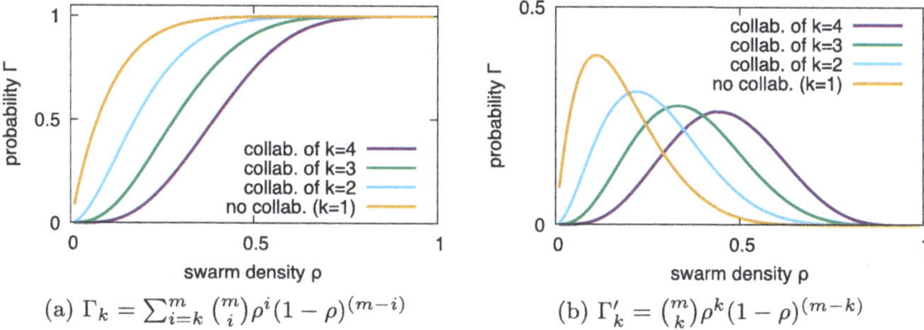

(a) $\Gamma_k = \sum_{i=k}^{m} \binom{m}{i} \rho^i (1-\rho)^{(m-i)}$ (b) $\Gamma'_k = \binom{m}{k} \rho^k (1-\rho)^{(m-k)}$

Fig. 2.4 Combinatorial explanation of chances for collaboration, collaboration probability Γ_k for $k \in \{1, 2, 3, 4\}$ and neighborhood size $m = 9$ (Moore neighborhood), and swarm density ρ; two scenarios: (**a**) without and (**b**) with interference. *Source* Hamann et al. [335]

the initial increase is steep and close to linear. We say that robots require low-order interactions. If robots require considerable help from other robots to perform the task, then the initial performance remains low for small sizes N and shows a curved (nonlinear) increase. We say that robots require high-order interactions. We give evidence for this conjecture through a simple analysis: a simple combinatorial argumentation (for more details and an empirical approach, see Hamann et al. [335]).

Our combinatorial consideration is the precondition for robot–robot collaboration: robots need to be in close proximity to each other. In swarm robotics, robot movement is often based on random motion (see Sect. 5.2.4). We consider the probability that collaboration among k robots takes place as a stochastic event proportional to k and swarm density ρ. Assuming a simple grid environment where collaboration takes place between neighboring robots, we can derive the probability of having at least k robots in Moore neighborhoods of 3×3, that is, $m = 9$ cells. Swarm density ρ indicates the independent probability of finding a robot in a given cell. The probability Γ_k of finding at least k robots in a Moore neighborhood of $m = 9$ cells corresponds to the cumulative distribution function of the Binomial distribution

$$\Gamma_k = \sum_{i=k}^{m} \binom{m}{i} \rho^i (1-\rho)^{(m-i)} . \tag{2.3}$$

In Fig. 2.4a, we show Γ_k as a function of ρ for $k \in \{1, 2, 3, 4\}$. As expected, the probability that at least k robots meet (our assumed precondition for collaboration) decreases by increasing k. Looking at the initial part of the curves, for low-density values, larger groups have a slow (nonlinear) increase. Instead, small groups (e.g., $k = 1$ or $k = 2$) have fast and almost linear increases. In Fig. 2.4a, we assume no interference between robots, thus values larger than k still allow for collaboration without overhead. In Fig. 2.4b, we assume that values larger than k would prohibit collaboration and only k robots could do useful work. The shown probability is

$$\Gamma'_k = \binom{m}{k} \rho^k (1-\rho)^{(m-k)} . \tag{2.4}$$

Despite the different shapes for high swarm densities, the initial part shows the same type of shapes for varying k as for Eq. 2.3.

2.4 Computational Models of Scalability

As mentioned above, parallel computing is a mature field that can provide us with basics for a theory of scalability. We will go through scalability in computational systems. Although these models were originally developed for phenomena in the context of computation, they are of interest for swarm robotics. There are many different fields of research that contributed such models but we focus on the perspective of computer science in this section. Scalability in computer science is mostly about the number of processors, CPUs (central processing units), or cores (independent processing units within a CPU).

To better understand the context, we quickly go through some of the history of parallel computing [359]. All the first computers that were developed, were single-processor computers. First efforts of scaling up were done by bit-level parallelism, that is, doubling the so-called "'computer word" size (natural unit of data of a certain processor design). It started from 4-bit microprocessors (e.g., Intel 4004, 1981) and continued doubling gave us 8-bit, 16-bit, 32-bit, and then 64-bit microprocessors. The next step was instruction-level parallelism by introducing the concept of pipelining (i.e., the processor executes different stages of multiple instructions simultaneously, similar to an assembly line). However, the interesting parallelism for our context is rather task parallelism where computer code is executed across several processors. For example, the Burroughs D825 (1962) was one of the first multiprocessing computers that allowed for task parallelism. The first massively parallel computer was the ILLIAC IV (1966) with 64 floating point units (not full CPUs). This new computer hardware required a theory of parallel computing. First key contributions are often attributed to Amdahl [21] and Gustafson [309]. These foundational models provided the basis for understanding computational speedup.

In parallel computing, scalability is often classified into two main categories: weak scaling and strong scaling. Weak scaling refers to increasing the workpackage size as more processors are added, ensuring that the computational load per processor remains constant. Weak scaling grows the total problem size and keeps the work per CPU constant as processors are added. In contrast, strong scaling refers to solving a fixed-size problem faster by adding more processors, aiming to reduce execution time. Strong scaling keeps the total problem size constant but CPUs receive proportionally smaller workpackages and we can possibly profit from memory effects. In practice, one might first add CPUs until efficiency drops, then increase the problem size to leverage available resources. In swarm robotics, we have no equivalent to that. We can add more and more robots N while optionally also adding operating space A but there is no equivalent to workpackages or workpackage size.

2.4.1 Speedup and Throughput

We add a few more definitions to improve our terminology about system-level performance. We take inspiration from parallel computing, a mature field that provides us with a well developed theory of scalability. Revisiting some of the above thoughts, we consider the time advantage of computing a problem in parallel instead of running the program on a single processor only. We assume that we can measure an average execution time T_{serial} for running on a single processor and an average execution time T_{parallel} for running on $p > 1$ processors. If the parallel approach is successful, we would expect $T_{\text{parallel}} < T_{\text{serial}}$.

We call the fraction $S = \frac{T_{\text{serial}}}{T_{\text{parallel}}}$ of these two times speedup, that is, the factor in time that we save when executing the program in parallel on several processors. For a proper definition we need to be explicit about the number of processors or robots. To be more specific, we define T_1 as the time required to complete the program or latency on a 1-core system (or one processor). We define T_p as the latency on a p-core system (with p integer and $p > 1$).

Speedup

We define as speedup

$$S_{\text{time}} = \frac{T_1}{T_p} . \tag{2.5}$$

If the considered task is inherently collaborative, then a single robot cannot solve it (e.g., think of a collective transport scenario). T_1 would be undefined or $T_1 = \infty$. So we should be careful whether sometimes it is more useful to define speedup with respect to the minimum number N_m of required robots to solve the task. We would then have $S = \frac{T_{N_m}}{T_p}$ [575].

In addition, we define another but similar concept: throughput. Throughput is the rate at which a system completes operations per time unit (i.e., jobs done per time unit) which indicates its productivity. This is a common concept from networks: the rate of successful data packet delivery.

Throughput

We define X_i as the throughput of an i-unit system (e.g., i robots) where X_i counts the completion of task-specific jobs.

For ease of use we abbreviate throughput in our equations by "'thruput." As speedup based on throughput we define

$$S_{\text{thruput}}(N) = \frac{X_N}{X_1}, \tag{2.6}$$

for the speedup of a robot swarm of swarm size N compared to a single robot's performance. Also here we have the problem that a single robot may not be able to solve the task which would give us $X_1 = 0$ and a division by zero. We could try to interpret $S_{\text{thruput}}(N) \to \infty$ or we rather go with the minimum number N_m of robots to solve the task as discussed above. We would get $S_{\text{thruput}}(N) = \frac{X_N}{X_{N_m}}$. We generally prefer to speak of throughput instead of processing time but ideally the speedup defined by time and the speedup defined by throughput are synonymous. Throughput seems more intuitive because we often do not have that explicit synchronization point in our robot swarms that would allow us to actually measure T_N.

We start our little journey through scalability models for computation systems starting with a few more definitions about performance metrics from parallel computing. In Sect. 2.4.2 we present the most prominent but also most basic models: Amdahl's Law and Gustafson's Law. These are phenomenological models of scalability, that is, there are no first principles that would allow us to derive them. Instead, they are creative insights by their authors. We discuss them to have a complete picture and also because they arguably provide the optimistic and pessimistic view on scalability of parallel systems.

In Sect. 2.4.3 we continue with one of the most fundamental and effective modeling approaches for computational systems: queuing theory. It is the classical approach from computer science to assess performance and processing times of systems where resources are shared among competing demands. Despite the high level of abstraction, queuing models have applications in swarm robotics as discussed below.

We conclude our discussion of models with computational background in Sect. 2.4.4 where we present Gunther's Universal Scalability Law (USL). It originates from distributed computing but also applies quite well to swarm robotics. This model importantly addresses the diminishing returns scenario, that is, it proposes a concave function and can model the situation of having a peak performance Π^* for a critical swarm density ρ_c.

2.4.2 *Amdahl's Law and Gustafson's Law*

To understand the limits of scalability in systems with both parallel and sequential (portion of computation that cannot be parallelized) components, we turn to Amdahl's law. It quantifies how the presence of a serial portion $\sigma \in [0, 1]$ constrains the overall speedup that can be achieved through parallelization, even as the number of processing units increases. Amdahl's law [21] predicts the maximum speedup for

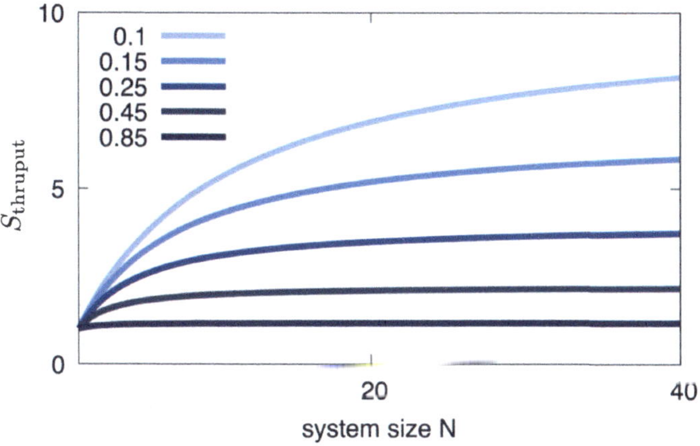

Fig. 2.5 Amdahl's law

tasks with a serial portion $\sigma \in [0, 1]$, highlighting parallel computing's limits by the necessary time for sequential parts. Throughput scales with

$$S_{\text{thruput}}(N) = \frac{N}{1 + \sigma(N-1)} . \tag{2.7}$$

See Fig. 2.5 for a plot of several settings of serial portion σ. Amdahl's law can be seen as the pessimistic perspective on parallel computing as it defines a kind of speed limit. The portion of a program that cannot be parallelized sets a fundamental upper bound on performance. Examples for the sequential portion of a program are: initializing data structures, aggregating results, handling dependencies between computations, or critical sections requiring synchronization. If we want to translate Amdahl's law to swarm robotics, we have to think of fundamental limits. There is no clear way of doing that but options could be communication overhead as a bottleneck or physical constraints leading to congestion.

Gustafson's law [309] suggests parallelization benefits increase with workload and processing power, allowing larger problems to be solved in the same time frame. Throughput scales with

$$S_{\text{thruput}}(N) = N + (1 - N)\sigma . \tag{2.8}$$

See Fig. 2.6 for a plot of several settings of σ. Gustafson's law can be seen as the optimistic perspective on parallel computing, as it emphasizes how increasing computational resources enable solving larger and more complex problems rather than just reducing execution time. In real-world applications, the sequential portion σ often is not constant but shrinks relative to the expanding problem size. Translating this finding into swarm robotics is again not possible in a direct way. Whether more robots

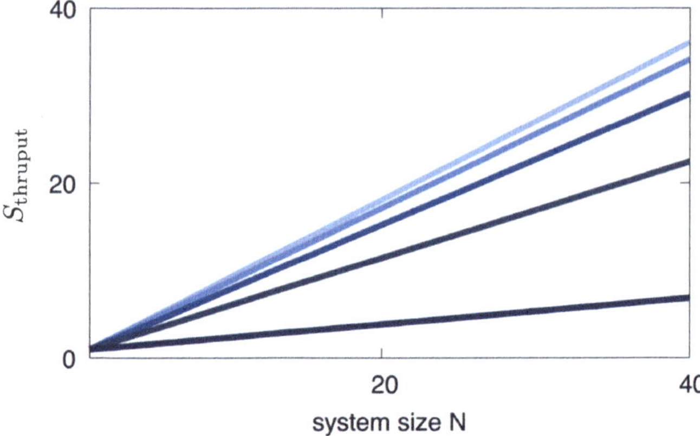

Fig. 2.6 Gustafson's law

can handle larger environments or more complex tasks, is not a given. The linear scaling $\Pi(N) \sim N$ as suggested by Gustafson seems indeed overly optimistic for increasing swarm density ρ in swarm robotics. Instead a saturation effect $\Pi(N) \rightarrow \Pi_{\text{const}}$ for large swarms N as defined by Amdahl seems more realistic. However, if we can afford to add space ΔA for each additional robot $\Delta N = +1$ and keep swarm density $\rho_{\text{const}} = \frac{N+\Delta N}{A+\Delta A}$ constant, then linear scaling seems feasible (non-shrinking marginal gain $\Delta\Pi(N) > 0, \forall N$). For a more optimistic take, also consider the possibility of super-linear scaling (see Sect. 2.4.5).

2.4.3 *Queuing Theory*

A simple modeling technique is queuing theory that helps to better formalize scalability and has applications to swarm robotics despite its simplicity. We provide a quick introduction to queuing theory. Queuing theory is the mathematical study of waiting lines or queues. It analyzes the behavior of queues in systems where resources are shared among competing demands to optimize performance, minimize delays, and ensure efficient use of resources. Queuing theory provides a framework to model, analyze, and predict the performance of these systems under varying conditions. Knowledge obtained by this kind of modeling may help to improve customer satisfaction (e.g., reduced waiting times). Queuing theory has many applications in computer science, for example, in network traffic management, operating systems, and database management. Despite its abstractness and the theory being tailored for computing systems, it can still be insightful and inspiring for swarm robotics as a fundamental method. Robot swarms operate in a highly distributed manner: rather than lining up in an orderly fashion or being served sequentially at a defined location, robots typically compete for access to shared resources in a less organized way.

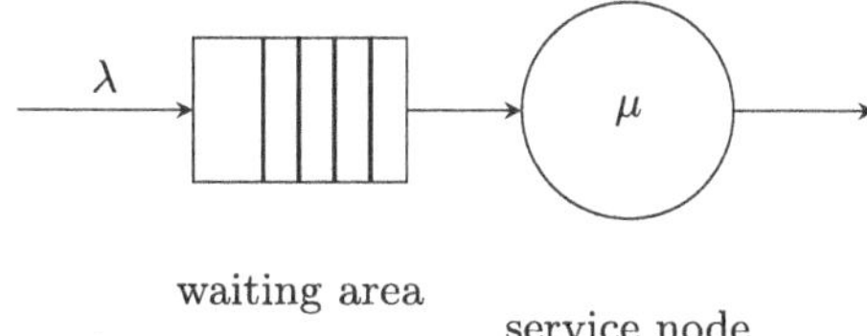

Fig. 2.7 A simple visualization of a queue for arrival rate λ and service rate μ

Queuing theory describes how jobs or tasks move through a system with limited resources (e.g., a server or more generally a "'machine") and how their arrival and processing are modeled mathematically. A job represents a task, request, or unit of work that requires processing. It enters the queue, waits its turn according to the so-called queue discipline, and is completed once it receives the required service (see Fig. 2.7). Often used queue disciplines are: FIFO (First In, First Out) that processes the first job that arrives first, LIFO (Last In, First Out) that processes the most recent job first, and the queue discipline called "priority" that processes jobs based on their importance. The arrival rate λ defines the average rate at which jobs arrive at the queue (new jobs per time unit). The service rate μ defines the average rate at which jobs can be served (completed jobs per time unit).

One can configure many different types of queues. To get a good overview and to order these options, there is a specialized notation system. Kendall's notation is a standard shorthand to classify queuing systems based on six parameters: arrival process (A), service process (S), number of servers (c), system capacity (K), population size (P), and queue discipline (D), formatted as $A/S/c/K/P/D$. A is the arrival process, for example, M for Markovian and Poisson or D for deterministic. S is the service time distribution, for example, M for Markovian and exponential or D for deterministic. c defines the number of machines (e.g., servers) in the queue. K is the system capacity, that is, the maximum number of entities that can stay in the system at the same time including those in service. P is the population size, that is, the total number of potential jobs in the world. D is the queue discipline as described above. However, K, P, and D are often omitted. The simplified and more frequently used form is $A/S/c$ where we assume infinite capacity ($K \mathrel{\widehat{=}} \infty$), infinite population size ($P \mathrel{\widehat{=}} \infty$), and FIFO as queue discipline (D is FIFO by default).

The term Markovian refers to a property of processes where the future state depends only on the current state and not on the history of how the system arrived at that state. Mathematically this can be expressed as

$$\begin{aligned} P(X(t+1)) &= P(x \mid X(t), X(t-1), \ldots, X(0)) \\ &= P(x \mid X(t)) . \end{aligned} \tag{2.9}$$

In other words, we do not care about what has happened one or more than one time steps before.

If we define the arrival process A as $A = M$ for Poisson, the incoming jobs are distributed according to a Poisson process. That is a specific type of stochastic process describing events occurring randomly at a rate λ over a continuous interval. The Poisson distribution is defined as

$$P_{\lambda}(k) = \frac{\lambda^k}{k!}\,\mathrm{e}^{-\lambda}\,. \tag{2.10}$$

The rate parameter λ describes the average rate of occurrence. The mean and the variance of a Poisson distribution are both equal to λ. For clarification we look at an example. With $\lambda = 3$ events/hour, the probability of having exactly k events in one hour is given by $P(k; \lambda) = (\lambda^k/k!)\mathrm{e}^{-\lambda}$. For example, for $k = 2$ events we get $P(2; 3) = (3^2/2!)\mathrm{e}^{-3} \approx 0.224$. The process is memoryless. The time that passes until the next event is independent of the time since the last event (i.e., Markovian). The process is homogeneous. Events occur at a constant rate λ over time. We have discrete events. Each event occurs independently.

If we define the service process S as $S = M$ for exponential, the time required to serve a job (service rate) follows an exponential distribution. The mean of an exponential distribution is $1/\mu$ and the variance is $1/\mu^2$. The exponential distribution is defined by the probability density function

$$f_{\mu}(t) = \mu\mathrm{e}^{-\mu t}, \quad t \geq 0\,. \tag{2.11}$$

The rate parameter μ represents the average number of events (or jobs served) per unit time. The exponential distribution is memoryless. The probability of a job completing in the next t units of time is independent of how long the job has already been in service. For example, with $\mu = 2$ jobs/second, the probability that a job takes more than 1 s to complete is given by $P(t > 1) = \mathrm{e}^{-\mu t} = \mathrm{e}^{-2\cdot 1} = \mathrm{e}^{-2} \approx 0.135$.

The most prominent and common queue is an M/M/1 queue.

M/M/1 Queue

An M/M/1 queue has a Poisson arrival process (first $M = A$). Its service time distribution is exponential (second $M = S$). The number of machines is set to $c = 1$.

So we get a basic model with a single server, first-come-first-served queue discipline, infinite system capacity, and unlimited population size. Relevant known system properties of an M/M/1 are the following. The average number of customers (or jobs) present in the system at the same time is $L = \lambda/(\mu - \lambda)$. The average time a customer spends in the system is $W = 1/(\mu - \lambda)$. We can also determine a probability of having n customers in the system at the same time by $P_n = \left(1 - \frac{\lambda}{\mu}\right)\frac{\lambda}{\mu^n}$. Machine (server) utilization is λ/μ, which is stable for $\frac{\lambda}{\mu} < 1$, critical for $\frac{\lambda}{\mu} = 1$, and instable for $\frac{\lambda}{\mu} > 1$. There is a clash of notation as textbooks often denote machine utilization by ρ but we have defined that as swarm density.

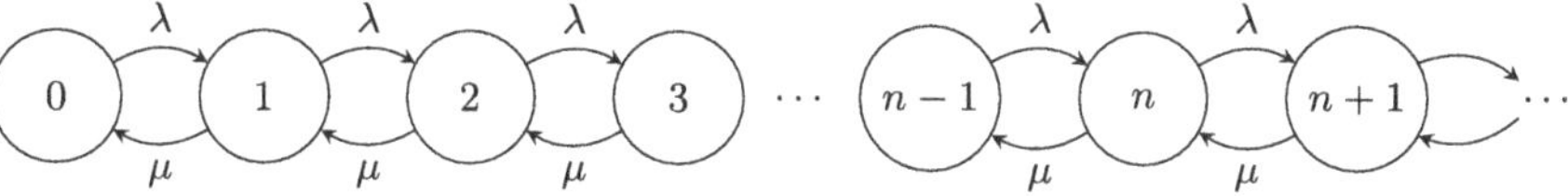

Fig. 2.8 State space representation of an M/M/1 queue with transitions labeled by the arrival rate (λ) and the service rate (μ). Each state represents the number of (waiting/currently served) robots in the system

M/M/c queues can be modeled as Markov chains because they have the memoryless property (see Fig. 2.8). A Markov chain is a stochastic process that models a sequence of states, where the probability of transitioning to the next state depends only on the current state. Transitions between states are defined by a transition probability matrix, which specifies the likelihood of moving from one state to another.

M/M/1 queue described as a continuous-time Markov chain

An M/M/1 queue is a stochastic process that can be described as a continuous-time Markov chain that we define as triple $(S, \mathbf{Q}, \varphi_0)$. S is a (finite or countable) state space (number of robots in the queue at any given time). $\mathbf{Q} = [q_{ij}]$ is the transition rate matrix (also known as infinitesimal generator matrix) with transition rate q_{ij} from state i to state j for $i \neq j$, birth rate $q_{i,i+1} = \lambda$ for all $i \geq 0$, death rate $q_{i,i-1} = \mu$ for all $i > 0$, and $q_{ii} = -\sum_{j \neq i} q_{ij}$ for $i > 0$, and $q_{00} = -\lambda$, $q_{ij} = 0$ otherwise.

This will immediately get clearer by checking this notation of a transition rate matrix:

$$Q = \begin{pmatrix} -\lambda & \lambda & & & 0 \\ \mu & -(\mu+\lambda) & \lambda & & \\ & \mu & -(\mu+\lambda) & \lambda & \\ & & \mu & -(\mu+\lambda) & \lambda \\ 0 & & & & \ddots \end{pmatrix}. \tag{2.12}$$

Each state (node) of the chain has at most two edges leading out. Either one job leaves or one job is added. There is no way that we do a jump of ± 2 or more. Note that we do not allow for remaining in a state (no self-loop, see Fig. 2.8) as defined by the rates $-(\mu + \lambda)$ that force to do the next step. We need to define an initial distribution φ_0 over the states (note another clash in our notation: distributions over Markov chain states are often notated as π, which we use here for individual performance). For example, we can simply set $\varphi_0 = [1, 0, 0, \dots]$, that is, the queue is initially empty. By modeling the queue as a Markov chain we get access to the rich set of methods of Markov theory. For example, we can try to answer the interesting question: What is the stationary distribution φ^* for a given $\mathbf{Q}$? This can be answered by solving a system of linear equations $\varphi^* \mathbf{Q} = 0$ (see Task 2.2).

To make this whole concept more accessible, we go through an empirical example. For an M/M/1 queue, we set the arrival time to $1/\lambda = 1.1$ time units and the service time to $1/\mu = 1.0$ time units. We are using the reciprocal, as speaking of average times for arrival and service may be more intuitive than speaking of rates. Also note that we have chosen an instable queue because $\frac{\lambda}{\mu} > 1$. This leads to a more interesting system behavior to observe. We initialize the queue to $\varphi_0 = [1, 0, 0, \dots]$. We can easily program a simulator that probabilistically models that queue and iterates over arrival and service events for many time units. We have done that for 1000 time units and we have measured jobs in the system, completions, and throughput rate (see Fig. 2.9). As we have $1/\lambda > 1/\mu$, there are more jobs coming in than we can serve. With maximal utilization the system can only process one job per time unit. This is what we observe in the throughput rate that settles around one as expected. Also completions arrive at close to 1000 as expected for 1000 simulated time units. Maybe the most interesting value are the measured jobs/customers in the system. They accumulate to around 100 after 1000 time units as expected but seem to fluctuate quite considerably throughout.

Queuing theory provides us with models that are simple and easy to handle. However, the enforced serialization as stereotypically represented by a queue is not justified for a typical swarm robotics scenario. Our robots operate in continuous space and do not need to line up in queues. However, there are scenarios where robots are forced to follow lines on the floor that form maybe a whole street and traffic system in the form of a regular grid for a Manhattan-style map. When navigating in a grid, robots may line up at crossings, for example. In Sect. 2.5.3 we will see that queuing theory can definitely help to better understand swarm robotics in certain cases despite its abstractness and simplicity.

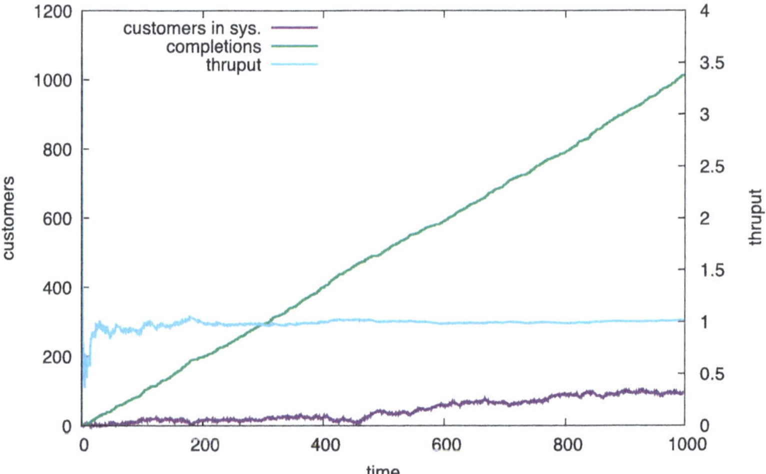

Fig. 2.9 Results from a numerical simulation of an M/M/1 queue with arrival time $1/\lambda = 1.1$ and service time $1/\mu = 1.0$. There are 1023 completions within the simulated 1000 time units. The resulting throughput rate is 1.023 customers per time step and we get a server utilization of 100%

2.4.4 *Universal Scalability Law (USL)*

There is a model for parallel processing performance by Gunther [305] that models the effects of contention and coherence delays in parallel systems (e.g., multiprocessor systems with shared memory). First we introduce the concept of capacity. Capacity C of a system in context of parallel computing is the maximum amount of work that the system can perform within specified performance constraints. Here, we only consider concurrency-related constraints, that is, we are concerned about the number of concurrent processes N that the system supports simultaneously.

Relative Capacity $C(N)$

We define relative capacity $C(N)$ as a dimensionless measure of a system's throughput at a given concurrency level N relative to its baseline throughput $S_{\text{thruput}}(N = 1)$ at a single processor (i.e., $C(N) = \frac{X_N}{X_1}$, $C(1) = \frac{X_1}{X_1} = 1$).

Relative capacity $C(N)$ is dimensionless and comparable regardless of raw throughput units. So it can certainly be understood as a swarm performance measure $\Pi(N) = C(N)$. Gunther [305] calls his model the Universal Scalability Law. The reasoning for this name is that the model is based on fundamental principles of queuing theory and resource contention. These principles are universal in the sense that any system with shared resources and parallel work will exhibit similar scalability patterns governed by the same underlying factors of concurrency, contention, and coherence costs. For a relative capacity $C(N)$ we define

$$C(N) = \frac{N}{1 + \alpha(N-1) + \beta N(N-1)}, \tag{2.13}$$

for a coefficient α that gives the degree of contention (interference) in the system and coefficient β that gives the lack of coherency in the distributed data. Contention occurs because resources are shared. Whenever the capacity of a shared resource is used completely and another process requests to use that resource, then the process has to wait. Contention increases with increasing system size, while keeping resources at the same capacity (e.g., increasing swarm size N while keeping operating area A constant). Lack of coherency occurs because processes, to a certain extent, operate locally. For example, they have local changes in their caches that are not immediately communicated to all other processes. Maintaining coherency is costly and the costs increase with increasing system size. In shared-memory multiprocessor machines, maintaining coherence typically involves cache coherency protocols, such as snooping or directory-based mechanisms, which ensure that all processors have a consistent view of memory [740]. These protocols require frequent communication and coordination between processors, and their overhead grows with system size. This is similar to overheads in swarm robotics due to self-organized coordination of many robots with high swarm density.

Gunther identifies four qualitatively different speedup situations for relative capacity C over number of processors N:

a. Linear scaling: If contention and lack of coherency are negligible, then we get "equal bang for the buck" [305] and have a linear speedup ($\alpha = 0$, $\beta = 0$, Fig. 2.10a).
b. Sublinear scaling: If there is a cost for sharing resources in the form of contention, then we have a sublinear speedup ($\alpha > 0$, $\beta = 0$, Fig. 2.10b).
c. Diminishing returns: If there is an increased negative influence due to contention, then the speedup clearly levels off ($\alpha \gg 0$, $\beta = 0$, Fig. 2.10c).
d. Retrograde performance: If in addition there is also an influence of incoherency, then there exists a peak speedup and for bigger system sizes the speedup decreases ($\alpha \gg 0$, $\beta > 0$, Fig. 2.10d).

In the original work of Gunther [305], superlinear performance increases are basically not allowed. In a later work [307], superlinear speedups are discussed and negative contention coefficients $\alpha < 0$ are allowed to model that situation (see Fig. 2.10e). While contention $\alpha > 0$ refers to capacity consumption due to sublinear scalability, $\alpha < 0$ refers to a capacity boost due to superlinear scalability. In parallel computing, superlinear speedups can occur due to some interplay between problem size per computing unit and available memory (see Sect. 2.4.5).

In the context of swarm robotics we can interpret contention as interference between robots due to shared resources, such as the area around a base station that robots need to pass in possibly crowded situations or in general shared space. Following this interpretation, collision avoidance is a waiting loop due to the shared resource *space* being currently unavailable. That is intuitive and similar to an airplane flying a holding pattern because the resource *runway* is currently in use and should certainly not be shared. Since superlinear scaling requires a negative contention coefficient ($\alpha < 0$), it becomes an intriguing thought experiment to consider what negative contention might represent in swarm robotics. It may seem intuitive to perceive it as a form of *inverse resource conflict*. For example, in collective transport, additional robots sharing the burden can generate synergistic effects in contrast to hindering one another in high swarm density situations.

Incoherency is more difficult to interpret as an analogy to swarm robotics. Incoherency refers to the performance loss caused by inconsistent information across the swarm, overhead due to limited communication of information (e.g., delays), or due to imperfect synchrony. As a result robots may act on outdated or divergent information. Incoherency concerns the alignment of internal state and decision-making across distributed robots. Just as coherency protocols (e.g., snooping or directory-based) coordinate updates across processors to ensure a consistent global view, swarm robots need to align their internal states and decision-making despite limitations in communication or partial observability. In both cases, there is a challenge of propagating updates efficiently and ensuring consistency without introducing excessive coordination overhead.

While Gunther assumes that there cannot be a system-wide deadlock situation due to contention only (speedup monotonically increases with increasing α), that

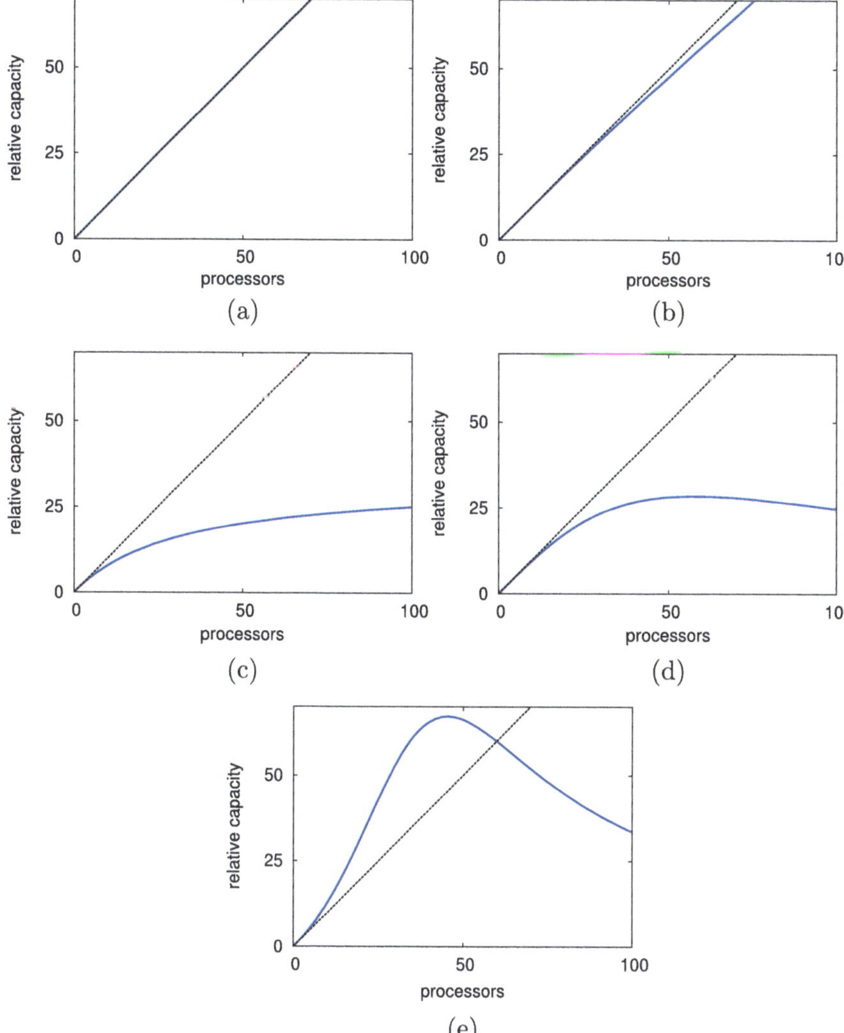

Fig. 2.10 Universal Scalability Law following Gunther [305], four standard situations and superlinear speedup [307] depending on parameters α (degree of contention) and β (lack of coherency). **a** Linear speedup, $\alpha = 0$, $\beta = 0$ **b** Sublinear speedup, $\alpha = 0.001$, $\beta = 0$ **c** Diminishing returns, $\alpha = 0.03$, $\beta = 0$ **d** Retrograde performance, $\alpha = 7\times10^{-4}$, $\beta = 3\times10^{-4}$ **e** Superlinear, $\alpha = -0.03$, $\beta = 5\times10^{-4}$

could occur in a swarm robotics system. For example, the swarm density could be too high, such that all robots permanently try to avoid collisions resulting in zero performance.

2.4.5 *Superlinear Scalability: A Dream Comes True?*

Superlinear scaling means we experience a swarm performance increase of $\Pi(N) > N\Pi(1)$. This may seem quite peculiar. A superlinear performance increase means not only that the overall performance increases with increasing swarm size, but also the individual performance π of each robot increases with increasing team size. This was observed, for example, in wasps [384]. In their study of the social wasp *Polybia occidentalis*, Jeanne and Nordheim showed that both total and per capita productivity increased significantly with colony size during early colony development. They attribute this superlinear scaling effect to reduced queuing delays among workers in larger groups, which allows individuals to work more efficiently. However, that does usually not correspond well to our human everyday experience. If a team is struggling to coordinate and collaborate, then adding more persons to the team typically worsens the problem instead of improving the situation. Indeed superlinear speedups seem special and they were frequently discussed and studied [236, 307, 310, 354].

There even exists a proof showing the impossibility of superlinear speedups but it assumes fixed problem size [236]. Their analysis suggests that observed superlinear speedup often results from effects, such as improved cache utilization or reduced overhead in parallel execution, instead of an inherent advantage of parallelism as such. In parallel computing, the concept of superlinear scalability is typically considered rare and of limited practical relevance, and is sometimes even dismissed as non-existent by definition. For example, in the book by Grama et al. [294] they accept that superlinear speedups can be observed but perceive them as a form of artifact or "speedup anomaly." More specifically, they define "acceleration anomalies" as instances where $\Pi(N) > N\Pi(1)$, and "deceleration anomalies" as those where $\Pi(N) < N$. Basically, defining everything an anomaly that does not obey the supposedly ideal linear scaling $\Pi(N) = N\Pi(1)$. This seems a common assumption in parallel computing and, for example, has seemingly also influenced Gunther [305] to not include it in his model initially (see Sect. 2.4.4).

An interesting take is that of Helmbold and McDowell [354]. Inspired by Miya [562] they define a sophisticated terminology. Helmbold and McDowell [354] define superlinear speedup as $\lim_{N\to\infty} \Pi(N) = \infty$, a definition that we do not adopt here. This indicates an ever increasing speedup with growing system size. Certainly not a realistic assumption. Closer to our definition, they introduce the notion of "linear superunitary speedup," defined as $\lim_{N\to\infty} \frac{\Pi(N)}{N}$ being finite and greater than 1. They define sublinear speedup as the case where $\lim_{N\to\infty} \frac{\Pi(N)}{N} = 0$, reflecting a scenario in which overhead more realistically grows unboundedly with increasing system size. They describe six scenarios of superlinear scalability (i.e., "linear superunitary speedup" in their terms). We focus here on the three most relevant cases: (1) Superlinaer scaling is impossible if we face, for example, monotonically increasing overhead in the parallel implementation. In reference to our above discussion and notation that would correspond roughly to $\Pi(N) = \Pi'(N) - \gamma(N)$ with $\gamma(N) \geq N$. (2) Superlinear scaling due to cache size, which we discuss in detail below. (3) Superlinear scaling can occur in randomized algorithms where multiple simultaneous attempts can reduce expected execution time compared to sequential approaches. This could actually be a likely explanation of some of the more difficult to analyze cases of superlinear scaling in swarm robotics. Imagine a scenario where robots need to collaborate in teams of two but meeting a second robot is a matter of chance. With more robots in the system and for sure with an increased swarm density, robot–robot encounters will increase which can be seen as an analogy to exploiting variance in parallel implementations of randomized algorithms (see Sect. 2.3.3).

A first possible cause for observing superlinear scalability in parallel computing can be that the serial implementation is at disadvantage compared to the parallel implementation because of memory hierarchy. Each processor may have a cache of a certain size. For a given problem size s (say in bytes) we distribute the work as packages to each processor and in the parallel implementation with N processors these work packages will be smaller, say, s/N. It may happen that for the given computer architecture the s/N-size work packages happen to fit in the cache while the s-sized work packages don't fit. As a consequence the serial implementation will run much slower, possibly causing superlinear scalability $\Pi(N) > N\Pi(1)$. For example, Grama et al. [294] just "disregard superlinear speedup due to hierarchical memory." More generally, one can argue that what might appear as superlinear scalability is actually more efficient utilization of existing resources. For example, adding more processors might lead to better utilization of caches or memory bandwidth, which improves performance but doesn't necessarily mean the system is scaling superlinearly. Instead, it's argued that the system is simply approaching its optimal utilization of resources.

Grama et al. [294] also discuss a second possible cause of superlinear scalability: exploratory decomposition. The algorithm may be required to search a tree data structure. For a given instance of the problem, the serial implementation may be required to search almost the whole tree to find the solution. However, a parallel implementation that assigned subtrees to processors may speed up superlinearly because a processor may reach the solving node in its assigned subtree quickly. If all other processes on the other processors are aborted immediately, then the parallel implementation may need to process even less nodes in total than the serial implementation.

Grama et al. [294] define W as "the amount of work done by a single processor, and W_p [as] the total amount of work done by p processors"; and "a search overhead factor as the ratio of the work done by the parallel formulation to that done by the sequential formulation" as $\frac{W_p}{W}$. This gives us an "upper bound on speedup for the parallel system" of $p\frac{W}{W_p}$. In some cases we may achieve a search overhead factor of $\frac{W_p}{W} < 1$ which would give us a superlinear speedup. However, they argue that if "the search overhead factor is less than one on the average, then it indicates that the serial search algorithm is not the fastest algorithm for solving the problem." In that case, superlinear speedup could only exist in an unlikely scenario where the problem instances for a given scenario are not uniformly distributed (e.g., solving nodes in the tree always appear in the parts of the tree that are traversed late by the serial implementation).

While superlinear speedups are rather infrequently observed in parallel computing, they are frequently observed in swarm systems. When observed, superlinearity is often a discrete physical effect (i.e., too few robots cannot succeed but given a minimum number of robots N_m they can), such as a robot group being able to form a bucket brigade [473, 636], robots that manage to overcome a gap in a team [790], or pass a step [300, 575, 601].

Superlinear scalability holds significant potential, providing strong motivation to investigate its occurrence across different domains and to explore how it can be systematically induced. A more subtle effect of superlinear scalability is shown, for example, in the stick-pulling scenario [373] as discussed in Sect. 5.3.6. Requiring two instead of one robot to pull a stick is again the physical effect as discussed above, however, there is another superlinear scalability effect. Ijspeert et al. [373] report that also for swarm sizes $N \in \{2, 3, \ldots, 6\}$ the individual robot performance (or efficiency) increases linearly which means we have superlinear swarm performance scaling. This part cannot be explained easily by a physical effect but has to be linked to robot–robot encounter probabilities when robots wait for other robots to collaborate (see Sect. 5.3.6).

Superlinear scalability is a fascinating phenomenon in swarm robotics. While often viewed as an anomaly in parallel computing, in swarms it is tied to collaborative opportunities. We still need to better understand the more sophisticated effects of superlinear scalability and how to design for it.

2.5 Swarm Robotics Models of Scalability

There are a few scalability models that were designed for applications in swarm robotics but they are not necessarily limited to that use. All of these models focus on concave functions which again shows how important that feature seems to be in swarm robotics. Hayes [345] was among the first to propose a performance function for swarm robotics for a specific search task. He provides a cost function of the form

$$\frac{1}{\Pi(N)} = \frac{a + bN^2}{N}, \tag{2.14}$$

for constants a defining costs depending on task duration and b defining energy and initialization costs that scale by group size. This is an amazingly simple formula that provides a good model of scalability (see Fig. 2.11). Another important early work

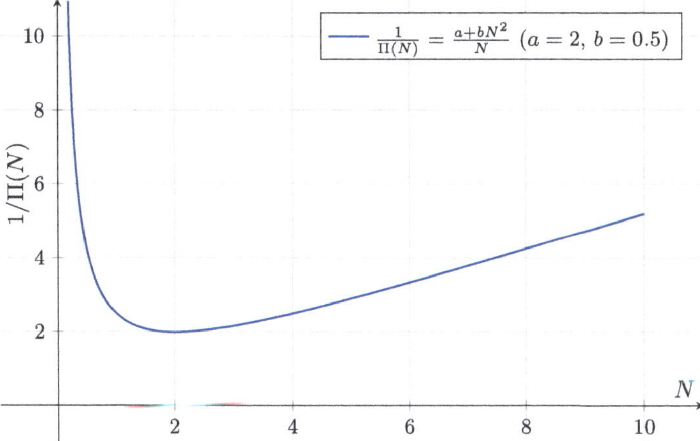

Fig. 2.11 Plot of simple swarm scalability model as a cost function by Hayes [345] for $a = 2$ and $b = 0.5$

is that of Rosenfeld et al. [682]. They study why performance in a swarm can drop due to interference, however, in a purely empirical approach using a robot simulator.

We start in Sect. 2.5.1 with a simple phenomenological model of scalability in swarm robotics. This modeling approach focuses on the positive effect of creating opportunities for cooperation and its negative counter-effect of increased competition for shared resources.

In the following Secs. 2.5.2 and 2.5.3, we present a recent model that discusses an important challenge of swarm performance. The swarm density ρ^* that shows in average the best performance Π^* may be critical in a sense that systems at that density ρ^* may also completely fail over longer time periods of operation ($\lim_{t\rightarrow\infty} \Pi(\rho^*, t) = 0$).

2.5.1 Swarm Performance Model

We introduce a phenomenological model of system performance in robot swarms that I proposed myself [319]. The main idea of this model is that opportunities for cooperation are countered by competition for shared resources. A swarm system has a theoretical ideal performance $\Pi'(N)$ that could be achieved through perfect cooperation without any interference or other overhead. This theoretical best swarm effort forms the first component of our model. The second component is the counterforce, that is, the overhead due to interference. For example, robots need to compete for shared resources of limited capacity (e.g., space) and accessing them in situations close to capacity generates overhead.

We define the general shape of swarm performance Π by

$$\Pi(N) = \frac{N}{e^N}, \tag{2.15}$$

which is roughly $0.37N^{1-N}$. With all relevant constants we get

$$\Pi(N) = \mathcal{C}(N)(\mathcal{I}(N) - d) \tag{2.16}$$
$$= a_1 N^b a_2 \exp(cN), \tag{2.17}$$

for parameters $c < 0$, $a_1, a_2 > 0$, $b > 0$, and $d \geq 0$ [319]. Parameter d is used to set the limit $\Pi(N \to \infty) = I(N \to \infty) - d = 0$. Furthermore, we refine our definition of $\Pi(N)$ by introducing two components, $\mathcal{C}$ and $\mathcal{I}$. We define a theoretical swarm effort that could be reached without any negative feedback by the cooperation function

$$\mathcal{C}(N) = a_1 N^b. \tag{2.18}$$

The same formula was used by Breder [106] in the context of cohesiveness within fish schools and by Bjerknes and Winfield [80] to model swarm velocity in emergent taxis. They used parameters of $b < 1$ while we also allow $b > 1$. Potentially that is a major difference because $b < 1$ represents a sublinear performance increase due to cooperation while $b > 1$ represents a superlinear increase. We define the interference function by

$$\mathcal{I}(N) = a_2 \exp(cN) + d. \tag{2.19}$$

This can be interpreted as the theoretical swarm performance that could be achieved without any cooperation (i.e., without positive feedback). Nonlinear effects that decrease efficiency with increasing swarm size are plausible due to negative feedback processes such as the collision avoidance behavior of one robot that triggers the collision avoidance behavior of several others in high-density situations. Still, there are many options of available nonlinear functions but best results were obtained with exponential functions. Also Lerman and Galstyan [479, Fig. 10b] report an exponentially decreasing efficiency per robot in a foraging task.

2.5.2 *The SGF Model: Solo, Grupo, and Fermo*

A strength of the above models of scalability is their simplicity, which is helpful when handling them. However, their contribution to a deeper understanding of the underlying processes that result in these system properties is limited. In that sense these models are rather phenomenological, that is, they primarily are based on observed phenomena, rather than deriving their equations from fundamental principles or first-principle laws. This would be a beautiful approach but it is also not clear how that can be achieved. Connecting individual robot behavior with system-level properties is a known challenge in swarm robotics as discussed later, for example, in Sect. 4.4.

For scalability modeling I have tried myself together with Andreagiovanni Reina to design a model that goes beyond merely being phenomenological. This effort was motivated in part by writing the first edition of this book. When I had revisited the basics of scalability in swarm robotics, I was reminded that there is a knowledge gap in the literature that needs work. For sure, our proposal is just a first step, but our effort produced what we call here the "SGF model" [322]. It is a continuous-time population model. The main idea is that for any swarm robotics system, we can assign any robot any of three states anytime: *solo*, *grupo*, or *fermo*. These are Esperanto words that we chose for fun and to make them distinguishable from other terminology. State *solo* refers to the robot that currently operates by itself. There are no other robots around. It cannot cooperate or collaborate, but there is also no interference. Depending on the task, this may be an efficient state for the robot. State *grupo* refers to the robot that currently operates in a group. It is in contact with other robots, which may help to cooperate or collaborate but it may also cause some interference from time to time (e.g., collision avoidance actions required due to limited space or collisions in radio communication). Depending on the task, this may be an efficient state for the robot, for example, if the task requires cooperation or collaboration. State *fermo* refers to the robot that currently is stuck. There are too many robots around such that all shared resources (e.g., space or radio channel) are depleted and the robot cannot do anything useful. The robot is stuck in a traffic jam or deadlock and is useless for now.

The dynamics of the system is based on rates of robots making transitions from *solo* to *grupo* or back and from *grupo* to *fermo* or back (see Fig. 2.12). Note that

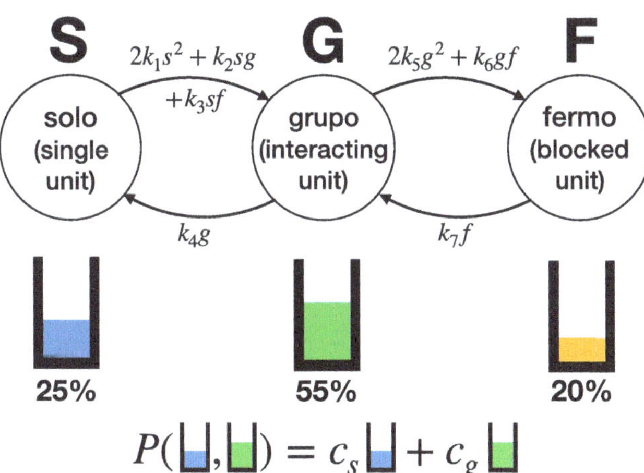

Fig. 2.12 Schematic of the SGF model indicating transition rates, the population model aspect giving swarm fractions in states *solo* (S), *grupo* (G), or *fermo* (F), and an example performance function P. *Source* Hamann and Reina [322]

there are no direct transitions between *solo* and *fermo*. Generally, the model can be interpreted both at an individual (microscopic) and a swarm (macroscopic) level. We introduce notation for the three sates: S for *solo*, G for *grupo*, and F for *fermo*.

The transitions and their rates are motivated by robot–robot encounters and in particular the simplest one, that is, in groups of only two (i.e., the considered robot and a robot it is encountering). We can now go through all combinatorial possibilities SS, SG, SF, GG, and GF, while ignoring ordering (e.g., SG and GS are not treated differently). We ignore FF because robots in *fermo* encountering each other will stay in *fermo*. Based on the minimal robot–robot interaction (a group of two) and almost mechanistic considerations, we can intuitively define what results from any of these robot–robot encounters. To derive the model, we follow a rate equation approach (for details about this method see Sect. 7.4.1). For each robot–robot encounter that results in a state transition, we notate a reaction equation and get

$$2S \xrightarrow{k_1} 2G\,, \tag{2.20}$$

$$S + G \xrightarrow{k_2} 2G\,, \tag{2.21}$$

$$S + F \xrightarrow{k_3} G + F\,, \tag{2.22}$$

$$2G \xrightarrow{k_5} 2F\,, \tag{2.23}$$

$$G + F \xrightarrow{k_6} 2F\,. \tag{2.24}$$

We do not need to explicitly model reactions that don't cause change. However, for now this is a one-way street as we are missing the "backlinks" (see Fig. 2.12). How will a robot ever go from *fermo* back to *grupo* or from *grupo* back to *solo*? This is only possible if a robot is able to break free from the crowd and claim some additional space around itself. Obviously, this cannot occur during an encounter with another robot; in fact, quite the opposite, it must happen alone. Unfortunately, the exact nature of this underlying process must remain unclear here and we cannot directly model it in a generic way. On the positive side, these backlinks are mathematically simple. We define a flow of robots going from *grupo* to *solo* and from *fermo* to *grupo* proportional to how many robots are in the respective state. We get

$$G \xrightarrow{k_4} S\,, \tag{2.25}$$

$$F \xrightarrow{k_7} G\,. \tag{2.26}$$

We interpret these flows as capturing the ongoing tendency of robots to escape from their *fermo* or *grupo* cluster and re-enter individual activity.

These seven reaction equations are a microscopic model of how a robot transitions between the three defined states. To obtain a population model of the swarm, that is, a macroscopic model, these reaction equations can directly be translated into rate equations that form a system of ODEs. Through van Kampen expansion [812], the reaction Eqs. 2.20 through 2.26 can be transferred borrowing the methodology from statistical physics and chemical kinetics, which relies on two main assumptions: the system is well-mixed and sufficiently large. We get

$$\begin{cases} \frac{ds}{dt} = -2k_1 s^2 - k_2 sg - k_3 sf + k_4 g \\ \frac{dg}{dt} = 2k_1 s^2 + k_2 sg + k_3 sf - k_4 g - 2k_5 g^2 - k_6 gf + k_7 f \\ \frac{df}{dt} = 2k_5 g^2 + k_6 gf - k_7 f \,, \end{cases} \tag{2.27}$$

where the system variables s, g, and f give the numbers of robots in states *solo* S, *grupo* G, and *fermo* F. If we do not allow robots to come and leave during the considered time interval, then we have robot conservation:

$$N = s + g + f \,, \tag{2.28}$$

and we can get rid of one of the ODEs in Eq. 2.27. For example, we can leave out $\frac{dg}{dt}$ and use $g = N - s - f$ instead.

To make use of this model, we need to specify the seven rate coefficients k_i and we usually define an initial value problem. We define an initial system state, for example, $s(0) = N$, $g(0) = 0$, and $f(0) = 0$. When we choose to solve this problem numerically, we obtain the temporal evolution of this population model and an equilibrium state that it converges to. We are interested in the stable fixed points (s^*, g^*, f^*) because we interpret them as a prediction of how many robots will tend to stay in either of the three states. We can calculate a series of these fixed points for different initializations $s(0) = N$ with $N \in \{1, 2, \dots\}$. These fixed points can be plotted to show the scalability of the system as predicted by the SGF model (see Fig. 2.13). For a given task, we can define a function that defines what the desired states are. Do we want the robots to work exclusively by themselves (*solo*), in teams (*grupo*), or both (combinations of *solo* and *grupo*)? We do not desire robots to be in *fermo* as that is by definition an undesired state without any productivity (e.g., robot in traffic jam or deadlock due to a depleted shared resource).

To connect the resulting swarm fractions s^*, g^*, f^* to an actual estimate of swarm performance, we need to define a function that assigns score to being in state *solo* or *grupo*. We define

$$\Pi(N) = c_s s^* + c_g g^* \,, \tag{2.29}$$

where the contribution coefficients c_s and c_g define the contributions to a specific task by robots in state S and robots in state G. For $c_g = 0$ and $0 < c_s \leq 1$, we have a scenario that requires standard parallelization, that is, robots are not required to interact or cooperate. For $c_s = 0$ and $0 < c_g \leq 1$, we have a scenario that requires

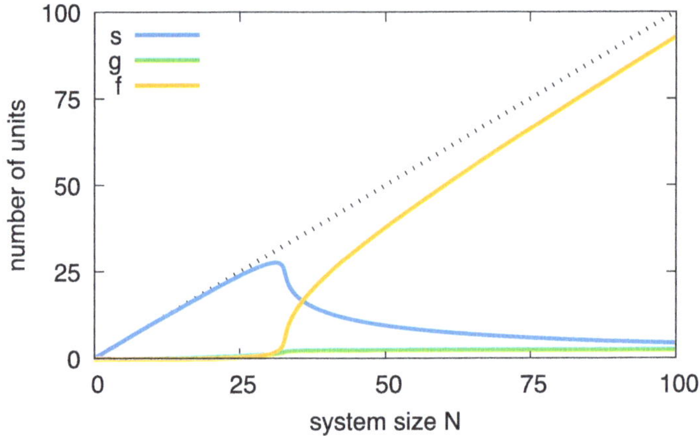

Fig. 2.13 Example data for the scalability as predicted by the SGF model (series of fixed points s^*, g^*, f^*) for $N \in \{1, 2, \ldots, 100\}$). Parameter setting is: $k_1 = 0.005$ $k_2 = 0.1$, $k_3 = 0.06$, $k_4 = 10$, $k_5 = 0.15$, $k_6 = 0.3$, $k_7 = 0.8$. *Source* Hamann and Reina [322]

cooperation. For $c_s = 0$ and $c_g > 1$, we have a scenario that requires collaboration and has superlinear performance.

A disadvantage of this model is its complexity. We require seven rate coefficients k_i that we need to determine in an application. This introduces a high number of degrees of freedom, and the resulting fitted parameter sets may not always be meaningful or interpretable. Dyson [217] famously quoted Enrico Fermi remembering John von Neumann saying "with four parameters I can fit an elephant, and with five I can make him wiggle his trunk." Who knows what our elephant can do with seven parameters? In an application we also need an interpretation of actual robot states and how they relate or translate to *solo*, *grupo*, and *fermo*. This could be done based on the robots' internal state but probably requires also an interpretation of their environment possibly on different timescales; for example: Is the robot currently blocked? Is it in a deadlock?

An advantage of this model is that it is not fully phenomenological but starts from an individual robot behavior, that is, observable microscopic behaviors can be used to estimate the model. These first principles may be rather simple and still have some ad hoc character but this goes beyond the above discussed models of USL, Gustafson's law, etc. Notably, these classical models can also be derived as special cases within this framework.

The above scalability models occur now as special cases of the SGF model. We get Amdahl's law for $k_1 > 0$, $k_2 > 0$, and $k_4 > 0$, while keeping $k_i = 0$ for $i \in \{3, 5, 6, 7\}$, for details see Hamann and Reina [322]. For the special case of choosing $k_2 = 2k_1$ we get for the stable fixed point (s^*, g^*, f^*) as a result

$$S_{\text{thruput}}(N) = \frac{N}{1 + \frac{k_2}{k_4} N} . \tag{2.30}$$

This equation has the same form as Amdahl's law (Eq. 2.7) suggesting that we get for the serial part parameter $\sigma = \frac{k_2}{k_4}$. The SGF model can describe also Gustafson's law by setting rates $k_1 > 0$, $k_4 > 0$, $k_4 > k_1$, and $k_i = 0$ for $i \in \{2, 3, 5, 6, 7\}$, for details see Hamann and Reina [322]. We obtain

$$S_{\text{thruput}}(N) = \frac{\sqrt{k_4^2 + 8k_1k_4N} - k_4}{4k_1}. \tag{2.31}$$

Even a special case of the USL can be obtained by fixing the rates $k_2 = k_3 = k_5 = k_6 = 2k_1$, $k_4 = k_7$, and by applying an approximation, for details see Hamann and Reina [322]. We get

$$S_{\text{thruput}}(N) = \frac{N}{\frac{k_2^2}{k_4^2}N^2 + \frac{k_2}{k_4}N + 1}. \tag{2.32}$$

We can translate the parameters via $\alpha = \frac{k_2}{k_4}$ and $\beta = \frac{k_2^2}{k_4^2} = \sigma^2 = \alpha^2$. Note that this is a restrictive special case of the USL. Also, Gunther et al. [307] allow negative contention parameters α. Interestingly, as we have $\alpha = \frac{k_2^2}{k_4^2}$ here, we cannot achieve negative values.

The SGF model is a step towards scalability models for swarm robotics that are well founded on microscopic and mechanistic first principles. However, it comes with a lot of complexity in both number of parameters and mathematical efforts required to solve the system of equations. As said, there is still a lot of work to do in this field and everybody is invited to contribute. We continue in the next section with a specific and fundamental insight that the SGF model provides.

2.5.3 Two-Phase Performance: Living on the Edge

Almost all literature in swarm robotics on system performance and scalability focuses on mean values of performance $\bar{\Pi}(N)$. Generally, that is a valid approach but in what conditions could that be problematic? $\bar{\Pi}(N)$ is a useful and meaningful observable if it is normally distributed: $\Pi(N) \sim \mathcal{N}(\bar{\Pi}, \sigma^2)$. It may be misleading, however, if the underlying distribution is bi- or multi-modal. One would obtain a bi-modal distribution of performances in swarm robotics, if independent runs of the system would sometimes show low performance (i.e., first peak in the distribution of Π) and sometimes high performance (i.e., a second peak in the distribution of Π).

Building on earlier work [322], we conducted a collaborative study to examine unusual features observed in one of the figures. In Fig. 2.13, the slope of $s(N)$ near $N \approx 33$ appears unusually steep, almost like a sudden jump. Also consider that this plot represents fixed points s^* that are determined numerically. This raises the possibility that the sharp change is a numerical artifact. However, it could also reflect a genuine discontinuity in the function $s(N)$ around $N \approx 33$, which is mathematically possible in systems of ODEs. To investigate further, we tested a stochastic simulation algorithm (Gillespie algorithm) that is often able to reproduce finite-size and noise effects, while the ODE system is representing mean values. We found that the rate equations of the SGF model allow for two phases of performance Π to coexist. That means that the system set to $N = 33$ as in Fig. 2.13 may dwell in either of two system states of $s^* \approx 30$ or $s^* \approx 10$. For $c_s = 1$ and $c_g = 0$ we have $\Pi(N) = s^*$. So we found a bimodal distribution of swarm performance with $\Pi(N = 33) \approx 30$ or $\Pi(N = 33) \approx 10$, with a mean of $\bar{\Pi}(N = 33) = 20$. That mean value is actually almost never observed in the system because it lies between the two peaks in the performance distribution.

In addition, there was also the known result from Hamann [319] that clearly indicated a bimodal distribution for swarm speeds in the emergent taxis scenarios. The question was not whether bimodal swarm performance distributions exist but how common they are, whether there is a good way of modeling them, and what the consequences of that finding are.

We found two more scenarios that clearly show a bimodal distribution. One scenario was the simulation of a warehouse solution for modern online fashion retailers aiming to maximize economies of scale. It uses robotic storage units forming a swarm of up to 10^5 low-cost, autonomous "hangers" on a powered rail grid, each carrying a bag as a storage unit. Combining global messaging with local coordination at rail intersections supports decentralized control, making it a hybrid system (combination of centralized and decentralized elements) that has potential to scale (see Kuckling et al. [451] for more details). The other scenario was an object clustering task, where a swarm of robots collects randomly placed objects autonomously. An external scalar field is used to guide robots and nudge objects toward the field's global minimum [817]. Both of these additional scenarios showed that interesting two-phase swarm performance, which clearly indicated that it is a rather common system property in swarm robotics.

Based on these findings we assume that it is a common feature in systems of swarm robotics that we have bimodal swarm performance distributions (common but probably not omnipresent). We call them two phases: an upper phase of high system performance $\Pi \approx \Pi_{\text{opt}}$ that is often near optimal and a low phase of minimal performance $\Pi \approx 0$. Between the two-phases ($0 < \Pi < \Pi_{\text{opt}}$), there is a region in the performance distribution that is barely populated. This poses a potentially dangerous scenario. If we add one (or a few) more robots to a close-to-optimal swarm system $\Pi(N) \approx \Pi_{\text{opt}}$, then there seems to be a chance that the system does not only not scale well but it may break down immediately showing low performance $\Pi(N + 1) \approx 0$. The well performing system can be in a critical state close to a tipping point and adding a few more robots may push it over the cliff.

We used the above SGF model (Sect. 2.5.2) to study this potentially challenging condition. The results by Kuckling et al. [451] show that this situation does indeed exist and that we can also model it. We performed a complete analysis of the system using methods of nonlinear dynamics and identified a specific type of bifurcation. A bifurcation in a dynamical system, such as our swarm system described by the time-dependent swarm performance $\Pi(N, t)$, occurs when a small change in a system parameter (in this case, the swarm size N) causes a sudden, qualitative change in the system's long-term behavior. By long-term behavior, we mean the system state to which the swarm settles over time, denoted as $\hat{\Pi}(N) = \lim_{t\to\infty} \Pi(N, t)$. In nonlinear dynamics, this is called the asymptotic behavior and in the SGF model it is the equilibrium state (Sect. 2.5.2). In this case, the system exhibits a so-called cusp bifurcation. This type of bifurcation arises when two control parameters, in our case, the swarm size N and a rate coefficient of the SGF model, jointly influence the system's behavior, leading to different types of bifurcations depending on their values. It is named after the cusp-shaped structure of its bifurcation diagram, resembling the pointed tip of a breaking wave at the beach, see Fig. 2.14.

A robot swarm system may experiences a bifurcation, making its performance highly sensitive to small changes in swarm size N. There are also hysteresis effects that would complicate to adjust the swarm size during operation. Hysteresis occurs when a system's response depends on its history, not just its current conditions. This memory effect can cause the system to remain in a given state even as inputs shift slightly. As a result, changes that would normally alter the system's behavior may need to be more extreme to overcome its previous state. Here that means that

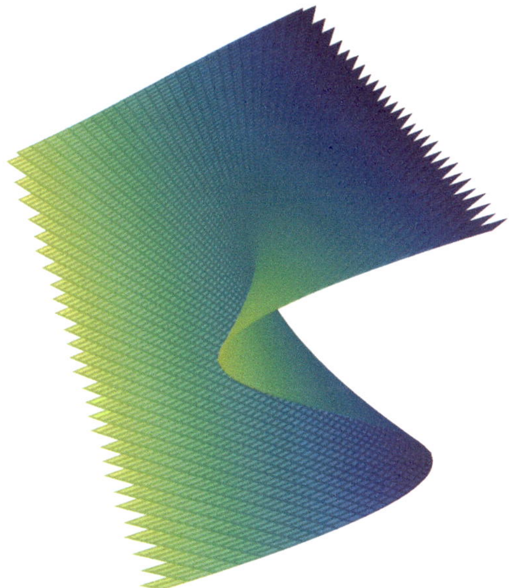

Fig. 2.14 Schematic representation of a cusp bifurcation, in our context here: the upper part represents the high performing ("upper phase") and the lower part the low performing swarm configuration ("lower phase"), they coexist for a certain interval of swarm sizes (horizontally), and navigating from one phase to another happens suddenly in jumps (i.e., a danger of suddenly breaking down swarm performance). The depth of the plot represents a parametric dimension (e.g., here for example a parameter k_i of the SGF model). The middle part of the s-shape represents an instable fixed point, so the middle ground also doesn't come for a rescue

after having added one more robot ($N + 1$) and having observed a breakdown to low performance, we may need to remove more robots than one to get back to the high-performance phase.

For swarm robotics systems optimized for peak performance, this situation can place them precariously close to sudden, catastrophic breakdowns. Extended transients might conceal this vulnerability or potentially allow the swarm to maintain high performance for the entire mission duration.

2.6 Analogies from Other Fields

One may be tempted to disregard a possible deeper meaning of these scalability models and results. Definitely, it is fine to consider the methods presented in this chapter exclusively within applications of swarm robotics. However, there seems to be a fundamental relationship between swarm size and swarm performance that holds across a much wider variety of otherwise intriguingly diverse systems. We observe similar performance-over-density curves across different disciplines, such as biology, electrochemistry, and modeling of engineered transport systems. Common to all of these systems is that they rely on some form of material transport (e.g., sugar, ions, vehicles). This indicates that our investigations of scalability in swarm robotics could be studied as a more general phenomenon of material transport and possibly be modeled in one fundamental approach. If required, this speculation can serve as an additional motivator to push for scalability models also in swarm robotics. Next, we give a few examples from different fields. They are related on an abstract level and share features and modeling properties with robot swarms.

2.6.1 Traffic and Flow

2.6.1.1 Transportation Forecasting

One of the most prominent examples for scalability and its modeling may be vehicle traffic as observed, for example, on a highway. There is a long history of developing and studying traffic flow models. The purpose of transportation forecasting is to better understand how congestion forms and how to predict it. A good overview is given by Helbing [349] as well as by Wageningen-Kessels et al. [813] and Li [488].

The most basic vehicle traffic model is the so-called fundamental diagram of traffic flow that plots flow against density. Flow is the rate at which vehicles pass a certain point on the road, for example, measured in vehicles per hour. Density is the number of vehicles per unit length of the road, for example, measured in vehicles per kilometer. In our context here, we could speak of throughput (see Sect. 2.4.1) and swarm density (see Sect. 2.2). The fundamental diagram captures the tradeoff between how many vehicles are on the road and how efficiently they can travel given

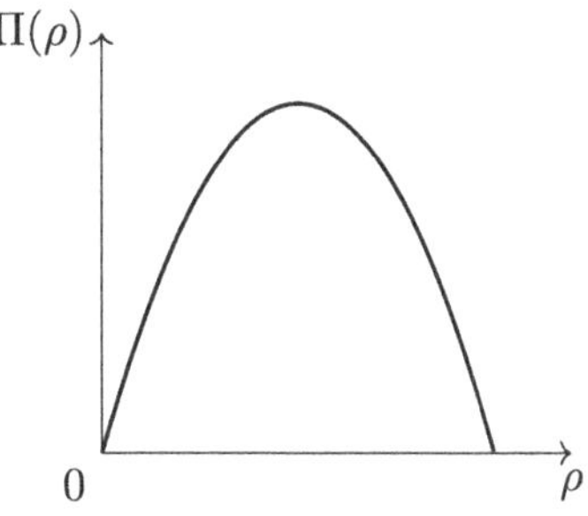

Fig. 2.15 Fundamental diagram of traffic flow $\Pi(\rho)$ (e.g., vehicles per hour) over vehicle density (e.g., vehicles per kilometer) according to early influential works of Greenshields et al. [296]

the limited shared resource of space. Following the early model of Greenshields et al. [296] we have

$$\Pi(\rho) = \rho(1 - \rho), \tag{2.33}$$

for vehicle density ρ and flow Π. Note that we have a concave function (see Sect. 2.3.1). The full model actually includes constants: $\Pi(\rho) = v_f \rho(1 - \frac{\rho}{\rho_{\max}})$ for free speed v_f and maximal density $\rho_{\max}$. Key findings of the fundamental diagram of traffic are: (1) At low traffic density, adding more vehicles increases flow due to minimal interference (see Fig. 2.15). (2) As density rises, vehicles slow each other down, eventually reaching a peak flow (road capacity) at a critical density ρ_c. For $\rho > \rho_c$ flow is reduced. (3) For a given flow $\Pi \neq \Pi_{\max}$, we get two possible densities ρ_1 and ρ_2 that implement that flow. In vehicle traffic, density can be influenced by speed limits. Hence, a traffic engineer, at least in the ideal situation, has at least two options to design the traffic flow. Option 1: Operate at lower density ρ_1 with higher vehicle speeds and more spacing. Option 2: Operate at higher density $\rho_2 > \rho_1$ with lower vehicle speeds and reduced spacing.

2.6.1.2 Traffic Jams in Ants?

While it is maybe expected that humans struggle to organize their vehicle traffic, also ants have to deal with physical interference sometimes although they have fantastic crawling and climbing abilities. The study by Dussutour et al. [215] on optimal traffic organization in ants demonstrates how the black garden ant (*Lasius niger*) efficiently manages crowding on trails. Ants use pheromone trails to direct traffic while foraging. However, when trails get overcrowded, interactions like collisions can slow down the group's overall efficiency. In their setup, ants travel between their nest and food across a diamond-shaped bridge with two branches: a wide and a narrow branch. At low density, most ants choose a single branch, optimizing based on pheromone concentration. In high-density conditions, ants distribute themselves more evenly across both branches and show lane splitting. This transition to symmetrical traffic is an effect of self-organization. When ants collide on a narrow trail, ants coming from the food source (returning to the nest) are more likely to push ants going toward the food source. This helps to distribute traffic more evenly across the two branches. Dussutour et al. [215] focus in their modeling approach on a temporal model of flow including dynamic, feedback-driven contention and a load-balancing mechanism that reduce the risk of retrograde scaling (i.e., no decrease in swarm performance for increasing swarm density, rather saturation of swarm performance).

In a similar study, Poissonnier et al. [643] investigate how Argentine ants manage traffic flow under extreme density. They found that ant flow saturates but remains constant even at high densities. They introduce a stepwise linear two-phase flow model. They define the flow of ants (ants per unit time) for a density ρ as

$$\Pi(\rho) = \begin{cases} \rho v & \text{if } \rho \leq \rho_{\text{jam}} \\ \rho_{\text{jam}} v & \text{if } \rho > \rho_{\text{jam}} \end{cases}, \tag{2.34}$$

for mean walking speed v of ants under low-to-moderate densities and $\rho_{\text{jam}} v$ is the saturation flow, which is the maximum throughput of the system. At low densities, ant flow increases linearly with density (similar to Greenshields' model). At high densities, flow saturates, but crucially does not degrade, indicating remarkable resilience to congestion. Microscopically, ants dynamically adapt their speed and interaction time, accelerating at intermediate densities and avoiding entering overcrowded paths. It is impressive that ants manage to maintain a saturated flow under these high-density conditions. Eq. 2.34 can be seen as an approximation to Amdahl's law or the diminishing returns case (e.g., the Universal Scalability Law with $\alpha \gg 0$, $\beta = 0$).

2.6.1.3 Optimal Concentrations in Transport Systems

Particularly relevant to our view of scalability as a fundamental and general phenomenon are established models of transport systems. These models are already less specialized and can be applied across a wide range of domains. A clever modeling approach can, for example, connect biological results on plants with results from studying engineered systems. The study of Jensen et al. [386] illustrates how simple principles (e.g., balancing flow and resistance) scale across diverse biological (animals and plants) but also engineered systems (vehicle traffic). For example, they study sugar transport in plants (vascular flow) for the efficient transport of phloem (nutrients and sugars). They also check data for drinking from a tube and blood flow in vertebrates. Key concepts that they combine in their model include viscosity-induced impedance as resistance that limits the rate of material transport and substance concentration in the role of determining the efficiency of transport. Translated into our context, viscosity is a measure of interference in the system and concentrations represent swarm densities. They obtain a concave function of flow over concentration again (cf. Sect. 2.3.1) that defines optimal concentrations for all studied transport systems. The normalized flow Π over normalized concentration or density ρ is defined as a function in the form of $\Pi(\rho) = \rho(1 - \rho)$ (Greenshields,

Eq. 2.33). The normalized impedance is in the shape of $\sqrt{1/(1-\rho)}$ similar to the interference function by Eq. 2.19.

Many transport systems, natural or artificial, seem to obey similar laws. This supports our speculation that scalability could be studied as a general material transport problem. A unified theoretical approach to scalability could be developed starting from models of transport systems.

2.6.2 *Biology, Chemistry, and Human Groups*

In the following, we give three examples, one from biology, chemistry, and humans each.

2.6.2.1 Optimal Group Size in Biology

In biology, scalability as such is rarely discussed. Probably for at least two reasons. Most animal systems are well adapted and prevent scalability issues before they pose a problem (as the example of ants above in Sect. 2.6.1.2). Hence, biologists are not motivated to search for scalability models. The other reason could also be the lack of data, which is much harder to obtain in biological systems.

The arguably most important subject for scalability in biology is the group-size paradox [445, 733]. This is an idea from the perspective of behavioral ecology and evolutionary biology on expected group sizes across all social species. While a particular group size N^* may be optimal for the swarm (highest fitness for all swarm members), it does not guarantee stability of the group size. Stability can be understood here roughly in the terminology of game theory. An (optimal) group size is not stable if it is attractive for single individuals to join the group and consequently increase group size.

The concept of Sibly [733] is that stable (and hence observed) group sizes are generally larger than the optimal group size ($N_{\text{eff}} > N^*$). It may be beneficial for individuals who are alone or in smaller groups to join an already optimal-sized group and hence increase its size further. Say, individuals arrive sequentially and select among available resource sites. We assume that their choices are guided by an individual fitness function $f(N)$ for group size N (see Fig. 2.16, cf. individual performance π). We also assume free entry conditions, that is, animals are free to choose whether to join a group without being excluded by others or by physical constraints. Note that in biological publications the actual fitness function $f(N)$ is usually not made explicit mathematically. The first individual chooses a random site. The second individual chooses to join the first one because it is beneficial to be in a group. Subsequent individuals continue to make the same choice of joining this group as long as the fitness benefit of joining the group (marginal gain) outweighs the fitness of being solitary, that is, $f(N+1) > f(1)$. Individuals benefit from joining a group even if its size N exceeds the optimal group size N^*. If for N_{eff} we have

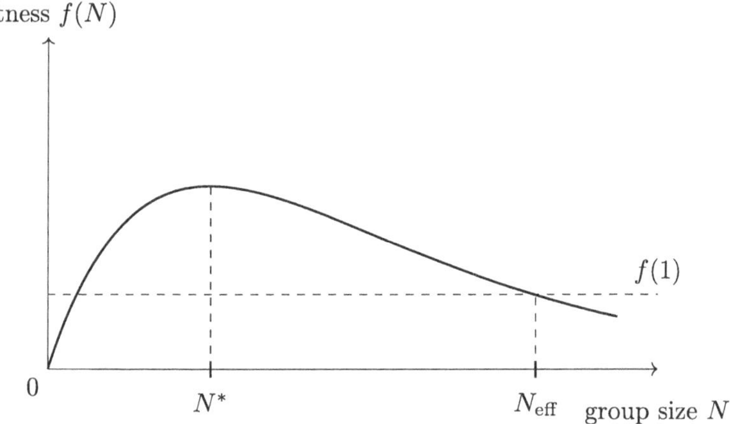

Fig. 2.16 Group-size paradox: This qualitative graph shows individual fitness f for group members over group size $N \geq 1$. There is an optimal but instable group size N^* and effective group size N_{eff} with $f(N_{\text{eff}}) \approx f(1)$. The idea is that animals keep joining the group as long as the fitness of the increased group is bigger than living alone: $f(N+1) > f(1)$. Shown is here an arbitrary choice of $f(N) = 3N \exp(-0.5N)$

$f(N_{\text{eff}}) \geq f(1)$ and $f(N_{\text{eff}}+1) < f(1)$, then all groups will theoretically stabilize at an effective group size N_{eff}. This interpretation, however, raises a paradox: groups stabilize at a size where membership offers almost no advantage over being solitary, challenging the evolutionary basis for grouping under the strong assumption of so-called free entry conditions.

This needs an explanation that reconciles the paradox by considering additional ecological and evolutionary factors. The modeling assumptions above are too simple as we ignore that resource availability and competition change with group size. Larger groups may experience diminishing returns due to intra-group competition. The above reasoning assumes free entry conditions as individuals can selfishly join a group without restriction.

We assume that anyone can join, that there are no movement costs (switching groups may have a cost), and also perfect knowledge about the resulting fitness. All these assumptions seem rather unrealistic.

A related concept is the Allee effect [13, 237, 747]. Individual fitness may actually increase with group size when groups are very small (see Sect. 2.3.3). Mechanisms, such as cooperative defense, predator detection, or collective foraging, can cause individuals to perform poorly in isolation but better once a minimal group size is reached. In terms of the fitness function $f(N)$, this means that $f(1)$ may be lower than $f(N)$ for small $N > 1$, producing an upward slope before competition effects dominate. The Allee effect complements Sibly's argument by emphasizing that below a critical group size, individuals may struggle to survive or reproduce, so the incentive to join groups is not just advantageous but essential.

2.6.2.2 Chemistry and Conductivity in Electrolytes

Our next example is from chemistry and may be unexpected. A quite specific but relevant example for transport in chemistry is the conductivity in electrolytes. An electrolyte (e.g., table salt) that is dissolved in a solvent (e.g., water) separates into cations (positively charged ions) and anions (negatively charged ions). Once they are subject to an electric potential, anions move towards the anode (positively charged) and the cations move towards the cathode (negatively charged). The conductivity depends on temperature but more importantly on the electrolyte's concentration. This concentration is the equivalent to our robot density ρ. In literature on chemistry, the conductivity is often given as molar conductivity Λ_m that is the conductivity normalized by the concentration [875]. Here we avoid normalizing conductivity by concentration and discuss instead absolute conductivity. For conductivity in strong electrolytes (i.e., fully dissociated, all molecules separated into cations and anions) we get according to Kohlrausch's law of independent ionic migration a (molar) conductivity of

$$\Pi(N) = aN - bN\sqrt{N}, \tag{2.35}$$

for constants a (limiting molar conductivity) and b (Kohlrausch constant). This is also a concave function (cf. Sect. 2.3.1) that is indeed similar to performance measurements in swarm robotics.

Why might models of ion transport in electrolytes possibly show similarities to models of congestion issues in swarm robotics? In an electrolyte, positive and negative ions move in opposite directions under an electric field. In a swarm, robots move in various directions, often interfering with each other. At low concentrations, ions move freely, similar to robots in a sparse swarm. However, as ion concentration increases, ion–ion interactions become more frequent (creating resistance, lowering conductivity), impeding their mobility. In high-density robot swarms, frequent collisions or interference among robots reduce overall mobility and efficiency.

It is interesting to check the history of science for models of conductivity in electrolytes. Kohlrausch's law of independent ionic migration was published around 1900. Conductivity models were focused on low concentrations and even linear models were suggested. However, with increasing concentrations all simple modeling approaches break down. For example, for higher concentrations there are electrostatic forces that slow down the motion of ions. There were a number of efforts in physical chemistry to fix these oversimplified models. For example, the Debye–Hückel theory (1923) where each ion is taken to have an ionic atmosphere made up of ions of the opposite charge and ionic strength is proportional to concentration. Another important contribution is that of Fuoss and Onsager [262, 606] providing the Fuoss–Onsager equation (between 1932 and 1955) that we simplify here to

$$\Pi(N) = aN - bN\sqrt{N} + cN\ln(N) + dN^{5/2}. \tag{2.36}$$

A slightly simpler model is provided by Kay [410] (1960) given by

$$\Pi(N) = aN - bN\sqrt{N} - cN^2 . \tag{2.37}$$

All of these scalability models are close to our observations in swarm robotics. The connection is the underlying transport phenomenon. Ions travel through space. With increasing concentration (i.e., increasing swarm size), they increasingly interfere with other particles; slowing them down. This is similar to our robots spending more and more time on collision avoidance actions with increasing swarm density.

2.6.2.3 Motivation in Human Groups

To conclude, let us turn to human groups. Also non-physical interference can increase the overhead due to high swarm densities. An example from the performance of human groups is the so-called Ringelmann effect which describes the loss of motivation of individuals working in groups with increasing group size. A nonlinear decrease of individual performance with increasing group size is reported by Ingham et al. [375] (see also Kennedy and Eberhart [419, p. 236]). Ringelmann [676] himself reported, for example, empirical data for an unspecified physical task (see Fig. 2.17) showing throughput speedup of $\Pi_{\text{thruput}}(N = 8) \approx 4$ and hence a decrease in individual performance from $\pi(N = 1) = 1$ to $\pi(N = 8) \approx 0.5$. Overall, the data in Fig. 2.17 shows diminishing returns for relative group performance Π and linearly decreasing relative individual performance π.

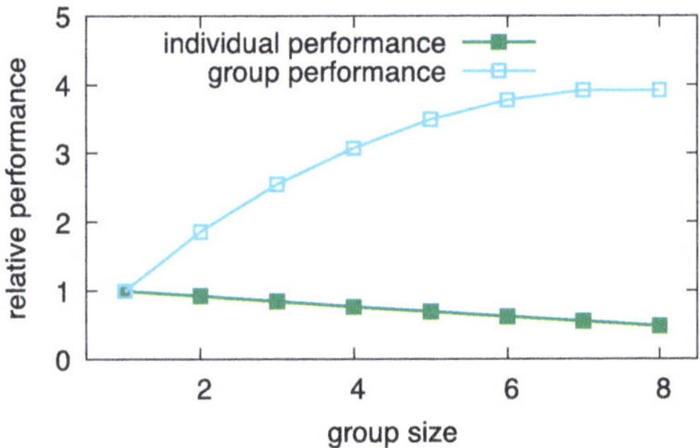

Fig. 2.17 Scalability in human groups according to Ringelmann [676] in an unspecified physical task, relative performance Π shows diminishing returns as a function of group size N and individual performance π reduces linearly

2.7 Implications for Swarm Design

We end this chapter by raising questions and pointing to a couple of specific scalability aspects that could prove to be of interest in the future. Inspired by Weinstock and Goodenough [855] we can ask questions, such as: If a system fails to scale, what caused it? Is scalability inherently tied to a tradeoff, meaning that achieving scalability may require sacrificing something else, such as performance or by paying another cost? What are scalability assurance methods? How to do a scalability audit? Weinstock and Goodenough [855] focus on computing systems but their take on whether testing for scalability is actually feasible is interesting and may also translate to swarm robotics: "You can't test for scalability in the sense that it may be impossible even to simulate the effect of increased workload and increased capacity." Usually in robotics, we are quite eager to simulate systems and we are also ready to use sophisticated software to simulate the physics of robots and other complexities; however, the feasibility of testing for scalability can indeed be challenging. We discuss this shortly below as meta-scalability (see Sect. 2.7.3).

We focus on several implementation details. In real-world deployments, swarm size and density may change dynamically. This requires adaptive mechanisms to preserve performance. We discuss that as online scalability. Can a robot swarm monitor its own scalability and respond to signs of degradation? We address that as robust scalability and congestion control. A critical enabling technology for scalability is communication. We quickly discuss some available options as of today.

2.7.1 *Online Scalability*

Scalability in swarm robotics is commonly treated as an offline property. Following this perspective, we assume that the initial number of robots N remains fixed throughout a mission, with no additions or removals (e.g., due to failing robots) once the task has begun. In contrast, a few studies have examined what could be called online scalability. We can attempt to implement a distributed self-organizing mechanism that enables a robot team to either dynamically regulate swarm size N at runtime autonomously (i.e., actively change swarm density ρ) or at least aspects of the robot behavior to adapt to changing swarm densities (i.e., react to changes that are external).

For example, Mayya et al. [532] propose a decentralized strategy for reducing interference in swarms performing distributed object collection tasks. A robot independently decides whether to continue participating or to withdraw from the task, guided solely by encounters of other robots and without explicit communication. Robot experiments show that this mechanism effectively governs swarm density and improves overall swarm performance.

Similarly, Wahby et al. [840] investigate how a robot swarm can adapt to external factors that influence its density. Their approach leverages localized information-sharing, allowing individual robots to detect shifts in density and collectively adjust their behavior to maintain optimal swarm performance. Swarm robot experiments confirm that this form of swarm-wide density awareness enhances adaptability, offering a practical solution for environments where swarm density may fluctuate over time due to external effects.

Both of these general strategies, actively modulating swarm density autonomously by the swarm itself and adapting robot behaviors to dynamic swarm densities, haven't received the attention in research that they deserve. In applications, robot swarms will face dynamic swarm densities and we need to prepare for them.

2.7.2 Robust Scalability and Congestion Control

An under-explored theme in scalability for swarm robotics is how to ensure scalability and how to detect issues concerning scalability at runtime. In other words, how to create swarm-awareness of how well the system scales dynamically online after deployment. Surely many swarm algorithms touch this challenge by trying to regulate flows of robots and by cleverly avoiding collisions, etc. However, rarely there is an effort in trying to measure and adapt for scalability with the exception of the two above mentioned papers [532, 840].

Marcolino and Chaimowicz [514] address the problem of congestion in swarm navigation when many robots simultaneously attempt to reach the same target. They propose a distributed coordination algorithm based on a probabilistic finite state machine, where robots switch between states (waiting, locked, or impatient) to regulate access to a target region. Using only local sensing and communication, robots self-organize so that only a few approach the target at a time, while others temporarily wait. The method is analyzed mathematically and validated in both simulations and experiments with up to $N = 12$ e-puck robots. Results show that the approach reduces congestion, smooths navigation, and scales well with increasing swarm size. Similar approaches include those by Soriano Marcolino et al. [739] and dos Passos et al. [206].

What is found more frequently in the literature is the (offline) analysis of swarm density on different systems and scenarios. For example, Kwa et al. [457] investigate how agent density influences the performance of robot swarms in collective target tracking tasks. They identify a characteristic transition phase in swarm density: at low densities, swarms fail to track targets due to insufficient cooperation, while at high densities, performance degrades due to interference and motility-induced phase separation.

As a second example, Khaluf et al. [429] investigate how agent density affects the scalability of collective-decision-making in swarms. They show that sparse swarms relying on noise-induced bistability do not scale well because their decision dynamics deteriorate with increasing size. A key finding is the identification of a critical density threshold, beyond which collective-decision-making transitions from unistable (undecided) to bistable (decided) behavior.

We may also draw useful insights from other fields. A prominent example is transportation, where various mechanisms are employed to manage congestion and optimize flow, such as traffic lights, metered ramps, congestion pricing (tolls that increase with traffic), and smart routing applications. Similarly, air traffic control relies on multiple constraints, including flight scheduling, holding patterns, and slot management at airports.

A field that may provide more readily applicable methods, however, is likely Internet congestion control in the form of Transmission Control Protocol (TCP) congestion control [408, 656], and congestion control in datacenters [452, 491]. Congestion control defines methods to detect or anticipate congestion and methods to adjust the protocol accordingly. Signals for congestion in a network include packet loss possibly because router queues are overflowing, increasing delay because packet waiting times increase, and explicit messaging by routers or switches that notify about congestion. Work or "high density" in a network is generated by senders. They can adjust their data transmission rates. In Additive Increase/Multiplicative Decrease (AIMD) a sender gradually increases its sending rate until it detects congestion, then sharply decreases it to alleviate network pressure. In Congestion Avoidance, the sender carefully increments its sending rate more slowly after having detected congestion to find a stable operating point that doesn't overwhelm the network.

Another relevant contribution is that of Aguilar et al. [5] as they investigate how groups of fire ants optimize excavation performance in confined tunnels by individual idleness and selective retreating. They apply their findings to a group of mobile robots. These strategies minimize clogging and improve traffic flow. Unequal workload distributions and occasional retreating behavior help to prevent persistent clusters and reduce traffic congestion.

Asadi et al. [37] study multi-agent pickup and delivery (MAPD) which combines multi-agent path finding (MAPF) with a sequence of pickup and delivery tasks. They offer an online algorithm for the MAPD problem that combines local collision detection with a deep-learning-based congestion prediction model to improve efficiency. Simulation results show that the method significantly enhances throughput and scalability.

Whether these methods can be directly transferred to swarm robotics remains a research question. Some strategies, however, suggest themselves, for example, detecting congestion, slowing robots down, or increasing inter-robot spacing to minimize interference.

2.7.3 Meta-Scalability

It seems clear that testing for scalability is potentially challenging. In the extreme case, we would need to buy many robots and test the scalability of our robot swarm in hardware. That is not how we do engineering, that is, to build something and test it once it is physically existent. Even if we turn to simulations, the question is raised whether our simulator scales. That is what I call "meta-scalability": the models and tools to implement, test, ensure, and study scalability need to be scalable themselves. In the case of swarm robotics, this tool is most prominently our robot simulator. Indeed, that is a challenge. There are only few robot simulators on the market that properly scale as discussed in literature [633, 679, 820]. Basic constraints in terms of computation complexity are given by calculating pairwise interactions that indicates time complexity of $\mathcal{O}(N^2)$. However, using spatial data structures (e.g., quad-trees) can reduce the complexity to $\mathcal{O}(N \log N)$ by limiting checks to nearby robots. A good way of improving meta-scalability can be hardware-in-the-Loop (HIL) simulations. We can combine scalable simulators with real-world hardware. Hardware in the loop setups can help to parallelize also computational load in hardware and to narrow the reality gap. Another aspect is whether it is always possible to communicate the results of complex observations of the swarm to a human designer. This may sometimes require sophisticated methods of processing, filtering, and visualizing data [438].

2.7.4 Scalability in Communication Technology

Specific technological aspects contribute directly to scalability in swarm robotics. Here we focus on the scalability challenge of communication, which is often a limiting technical factor in implementations of swarm robotics. In swarm robotics, we restrict communication to neighbors only to avoid bottlenecks. Still, the range of this 1-hop communication is a design parameter and a clever choice of technology is required. In standard approaches of swarm robotics, strict constraints on maximum latency or similar timing aspects are typically absent. However, such constraints may become critical when applying methods from distributed control theory [377, 489, 603].

Wi-Fi (IEEE 802.11 Standards). Wi-Fi is a widely adopted wireless communication technology known for its high data rates and broad compatibility with various devices. In swarm robotics, Wi-Fi could be an option if we need to transmit a lot of data, for example, video streams. However, Wi-Fi tends to have significant scalability issues for high numbers of robots.

Bluetooth and Bluetooth Low Energy (BLE). Bluetooth, particularly its Low Energy variant, is an interesting choice for short-range, low-power communication in swarm robotics. Bluetooth Mesh networking allows robots to relay messages. Bluetooth nowadays is very plastic and can be customized for specific use cases. Still, range and data rates are rather limited. BLE in general offers only moderate scalability through mesh networking.

Zigbee (IEEE 802.15.4). Zigbee is a low-power, low-data-rate wireless communication standard for mesh networks. It has limited bandwidth and managing large mesh networks can become complex. Zigbee is supposed to scale effectively due to its mesh networking capabilities.

Infrared (IR) Communication. Infrared (IR) communication utilizes light in the infrared spectrum. In swarm robotics, IR is commonly used for proximity sensing and can in addition also be used for short-range (cm-range), line-of-sight communication. IR communication scales easily because of its short range.

Cellular Networks. Cellular network technology offers high bandwidth, low latency, and extensive coverage. In swarm robotics, it is relevant for outdoor use when large-scale swarms operate over wide areas. However, it relies on existing cellular infrastructure. It is supposed to be highly scalable supporting thousands of connected robots.

Ultra-Wideband (UWB) Communication. Ultra-Wideband is a wireless technology that transmits data using very short pulses across a wide frequency spectrum, offering also precise localization in addition to communication. In swarm robotics, UWB is an excellent choice when accurate (relative) positioning is required (e.g., formation control). The complexity of implementing UWB systems may require specialized expertise. UWB is moderately scalable.

2.7.5 Deployment Scalability

The challenge of deployment scalability is easily summarized by the following consideration. In lab experiments as presented by Rubenstein et al. [686], one needs to deploy up to $N = 1000$ robots. In general that would require to prepare them possibly by uploading the newest version of the controller software, charging them, testing them, and placing them in their initial position for the experiment. If that would require only $t_{\text{prep}} = 3$ minutes per robot, this would sum up to $T_{\text{prep}} = N t_{\text{prep}} =$ 50 hours. Say, an experiment (aka mission) lasts $T_{\text{mssn}} = 25$ hours, then we would get an effective preparation-to-mission-length ratio[2] of $\frac{T_{\text{mssn}}}{T_{\text{prep}}} = \frac{25}{50} = 0.5$. That could be tolerable in an academic environment and given we have $N = 1000$ robots at work. However, for outdoor robot swarms with rather complex robot hardware and bigger sizes, handling the deployment becomes a scalability issue. Realistic values could then be a preparation time per robot of $t_{\text{prep}} = 30$ minutes, for a swarm of $N = 10$ robots, and a mission time of $T_{\text{mssn}} = 4$ hours. That would give us a ratio of $\frac{T_{\text{mssn}}}{T_{\text{prep}}} = 0.8$ that we may not want to tolerate especially in a commercial context. First and foremost, this is an engineering problem of simplifying the deployment as much as possible (reducing t_{prep}) It can be countered by trying to increase the mission time (increasing T_{mssn}), which means to push towards long-term autonomy [221].

Deployment scalability, particularly in terms of hardware handling, should be recognized as a key challenge in swarm robotics. Even the most scalable algorithms offer limited value if we cannot deploy large numbers of robots efficiently and within realistic timeframes. Possibly, the deployment of robots can also be automatized. In

[2] This idea was proposed by Dr. Michael Benjamin during a personal conversation.

summary, we need methods for "swarm logistics" [144] including rapid mass deployment of large-scale swarms of mobile robots in the wild but also maintenance (e.g., charging infrastructure, storage solutions, integrated transport systems, recovery system) and safety protocols (e.g., human override). At the same time, we need to extend mission times by pushing towards long-term autonomy, for example, by improving hardware and software reliability, fault tolerance, durability, weather resistance, and energy efficiency.

2.8 Further Reading

The literature on this subject is a bit scattered across fields. An interesting read is certainly Gunther [306] to get a rather hands-on idea for how retrograde performance can be modeled using the USL and measured in distributed computing. Another option to understand the engineering perspective on scalability is the technical note by Weinstock and Goodenough [855]. They also discuss a "scalability audit" including scaling strategies and scalability assurance methods. To understand the connection between computing and robotics, read our paper [322]. A good intuition about scalability of robot swarms guided by empirical results is provided by Rosenfeld et al. [682]. The book of Valentini [800] contains a short section on scalability in collective-decision-making. Similarly, Almansoori et al. [17] study scalability in evolved algorithms for collective-decision-making. For those interested in scalability of human–swarm interaction, the paper by Olsen Jr and Wood [605] provides fundamental ideas (more details in Sect. 6.2.1).

2.9 Exercises

Exercise 2.1 (*Scaling of a simple computing system*) We create a simple model of a computer system. The following is a simple example from queuing theory. We assume that new jobs for the computer system come in with a constant rate of λ and are added to a waiting list. We also assume that the process is without memory, that is, incoming jobs are statistically independent from each other. Therefore, we model the incoming jobs as a Poisson process. The probability that i jobs arrive within a given time interval is given by

$$P(J = i) = \frac{e^{-\lambda}\lambda^i}{i!} \quad . \tag{2.38}$$

a. Plot $P(J = i)$ for a reasonable interval of J and $\alpha \in \{0.01, 0.1, 0.5, 1\}$.
b. Implement a program that samples numbers of incoming jobs from $P(X = i)$.
c. Implement a model that iterates over these two phases: calculate and administrate new incoming jobs, then operate on the current job (one at a time). Generate a sample of 2000 time steps for $\lambda = 0.1$ and a processing duration of 4 steps per job. What is the average length of the waiting list?
d. Change your program such that you can average the waiting time list length over many samples (independent runs of your model over 2000 time steps each). Determine the average list length for rates $\lambda \in [0.005, 0.25]$ in steps of 0.005 based on 200 samples and plot it.
e. Do the same for a processing duration of only 2 steps per job with rates $\lambda \in [0.005, 0.5]$ in steps of 0.005. Compare the two plots.

Exercise 2.2 (*Compute the stationary distribution of a finite M/M/1 queue*) Consider a simple M/M/1 queue modeled as a continuous-time Markov chain with a finite number of states $n = 5$ (i.e., a truncated queue with at most four customers in the system). Let the arrival rate be $\lambda = 2$ and the service rate be $\mu = 3$. We define the transition rate matrix $\mathbf{Q}$ for this finite system with five states $S = \{0, 1, 2, 3, 4\}$, where state i denotes that there are i customers in the system (either waiting or being served):

$$\mathbf{Q} = \begin{pmatrix} -2 & 2 & 0 & 0 & 0 \\ 3 & -5 & 2 & 0 & 0 \\ 0 & 3 & -5 & 2 & 0 \\ 0 & 0 & 3 & -5 & 2 \\ 0 & 0 & 0 & 3 & -3 \end{pmatrix}. \tag{2.39}$$

We want to find the stationary distribution $\boldsymbol{\varphi} = (\varphi_0, \varphi_1, \varphi_2, \varphi_3, \varphi_4)$. The stationary distribution gives the long-term probability of being in a certain state. That is:

$$\boldsymbol{\varphi} \cdot \mathbf{Q} = \mathbf{0} \quad \text{and} \quad \sum_{i=0}^{4} \varphi_i = 1. \tag{2.40}$$

Each φ_i represents the probability that the system is in state i in the long run. We assume that such a distribution exists and is unique (which it is for this ergodic, finite chain).

a. Set up the system of linear equations: You need to solve the equation $\boldsymbol{\varphi} \cdot \mathbf{Q} = \mathbf{0}$. This gives five equations (one per state), but they are linearly dependent. So, instead of using all five, use only four equations and add the normalization condition $\sum \varphi_i = 1$ as the fifth.
b. Numerically solve the system: You can solve this using Python (e.g., NumPy), MATLAB, or even by hand with a calculator or spreadsheet if needed.
c. Interpret the result.

Exercise 2.3 (*Scaling of cache coherence*) In computing systems with memory hierarchy, cache coherence refers to the consistency of data stored as copies in different caches of a multiprocessor system. When multiple processors attempt to read and write data stored in shared memory, cache coherence protocols ensure that all processors have a consistent view of the data. Without such mechanisms, processors might work with outdated or incorrect data, leading to errors in computation. Maintaining cache coherency is a scalability challenge because if more processors are added, contention for shared resources and the overhead of synchronizing cache states both grow.

Program a simple software to simulate a parallel system where a fixed number W of computational tasks are distributed across a varying number of processors N. Each task represents a memory operation that can result in one of three outcomes: cache hit, cache miss, or memory page load. A cache hit occurs when the required data is already present in the cache. A cache miss happens when the data must be fetched from main memory, causing a moderate delay. Worst case is a memory page load, where a memory page must be loaded from disk. Your task is to measure the system's performance in terms of throughput X (number of tasks completed per unit of time) and speedup.

We simulate task execution times in the following way: a cache hit lasts only 1 time unit, cache miss lasts 5 time units, and a memory page load lasts 10 time units. We define a fixed task pool of size $W = 10^4$. We model cache coherency probability by

$$P_c(N) = 1 - \frac{1}{1 + \exp(-0.02(N - 150))} . \quad (2.41)$$

This directly models the probability of a cache hit $P_{\text{hit}} = P_c$. The probability of a cache miss is $P_{\text{miss}} = 0.8(1 - P_c)$. As a result we have a probability of a page load of $P_{\text{load}} = 0.2(1 - P_c)$. We independently simulate processor numbers $N = \{1, 2, 4, 8, 16, 32, 64, 128, 256\}$.

a. Simulate everything as described above.
b. Instead of a constant cache miss probability $P_{\text{miss}} = 0.8(1 - P_c)$, we introduce also a function for that: $P_{\text{miss}}(N) = 1 - (1 + \exp(-0.02(N - 150)))^{-1}$. Simulate everything again.
c. Instead of constant durations for cache hit, cache miss, and memory page load, we scale them linearly with N: a cache hit lasts $(0.01N + 1)$ time units, a cache miss lasts $5(0.01N + 1)$ time units, and a memory page load lasts $10(0.01N + 1)$ time units. Simulate everything again.
d. Use your three data sets from all three sub-tasks above and fit α and β of the USL to it.

Exercise 2.4 (*Superlinear speedup*) In the following we try to implement a simple simulation of a computer system that shows a superlinear speedup as modeled by the Universal Scalability Law for $\alpha < 0, \beta > 0$ (see Fig. 2.18). To do so, we choose to simulate influences of cache coherence and cache hits in a simplistic way.

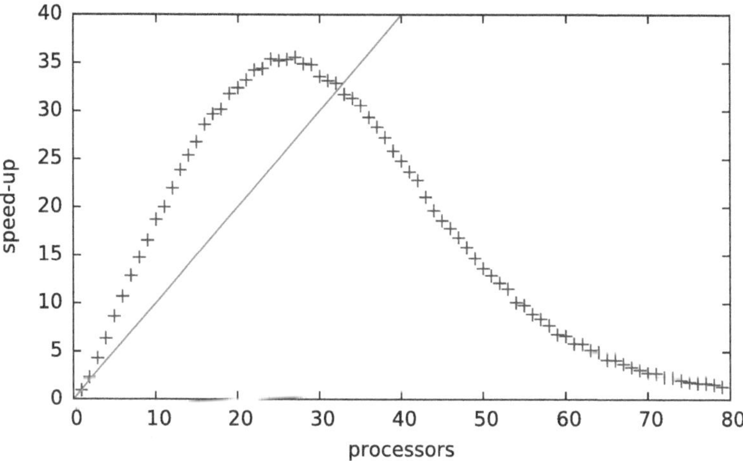

Fig. 2.18 Universal Scalability Law by Gunther [305]

a. Implement a program that iterates over $t = 1000$ time steps. In each time step simulate with simple sequential code the parallel execution of N processors. Say the problem size is 100 and each processor gets a piece of it of size $s = \frac{100}{N}$. Each processor should do one of three different tasks in each time step.
   Either it executes a task

   (1) to obtain cache coherency with probability

$$P_c(p) = \frac{1}{1 + \exp(-0.1(p - 30))}, \tag{2.42}$$

   depending on the number of processors N;
   or if the cache is currently coherent $(1 - P_c(N))$ it completes

   (2) one job of the actual task (cache hit) with probability

$$P_h(s) = \frac{1}{2}\left(1 - \frac{1}{1 + \exp(-0.05(s - 15))}\right) + \frac{1}{4}, \tag{2.43}$$

   for each processor's piece of the problem of size s;
   or it executes a task to

   (3) load a page from memory (cache coherent but cache miss) with probability

$$1 - P_h(s). \tag{2.44}$$

   For each execution of task 2 (completion of one job of the actual task) increment a variable that measures the performance.

b. Use your code to simulate different system sizes $N \in \{1, 2, \ldots, 80\}$. Adapt the size of each processor's piece of the problem $s = \frac{100}{N}$ accordingly. Keep track of the obtained performance (number of completed jobs) and calculate the speedup for all settings (performance normalized by the performance for $p = 1$). Plot a diagram of speedup over number of processors. Play around with probabilities $P_c(N)$ and $P_h(s)$. What is necessary to achieve a non-linear speedup?

Exercise 2.5 (*Swarm performance depending on opportunities for cooperation*) Show how the swarm performance scales with the number of robots under certain assumptions about the cooperation likelihood as a function of swarm density. You will model the probability that at least k robots successfully cooperate at a task site and explore how this affects global swarm performance as the swarm grows. A swarm with N robots operates in a 2D arena of area A. There are S cooperative task sites (i.e., specific areas associated with a task that require a collaborative effort to be completed), each of which requires at least k robots to succeed. At each site, an average of m robots are present (this depends on swarm density ρ). Robots are uniformly distributed in the arena. At each site, the number of nearby robots m grows linearly with swarm size N. We get $m = \beta N$. For example, we can choose $\beta = \frac{1}{S}$. The collaboration of robots succeeds at a site if at least k out of m nearby robots are available. Derive the probability $\Gamma_k(m, \rho)$ that at least k out of m robots are available to cooperate at a site. Model the swarm performance $\Pi(N)$. Let $\Pi(N)$ be the expected number of successful cooperative actions per time unit: $\Pi(N) = S\Gamma_k(m, \rho)$. Analyze how $\Pi(N)$ changes as a function of N.

a. Calculate $\Pi(N)$ for a specific case with reasonable parameters, for example: $k = 2$ (e.g., stick-pulling task), $\rho = 0.3$ $\beta = 0.01$, $S = 10$, and $N \in [1, 200]$. Plot $\Pi(N)$ and as reference $L(N) = \frac{N}{2}\Pi(2)$.
b. Explore an alternative assumption: $S \sim N$. Now assume that the number of task sites grows linearly with swarm size: $S = \alpha N$. Analyze what can be observed for the swarm performance $\Pi(N)$.

Exercise 2.6 (*Fitting models*) Find a paper on swarm robotics that contains an empirical approach to swarm performance $\Pi(N)$ (i.e., any plot of system-level performance over number of robots N or robot density ρ). A good example is the paper of Rosenfeld et al. [682]. Extract the data (get help from graph digitizer software tools). Select a number of models from this chapter that provide a function in analytical form (e.g., Eqs. 2.37, 2.13, 2.14). Choose a fitting tool to automatically fit the data to the model. Try to get a feeling for which models fit well to what kind of curves.

a. Identify plots showing swarm performance over swarm size (e.g., time to complete task vs. number of robots). If the data is not in tabular form: use graph digitization software to extract numerical values from plots.
b. Structure your dataset appropriately. Normalize performance if needed.
c. Fit Amdahl's law via σ by minimizing squared error using a tool of your choice. Plot your fit and interpret the fitted parameter σ.

d. Fit Gunther's USL via α and β. Plot your fit and interpret the fitted parameters. For example, do you obtain negative contention $\alpha < 0$? This would indicate a superlinear speedup.
e. Proceed with other models.

Exercise 2.7 (*Foraging simulation and group size analysis*) The idea is to simulate a foraging scenario roughly following Beauchamp and Fernández-Juricic [64] to analyze how individual decision-making influences group sizes under different conditions. Implement a 30×30 grid-based environment with 40 randomly distributed food patches of varying quality. Each patch provides a fixed intake rate to solitary foragers (e.g., between 0.25 and 2 items per step). Foragers, initially placed randomly, follow movement and decision rules. A population of $N \in \{50, 100, \ldots, 500\}$ foragers shares food at a patch according to a peaked fitness function, where intake increases with group size and decreases thereafter. In renewing habitats, patches replenish after depletion; in non-renewing habitats, they do not. Foragers move randomly or leave patches based on two models. (1) Null model: Foragers leave only when patches are depleted. (2) Learning model: Foragers leave if their intake rate falls below their estimate of habitat quality H_t, updated using:

$$H_{t+1} = \alpha I_t + (1 - \alpha) H_t \,, \tag{2.45}$$

where I_t is the current intake, and α $(0 < \alpha < 1)$ is a memory factor. Initialize the environment and simulate for 2000 time steps for both models under varying conditions. Record group sizes at each patch. Compute and compare the group-size distributions across both models for different population sizes and patch renewal conditions. Do these agents achieve a good load balancing across food patches? Does the learning-based model allow for larger populations?

Chapter 3
Short Introduction to Robotics

"I can't define a robot, but I know one when I see one."
—*Joseph Engelberger*

"A robot must know it is a robot"
—*Nikola Kesarovski, The Fifth Law of Robotics*

Abstract This is a little crash course in robotics for the case you haven't heard so much about robotics yet. This short introduction to mobile robotics starts from a general perspective and quickly introduces fundamental concepts, such as sensors, actuators, and kinematics. We continue with a short introduction to open-loop and closed-loop control. Agent models, behavior-based robotics, and potential field control are introduced as control options particularly interesting for swarm robots. We conclude with a presentation of several hardware platforms dedicated to swarm robotics, such as the s-bot, the I-SWARM robot, Alice, and the Kilobot, among others.

3.1 Introduction

3.1.1 Learning Objectives

This chapter offers a concise introduction to robotics, moving from general principles to core concepts and applications in swarm robotics. After completing the chapter, the reader should be able to:

- Define a robot and identify its essential components (manipulators, sensors, actuators, controller, software).
- Differentiate between robot classes (manipulators, mobile robots, hybrids) and their relevance for swarms.
- Describe main robot components, including body, joints (ball, hinge, slider), effectors, and actuators (e.g., motors, differential drive).

H. Hamann, *Swarm Robotics*,
https://doi.org/10.1007/978-3-032-10584-4_3

- Explain Degrees of Freedom (DOF) and their role in classifying robots.
- Distinguish between open- and closed-loop control, and understand the basics of PID feedback control.
- Discuss control paradigms important for swarms: reflex agents, behavior-based robotics, and potential field control.
- Recognize the challenges of odometry (dead reckoning), error sources, and the role of calibration.
- Outline the principles of SLAM and sensor fusion for reliable state estimation.
- Differentiate between forward and inverse kinematics and their applications.
- Identify key swarm robot platforms (s-bot, I-SWARM, Alice, Kilobot, Thymio II, TurtleBot 4) and their distinctive features.
- Appreciate active matter systems and how collective behavior can emerge without traditional robot components.

3.1.2 Guiding Questions

- How do different definitions of a robot (autonomous machine, reprogrammable manipulator, self-controlling system) shape our understanding of robots, especially in swarm contexts?
- What are the practical implications of Degrees of Freedom for functionality and control complexity?
- Compare active and passive sensors: what are their advantages and drawbacks, and how does interference affect active sensors in multi-robot systems?
- Why is closed-loop control generally preferred over open-loop control in mobile robots, and how does PID control minimize errors in dynamic environments?
- What are the main challenges of odometry, and how do calibration and sensor fusion help overcome them?
- How does Simultaneous Localization and Mapping (SLAM) enable operation in unknown environments, and what map representations are possible?
- Which reactive control paradigms (reflex agents, behavior-based robotics, potential fields) are effective for swarm robots, and how do they enable self-organization?
- Drawing on biology (e.g., ants), what navigation strategies (route following, path integration, map-like navigation) are relevant for swarm robots?
- How do design choices (size, locomotion, power, communication, processing) affect the suitability of swarm robot platforms such as Kilobot, s-bot, or CapBot?
- How do active matter systems challenge conventional robot definitions, and what insights do they provide into self-organization and collective behavior?

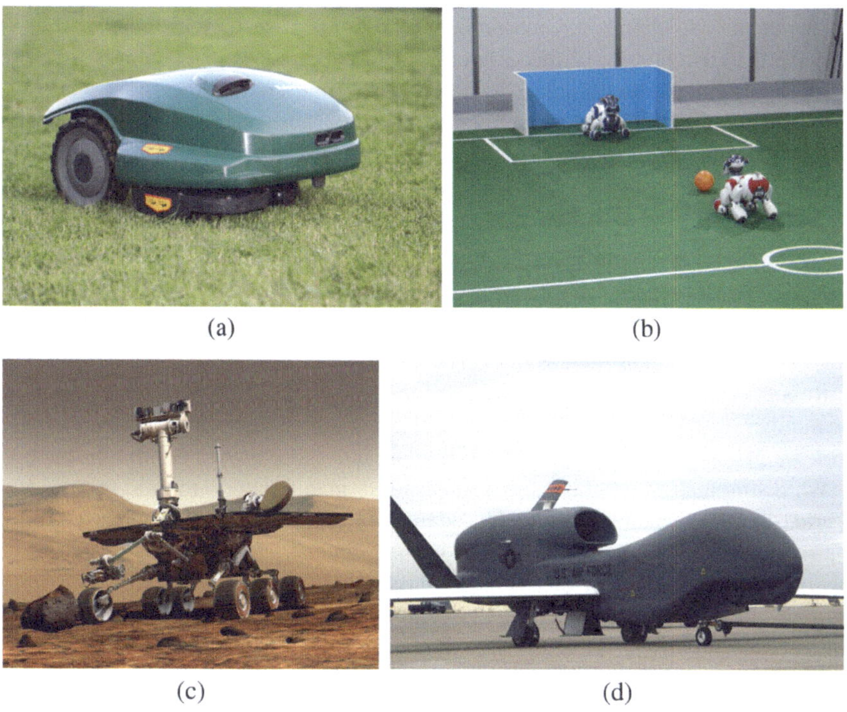

Fig. 3.1 Example robots. **a** Robomow® RM 400 (photo by Holger Casselmann, CC BY-SA 3.0). **b** Sony® Aibos at Robocup 2005 (photo by Alex North, CC BY-SA 2.0). **c** NASA Mars Exploration Rover (photo by NASA, public domain). **d** RQ-4 Global Hawk (public domain)

3.1.3 On Robots

Before we start with the actual subject of robot swarms, we have to go through some basic concepts of mobile robotics. We start with discussing what a single robot needs, and at the end of the chapter, we have a look at some examples of robots specifically designed for swarm robotics.

Similarly to other complex issues, such as intelligence or life, the concept of a robot is not easily defined. One can say that a robot is a computer-controlled, autonomous machine built for real-world functions (see Fig. 3.1 for examples). However, it is not clear whether that definition is sufficient. The *Robot Institute of America* defines a robot as "a reprogrammable, multifunction manipulator" (1979)[1] and the *European Common Market* defines a robot as "an independently acting and self-controlling machine." When we define a robot based on what it needs to be complete, we can say that a robot needs manipulators to change either its own position within the

[1] New Guide to Science by Isaac Asimov, 1984.

environment or to change its environment. A robot also needs sensors to (partially) perceive the current state of the environment, it needs a computer to process these inputs, and it needs to calculate actuator values that control the robot's motors. Usually, one adds that the robot should be a multipurpose machine, that is, it can be applied not only for exactly one task but also for many different tasks.

Robot

A robot is a computer-controlled, autonomous machine equipped with sensors, actuators, and a controller to manipulate and interact with its environment, typically designed as a reprogrammable and multipurpose system.

Robots can also be defined in terms of their degrees of freedom (DOF), that is, the number of independent joint variables that determine their possible motions (see Sect. 3.2.2) and thus their ability to manipulate objects or navigate through space.

We distinguish three classes of robots: mere manipulators, mobile robots, and hybrid robots. The manipulators are basically robot arms. The typical industrial robot is a robot arm and is commonly found in many industrial settings. These kinds of robots are very popular and have been around for many decades. However, for swarm robotics, they have little or no relevance. In fact, they are rather boring machines, often even without sensors.

Mobile robots are our primary focus. They are uncrewed vehicles that are capable of locomotion. An example is vacuum cleaning robots. Once a robot is mobile, it definitely needs sensors to make sure it is not endangering others or itself. Proximity sensors, which measure distances to nearby objects, are helpful here. Almost all robots used in swarm robotics are mobile robots.

The third class is on hybrid robots. These are mobile robots that have manipulators. The stars of hybrid robots are humanoid robots, that is, robots with arms and legs similar to human beings. Their drawback is that they are relatively complex machines, and hence they are expensive. Hybrid robots are not of much importance in swarm robotics yet.

3.2 Components

Next, we examine the various components of robots. Starting from a body (some basic static structure), a robot has effectors, actuators, sensors, a controller, and software. The layperson may forget about the existence of software because it can't be seen, but, as you know, it is the essential part of a robot. Other components are joints and additional structural parts.

3.2.1 Body and Joints

The body of a robot can be understood as a graph of links and joints. The links are parts that provide structural and physical properties. Joints connect these parts by constraining the spatial relation of two or more links. Once a joint connects two links, these two links can no longer move entirely independently of each other. What kind of movements are still possible without causing damage depends on the type of joint. There are many different types, and we just mention three here. There is a ball joint that allows rotation in three dimensions. Remember that our world is three-dimensional and three axes determine positions. An object can be moved up and down, left and right, forward and backward. In addition, objects can be oriented in different ways by rotating them around the three axes. From your flight simulation game, you may know the three ways of rotation: yawning, pitching, and rolling. So naturally, there are six so-called degrees of freedom, and a joint restricts this set of possibilities. So the ball joint allows rotation in any of the three axes. An example in your body is the shoulder joint. A hinge joint allows only for rotation around one axis. Examples in your body are the joints in your fingers and toes, while the knee actually allows some rotation (however not much as many footballers learn the hard way). A slider joint allows a translation along one axis. For some reason, the slider joint is not popular in nature. We don't have a slider joint in our body, although it may be handy in certain situations.

3.2.2 Degrees of Freedom and Pose

When you want to talk to a roboticist, make sure you have understood the concept of degrees of freedom (DOF). It is a popular approach to classify robots (especially manipulators) according to the total number of DOFs that they allow. For that approach, the number of DOF is summed over all joints. As a rule of thumb (pun intended), you can use a three-finger rule. Your thumb, first, and second finger point in the three directions of possible translations (see Fig. 3.2). In addition, imagine the three rotations around each of them. How many DOF can you identify in your arm?

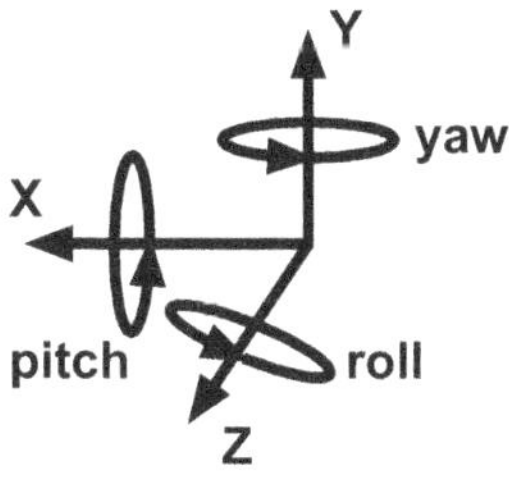

Fig. 3.2 Six degrees of freedom (DOF)

A healthy human arm is considered to have seven DOF: shoulder pitch, shoulder roll, arm yaw, elbow pitch, wrist pitch, wrist yaw, and wrist roll.

Closely related to the concept of degrees of freedom is the notion of a robot's pose.

Pose

Pose is the complete description of a robot's position and orientation in space.

In three-dimensional space, the pose is therefore specified by three translational coordinates (x, y, z) and three rotational coordinates (roll, pitch, yaw). For mobile robots moving on a plane, the pose is typically reduced to two position coordinates (x, y) and one orientation angle (heading). While DOF indicate how many independent ways a robot can move, the pose specifies its actual configuration in the environment at a given moment.

3.2.3 *Effector*

A robot's effector (also end-effector) is the component that is used to accomplish some desired physical function. The effector in a complex robot can be a hand to grip objects, but more commonly would be a torch, wheels, or legs. Hence, the effector implements the manipulation task that we determined in the definition of robots. An effector, even if it is just a wheel, helps a robot to manipulate the environment or to reposition itself within the environment.

3.2.4 *Actuator*

The actuators can be considered as the robot's muscles. In the standard case of a mobile robot, they are electric motors because they are cheap and easy to control. Many industrial robots, however, are driven by pneumatic or hydraulic systems. More fancy actuators can be piezoelectric motors or any other device that allows the robot to apply a physical force on command.

The most common actuation system in mobile robotics is the differential drive. Many robots just have two wheels, but a distinct motor controls both of them. By turning the motors at different speeds, the robot can turn or even spin on the spot.

3.2.5 *Sensor*

Maybe the most interesting type of component is the sensor. They are essential to make a robot different from a computer, as sensors allow the robot to perceive its environment. A robot doesn't have to wait for explicit user input like a computer; instead, it can observe its environment and react autonomously to changes. Sensors come in two categories: active and passive sensors.

Active Sensors

An active sensor derives information from a reaction of the environment to a previous action of the robot. Examples are bumpers, infrared, and sonar sensors. The bumper detects an object by having the environment push it into the robot. It is considered active because here we think of mobile robots that actively move. However, one could argue that a moving object could push the bumper even if the robot is static. The case is clearer for infrared and sonar sensors. They functionally rely on the concept of first sending out a signal, light or sound, respectively, and then waiting for the echo. By analyzing this received signal, the robot can determine certain properties of the environment. In most cases, infrared and sonar sensors are used to measure distances to nearby objects.

Infrared and sonar sensors belong to the class of range finders. These are sensors that are used to determine distances. Other examples are radar, lasers, whiskers, and even GPS. The problem with active sensors used as range finders is interference. Once you have too many sensors operating in a small area, they interfere with each other. Then a sensor cannot distinguish between echoes of its own signal and signals from other sensors. One could have too many sensors in one place because there are many on one robot or, typical for swarm robotics, too many robots with active sensors in one place. Interference is a challenging problem, and there is no simple solution. However, one option is to limit the range of the used sensors, hence scaling the problem down. The range of an active sensor can be restricted by limiting the transmission power (smaller amplitudes in the light/sound signal).

Passive Sensors

Imaging sensors are typical examples of passive sensors, as they, of course, rely on ambient light. They create a visual representation of the robot's environment. Besides usual cameras, also uses stereo vision systems to add depth information. Using vision is not very common in swarm robotics because it typically requires a lot of computations. Often, swarm robots, however, do not have very powerful microprocessors, and the computational load roughly increases with each additional pixel. In some cases, swarm robots have low-resolution cameras with a low number of pixels or even only a single line of pixels.

Other sensors are specialized in obtaining depth information. Time-of-Flight cameras measure depth by calculating the time it takes for emitted light to bounce off objects and return to the sensor. They provide dense, real-time 3D information and are commonly used for obstacle avoidance, mapping, and gesture recognition.

Environmental sensors measure physical or chemical properties of the robot's surroundings, such as temperature, humidity, pressure, or gas concentration. These sensors are essential for applications such as environmental monitoring, agriculture, or plum detection, where understanding ambient conditions is crucial for informed decision-making.

Example Perception-to-Action Pipeline for Vision-Based Approach

Mezey et al. [555] propose collective motion of ground robots based on vision only. A vision-based approach begins with each robot capturing raw images of its surroundings using an onboard camera. This raw visual input forms the basis of local perception and is the only source of information available to the robot in this example.

The captured image is first preprocessed to correct distortions (e.g., unwarping a fisheye lens). Visual features, such as the presence of other robots, are extracted using image processing and machine learning techniques. A common method involves object detection algorithms, such as convolutional neural networks (CNNs), which identify and localize relevant agents in the image.

Note that CNNs are a common choice for detecting and localizing objects in camera images. Still, they are computationally expensive to evaluate, particularly on small mobile robots with limited processing power. To address this constraint, the system described in this example utilizes a lightweight onboard processor (e.g., a Raspberry Pi) in conjunction with a dedicated neural inference accelerator, such as a tensor processing unit (TPU), to efficiently execute the CNN in real time.

This hardware setup allows the robot to process camera images locally at a reasonable frame rate, extract bounding boxes of visible peers, and convert this information into a simplified visual projection field. By offloading the most intensive part of the visual processing (CNN inference) to a specialized accelerator, the system achieves fast and reliable detection suitable for real-time decentralized control.

From this processed image, the robot constructs a compact representation of the visual scene. In this specific case, by Mezey et al. [555], it is a one-dimensional visual projection field that encodes where other robots are visible within the field of view. In cases of partial visibility (e.g., at image boundaries), heuristic or learned models can reconstruct the likely full presence of neighboring agents.

This abstracted visual input is then passed into a control model that governs the robot's behavior, such as adjusting its speed or turning rate. The output of this model directly drives the robot's actuators, enabling it to align, avoid collisions, or maintain cohesion with its neighbors.

This perception-to-action pipeline exemplifies how raw camera data can be transformed into meaningful control signals, enabling collective behavior without the need for global positioning systems, direct communication, or shared maps.

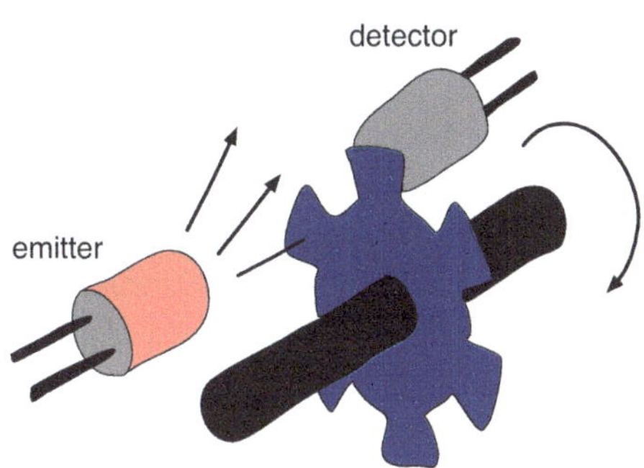

Fig. 3.3 A shaft decoder as example of a proprioceptive sensors

Sensor Fusion

One can also combine multiple sensors, ideally of different types, to improve reliability and reduce uncertainty. For example, combining data from inertial measurement units (IMU) with wheel encoders and GPS can enhance the overall state estimation of a robot, compensating for the individual weaknesses of sensor types. This is known as sensor fusion and is a common approach in robotics.

Proprioceptive Sensors

Another class of sensors are proprioceptive sensors. These are sensors that gather information about the robot's internal state, such as joint angles, wheel rotations, acceleration, or battery level. Proprioceptive sensors look inside the robot instead of monitoring the environment.

Proprioceptive Sensor

A proprioceptive sensor measures a robot's internal state, providing information about the robot itself rather than the external environment.

A typical example is a shaft decoder (see Fig. 3.3) that counts the revolutions of an axis. It is built based on a light barrier and a wheel with holes on the axis that sometimes blocks the light barrier and sometimes does not, depending on its angular position. By counting the alternating phases of light and no light, the position of the axis and how often it turned can be detected. This gives valuable information for configuring the robot and is used in the robot's odometry.

3.3 Odometry

Odometry is an important field within robotics that works on a problem that we as human beings cannot fully connect to.

> **Odometry**
>
> Odometry is the process of estimating a robot's position and orientation from a known previous position by integrating motion information from internal sensors over time.

In contrast, dead reckoning refers more broadly to estimating position from past motion without necessarily relying on dedicated sensors, making odometry the robotics-specific form of dead reckoning based on proprioceptive measurements. The underlying idea is that the robot always knows where it is. Usually, whenever I would ask you where you are, you could give me a pretty straightforward answer, even specifying the estimated distance to the next door and the room's walls if necessary. For a robot, that is not a simple task. Remember, a robot may not have vision and only a few range finders. If you like, a robot is somehow locked in a tin can and knows only a little about its environment. If you had to close your eyes while wearing a noise-canceling headset, not being allowed to move your arms, and only being told distances to the next object to your front and back, navigation would also become tricky for you.

You can view odometry as a form of dead reckoning, that is, the art of determining your position based on a known initial position and subsequent headings and speeds. Probably you remember stereotypical scenes from some submarine movie, where the captain is doing precisely that on a nautical chart (yet another example of people locked in a tin can). You may immediately guess that uncertainty is an issue in odometry. Errors in determining the robot's correct speed and heading will sum up, and the determined position may be different from the robot's actual position. Even if the robot has perfect information about its initial position and measures the revolutions of its axes with proprioceptive sensors, errors will occur, sum up, and the robot will determine an incorrect position after a while. Please note that GPS is not always the simple answer here because we may need higher accuracy to navigate safely, or we may operate indoors.

3.3.1 Non-systematic Errors, Systematic Errors, and Calibration

In odometry, there are two classes of errors depending on their source. There are non-systematic errors that can usually not be measured and hence cannot be included in a model. Uneven friction of the surface, wheel slippage, bumps, and uneven floors can cause these errors. Differences between the model cause systematic errors, that the robot has of itself, and the actual behavior of the robot. These systematic errors can be measured and then incorporated into the model. We call these corrections of the model calibration. Another unexpectedly difficult task in robotics is to make robots go straight. Especially in swarm robotics, where we use small and simple components and the robots are lightweight, making robots go straight is nontrivial. Calibration can help here to compensate for nonsymmetrical features in the hardware. Typical examples are unequal wheel diameters, uncertainty about the wheel base (distance between axes), or just faulty axle bearings.

The uncertainty about the robot's position grows over time due to systematic and non-systematic errors. In classical robotics, standard techniques such as Bayesian filters, particle filters, and Kalman filters are used. Here we don't go into these details because they are out of this book's scope. Instead, we have a quick look at map-making and then make a little excursion into how animals (here: ants) solve the problem of localization and navigation.

3.3.2 The Art of Map Making

The one moment when we as human beings also have trouble in locating ourselves is when we have to navigate a new city. For that purpose, we have navigation systems and smartphones with map apps. In former times, we would have used an actual map on paper. Hundreds of years ago, map-making was a true art because a lot of information about regions, countries, or whole continents was just missing (see Fig. 3.4). Robots face similar challenges whenever they are deployed in a new area. Instead of relying on odometry exclusively, they can also build a map on the fly. This is another popular field within robotics, called simultaneous localization and mapping or short SLAM. There are many different ways of representing the environment in the form of a map. Examples are maps based on a grid (filling grid cells where objects and walls were detected), landmarks (e.g., features that are recognizable by vision), or a laser scan (i.e., processing many detected points in 3D).

3.3.3 Excursion: Homing in Ants

In swarm robotics, we often take inspiration from biological systems. We do that not as a purpose of its own, but to get novel design ideas of how we can engineer robots. Swarm robots need to be cheap and straightforward, and relatively simple animals, such as insects, can be a true inspiration. So how do ants solve the problem of localization and navigation? They face the same problem as they need to navigate, for example, from a feeding place back home to their nest to transport food to the colony. Wolf [872] considers three methods of how ants can do that (see Fig. 3.5). In the case of route following, an ant can drop pheromones on its way to the food

Fig. 3.4 The art of map-making: sea map of Portugal by Lucas Janszoon Waghenaer (1584)

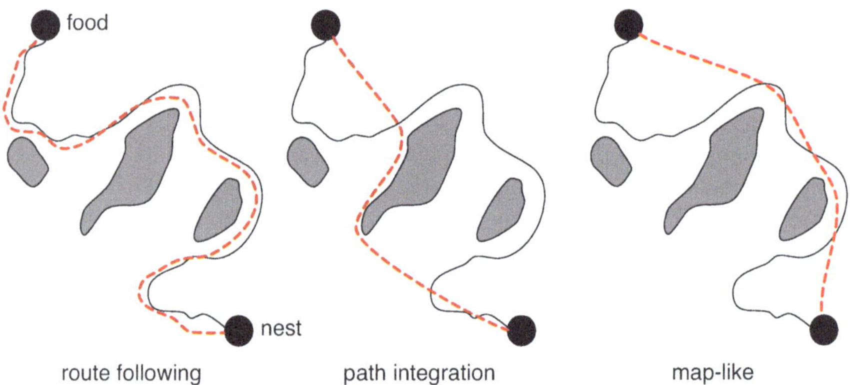

Fig. 3.5 Trajectories of homing ants from a food place to the nest for three methods: route following (e.g., by pheromones), path integration (e.g., by the terrestrial magnetic field and the number of steps), and map-like navigation (e.g., by remembering landmarks) [872]

and then follow them back to the nest. In the case of path integration, an ant can remember the general heading and the number of steps toward the nest with help, for example, from the terrestrial magnetic field. In the case of map-like navigation, an ant can remember several landmarks on the way and navigate back by approaching each of them sequentially. For each of these methods, at least one ant species exists that navigates that way.

If you are wondering how an ant could possibly count its steps, then you can, of course, take the skeptic position and doubt it. However, biologists have experimented to test this hypothesis. The experiment is rather cruel, but in most countries (e.g., within the European Union) there is no legislation protecting insects in the context of scientific experiments. Wittlinger et al. [871] have changed the step length of a group of desert ants (*Cataglyphis*, that is the ant in question) by cutting off parts of their legs (i.e., shortening their step length) and extending the legs of another group of ants (i.e., extending their step length). When these ants were left to navigate back to their nest, indeed, those with the longer step length passed the nest, and those with the shorter step length stopped short. Hence, this proves that the desert ant navigates by path integration based on step counting. It seems reasonable that they cannot actually count their steps, but they can obviously estimate the number well enough.

If these methods were found by natural selection after many, many generations of ants, then they are probably also useful to be applied in swarm robots. Only, we would count revolutions of axes instead of counting steps, as long as we usually don't have legged swarm robots.

3.4 Kinematics

The calculations of a robot's position by using odometry are an example of kinematics. Kinematics is the study of motion without regard to the causing forces.

Kinematics

Kinematics is the study of motion, excluding the forces or torques that cause it, and focusing instead on the geometric and temporal relationships between the positions, velocities, and accelerations of a robot's components.

Accelerations are described purely as changes in velocity over time, without reference to the underlying forces or torques that produce them. Hence, it is a kind of simplified physics that only pays attention to the time-based and geometric constraints. Kinematics is yet another popular field within robotics because it is especially relevant for robot arms and industrial robots. There are two kinds of kinematics: forward kinematics and inverse kinematics.

3.4.1 Forward Kinematics

The problem of forward kinematics is stated in the following way. I give you the starting configuration of the robot arm, including the angles of all its joints and some desired changes to some joints. Your task is then to calculate the new configuration.

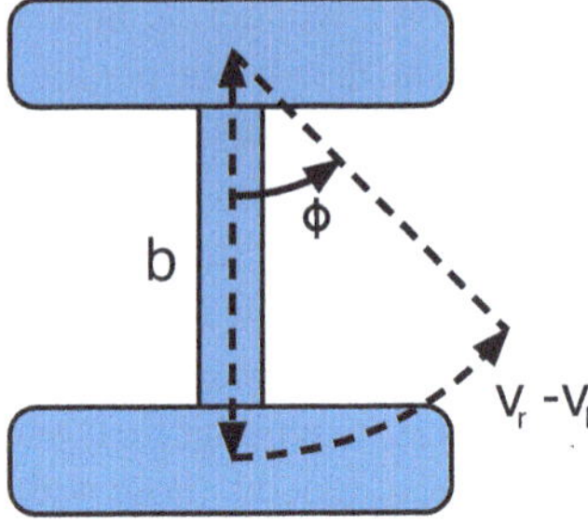

Fig. 3.6 Kinematics of the differential drive

Note, this is a typical setting for industrial robots. So you would need to calculate the position and orientation of the end-effector and all the joints.

Forward Kinematics

Forward kinematics is the process of computing a robot's position and orientation in space from its control variables (e.g., joint angles, wheel rotations, or steering angles), based on its geometric and kinematic model. In manipulators, this provides you with the end-effector pose, while in mobile robots, it gives the robot's pose in the environment.

A typical case for a swarm robot is the kinematics of differential steering (i.e., two wheels, each driven individually, see Fig. 3.6). The robot system is described by three variables: position of the robot's center of mass (x, y) and the robot's heading ϕ. In a problem of forward kinematics, I would give you the current robot position (x_0, y_0), its heading ϕ_0, and a desired change in terms of wheel speeds v_r for the right wheel and v_l for the left wheel. Given also the robot's track b (distance between wheels that share an axis), you can calculate the robot's speed m and write down the appropriate differential equations

$$\frac{dx}{dt} = m(t)\cos(\phi(t)), \tag{3.1}$$

$$\frac{dx}{dt} = ((v_r + v_l)/2)\cos(\phi(t)), \tag{3.2}$$

$$\frac{dy}{dt} = m(t)\sin(\phi(t)), \tag{3.3}$$

$$\frac{dy}{dt} = ((v_r + v_l)/2)\sin(\phi(t)), \tag{3.4}$$

$$\frac{d\phi}{dt} = (v_r - v_l)/b. \tag{3.5}$$

dx/dt is the first component of speed (in 2D) and dy/dt is the second component of speed. $d\phi/dt$ is the arc change over the radius (angular speed). To get nice functions of time, we integrate them all and get

$$\phi(t) = (v_r - v_l)t/b + \phi_0, \tag{3.6}$$

$$x(t) = x_0 + \frac{b(v_r + v_l)}{2(v_r - v_l)}(\sin((v_r - v_l)t/b + \phi_0) - \sin(\phi_0)), \tag{3.7}$$

$$y(t) = y_0 - \frac{b(v_r + v_l)}{2(v_r - v_l)}(\cos((v_r - v_l)t/b + \phi_0) - \cos(\phi_0)). \tag{3.8}$$

The turn radius for a circular trajectory is given by $\frac{b}{2}\frac{v_r+v_l}{v_r-v_l}$.

We should not unquestioningly trust these mathematical formulas but check for cases where we may have issues. There is the term $v_r - v_l$ in the denominator, which means we have an asymptote (singularity) for $v_r - v_l \rightarrow 0$. For this case, special handling is required. Fortunately, $v_r - v_l \rightarrow 0$ implies that the robot is going straight, and we could use another, simpler model.

3.4.2 *Inverse Kinematics*

The inverse kinematics problem is to calculate the appropriate parameters for all joints for a given position of the end-effector.

Inverse Kinematics

Inverse kinematics is the process of determining the control variables (e.g., joint angles, wheel rotations, or steering angles) required for a robot to achieve a desired position and orientation in space, based on its geometric and kinematic model. In manipulators, this means finding joint configurations that place the end-effector at a target pose. At the same time, in mobile robots, it corresponds to computing the control inputs needed to reach a specified pose in the environment.

For example, this could arise as a problem in situations as shown in Fig. 3.7. It is important to avoid collisions, not only of the robot arm with other structures but also of the robot arm with itself. This is mathematically a complex problem because there are incorrect configurations, singularities, and multiple possible solutions for specific issues. Also, inverse kinematics is an intensively worked on field in robotics.

3.5 Control

Finally, we must discuss robot controllers. This is probably the most exciting component of a robot because it is basically the robot's brain. We all want intelligent robots, but obviously, it is a challenge to program robots to behave intelligently in almost all situations. The controller directs the robot on how to move and act. In

Fig. 3.7 The so-called "Canadarm2" robot arm of the International Space Station (photo by NASA, public domain)

classical robotics, there are two main controller paradigms: open-loop control and closed-loop control.

Open-loop Control

Open-loop control executes the robot's movement without paying attention to any feedback, that is, there is no chance to make corrections.

Closed-loop Control

Closed-loop control is a control strategy that continuously compares its output with a desired reference (setpoint) and adjusts its actions using feedback to minimize error. Hence, with closed-loop control, we can compensate for errors.

If we want to control a mobile robot to make it drive parallel to a wall with an open-loop approach, then it is likely to veer off the path. The movement itself can be noisy; there could be slippage, model inaccuracies, or bumps. Open-loop control can hence not be applied in most cases when working with mobile robots. For industrial robots, open-loop control is more often an option. With a closed-loop control we can make use of a proximity sensor. We could position the sensor at the robot's side, facing the wall. If the robot veers away or toward the wall, we can detect that and compensate for it.

Fig. 3.8 5×10^3 sampled trajectories of a simple and deliberately exaggerated differential-drive robot model to illustrate how even small actuation noise can cause significant drift over time, highlighting the robot's struggle to move in a straight line. The robot is initially positioned on the left and is commanded to move in a straight line (horizontally) to the right

3.5.1 Trajectory Error Compensation

When trying to keep a robot (or its end-effector) on a particular path, it will usually veer off eventually. Hence, we need a controller that compensates for that error. You can also think of the cruise control of a car and how you would program a controller to change the car's acceleration appropriately.

For example, most simple mobile robots actually struggle to move in a straight line. Many small mobile robots are differential-drive robots, meaning they have two independently driven wheels. However, even minor differences in wheel speed, motor calibration, or surface conditions can cause the robot to deviate from its intended path gradually. That is a phenomenon known as drift. These errors accumulate over time, leading to increasingly inaccurate motion if left uncorrected. See Fig. 3.8 for a simple but powerful visualization. A simple simulation models a differential-drive robot attempting to move in a straight line along the x-axis. The robot has two independently driven wheels, and at each time step, both wheel velocities are affected by zero-mean Gaussian noise with a slight standard deviation. This models the typical imperfections found in low-cost mobile robots, such as uneven motor responses. The noise accumulates over time, introducing small asymmetries in the wheel speeds of the left and right wheels that lead to deviations in the robot's orientation. As a result, even though the robot is commanded to move forward, its actual trajectory begins to drift away from the intended path. Figure 3.8 is based on this simulation and shows 5,000 sample trajectories of the robot under identical control conditions but with different noise realizations. All trajectories begin at the origin, but spread laterally, forming a fan-like pattern. This spread represents the range of possible lateral displacements caused by the accumulation of minor, random orientation errors. Some

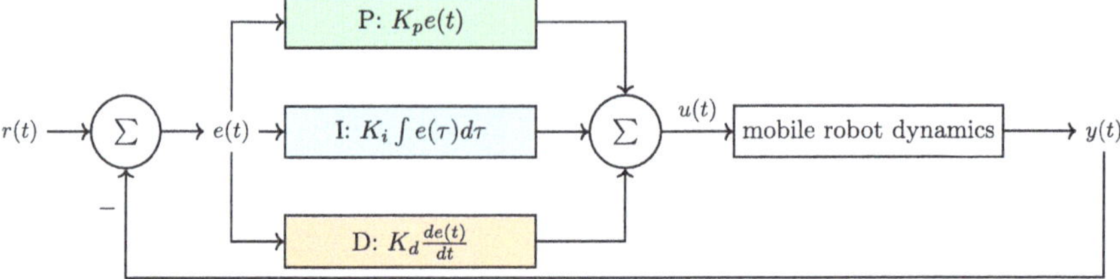

Fig. 3.9 Proportional–integral–derivative controller, short PID controller

trajectories veer to the left, others to the right, and a few remain aligned with the x-axis. To compensate for this drift and maintain a straight trajectory, continuous feedback control is typically required to monitor and adjust the robot's movement.

While there is an obvious case for reducing these errors through calibration and control methods, in swarm robotics, we can consider an unconventional option. Given that we have a population of robots, we could study whether the robots' imperfections are distributed in a way that allows them to average away. Raoufi et al. [663] explore the often-overlooked phenomenon of individuality in robotic swarms. Using the Kilobot platform as a case study, they demonstrate that such heterogeneity, which is typically treated as noise or error, is in fact a persistent and measurable feature. Variations in motor calibration, sensor sensitivity, and even internal timing lead to consistent differences in how robots move, perceive their environment, and synchronize with one another. Through experiments in phototaxis and random walk behaviors, Raoufi et al. [663] find that these differences can significantly affect performance and, in some cases, improve it compared to idealized, perfectly calibrated robots. Recognizing and modeling individuality can reduce uncertainty in swarm models and open up new design opportunities. Rather than treating such variation as a flaw, swarm roboticists should consider it a potential resource. This perspective is unique in the field, as most engineering approaches still aim to suppress variation. We now continue with the classical approach, which focuses on reducing errors through feedback mechanisms and calibration.

Control theory is the classical approach to control machines and generally engineered systems. The key is feedback control. It is a fundamental concept in control theory, where the output of a system is measured and used to adjust the system's input, automatically and continuously, to achieve a desired goal. Feedback control compares what you want the system to do (setpoint) with what it's actually doing (output), and adjusts the input accordingly to reduce the error. The setpoint $r(t)$ is the desired target value that the system (i.e., robot) should achieve. The output $y(t)$ of the system is here the mobile robot's state or current motion. The error $e(t)$ is the difference between setpoint and output: $e(t) = r(t) - y(t)$. The control signal $u(t)$ is what we calculate to be sent to the robot's actuator to control the robot's behavior.

There is a classical control approach called PID control (see Fig. 3.9). P, I, and D are components of a controller that can be combined for the best results, and they are distinguished by their mathematical property.

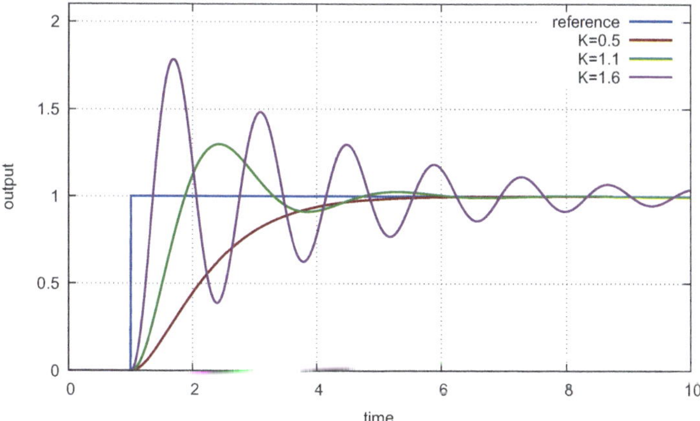

Fig. 3.10 Evolution of the error over time for different parameters. The horizontal axis represents time, and the vertical axis represents the robot's position. The desired position suddenly changes at $t = 1$. A too aggressive controller overshoots and oscillates, and a too slow controller compensates without overshooting but takes too long

PID Control

PID control (Proportional–Integral–Derivative control) is a feedback control method that computes a control signal as the sum of three terms: a proportional term responding to the current error, an integral term correcting accumulated past errors, and a derivative term anticipating future error trends.

The full equation to calculate the control signal of a PID control is given by

$$u(t) = K_p e(t) + K_i \int_0^t e(\tau)\, d\tau + K_d \frac{de(t)}{dt}, \tag{3.9}$$

with error signal $e(t)$ at time t, proportional gain K_p, integral gain K_i, and derivative gain K_d. The P-component $K_P e(t)$ provides a force in negative proportion to the measured error. The D-component $K_D \frac{de(t)}{dt}$ adds a force that is proportional to the first derivative of the measured error. Hence, it measures the speed of change of the error. For example, an error could get quickly bigger or not change at all. The I-component $K_I \int_0^t e(\tau)\, d\tau$ adds a force that is proportional to the integral of the measured error. Hence, it sums up the past errors and is big if the error was big for quite a while in the past. Combining all three components gives you a PID controller that is well prepared for many different dynamics in the error. Depending on certain parameters you can configure a PID controller (see Fig. 3.10) to ensure it is neither too aggressive (oscillations) nor too slow.

Control theory is applicable in situations where swarm robots can calculate an error $e(t) = r(t) - y(t)$ in a distributed manner. However, the setpoint often refers to a global feature of the swarm that an individual robot cannot easily monitor. So, unfortunately, we cannot easily apply this concept in swarm robotics, at least not in the actual challenging situations. There are more complex challenges in mobile robotics, such as deciding when to explore the environment and when to exploit what is already known. Should your robot try to pass the next door to check another room, or should it stay in the current room? Many of such questions cannot be resolved with a PID controller. Hence, it is obvious that we need additional approaches.

3.5.2 Controllers for Swarm Robots and Agent Models

Not only is the hardware of swarm robots supposed to be kept simple but also their software, that is, their control. The idea is to rely instead on processes of self-organization as mentioned before. By collaborating, the robots are still able to solve complex tasks that an individual would not be able to solve with their simple controller. The standard control approach in swarm robotics is hence based on reactive control. Russell and Norvig [690] introduce classes of agent models. The simplest model is that of the simple reflex agent.

Simple Reflex Agent

A simple reflex agent has only simple condition–action rules (if–then rules) that determine how it behaves.

A condition-action rule can, for example, be: "if the sensor value of proximity sensor 5 is below 33, then make a left turn." This is a purely reactive behavior because the robot directly reacts to its current perception. A history of perceptions is not considered. The second agent model in the hierarchy of Russell and Norvig [690] is the model-based reflex agent.

Model-based Reflex Agent

The model-based reflex agent has only simple condition-action rules, but has an internal state that allows the robot to potentially react differently to the very same perception.

The behavior of the robot is still determined by condition-action rules while considering a history of perceptions. The internal state of a model-based reflex agent can be as simple as in a finite state automaton, or it can be more complex in form, for example, of an array of floating point values used to store previous perceptions.

The concept of equipping robots with reactive controllers only is related to the so-called behavior-based robotics by Arkin [28] and the concept of subsumption architecture by Brooks [109, 110]. Also, from the point of view of biology considering, for example, social insects, reactive behavior seems plausible [91].

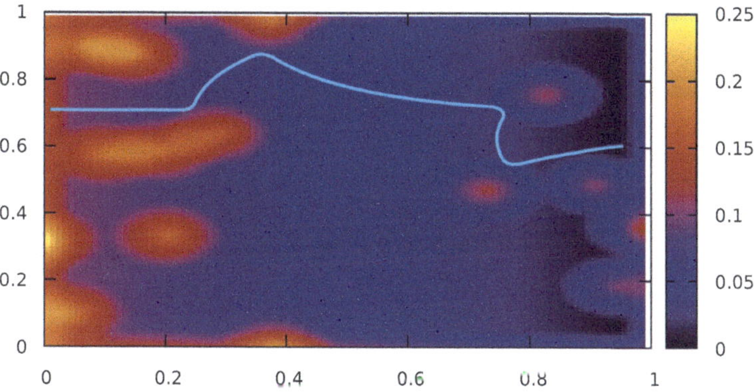

Fig. 3.11 Example trajectory of a robot in 2D space controlled by a potential field (colors encode height of the potential field)

The idea of reactive control is easily understood by the example of so-called potential field control.

Potential Field Control

Potential field control is a reactive navigation method. The robot moves according to an artificial potential field: the goal generates an attractive potential that pulls the robot closer, while obstacles generate repulsive potentials that push it away. The resulting control signal is the sum of these virtual forces.

Using potential field control, the problem of collision-free navigation is translated into a virtual physical world. The potential field is constructed based on the robot's perception and based on the defined task. As an example, the robot in Fig. 3.11 is supposed to navigate from the left to the right. This is implemented by an overall slope in the potential field with the lower end on the right-hand side. Cones represent obstacles. The navigation of the robot can be imagined as similar to a marble rolling down the slope while being repelled from the cones. This is a reactive behavior because the robot does not plan its actions ("Where will I end up when I do this now?") and the robot's actions are directly determined by the currently present slope of the potential field. However, depending on the implementation, past actions may influence the robot via its current acceleration, which may integrate over past influences.

In swarm robotics, there is a lot of research using methods of machine learning [690], such as evolutionary swarm robotics [201, 787]. This is the application of evolutionary algorithms [223] to swarm robotics. Usually, the weights of artificial neural networks are optimized using evolutionary computation to teach a swarm to solve a given task. However, there is also a lot of research where the controllers, for example, in the form of finite state machines, are simply hand-coded. The complexity of both the swarm behavior itself and the challenge of finding the right controller has its origin in the self-organized interactions between robots and between robots and the environment. Even though the controllers themselves are simple, finding them is difficult. We will discuss that in depth later (see Chap. 7).

3.6 Swarm Robot Hardware

Finally, we have a closer look at a few selected examples of swarm robot hardware. The most common design is a relatively small mobile robot (say, diameters between 4 and 12 cm) with a differential-drive and infrared sensors for proximity sensing; however, there are also some special designs.

3.6.1 Mobile Robots for the Lab

S-Bot

One of the most iconic hardware designs is the "s-bot" (see Fig. 3.12) of the swarm-bots project [574]. By today, it is a bit outdated, but back in 2004 it was an amazing approach to swarm robotics and robotics. It is a considerably small robot of 12 cm diameter, 15 cm height, and a weight of 660 g. Its two Li-ion batteries allowed energy autonomy for more than 1 h. The robot's controller is based on a 400 MHz custom XScale CPU board with 64 MB of RAM and 32 MB of flash memory. The robot

Fig. 3.12 Five s-bots (12 cm diameter) showing their ability of physically connecting to each other [574] (photo by Francesco Mondada and Michael Bonani)

was accessed by Wi-Fi, which is a great feature for debugging. The robot's actuators are two treels (i.e., a combination of wheels and tracks), turret rotation, rigid gripper elevation, and a rigid gripper. The original design included a 3-DOF side arm and a side arm gripper, which were later left out. The robot has eight RGB light-emitting diodes around the turret and red light-emitting diodes in the grippers. The robot has an amazing number of sensors: 15 infrared sensors around the turret, four infrared sensors below the robot, position sensors on all degrees of freedom except the gripper, force and speed sensors on all major degrees of freedom, two humidity sensors, two temperature sensors, eight ambient light sensors around the turret, four accelerometers, which allow three-dimensional orientation, a 640×480 camera sensor with a custom optic based on a spherical mirror that provides omnidirectional vision, four microphones, two-axis structure deformation sensors, and an optical barrier in grippers. Many experiments were reported using this robot, including experiments in evolutionary swarm robotics [201, 787]. The basic design of the s-bot was later redesigned in different versions called "marXbot" [92] and foot-bot within the Swarmanoid project [196].

I-SWARM

Another big step forward in swarm robotics hardware research was the I-SWARM project [728] as described in Sect. 1.7.1. The extremely ambitious objective of I-SWARM was to develop a tiny swarm robot of size $1 \times 1 \times 1\,\text{mm}^3$. Although this constraint was later relaxed a bit to $3 \times 3 \times 3\,\text{mm}^3$, it is obvious that such a design comes with many challenges. The robot (see Fig. 3.13) has no battery but is powered by a highly efficient solar cell. The robot has no wheels; instead, it is locomoted by three vibrating piezoelectric legs (slip-stick locomotion). Robot–robot communication is implemented by four infrared transceivers. The robot has a vibrating piezoelectric needle as a sensor, and the onboard electronics are a reprogrammable ASIC (i.e., custom chip design). The final robot was not fully functional due to issues in the ASIC because there was not enough funding to do several design iterations. Still, it was a great achievement, and the knowledge is there now. Any company with enough funding could start building this tiny robot. Potential applications could be cleaning and monitoring tasks in hard-to-reach parts of machines.

Alice

The microrobot "Alice" is another example of a relatively small swarm robot (see Fig. 3.14). It was developed by Caprari et al. [120] and has a size of $22 \times 21 \times 20\,\text{mm}^3$ or $9.24\,\text{cm}^3$. It has a PIC microcontroller, a NiMH rechargeable battery, and flexible PCBs. A true exception was the two "Swatch" motors provided by the watchmaker Swatch. At that time, small electromotors were rare.

Fig. 3.13 The tiny swarm robot developed by Seyfried et al. [728] in the I-SWARM project (size $3 \times 3 \times 3\,\text{mm}^3$)

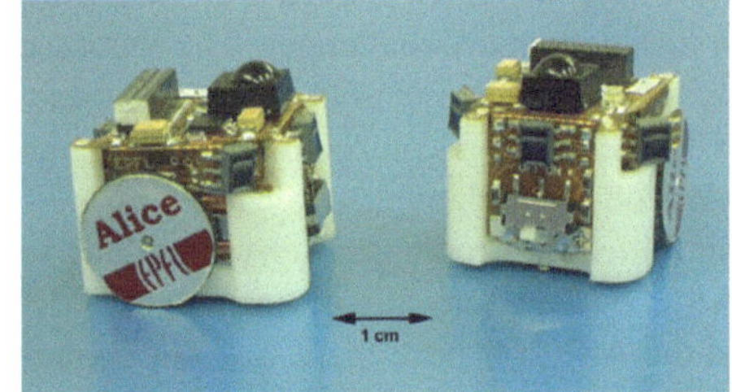

Fig. 3.14 The Alice microrobot, size of $22{\times}21{\times}20\,\text{mm}^3$ [120] (photo by Gilles Caprari)

Kilobot

A swarm robot that is currently popular in swarm robotics research is the "Kilobot" [685, 779]. It is also relatively small with a diameter of 33 mm and a height of 34 mm (see Fig. 3.15). The overall design of the Kilobot seems to be inspired by the I-SWARM robot, because it also has vibrating legs instead of wheels (slip-stick locomotion). The controller is an Atmega 328 (8bit, 8MHz), it has 32 KB flash memory, 1 KB EEPROM, and 2 KB SRAM. The robot has a Li-Ion battery. Communication of up to 7 cm is done via infrared light that is reflected off the ground surface (the infrared transceiver points downwards). Distance information of neighboring Kilobots can be estimated via signal strength. The robot has an RGB LED, an ambient

Fig. 3.15 The popular Kilobot (diameter of 33 mm) [685]

light sensor, and two vibration motors. They are independently controllable, hence, implementing a differential drive. This robot was used in the biggest robot experiment reported so far [686]. The experiment was done with 1024 robots. Hence, the name Kilobot was seemingly given in anticipation of that experiment.

One disadvantage of the Kilobot is that it is very slow due to the vibrating leg technique. Another issue not directly related to the robot is the difficulty of handling such large numbers of robots in experiments. All robots need to work reliably, but, for example, for recharging, they need to be handled. If a human operator has to handle them manually and maybe even calibrate them, then handling 1000 robots comes with a lot of overhead. Even if each of the 1000 robots only needs attention for two minutes, it will take more than 33 h to finish this task.

Thymio II

The "Thymio II" robot [675] was initially designed to introduce children of varying ages to robotics and programming (see Fig. 3.16). It is used to attract children to robotics, which is an excellent service to both the public and the robotics community. However, meanwhile, it is also quite popular in swarm robotics research. Thymio II includes preprogrammed behaviors for beginners, a visual programming interface, and a more advanced text-based language. The Thymio II is a differential-drive mobile robot equipped with seven horizontal infrared proximity sensors around its circumference (five at the front and two at the back), as well as two IR ground sensors, which enable it to detect obstacles and surface conditions. It can drive at

Fig. 3.16 The Thymio II robot, here extended with a Raspberry Pi, RGB color sensor, and an additional battery [235]

speeds of up to 20 cm/s, although in practice, the speed is often limited to reduce motor wear. The Thymio II is usually extended for use in research. Common is a Raspberry Pi connected through USB and the D-Bus interface, which allows running Python scripts and incorporating additional sensors (e.g., light sensors) [399, 840]. The Thymio II robot's chassis also has LEGO® attachment points, making it easy to mount custom hardware.

TurtleBot 4

The "TurtleBot4" robot was developed as an affordable and extensible platform to support education and research in mobile robotics and autonomous systems. It is part of the widely used TurtleBot series, which has played a key role in popularizing access to ROS (Robot Operating System, see Sect. 6.4.1) and robot development platforms (see Fig. 3.17). It is primarily used in teaching for topics such as SLAM and navigation, as well as in swarm robotics research. The TurtleBot4 is a differential-drive mobile robot built on a compact base (iRobot® Create3), equipped with a 2D LiDAR (Hokuyo or RPLIDAR), an IMU, and a front-facing RGB-D camera (OAK-D Lite or similar), enabling it to perform obstacle detection, mapping, and visual perception tasks. The onboard computer is a RaspberryPi4, running ROS 2. TurtleBot 4 supports USB and GPIO interfaces for custom extensions, and its top-mounted plate offers space for mounting additional sensors or communication modules.

Fig. 3.17 Four TurtleBot4 robots (photo by Elisabeth Böker, Centre for the Advanced Study of Collective Behaviour, Konstanz)

Fig. 3.18 The open platform robot Colias by Arvin et al. [33] (diameter of 4 cm, photo by Farshad Arvin)

Other Swarm Robots for the Lab

There are even more swarm robots. A class of swarm robots was designed explicitly for education, which doesn't mean they are not used in research as well. An example is the *e-puck* robot [576]. It is a small mobile robot with a differential drive and infrared sensors for proximity sensing and communication. Another robot design for education is the *Pheeno* robot [865]. It has a modular design, and the focus seems to be on a more classical robotics approach, including vision and gripping. The robot *Colias* (see Fig. 3.18) was designed by Arvin et al. [33] as a low-cost, open platform with a diameter of 4 cm, a relatively high maximum speed of 35 cm/s, and the typical infrared sensors.

Other robot platforms follow the standard differential-drive concept are *r-one* [537], *Wanda* [425], and *Jasmine* [383]. One of the earliest robot designs in swarm

Fig. 3.19 A six-legged swarm robot with a diameter of 10 cm [230] (photo by Wilfried Elmenreich)

robotics was the *Khepera* robot by Mondada et al. [573], which is barely used anymore. The recently reported GRITSBot by Pickem et al. [630] was designed as an inexpensive swarm robot "to lower the entrance barrier." A robot that is unique because it neither relies on wheels nor vibrating legs is the robot reported by Elmenreich et al. [230] (see Fig. 3.19). It is a six-legged robot with hexapod locomotion based on a platform of a remote-controlled toy. The Droplet robot by Farrow et al. [239] is another robot based on slip-stick locomotion. It has omnidirectional motion based on six linear directions, and it can turn in place.

Liu et al. [495] introduce *CapBot*, a battery-free swarm robotics platform that replaces conventional rechargeable batteries with supercapacitors. Unlike batteries, which are slow to charge, degrade after a few years, and create environmental waste, supercapacitors allow CapBot to fully recharge in about 16 s, provide more than 50 min of operation, and support rapid peer-to-peer energy transfer (trophallaxis) in under 20 s (cf. Schmickl and Crailsheim [709]). The robot is built from low-cost, commercial-off-the-shelf components, equipped with omnidirectional wheels, Bluetooth Low Energy networking, and an accurate energy meter. This robot shows an alternative to the virtually omnipresent batteries and has potential for better sustainability in swarm robotics (also see Sect. 6.9).

3.6.2 Modular Robotics

Other special platforms relevant to swarm robotics include some approaches from modular robotics, where autonomous robot modules can physically not only attach to each other but also operate as an independent entity, as done in the projects SYMBRION and REPLICATOR [484].

Generally, there is a rich literature on self-reconfigurable modular robotics. A discussion of that would go beyond the scope of this book. Instead, here are just a few notable mentions. Christensen et al. [143] propose an online learning strategy for locomotion in modular robots. These robots essentially climb on top of each other, forming a physically aggregated and assembled super-robot. Zykov et al. [905] propose self-reproducing modules, which point to the dream of modular robotics as initially coined by von Neumann [834].

A great modern modular robotics approach is that of Devlin et al. [186]. They present robots that mimic the dynamic physical behavior of living tissues by actively modulating internal forces to control both shape and mechanical strength. Inspired by biological processes, such as convergent extension and tissue fluidization, they have developed robots capable of topological rearrangements through the application of controlled inter-unit tangential forces. The system transitions between solid-like and fluid-like states, enabling self-shaping and adaptive load support.

Piranda and Bourgeois [639] introduce the design of a quasi-spherical robotic module intended as a building block for large-scale modular robots and programmable matter. The module can roll in any direction, enabling smooth locomotion, reconfiguration, and self-assembly. The authors discuss the mechanical design, actuation principles, and potential applications, highlighting how spherical symmetry supports scalability and robustness in swarm-like modular robotics. A survey from that perspective is given by Thalamy et al. [778].

3.6.3 Uncrewed Surface Vehicle (USV)

We focus here on autonomous sailboats that leverage renewable wind energy for long-endurance navigation, enabling applications such as oceanographic monitoring, carbon flux studies, and hurricane data collection (for more details see Sect. 6.7.1).

The boat, Onyx Pearl, introduced by Kedia et al. [412] measures 1,200 mm in length with a mast height of 1,600 mm. It integrates a Raspberry Pi 5 and an ESP32 microcontroller unit for processing and real-time computing. Equipped with a comprehensive suite of sensors and communication systems, it features an LCJ Capteurs CV7 ultrasonic wind sensor, RTK GPS with subcentimeter precision, a 9-DOF IMU MPU-9250, and a NORA-B120 Bluetooth 5.2 BLE module offering a range of up to 400 m for local boat-to-boat and broadcast communication. For long-range communication (up to 10 km line of sight), it employs an RFDesign UAV RFD900x telemetry system. The in-house-developed RTutilizesstem uses the ArduSimple simpleRTK2B platform, capable of operating in both Base and Rover configurations. RTCM correction data (standardized format for GNSS correction messages) is transmitted to the rovers via a unidirectional point-to-multipoint communication setup using the Digi XBee SX 900 RF module.

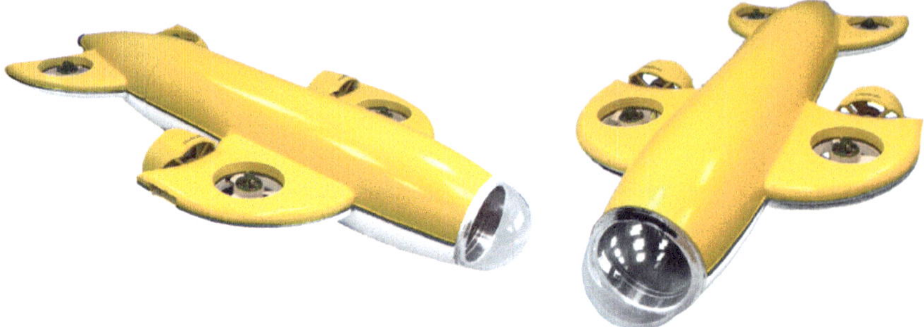

Fig. 3.20 The autonomous underwater vehicle MONSUN [608]

Sun et al. [761] discuss the development of autonomous sailboat sails. They cover the traditional flexible sail and advanced rigid wingsails, including specialized configurations like flap and self-trimming wingsails.

3.6.4 Autonomous Underwater Vehicles (AUV)

There are also a few projects doing underwater swarm robotics, that is, they develop autonomous underwater vehicles (AUV). Examples are the robots developed in the project CoCoRo [707] and the AUV called MONSUN (see Fig. 3.20) [608]. Especially, communication between robots is challenging underwater. Radio cannot be used because electromagnetic waves do not propagate far in water. While CoCoRo uses visible light for communication, MONSUN uses an underwater acoustic modem, which has a much bigger range than light. However, also dealing with acoustic waves underwater is challenging because they are reflected at the surface, possibly at confining walls, or even at water layers of different temperatures. Also, these underwater acoustic modems are usually rather expensive.

3.6.5 Flying Drones

Although there is currently a lot of research on flying drones (quadrocopters), few of the projects seem to focus on self-organizing drone swarms that do not make heavy use of global information (see works by the groups of Martin Saska and Gábor Vásárhelyi, e.g., [42, 49]; see Sect. 6.7.2). A reason may be that it is very challenging to control multiple quadrocopters within a small space, creating a lot of air turbulence, while having all of them rely on their onboard sensors exclusively. What can be found in the literature are combinations of robot swarms on the ground and flying robots. Examples are the work by Chaimowicz and Kumar [135] and the project Swarmanoid [196].

3.6.6 Active Matter Systems: "Robots" Out of Control

Following the definition of a robot strictly, the following systems can barely be seen as robots. However, they are significant contributions that (1) an engineer working on swarm robotics should be aware of and (2) maybe will transition into complete swarm robot systems with future technology. Active matter systems are collections of individual units that consume energy to produce motion [96, 185, 513] and they are studied mainly for their simplicity and controllability. These systems can exhibit collective behavior, such as flocking [173]. Examples of active matter systems include vibrated polar disks, colloidal rollers, motile colloids, and micro-swimmers. A motile colloid is a small particle capable of propelling itself in a fluid, rather than passively drifting [289]. They are "mindless robots" [96] as most of them do not possess any sensors, actuators, or control systems. They cannot be seen as autonomous robots.

Active matter encompasses a broad range of physically driven collectives that convert energy into motion, operating on vastly different scales and mechanisms of motion. There are micrometer-scale colloids and micro-swimmers propelled through fluids, where Brownian motion dominates and external fields (electric, vibrational, or optical) often supply the energy. At the other extreme are centimeter-sized, simple robots with onboard batteries, operating under dry friction and sometimes inertial effects, rather than viscous drag. All such "mindless robots" share core features: they have no explicit sensor-actuator loops or centralized control, but can exhibit collective behavior, such as flocking, clustering, and collective transport. Some units turn or reorient based purely on local cues (density gradients or indirect hydrodynamic interactions), effectively responding to neighbors even without sensors. Although most of these platforms cannot be called autonomous robots in the classical sense, they serve as minimal physical models for studying how large-scale coordination and emergent organization can arise from simple, local rules, a goal that also resonates with design principles in swarm robotics. Despite the distinct differences among active matter systems, they show similar collective behaviors. This suggests that insights gained in a specific active matter system may be broadly applicable.

Active matter systems consist of entities on different size scales: centimeter-sized [185], millimeter-sized [185], and micrometer-sized [70, 107, 357, 895]. While the centimeter-sized polar disks reported by Deseigne et al. [185] have their own onboard energy reservoir in the form of regular batteries, most of the other active matter systems require an external energy source. Typical sources of energy are external vibrations [185], electric fields [107, 895], and light [70, 357].

Colloids are substances containing small particles (from nanometers to micrometers in radius) dispersed throughout a continuous medium (e.g., liquid). Particles in a colloid are small enough not to settle under gravity. Zhang et al. [895] describe metal–dielectric Janus colloids that "turn towards regions of higher density." This turning behavior enables the colloids to respond to their environment and exhibit a form of taxis. Even without explicit sensors, these colloids manage to react to their environment. Similarly, Das et al. [173] describe colloids that align to each other by indirectly "sensing" and responding to their neighbors' positions and movements. Bricard et al. [107] study populations of self-propelled colloidal rollers that spontaneously organize into a collective motion. Long-range hydrodynamic interactions stabilize coherent motion and suppress large-scale density fluctuations, resulting in the formation of giant flocks in active matter.

Impressive are the swarm sizes N that can be achieved with these systems. They are ranging from tens [96], to hundreds and thousands [173, 185, 357], and even millions [107]. The complexity of the observed collective behaviors is, of course, limited but still astonishing given the simplicity of these "mindless robots." Observed behaviors include corralling (micro-swimmers spontaneously gather and confine passive particles) [70], flocking [173, 185], motility-induced phase separation (a form of clustering) [895], and aligned aggregates [70].

One of the most complex behaviors exhibited by active matter systems is demonstrated by the system reported by Heuthe et al. [357]. In a collective transport scenario (see Sect. 5.3.5), $N = 200$ micrometer-sized Janus microparticles controlled individually via laser spots, collectively rotate and transport a rod-shaped object.

Chvykov et al. [145] introduce "rattling" (measure of how strongly a swarm system responds to random perturbations or fluctuations induced by an external drive) as a unifying, predictive principle for understanding self-organization in active matter systems. They propose that in driven systems, low-rattling configurations (i.e., states with minimal drive-induced stochastic fluctuation) are preferentially selected over time. They demonstrate this concept both theoretically and experimentally using ensembles of simple shape-changing robots, which spontaneously form coordinated group behaviors depending on the external driving pattern.

Mayya et al. [531] explore how a swarm of minimalist vibration-driven robots (*brushbots*) can spontaneously form regions of high and low density. Inter-robot collisions cause density-dependent slowdowns in robot motion, which in turn trigger a phenomenon known as motility-induced phase separation (MIPS). By drawing a formal analogy between the *brushbot* swarm and a system of passive particles, the study demonstrates that robots can form stable clusters or patterns simply by colliding with one another, without requiring sensors or communication.

Despite their simplicity, these entities exhibit a vast repertoire of collective behaviors. By studying these simplified models, new insights may be gained about the fundamental principles governing self-organization. For example, the discovery of flocking by turning away [173] challenges conventional thinking about flocking and may lead to more unexpected collective behaviors. A key challenge is to develop models that bridge the gap between microscopic dynamics and macroscopic phenomena, and to push these insights into novel applications.

3.7 Further Reading

There are many good books on mobile robotics. Arkin's Behavior-Based Robotics [28] and Probabilistic Robotics by Thrun et al. [785] are two important books. Two more recent recommended books on mobile robotics are: Correll et al. [155] and Ben-Ari and Mondada [69]. The book by Floreano and Mattiussi [247] is not focused on robotics, but instead treats a rare combination of subjects, including behavior-based robotics, evolutionary computation, learning, artificial neural networks, self-organization, and swarm robotics. Complementary books include Robot Programming by Jones and Roth [393] and Braitenberg's Vehicles [99], although the latter focuses exclusively on behavioral aspects rather than robotics.

3.8 Exercises

Exercise 3.1 (*Kinematics of Differential Steering*) The equations for the kinematics of differential steering were given in Eqs. (3.1) to (3.8). Use Eqs. (3.2), (3.4), and (3.5) to integrate x, y, and ϕ numerically over a time interval $t \in [0, 3]$ for the following situation. The wheelbase and initial position (without units) are set to

$$\begin{aligned} b &= 0.05, \\ x_0 &= 0, \\ y_0 &= 0, \\ \phi_0 &= 0. \end{aligned}$$

The velocities for the first third of the time interval ($t \in [0, 1]$) are set to

$$\begin{aligned} v_r &= 1.0 \\ v_l &= 0.9. \end{aligned}$$

At time $t_1 = 1$ and for the time interval $[1, 3]$ the wheel velocities are changed to

$$\begin{aligned} v_r &= 0.9 \\ v_l &= 1.0. \end{aligned}$$

Discretize time in steps of $\Delta t = 0.01$.

(a) Calculate the robot's trajectory for the above setting using Eqs. (3.2), (3.4), and (3.5).
(b) Let's assume that we have an error of $v_l \pm 0.01$ for the left wheel velocity (e.g., due to errors in the wheels or odometry, etc.). Calculate two additional trajectories

for the whole time interval ($t \in [0, 3]$), one for $v_l + 0.01$ and one for $v_l - 0.01$, but using the integrated equations (3.6), (3.7), and (3.8) now.

(c) Plot all three trajectories in one diagram and compare them.

Exercise 3.2 (*Simple Sensor Fusion for State Estimation*) Simulate a mobile robot performing dead reckoning with noisy sensor data and fuse multiple noisy estimates (wheel encoders, IMU, and GPS) using weighted averaging or other intuitive methods to improve localization.

Define ground truth trajectory: A robot moves in a 2D arena. Generate a known path. We assume a ground truth trajectory defined over continuous time $t \in [0, T]$: $x(t) = t, \quad y(t) = \sin(0.2t)$ You discretize time using time steps of Δt, e.g., $\Delta t = 0.1$ s.

Wheel Encoders (Odometry): Odometry estimates position by integrating wheel velocities: Left and right wheel velocities are $v_l(t)$, $v_r(t)$. The wheelbase is b. We define a linear and angular velocity:

$$v(t) = \frac{v_l(t) + v_r(t)}{2}, \quad \omega(t) = \frac{v_r(t) - v_l(t)}{b} . \tag{3.10}$$

The position update in 2D for the differential drive is given by

$$x_{t+\Delta t}^{\text{odo}} = x_t^{\text{odo}} + v(t)\cos(\theta_t) \cdot \Delta t \tag{3.11}$$

$$y_{t+\Delta t}^{\text{odo}} = y_t^{\text{odo}} + v(t)\sin(\theta_t) \cdot \Delta t \tag{3.12}$$

$$\theta_{t+\Delta t} = \theta_t + \omega(t) \cdot \Delta t \tag{3.13}$$

We add a drift model. W model cumulative drift as a slow-growing additive offset:

$$x_{t+\Delta t}^{\text{odo, drift}} = x_{t+\Delta t}^{\text{odo}} + \varepsilon_x t, \quad y_{t+\Delta t}^{\text{odo, drift}} = y_{t+\Delta t}^{\text{odo}} + \varepsilon_y t , \tag{3.14}$$

where $\varepsilon_x, \varepsilon_y \sim \mathcal{N}(0, \sigma_{\text{drift}}^2)$ are small constants or drawn once at start.

IMU: Accelerometers measure linear acceleration. We derive it from ground truth:

$$a_x(t) = \frac{d^2x(t)}{dt^2} = 0 \tag{3.15}$$

$$a_y(t) = \frac{d^2y(t)}{dt^2} = -0.04\sin(0.2t) \tag{3.16}$$

We numerically integrate the acceleration to get the velocity and then the position:

$$v_x(t + \Delta t) = v_x(t) + a_x(t)\Delta t \tag{3.17}$$

$$x_{t+\Delta t}^{\text{imu}} = x_t^{\text{imu}} + v_x(t)\Delta t . \tag{3.18}$$

Similar for $y(t)$. We add noise to each acceleration reading:

$$\tilde{a}_x(t) = a_x(t) + \eta_x(t) + b_x, \quad \tilde{a}_y(t) = a_y(t) + \eta_y(t) + b_y, \tag{3.19}$$

where $\eta_x(t), \eta_y(t) \sim \mathcal{N}(0, \sigma^2)$ are Gaussian noise, $b_x, b_y \sim \mathcal{N}(\mu_b, \sigma_b^2)$ are constant biases per robot. This leads to slow drift over time, simulating the inaccuracy of an IMU.

GPS: GPS gives noisy position readings directly based on the ground truth, sampled at low frequency. We assume that the GPS update interval is $T_{\text{GPS}} = 1.0$ s. The simulated GPS position is

$$x_t^{\text{gps}} = x(t) + \delta_x(t), \quad y_t^{\text{gps}} = y(t) + \delta_y(t), \tag{3.20}$$

with $\delta_x(t), \delta_y(t) \sim \mathcal{N}(0, \sigma_{\text{gps}}^2)$ and for a consumer GPS, we assume $\sigma_{\text{gps}} = 1.0$ m. We update this reading only every T_{GPS}, retaining the last value otherwise.

You now have three position estimates: $\mathbf{p}^{\text{odo}}(t) = (x^{\text{odo}}(t), y^{\text{odo}}(t))$ from odometry, $\mathbf{p}^{\text{imu}}(t) = (x^{\text{imu}}(t), y^{\text{imu}}(t))$ from IMU, and $\mathbf{p}^{\text{gps}}(t) = (x^{\text{gps}}(t), y^{\text{gps}}(t))$ from GPS. We now combine these into a fused estimate $\mathbf{p}^{\text{fused}}(t)$.

(a) Option 1 is a weighted average (static weights). This is the most straightforward method of fusion. Let $w_{\text{odo}}, w_{\text{imu}}, w_{\text{gps}}$ be the weights such that: $w_{\text{odo}} + w_{\text{imu}} + w_{\text{gps}} = 1$. Then compute

$$\mathbf{p}^{\text{fused}}(t) = w_{\text{odo}}\mathbf{p}^{\text{odo}}(t) + w_{\text{imu}}\mathbf{p}^{\text{imu}}(t) + w_{\text{gps}}\mathbf{p}^{\text{gps}}(t)\ . \tag{3.21}$$

(b) Option 2 is a complementary filter (with memory). The complementary filter combines high-rate but drifting sources (such as odometry or IMU) with low-rate but accurate sources (like GPS). Use GPS as a correction at each GPS update:

$$\mathbf{p}^{\text{fused}}(t) = \alpha \cdot \left(\mathbf{p}^{\text{fused}}(t - \Delta t) + \Delta\mathbf{p}^{\text{odo}}(t)\right) + (1 - \alpha)\mathbf{p}^{\text{gps}}(t) \tag{3.22}$$

with $\Delta\mathbf{p}^{\text{odo}}(t) = \mathbf{p}^{\text{odo}}(t) - \mathbf{p}^{\text{odo}}(t - \Delta t)$ and $\alpha \in [0.7, 0.99]$ is a blending factor.

Exercise 3.3 (*Potential Field Control*) Implement a data structure for a potential field P (e.g., an array). Generate a potential field with a slope in one direction over the whole length (in Fig. 3.11, high potential on the left, low on the right), slopes leading away from the borders, and add a dozen obstacles in the form of local maxima within the potential field (e.g., cone-shaped, Gaussian—as you wish). Place the robot randomly at a position with high potential at one side (in Fig. 3.11, robot is started on the left).

(a) Generate trajectories for several potentials and a robot placements with a controller that reacts directly to the gradient of the potential field, that is, the slope in x- and y-direction of the potential field at the robot's current position directly determines the robot's velocity v. The robot's displacement in one integration step is

$$\Delta x(t) = v_x(t)\Delta t,$$
$$\Delta y(t) = v_y(t)\Delta t,$$

and the velocity is determined by the gradient of the potential field:

$$v_x(t) = \frac{\partial}{\partial x} P(x, y),$$
$$v_y(t) = \frac{\partial}{\partial y} P(x, y).$$

An approximation of the gradient is obtained easily by subtracting two neighboring values of the potential field in your data structure $P[\cdot, \cdot]$:

$$\frac{\partial}{\partial x} P(x, y) \approx P[i, j] - P[i + 1, j],$$

for appropriate indices i and j. You have to play around with parameters to make it work: choose an appropriate integration step Δt and/or nominal velocity v, optionally normalize the slope of the potential field to one and scale it by a factor up or down, etc.

(b) Generate trajectories for several potentials and a robot placements with a controller that reacts indirectly to the gradient of the potential field, that is, the slope in x- and y-direction of the potential field at the robot's current position only determines the acceleration of the robot. The robot's displacement is still determined by

$$\Delta x(t) = v_x(t)\Delta t,$$
$$\Delta y(t) = v_y(t)\Delta t,$$

but the robot's velocity is now determined by

$$\Delta v_x(t) = c v_x(t - \Delta t) + \frac{\partial}{\partial x} P(x, y),$$
$$\Delta v_y(t) = c v_y(t - \Delta t) + \frac{\partial}{\partial y} P(x, y),$$

for $0 < c \ll 1$ and followed by a normalization $\mathbf{v}/|\mathbf{v}| = (v_x, v_y)/|(v_x, v_y)|$ to prevent a too big increase of the velocity. That way, the robot can have a kind of memory of the former directions. Experiment with different parameters for the steepness of obstacles, normalize the acceleration vector, and implement a discount factor c that reduces the influence of former accelerations. What is a good setting such that the robot always (or at least most of the time) reaches the low values of the potential?

Exercise 3.4 (*Behaviors of a Single Robot*) Design a simple environment, which is rectangular, bounded by walls, and almost empty (maybe a few obstacles). Initially, position the robot randomly somewhere in your robot arena.

(a) Program a collision avoidance behavior—a robot that moves around while avoiding crashing into walls.
(b) Program a wall follower—a robot that moves along the walls without touching them (i.e., avoid the easy solution of letting the robot slide along the wall).
(c) Program an appropriate behavior for a vacuum cleaning robot. What are strategies to ensure that (almost) everything gets clean?
(d) Just play around a little and see what kind of behaviors you can generate.

Chapter 4
Short Journey Through Nearly Everything

"And what happens to that incredibly complex memory bank that remembers the whole system during these periods of 'swarming'?"

—Stanisław Lem, The Invincible

"We take off into the cosmos, ready for anything: for solitude, for hardship, for exhaustion, death. [...] A single world, our own, suffices us; but we can't accept it for what it is."

—Stanisław Lem, Solaris

Abstract We do a little example tour through many methods and ideas we are going to study in this book. This is a brief overview of the methodology that is relevant to designing swarm robot systems. We model a robot controller with a finite state machine for a collective-decision-making problem. We immediately face the typical challenge of distinguishing between microscopic information available to an individual robot and macroscopic information accessible only to an external observer. We continue with a simple macroscopic model of collective making and discuss whether it represents a self-organizing system.

4.1 Introduction

4.1.1 Learning Objectives

After reading this chapter, you should be able to:

- Understand the methodology of designing swarm robot systems and modeling controllers with finite state machines for collective decision-making.
- Distinguish between microscopic information available to individual robots and macroscopic information visible only to an observer, recognizing the micro–macro problem.

H. Hamann, *Swarm Robotics*,
https://doi.org/10.1007/978-3-032-10584-4_4

- Explain how robot–robot interactions drive state transitions and influence individual behavior.
- Explore simple macroscopic models of collective decision-making.

4.1.2 *Guiding Questions*

- How can robot controllers for collective decision-making be modeled with finite state machines?
- What is the micro–macro problem in swarm robotics, and how do microscopic (local) and macroscopic (global) information differ?
- How can simple macroscopic models of collective decision-making be described?

4.2 Finite State Machines as Robot Controllers

A robot controller can be represented by a finite state machine with states being associated with actions and transitions triggered, for example, by sensor input or timers. A state represents constant actuation for the time spent in that state. An example for a collision avoidance behavior modeled by a finite state machine is given in Fig. 4.1. The robot has a sensor to the left s_l and a sensor to the right s_r. Thresholds θ_l and θ_r determine when an object is too close (bigger values of s indicate closer objects). The turns are executed for a defined time until a timer is triggered.

A minimal example is shown in Fig. 4.2. The condition of the transition T can depend on a particular sensor value and a threshold, on a timer, on a received message, etc.

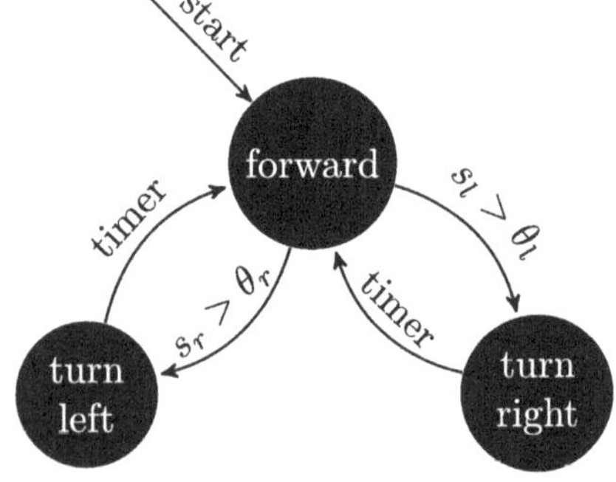

Fig. 4.1 Finite state machine for a collision avoidance behavior

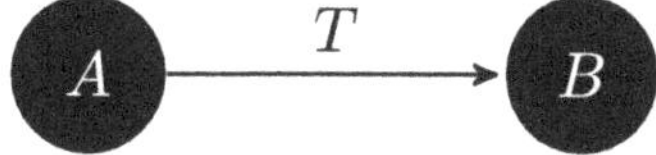

Fig. 4.2 Minimal example of a finite state machine

4.3 State Transitions Based on Robot–Robot Interactions

In the following, we focus on state transitions that depend on robot–robot interactions. In particular, we focus on situations when the states of neighboring robots determine each robot's state transitions and hence its behavior. Say there are only two possible states: A and B. Say the swarm size is N. Then we can define a variable a that counts robots currently in state A and a variable b that counts robots in state B. Both a and b reflect global knowledge about the swarm that is available to an observer but not to the robots themselves. An obvious condition is $N = a + b$. Furthermore, we can calculate the fraction of robots that are in state A by $\alpha = \frac{a}{N}$.

We assume that an agent can determine the internal state of all its neighbors. This could be implemented by explicit messaging or, for example, each robot could switch on a colored LED which encodes its internal state and which could be detected by vision. Furthermore, we assume that all robots also keep moving around, although we do not specify a particular purpose of doing so here. However, if the robots are in motion, then their neighborhoods are also dynamic.

Based on the unit disc model, we assume a sensor range r. For a robot R_0, all robots within distance r are within its neighborhood $\mathcal{N}$. For the situation shown in Fig. 4.3 we have $\mathcal{N} = \{R_1, R_2, R_3, R_4\}$. We assume that R_0 knows the states of all neighboring robots $\mathcal{N}$. In similarity to the above- defined variables, we can introduce variables $\hat{a}$ and $\hat{b}$ that give the number of neighboring robots of robot R_0 plus itself that are in state A and B, respectively. We have $|\mathcal{N}| + 1 = \hat{a} + \hat{b}$ and we can give a fraction of robots that are in state A by $\hat{\alpha} = \frac{\hat{a}}{|\mathcal{N}|+1}$.

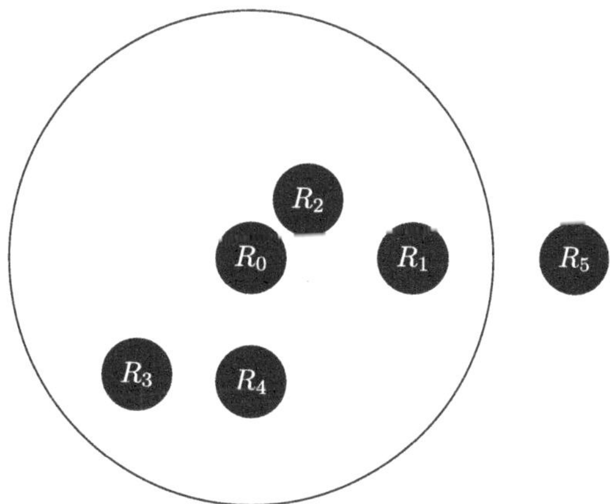

Fig. 4.3 Neighborhood of robot R_0 is $\mathcal{N} = \{R_1, R_2, R_3, R_4\}$

In the following, we assume that a robot's state transitions between A and B depend exclusively on $\hat{\alpha}$. For example, we could say if the robot is currently in state A and it measures $\hat{\alpha} < 0.5$ (i.e., there are more neighbors in state B than state A) then it switches to state B.

4.4 Early Micro–Macro Problems

We can interpret the swarm fraction $\hat{\alpha}$ as a measurement by robot R_0 of the actual current global situation, which is given by swarm fraction α. This is also called "local sampling" and will be discussed in detail later (see Sect. 7.3). Generally, we have $\alpha \neq \hat{\alpha}$. Under which conditions could we hope for $\alpha \approx \hat{\alpha}$? On average, and for a so-called well-mixed system, that is, a system without a bias, we can assume $\alpha \approx \hat{\alpha}$. However, the well-mixed assumption often does not hold, and the sampling error (i.e., variance in the robot's measurements) can have systematic effects and, hence, introduce bias as well. This complex of problems is already a small taste of the micro–macro problem that will be discussed in detail later. The main challenge of swarm robotics is to find connections between the microscopic level (here, local measurements of $\hat{\alpha}$) and the macroscopic level (here, the actual global situation α). We also speak of establishing a micro–macro link.

4.5 Minimal Example: Collective-Decision-Making

Next, we investigate a minimal example of collective-decision-making (more details later, see Chap. 8). The task in collective-decision-making is typically to find a consensus, that is, 100% of the robots in the swarm agree on a decision which could, for example, be to switch to the same internal state. Based on the above- defined result $\hat{\alpha}$ of locally sampling the neighborhood and a threshold of 0.5, we can define a finite state machine for this little collective-decision-making scenario (see Fig. 4.4).

These transition rules define what we call a majority rule because this approach tries to reinforce the current majority. If there are more close-by robots in state A, then the considered robot switches to A (otherwise it stays in B). If there are more close-by robots in state B, then the considered robot switches to B (otherwise it stays

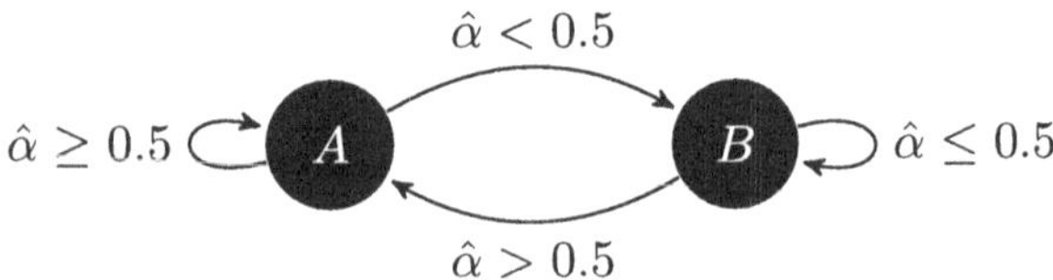

Fig. 4.4 Minimal example of collective decision-making based on the majority rule

in A). We hope that on average the local measurement $\hat{\alpha}$ gives a good approximation to the global state α (cf. micro–macro problem). As a consequence, the whole swarm should converge on a consensus of either $\alpha = 1$ or $\alpha = 0$.

4.6 Macroscopic Perspective

On the microscopic level, the situation is relatively straightforward. The measurement of $\hat{\alpha}$ is probabilistic, but the state switching behavior for a given $\hat{\alpha}$ is deterministic. If we want to determine the macroscopic effect of this microscopic behavior, then things get a bit more difficult. We have to look into the combinatorics of all possible neighborhoods. For simplicity, we restrict ourselves to a small neighborhood of $|\mathcal{N}| = 2$ and we make use of the above- defined true swarm fraction α.

The probability $P_{B\to A}$ to switch from B to A is given by

$$P_{B\to A}(\alpha) = (1-\alpha)\alpha^2, \tag{4.1}$$

because the considered robot has to be in state B (probability $1-\alpha$), we assume that its neighborhood is statistically independent, so we multiply by the probability that both ($|\mathcal{N}| = 2$) neighboring robots are in state A (otherwise the transition condition $\alpha > 0.5$ is not satisfied). According to combinatorics there would be $\binom{3}{1} = 3$ ways of arranging two A and one B (BAA, ABA, and AAB) but only the first one results in a switch with the considered robot in state B. A plot of Eq. (4.1) is shown in Fig. 4.5. For $\alpha > 0.66$, there are too few robots in state B that could potentially switch; that is why the probability is decreasing with increasing α. For $\alpha < 0.66$, there are too few robots in state A that could generate a local majority; that is why the probability is decreasing with decreasing α.

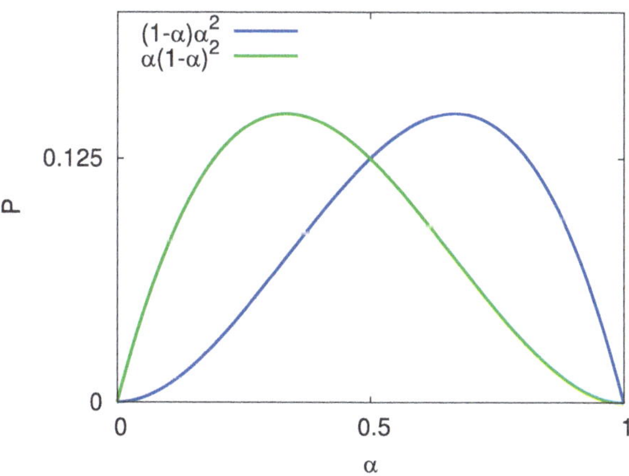

Fig. 4.5 Switching probabilities, Eq. (4.1)

The probability of switching from A to B is fully symmetric and defined by

$$P_{A\rightarrow B}(\alpha) = \alpha(1-\alpha)^2. \tag{4.2}$$

In all other situations, we observe no switch, and we get

$$P_{A\rightarrow A}(\alpha) = \alpha^3 + 2(1-\alpha)\alpha^2, \tag{4.3}$$

whereas the term $2(1-\alpha)\alpha^2$ accounts for the above-mentioned situations ABA, and AAB where we would have a majority of A but the considered robot is already in state A and hence does not switch to A. In symmetry, we get

$$P_{B\rightarrow B}(\alpha) = (1-\alpha)^3 + 2(1-\alpha)^2\alpha. \tag{4.4}$$

4.7 Expected Macroscopic Dynamics and Feedbacks

Now we would like to know the expected macroscopic dynamics of the system, that is, how does the swarm fraction α develop over time. For that, we need to introduce a representation of time. We define a time interval Δt that is long enough to observe the state transitions and short enough to observe not too many states transitions. We define the expected change $\Delta\alpha$ of α by using the above- defined probabilities of state transitions $P_{B\rightarrow A}$ and $P_{A\rightarrow B}(\alpha)$ and by weighting them according to the contribution of the state switches to the change of $\Delta\alpha$

$$\frac{\Delta\alpha(\alpha)}{\Delta t} = \frac{1}{N}((1-\alpha)\alpha^2) - \frac{1}{N}(\alpha(1-\alpha)^2). \tag{4.5}$$

The first term represents transitions $B \rightarrow A$, which contribute positively to α (i.e., generating more robots in state A). The second term represents transitions $A \rightarrow B$, which contribute negatively to α (i.e., generating more robots in state B). Factor $1/N$ accounts for an assumed switching rate of one robot per time step. Equation (4.5) is plotted in Fig. 4.6.

What is shown in Fig. 4.6 represents a feedback process. Whether it is positive or negative feedback is easily determined visually. The left half of the diagram represents minority situations in terms of state A because we have $\alpha < 0.5$ there. In the left half, we also have only negative values for $\Delta\alpha$. That means the minority of state A is even reinforced. Similarly, in the right half of the diagram, we have the majority of situations in terms of state A because $\alpha > 0.5$. In the right half, we also have only positive values for $\Delta\alpha$. That means the majority of state A is even reinforced. Hence, we have positive feedback.

Does this robot swarm qualify as a self-organizing system? It contains positive feedback. Does it also contain negative feedback? It is not obvious, but it does contain negative feedback. All real-world systems have limited resources, so also here the

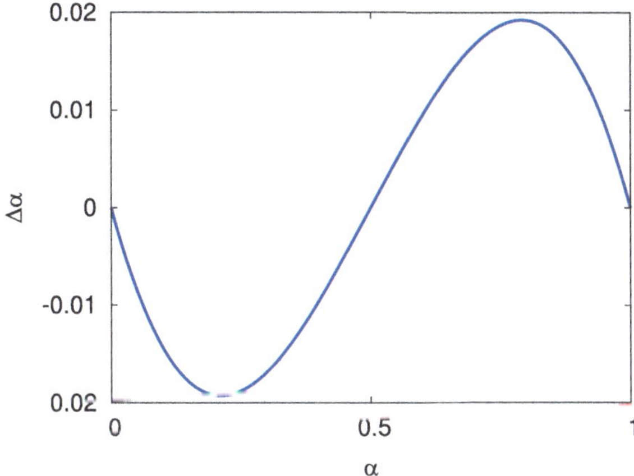

Fig. 4.6 Expected average change $\Delta\alpha$ of α

positive feedback is stopped once the minority is eaten up and no robots remain to switch to the majority (for $\alpha = 0$ and $\alpha = 1$). The system has initial fluctuations that determine whether we end up with 100% of the robots in state A or 100% in state B, and we have multiple interactions between many robots. The balance between exploitation and exploration is, however, biased towards exploitation. Once the swarm has converged on a consensus, it will never leave it. This cannot be good in certain situations, for example, in dynamic environments when the swarm should stay adaptive to new situations. Hence, this robot swarm is close to a self-organizing system but misses exploration.

How could exploration be included? We could allow robots to switch between states spontaneously. Say, in each time step, they have a 5% chance of a spontaneous switch. The above given expected change $\Delta\alpha$, Eq. (4.5), then changes to

$$\frac{\Delta\alpha(\alpha)}{\Delta t} = \frac{1}{N}((1-\alpha)\alpha^2) - 0.05\alpha - \frac{1}{N}(\alpha(1-\alpha)^2) + 0.05(1-\alpha). \tag{4.6}$$

While $\alpha = 0$ and $\alpha = 1$ were fixed points before, they are not anymore. For $\alpha = 0$, we have now $\frac{\Delta\alpha(\alpha)}{\Delta t} = 0.05$ and, for $\alpha = 1$, we have $\frac{\Delta\alpha(\alpha)}{\Delta t} = -0.05$. Hence, we have regions of negative feedback at the two boundaries for about $\alpha < 0.052$ and $\alpha > 0.948$. As a consequence, we always have a few robots in the opposite state, who can serve as explorers to check whether the other option has improved in utility. So then we have a self-organizing system.

4.8 Further Reading

This chapter is meant to give you a small taste of what we are going to discuss in this book. So you can keep reading, or if you want to know many more details about modeling collective-decision-making systems, then you can continue by reading

Valentini's book on the subject [800]. Vigelius et al. [825] study micro–macro models of collective-decision-making. Couzin et al. [160] explain how animals find good decisions. Biancalani et al. [79] describe a straightforward decision-making model that, however, most likely does not scale. Reina et al. [668] give a hint on how a software engineering approach for collective-decision-making in swarm robotics can look like.

4.9 Exercises

Exercise 4.1 (*Plot the Macroscopic Dynamic System Behavior*) Plot the expected change $\frac{\Delta\alpha(\alpha)}{\Delta t}$ with negative feedback as given in Eq. (4.6). How does it change when you choose different probabilities for spontaneous switching?

Exercise 4.2 (*Simulate Collective- Decision-Making*) Write a little simulation that simulates non-embodied agents moving randomly on a torus and switching states according to the rules we have defined in this chapter. Monitor the current global state α and plot its trajectory, that is, how it changes over time. Check also different probabilities for spontaneous switching again. Does it appear that the swarm changes from a large majority in favor of option A to option B and vice versa? How long does it take?

Chapter 5
Skills and Scenarios of Swarm Robotics

All this time we've been gazing upwards in anticipation of an alien species arriving from space, when intelligent life-forms have been with us all along, inhabiting part of the planet that we've never seriously attempted to explore.

—*Frank Schätzing, The Swarm*

From what you say, it follows that robots should be constructed quite differently from the way we've been doing it to be really universal: you'd have to start with tiny elementary building blocks, primary units, pseudo-cells that can replace each other, if necessary.

—*Stanisław Lem, The Invincible*

Abstract We provide a structured overview of fundamental skills and collective scenarios in swarm robotics, highlighting core models, methods, and applications. This chapter distinguishes between fundamental skills and more complex scenarios that form the building blocks of robotic collective intelligence. We categorize five domains of skills: physical (e.g., motion, collision avoidance), temporal (synchronization), numerical/logical (counting), computational (data processing), and minimal communication for information sharing. Individual skills, such as collision avoidance, ensure safety, while synchronization remains difficult without a global clock. Random motion strategies (e.g., Lévy flights) support efficient exploration. Collective behaviors are modeled through frameworks, such as those developed by Reynolds, Couzin, and Vicsek, which highlight complex dynamics and scale-free correlations. Scenarios progress from basic aggregation and dispersion to advanced tasks such as pattern formation, clustering, self-assembly, and collective construction. Further topics include collective transport with possible super-linear gains, as well as foraging, shepherding, and division of labor, where dynamic environments may yield counterintuitive effects such as "less is more."

H. Hamann, *Swarm Robotics*,
https://doi.org/10.1007/978-3-032-10584-4_5

5.1 Introduction

5.1.1 Learning Objectives

After reading this chapter, you should be able to:

- Categorize main skill domains: physical/spatial (e.g., motion, collision avoidance), temporal (synchronization without a global clock), numerical/logical (counting), computational (data processing), and minimal communication for information sharing.
- Explain key individual skills, such as collision avoidance for safety, synchronization challenges, and random motion strategies (e.g., Lévy flights) for exploration.
- Identify and analyze swarm scenarios ranging from aggregation and dispersion to pattern formation, clustering, self-assembly, and collective construction.
- Grasp the principles of collective motion (e.g., flocking) through models by Reynolds, Couzin, and Vicsek, including their interaction rules and emergent properties like criticality and scale-free correlations.
- Recognize advanced concepts such as collective transport foraging, shepherding, division of labor, and counterintuitive effects like "anti-robots" and "less is more."

5.1.2 Guiding Questions

- What are the core skill domains in swarm robotics (physical, temporal, numerical/logical, computational, communication), and what challenges do they pose (e.g., collision avoidance in dense swarms, synchronization without a global clock)?
- What roles do random motion strategies (Brownian motion, correlated random walks, Lévy flights) play in exploration, and how do environmental contexts favor different strategies?
- How do foundational models of collective motion (Reynolds, Couzin, Vicsek) differ in rules, emergent behaviors, and practical implementation on robots?
- What are the objectives and control strategies behind aggregation and dispersion, and how do they address the exploration–exploitation tradeoff in dynamic environments?
- How do complex scenarios like pattern formation, clustering, sorting, and self-assembly build on fundamental skills, and what bio-inspired mechanisms (e.g., stigmergy, reaction–diffusion, Brazil nut effect) support them?
- How can swarm size lead to super-linear performance gains in tasks such as collective transport and manipulation (e.g., the stick-pulling experiment)?
- What are the counterintuitive effects of "anti-robots" and "less is more," and how can reduced capability or deliberate disruption improve swarm adaptability?

- What are the key challenges of division of labor and task allocation, including partitioning, dynamic switching, and the use of local cues and probabilistic decision-making?
- How do catastrophic natural behaviors (e.g., circular milling in army ants) inform design principles for safe, robust swarm robotics and the avoidance of emergent traps?

We distinguish a set of skills and scenarios. This separation is partially artificial. What we consider skills could be seen as minor scenarios that rarely come as scenarios of their own. They are more likely to appear as building blocks in more complex scenarios.

5.2 Seven Skills

There are robot controllers and programming methods for swarm robots that are not directly tied to a specific task or scenario. These methods are somewhat more abstract, as we do not necessarily associate them with a swarm problem, and hence, they could easily be overlooked. As mentioned above, the distinction between skills and scenarios is, in part, arbitrary. Still, it can help to understand what could serve as building blocks on the swarm level. We also learn this way what we cannot get for free (e.g., a global clock) and understand some fundamental challenges.

We can categorize swarm robotics skills in five domains[1] that we will, however, not explore in all detail here.

1. Physical skills and spatial awareness include types of motion, determining the swarm's orientation (e.g., as in flocking, see Sect. 5.2.5), and basic maneuvers (e.g., collision avoidance, see Sect. 5.2.1). Different types of random motion are one of the most relevant examples (see Sect. 5.2.4). Localization and SLAM can be seen as elements of this category, but usually have a level of complexity that is far beyond other skills discussed here (see Sect. 6.5).
2. Temporal skills: Our swarms are missing a global clock and need to explicitly synchronize (see Sect. 5.2.2). Establishing a system-level sequencing (e.g., of tasks) also requires a considerable effort [269]. Even just establishing in the swarm that a collective decision-making process has been concluded is challenging and needs to be approached probabilistically.
3. Numerical and logical skills include counting how many robots are in a cluster of robots or in the swarm (see Sect. 5.2.3), basic arithmetic (in contrast to computation as stated below), and Boolean logic as discussed primarily in the context of collective decision-making (see Chap. 8).

---

[1] This is inspired by a conversation with Dr. Carlo Pinciroli.

4. Computational skills include algorithmic approaches to data processing and encoding. Intuitively, these are skills that involve selection, that is, making decisions. While such capabilities may appear fundamental and omnipresent, there are swarm methods that explicitly avoid computation (see Sect. 6.8.1).
5. Minimal communication skills: Our swarm robots often cooperate and collaborate, guided by the sharing of information through explicit communication. We avoid point-to-point communication to ensure scalability and often use simple methods, such as gossiping (see Sect. 5.2.6). In some cases, reducing communication, despite seeming counterproductive, can enhance swarm performance (see Sect. 5.2.7).

In the following, we discuss seven essential skills that are often overlooked. For example, collision avoidance is usually considered an embarrassingly simple problem; however, it can have a significant impact on a swarm's performance if not managed effectively. Similarly, synchronizing a robot swarm is a required skill that may be overlooked, for example, when implementing it by default in a simulation (i.e., executing the sense-act cycle for all robots in a synchronized manner). Random motion of robots may also be viewed as a simple issue, but it requires attention, especially for optimal exploration.

5.2.1 Collision Avoidance

Collision avoidance is arguably the "Hello World" problem of mobile robotics. Once the robot moves, we want it to avoid colliding with objects. This is already a single-robot challenge, but the situation becomes even more complex when multiple robots share the same space. Collision avoidance in swarm robotics is essential for ensuring that robots can operate safely and efficiently while maintaining their collective behaviors. Given we focus on decentralized control and local interactions, we usually cannot use methods of multi-agent path finding (MAPF) to compute collision-free paths [749]. Typical methods of MAPF use global information, such as global localization. Still, there are many variants of algorithms that implement collision avoidance. Given the fundamental importance of collision avoidance in single-robot systems, numerous algorithms have been developed to address this issue. Here we mention just a few, following Taylor and Nowzari [775]. A fundamental concept is that of potential fields (see Sect. 3.5.2). Obstacles generate virtual repulsive forces that push the robot away, preventing collisions. Gyroscopic steering is a collision avoidance method that adjusts a robot's trajectory by applying forces perpendicular to its velocity, rather than directly repelling it from obstacles. This helps to maintain smooth motion while avoiding abrupt stops. Control Barrier Certificates are a formal safety-critical control approach that ensures robots avoid collisions while maintaining their desired behavior. It enforces safety constraints by modifying a robot's control input only when necessary, making it a minimally invasive method. Optimal Reciprocal Collision Avoidance (ORCA) [808] is a decentralized collision avoidance algorithm for multi-robot systems. It builds on the Velocity Obstacle method, which defines a set of unsafe velocities that would lead to a collision within a given time horizon.

Taylor and Nowzari [775] study collision avoidance strategies and their impact on swarm behavior in robotic systems. They argue that simulations of swarm behaviors are often too simplified, especially when simulating robots without volume, and hence ignoring collisions. Using a ring formation swarming behavior as a test case, they analyze how different collision avoidance strategies impact swarm performance. Despite extensive tuning, all methods introduce unavoidable interference with the respective swarm behavior. Their findings suggest that collision avoidance should be co-designed with swarming behaviors rather than added later.

5.2.2 Synchronization and Desynchronization

Swarm systems are asynchronous systems. There is no central clock that everyone can access. Since our swarm robots lack a global clock, they must synchronize explicitly if required. Typical examples of synchronization phenomena [753] are acoustically coupled systems, such as rhythmic applause [594] and chirping crickets [869], visually coupled systems, such as flashing of fireflies [114], and mechanically coupled systems, such as coupled pendulums [660] and gait synchrony on bridges [755]. Suppose a group of humans needs to synchronize (e.g., gathering at the local pub). In that case, they usually use a global communication channel (i.e., a channel that all group members can easily access). Our robots, however, use local communication and can only speak to their neighbors. Hence, we need a distributed algorithm to synchronize the swarm globally. We will address that later, but first, let's consider a key aspect of hardware and then explore standard models for synchronization in the literature.

GPS-based Time Synchronization

One straightforward hardware solution is the use of GPS modules, which provide an external clock reference. Many off-the-shelf GPS units deliver a "one pulse per second" signal with microsecond-level accuracy, allowing robots to align their clocks to a global timescale. This approach has been widely adopted in outdoor robotics, but it is not always feasible indoors, underwater, or in GPS-denied environments. In such cases, robots must fall back to local synchronization mechanisms.

Mathematical Models of Synchronization Effects

A mathematical model of synchronization is the coupling in phase of two oscillators. For phases θ_1 and θ_2 of the two oscillators and their natural frequencies ω_1 and ω_2 (oscillator rate at which it oscillates in the absence of external influences), we get

$$\frac{d\theta_1}{dt} = \omega_1 + K \sin(\theta_2 - \theta_1) , \quad (5.1)$$

$$\frac{d\theta_2}{dt} = \omega_2 + K \sin(\theta_1 - \theta_2) , \quad (5.2)$$

for a coupling strength K. If the coupling K is sufficiently strong, the oscillators gradually synchronize, meaning their phase difference remains constant over time.

Timing and synchronization phenomena are modeled in physics with oscillators. There is a wide range of oscillator types. The simplest oscillator model is the harmonic oscillator, whose motion is described by $m\ddot{x} + kx = 0$ (where x represents the displacement, charge, or position, depending on the application). It is used to model, for example, mass-spring systems, simple pendulums (small angles), and inductor-capacitor (LC) circuits. The damped harmonic oscillator is described by $m\ddot{x} + b\dot{x} + kx = 0$ and models, for example, RLC circuits (resistor, inductor, capacitor) with resistive damping for filters and suspension bridges. Driven harmonic oscillators are described by $m\ddot{x} + b\dot{x} + kx = F_0 \cos(\omega t)$ and model, for example, a child on a swing. The Van der Pol oscillator is described by $\ddot{x} - \mu(1 - x^2)\dot{x} + x = 0$ (μ is a nonlinear damping coefficient) and models, for example, vacuum tubes, heartbeat rhythms, and brain wave patterns (see Fig. 5.2a). The Van der Pol oscillator is also similar to a sawtooth pacemaker. That is a timing mechanism, often observed in biological systems like fireflies or crickets, where an internal oscillator repeatedly builds up to a threshold and then rapidly resets to a baseline level, creating a repeating "sawtooth" waveform pattern.

In most cases, these oscillators are studied in isolation. However, there were early findings centuries ago that introduced the concept of coupled oscillators. For example, Christiaan Huygens found in 1665 that pendulums sitting on the same beam would synchronize [660]. Other examples include coupling between acoustic resonators and coupled oscillating electrical circuits. The standard simple model to study synchronization of oscillators is the Kuramoto model (see Sect. 8.5.7 for more details). In the standard Kuramoto model [455], there are $N > 1$ oscillators that are coupled, that is, their phases influence each other. Oscillator i has phase $\theta_i \in [0, 2\pi]$ and is updated by

$$\frac{d\theta_i}{dt} = \omega_i + \frac{K}{N} \sum_{j=1}^{N} \sin(\theta_j - \theta_i), \quad (5.3)$$

for a natural frequency ω_i and coupling strength K. The standard model has global coupling. Initially, all oscillators start with random phase $\theta_i \in [0, 2\pi]$. The Kuramoto model is critical and popular because it captures the essential synchronization

Fig. 5.1 Synchronized flash of fireflies (Photo by xenmate, CC BY 2.0)

behavior of any system of coupled (limit-cycle) oscillators, regardless of whether they are biological, electrical, or mechanical. A significant finding is that a large population of weakly coupled oscillators with distributed natural frequencies can spontaneously synchronize if the coupling strength exceeds a critical threshold. This indicates that synchronization is a universal and robust phenomenon. That is undoubtedly good news for swarm robotics.

Fireflies as Natural Synchronizers

An example of a biological system that shows synchronization is a population of fireflies (see Fig. 5.1). "Though most species of firefly are not generally known to synchronize in groups, there are some (for example, *Pteroptyx cribellata*, *Luciola pupilla*, and *Pteroptyx malaccae*) that have been observed to do so in certain settings."[2] Buck and Buck [114] study the synchronous flashing behavior of fireflies in Southeast Asia, highlighting how male fireflies congregate in specific trees and flash in near-perfect unison to attract female fireflies. The synchronization enhances signal visibility in dense swamp habitats, overcoming visual communication challenges. The behavior of the attracted females is driven by phototaxis (i.e., motion towards light). Buck et al. [113] demonstrate that firefly luminescence is directly con-

[2] http://ccl.northwestern.edu/netlogo/models/Fireflies.

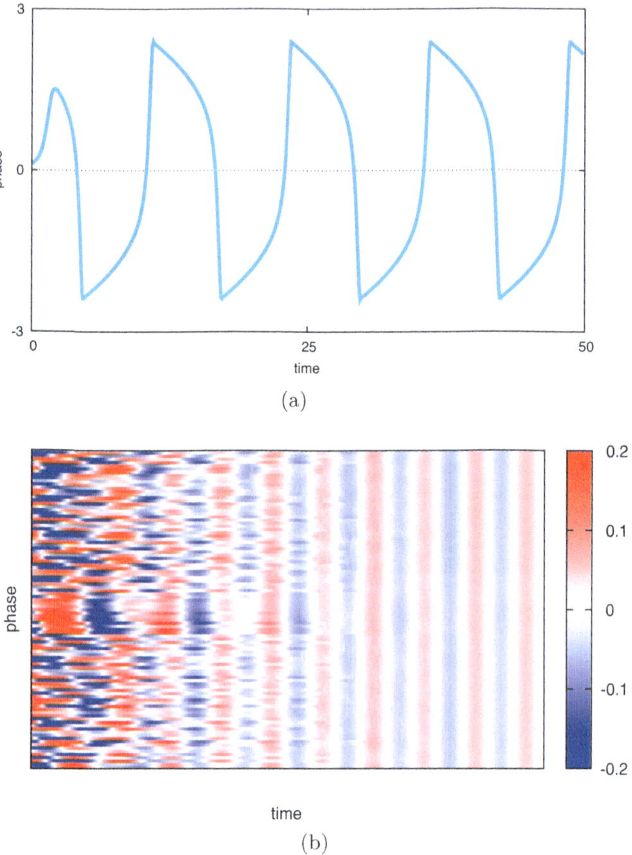

Fig. 5.2 **a** Regular Van der Pol oscillator, plot of phase over time. **b** Phase plot over time for all $N = 200$ pedestrians of the Millennium Bridge scenario following Belykh et al. [68]

trolled by neural signals, with consistent flash patterns and responses to electrical stimulation revealing a neuroeffector system akin to muscle physiology.

Chorusing-Based Synchronization, Swarmalator, and Sensor Networks

Holland et al. [363] introduced an approach based on chorusing, that is, a mechanism of coordinated signaling behavior inspired by fireflies and crickets. They built on a modification of a coupled oscillator model by Mirollo and Strogatz [560]. Each robot has a counter $c \in [0, c_{max}]$ that is increased by one in each time step. A robot can be in either of two phases: *refractory* phase for $0 < c \leq c_1$ or *listening* phase for $c_1 < c \leq c_{max}$. In the *listening* phase, a robot sends a signal with a given constant probability P or if it hears a signal from a neighbor. When $c > c_{max}$ is reached, the robot emits a signal. Sending a signal is followed by a counter reset $c = 0$ and the *refractory* phase where the robot is unable to send a signal and ignores neighbors. This is introduced to prevent waves of signals from continuously passing through the swarm. This whole procedure implements a phase advance interactive mechanism. By advancing its phase, a robot reduces the time lag between its own signal and the signal it detects, allowing for quicker swarm-level synchronization. This is in contrast to the phase delay mechanism, where robots delay their signal timing after detecting another signal. For swarm size N, a robot is expected to send a non-stimulated signal every $N - 1$ stimulated signals. This is based on the reasoning that in a synchronized swarm of N robots, each robot is expected to be the first to send a signal on a proportion of $\frac{1}{N}$ occasions (assuming period lengths of all robots follow the same distribution).

There are many other works on synchronization for swarm robots. Here, we only mention a few selected ones. A related concept to mere synchronization is treated with the so-called "Swarmalator," [612] a model of mobile oscillators whose locations affect their phase dynamics and vice versa (see Sect. 8.5.8). This is a model of swarm robots that synchronize depending on their location and also allow their motion to be influenced by the synchronization mechanism. The Swarmalator approach was implemented in real robot swarms by Barciś and Bettstetter [55].

Distributed clock synchronization is also relevant in the fields of sensor networks and ad-hoc networks; see, for example, Tyrrell et al. [798], Leidenfrost and Elmenreich [472], and Sivrikaya and Yener [735] for a survey.

Desynchronization

Also, the opposite of synchronization is nontrivial: "Desynchronization is the logical opposite of synchronization; instead of nodes attempting to perform periodic tasks at the same time, desynchronization occurs when each node performs its task as far away as possible from all other nodes" [178]. Degesys et al. [178] introduce an algorithm for desynchronization in wireless sensor networks, enabling nodes to evenly distribute their periodic transmissions without relying on a global clock. The algorithm supports collision-free Time Division Multiple Access (TDMA) scheduling and adapts to dynamic changes in node numbers.

Synchronization on the Millennium Bridge

As a final example, we examine a system that combines biomechanically inspired models of crowd dynamics, physical coupling of agents, pedestrian-induced lateral motion of bridges (also known as wobbling), and Van der Pol oscillators. A beautiful work uses the concept of coupled oscillators to model a phenomenon that was observed on the Millennium Bridge in London [68, 755]. Oscillations can amplify the seemingly insignificant force of a pedestrian's step into a powerful, destructive force capable of damaging or even collapsing bridges. A notable example occurred in 2000 during the opening ceremony of the Millennium Bridge in London, where synchronized footfalls caused unexpected swaying, highlighting the dramatic impact of resonance phenomena in structural engineering. Initially triggered by slight variations in pedestrians' walking patterns, the bridge began to exhibit lateral oscillations. In response, pedestrians unconsciously adjusted their steps to match the bridge's movement, further reinforcing and amplifying the swaying motion. With group sizes exceeding a certain threshold ($N_c = 165$ for the Millennium Bridge), the oscillations of the bridge are amplified, posing a threat to its structural integrity [68]. One approach to modeling such an oscillator system is the Van der Pol oscillator. In this model, each agent $i \in \{1, 2, \dots, N\}$ has a lateral gait phase x_i, natural gait frequency ω_i, and parameters λ_i and a that influence the pattern of movement. The behavior of the bridge is defined by its own lateral movement modeled via bridge position y, the coupling strength r between the bridge and the pedestrians, the natural frequency Ω of the bridge, and the dimensionless damping coefficient h. The mass of the bridge, the mass of the pedestrians, and the height of the center of mass of the pedestrians are considered in this approach, but are hidden in the parameters ω_i, Ω, h, a, and r. The following system of differential equations models the system. The lateral gait phase x_i of pedestrian i is modeled as a Van der Pol oscillator by

$$\frac{d^2x_i}{dt^2} = \lambda_i \left(\left(\frac{dx_i}{dt} \right)^2 + x_i^2 - a^2 \right) \frac{dx_i}{dt} - \omega_i^2 x_i - \frac{d^2y}{dt^2} \tag{5.4}$$

and the bridge's lateral motion is modeled by its position y that is updated by

$$\frac{d^2y}{dt^2} = -2h\frac{dy}{dt} - \Omega^2 y - r \sum_{i=1}^{N} \frac{d^2x_i}{dt^2} \ . \tag{5.5}$$

See Fig. 5.2b for a plot of all pedestrians' gait phases x_i. The plot shows a global synchronization effect after a short transient. The synchronization relies on global coupling via the bridge here. This is similar to the synchronization of pendulums coupled via a beam. We defer the local coupling for the Millennium Bridge scenario for a task (see Exercise 5.3).

5.2.3 Counting

Sometimes, the skill of counting may be required in swarm robotics. For example, a task may require a particular exact or minimum group size. A robot may need to estimate the size of its local cluster, or the swarm as a whole may need to determine its total size N. In swarm robotics, the swarm size N is considered global information, and we usually would want to avoid requiring a robot to know it. Strictly speaking, it does not scale as storing an integer does, which scales $\mathcal{O}(\log N)$ in terms of memory; however, that is rather a dogmatic argument (except for extremely restricted hardware, as it could be imagined for nanorobotics).

There is, maybe surprisingly, little literature on counting robots in swarms. An earlier paper is that of Melhuish et al. [544]. They build on their own previous work on chorusing as discussed above (see Sect. 5.2.2). So to establish the skill of counting, they also use the skill of synchronizing (see above in Sect. 5.2.2). They use synchronized signaling to form stable groups of a desired size and to coordinate collective departures and movements (convoying). Following the chorusing concept [363], robots synchronize and either send out stimulated signals because a neighbor sent a signal or non-stimulated signals because the considered robot initiated the signal itself. Swarm counting is implemented by letting each robot count the number of stimulated signals n between its non-stimulated signals. The number of counted stimulated signals n plus one is an estimate of swarm size: $n + 1 \approx N$. A better approach, of course, is to use a mean of a history of counts $\bar{n}$: $\bar{n} + 1 \approx N$. The approach strikes a balance between minimal communication and control, while achieving reliable group coordination. To aggregate robots of a certain group size G, each robot independently estimates the group size $\hat{G}$. If $\hat{G} \leq G$, the robot moves towards the group's center. If $\hat{G} > G$, the robot moves away. The approach is not fully scalable in terms of accuracy and time. In the form of a speed-accuracy tradeoff, accuracy depends on c_{max} relative to swarm size N. Bigger swarms require longer cycle lengths c_{max}.

Brambilla et al. [101] build on the work of Melhuish et al. [544] and address the issue of noisy and unstable size estimates. Their main idea for reducing noise in the approach is that robots that recently emitted a non-stimulated signal should have a lower chance of emitting again. They introduce a third phase: *stimulated only* phase. During *stimulated only* phase, "a robot can emit a stimulated signal, but has zero probability ($P = 0$) of emitting a non-stimulated one" [101]. Their signaling protocol reduces estimation error by imposing an implicit order on signal emissions, which improves both stability and agreement across the swarm. Their algorithm shows robustness and adaptability to dynamic changes in swarm size, as validated through extensive simulations. It seems plausible that this approach is also not fully scalable in terms of accuracy. Brambilla et al. [101] report only simulations for $N = 25$ and 50.

Similarly to the above approaches, Varughese et al. [820] introduce a communication framework following a simple state-based mechanism (inactive, active, refractory). Robots send signals that propagate through the swarm in wave-like patterns. A robot initially sets its counter c to a random value $c \subset [0, c_{max}]$. When $c \geq c_{max}$, the robot sends a signal and resets its counter randomly $c \in [0, c_{max}]$. Note that there is no explicit refractory phase in this case. Each time a robot receives a signal, it increases another counter n and relays the signal. When a robot sends a signal, it also stores its current counter n as an estimate of swarm size and resets it $n = 0$. As above, we have $\bar{n} + 1 \approx N$. This approach tends to underestimate swarm size. When multiple robots initiate signal waves simultaneously, the waves overlap, causing swarm members to detect them as a single wave. Robots count them as one event and underestimate the total. The estimate improves for longer cycle lengths c_{max}

relative to swarm size N. Accuracy can be achieved with time. Knowing only the order of magnitude of N is enough to choose sufficiently long $c_{\max}$ [819]. For online scaling, the cycle length must also be adapted online.

Wang and Rubenstein [846] leverage a probabilistic sampling method, where each robot i generates a random number r_i uniformly distributed $r_i \in [0, 1]$, and the swarm collectively determines the maximum random number $r_{\max} = \max_i r_i$ to estimate the swarm size. For larger swarms N, the maximum $r_{\max}$ will be closer to one. As above, it is recommended to work with averages $\bar{r}_{\max}$, here for statistically independent stochastic experiments. The probability density function of $r_{\max}$ is $f(x) = Nx^{N-1}$ for $x \in [0, 1]$. The expected value is $\mathbb{E}(r_{\max}) = \bar{r}_{\max} = \frac{N}{N+1}$. As estimated swarm size, we get

$$N \approx \frac{\bar{r}_{\max}}{1 - \bar{r}_{\max}} . \tag{5.6}$$

However, we need to know how to obtain maximum $r_{\max}$ in a swarm. The robots share their random numbers r_i with their neighbors. Each robot keeps track of the largest random number $\hat{r}_{\max}$ it has received so far, including its own. If a robot receives a larger number than its current maximum ($r > \hat{r}_{\max}$), it updates its value and propagates the new maximum number to its neighbors. To ensure accuracy, each robot operates in discrete rounds, where it listens, compares, and updates its maximum value before advancing to the next round ("listen-think-talk"). Theoretical analysis guarantees the algorithm's accuracy and convergence rate, showing that error decreases linearly with the number of sampling rounds, regardless of swarm size N. The method is validated through simulations and real-world experiments with Kilobot swarms, demonstrating its robustness, accuracy, and adaptability. Wang and Rubenstein [846] report that their approach is scalable as indicated by their simulation results for up to $N = 10^5$ and a fixed number of 10^3 rounds. However, for this approach, incomplete information due to a too short time for collectively determining the maximum value would result in an underestimation of the swarm size.

Saha et al. [692] presents the aggregate-and-broadcast algorithm, a deterministic and decentralized method for counting the number of nodes in a network when an upper bound on size is known. Every node in a connected network can compute the exact total number of nodes using only local communication and minimal memory. The algorithm works by progressively pruning the network into regional hubs that aggregate counts, which are then broadcast across the network. Compared to previous methods, Saha et al. [692] significantly reduce communication overhead and memory usage. Hence, the approach is a good option for our resource-constrained robot swarms.

Generally, the problem and methods of estimating swarm size N are related to the issue of online scalability as discussed in Sect. 2.7.1. For example, in the approach of Wahby et al. [840], the robots estimate swarm density ρ, which is directly connected to swarm size via operation area size A: $\rho = \frac{N}{A}$. There is also a connection to quorum sensing as observed in bacteria, where swarms collectively detect and respond to a threshold number or density of individuals, triggering collective behavior (see Sect. 1.4).

5.2.4 *Random Motion*

How can random motion possibly be a skill? Even an impaired person can perform a random walk [903], so why should it be challenging to implement on robots? First, random motion can play a crucial role in exploration. If our robots know little about their environment, they can't do better than search randomly. This is also seen in foraging animals and studied widely [149, 381]. Second, there are several ways to implement random motion. A particular chosen approach needs to be clearly specified and depending on the scenario specific movement strategies will perform better than others [616]. Also note that the following models are exclusively focused on motion, whereas in multi-robot search or exploration, we would have the option of coordinating the robots through communication. Random walks are also studied in discrete space and on networks [520]. Note that some of the following methods are also used for time series generation in benchmarking for machine learning.

In a prominent paper on swarm robotics that focuses on random walks, Dimidov et al. [188] study correlated random walks and Lévy flights in simulations and with a Kilobot swarm. They find that correlated random walks with high persistence (high correlation in headings between two consecutive time steps) are the most effective strategy in bounded areas, whereas Lévy flights are more advantageous in unbounded space.

The latter paper by Kegeleirs et al. [414] also focuses on random walks in swarm robotics. They study Brownian motion, correlated random walk, Lévy walk, and Lévy taxis (added directional bias in response to an external gradient).

Brownian Motion and Wiener Process

Historically, the first mathematical models of random walks appeared in the context of Brownian motion [227, 799, 835]. The Wiener process mathematically describes Brownian motion. The Wiener process is a continuous-time stochastic process W_t. We initialize to $W_0 = 0$. The distribution of any increment $W_{t+v} - W_t$ depends only on the length of the time interval v. For $0 \leq v < t$, the increment $W_t - W_v$ has a normal distribution with mean $\mu = 0$ and variance $\sigma^2 = t - s$:

$$W_t - W_v \sim \mathcal{N}(0, t - v). \tag{5.7}$$

It may be somewhat unsatisfying that, despite being a continuous-time process, we define it by referring to a time increment v. It is essential to recognize that the time increment v is not a fixed time step, but rather an arbitrary real number indicating the distance we extend beyond t in our observation. It is simply the parameter that measures the length of the time interval over which we consider the change in the Brownian path. There is no easy way to describe it as an ordinary differential equation, as the Wiener process is almost surely nowhere differentiable. An alternative are stochastic differential equations (SDE) with the interpretation in terms of the Itô calculus, such as the Langevin equation (see Sect. 7.4.2 for more details). Also note that the Wiener process is self-similar, that is, for every $c > 0$ the process $V_t = (1/\sqrt{c})W_{ct}$ is another Wiener process (you can "zoom in and out" in the time dimension as you like).

Random Walks in Discrete Time

In that sense, the Wiener process is more of a mathematical construct than something that can be truly implemented in our robots. We instead focus on random walks in continuous space with discrete time, that is, a sequence of positions $\mathbf{x}_t \in \mathbb{R}^n$ evaluated at discrete time steps $t \in \{0, 1, 2, \ldots\}$. The update rule from time t to $t + 1$ is

$$\mathbf{x}_{t+1} = \mathbf{x}_t + s_t\, \mathbf{u}(\theta_t)\ , \tag{5.8}$$

for step size s_t, heading θ_t, and $\mathbf{u}(\theta_t)$ is the unit vector in the direction θ_t. For $n = 2$ dimensions, we simply get

$$\mathbf{u}(\theta_t) = \begin{pmatrix} \cos(\theta_t) \\ \sin(\theta_t) \end{pmatrix}\ . \tag{5.9}$$

The heading θ_t is drawn from a probability distribution. The step size s_t can be deterministic or also drawn from a probability distribution, which adds another layer of variability. The next heading θ_{t+1} can be independent of the current heading θ_t or dependent. Depending on how θ_{t+1} relates to θ_t, the random walk can be uncorrelated or correlated.

Uncorrelated Random Walk

In an uncorrelated or memoryless random walk, the heading θ_{t+1} is chosen independently of θ_t. For example, θ_{t+1} can be drawn from a uniform distribution over $[0, 2\pi)$, that is, with no directional preference or memory of the previous heading. For the uncorrelated random walk, we define

$$\theta_{t+1} \sim U[0, 2\pi) \tag{5.10}$$

and a constant step size $\forall t : s_t = s$. An uncorrelated random walk tends to spread out in all directions over time. In two dimensions, the mean distance from the origin grows on the order of $\sqrt{t}$. Sometimes the term "isotropic random walk" is used. It refers to a random walk that has no preferred direction for any given step and does not specify dependencies between two consecutive time steps. In 2D, isotropy usually refers to drawing a heading θ from $U[0, 2\pi)$.

Correlated Random Walk

In a correlated random walk, θ_{t+1} depends on θ_t. This dependence introduces memory or persistence in direction. There are several options for implementing the correlation, and there are options of probability distributions to draw from. We can start by assuming a Gaussian (or normal) distribution. For the correlated random walk, we define

$$\theta_{t+1} \sim \mathcal{N}(\mu = \theta_t, \sigma^2) \tag{5.11}$$

and a constant step size $\forall t : s_t = s$. This notation is, however, ambiguous as the Gaussian distribution is unbounded and defined on $(-\infty, +\infty)$. The more precise definition, hence, uses a wrapped normal distribution as defined in directional statistics. The probability density function of the wrapped normal distribution is defined as

$$f(\beta; \mu, \sigma) = \frac{1}{\sigma\sqrt{2\pi}} \sum_{k=-\infty}^{\infty} \exp\left[\frac{-(\beta - \mu + 2\pi k)^2}{2\sigma^2}\right], \tag{5.12}$$

which gives us a normal distribution defined on $\beta \in [-\pi, \pi)$. For standard deviation $\sigma = \infty$, we get a uniform distribution (i.e., uncorrelated random walk), and for any finite value, it gives us a normal distribution and a correlated random walk. Another alternative is the von Mises distribution, also known as the circular normal distribution.

We can also use a random turn model and sample a turn angle ϕ_t relative to the current heading θ_t instead of sampling the next heading directly. We define $\theta_{t+1} = \theta_t + \phi_t$ and draw the turn angle, for example, from a narrower uniform distribution $\phi_t \sim U(-0.5\pi, 0.5\pi)$ or again a wrapped normal distribution. The next heading θ_{t+1} is still correlated.

Correlated random walks are widely used in mobile robotics and swarm robotics. In most application scenarios, the robot is supposed to explore a particular area. A random walk is a sufficient option but the robot still needs to be efficient by covering some distance. The correlated random walk appears to be a good compromise and is relatively easy to implement.

Lévy Flight and Lévy Walk

A Lévy flight is a special type of random walk known for its heavy-tailed step size distribution (extremely large values occur more often than in light-tailed distributions, such as a normal distribution). It can be used as a model for foraging

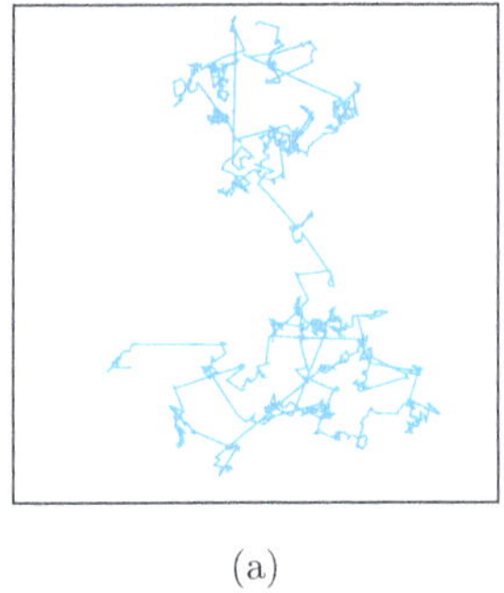

(a)

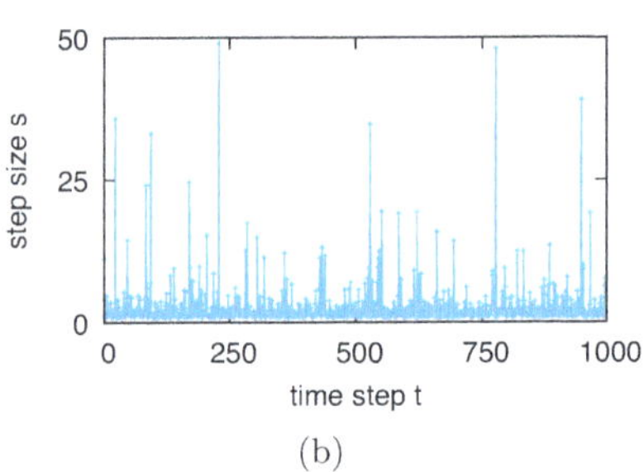

(b)

Fig. 5.3 Levy flight as a form of random motion ($\alpha = 1.5$, $s_{\min} = 1$, 10^3 simulated time steps). **a** Typical trajectory clearly showing "cluster-and-jump" dynamics. **b** Step size s_t over time steps showing the rare but wide jumps due to the heavy-tailed probability distribution of step sizes

animals in sparse environments, that is, food patches are scattered and distributed over wide areas [832]. A typical choice of distribution for the step size is a power-law (or Pareto) distribution:

$$s_{t+1} \sim PL(\alpha, s_{\min}) , \tag{5.13}$$

for a probability density function of $p_{PL}(s) = \frac{\alpha-1}{s_{\min}} \left(\frac{s}{s_{\min}}\right)^{-\alpha}$. Such distributions allow for occasional large jumps, resulting in movement patterns that exhibit "cluster-and-jump" dynamics (see Fig. 5.3). In typical definitions, the direction of each jump is chosen uniformly at random $\theta_{t+1} \sim U[0, 2\pi)$, but one could also use heading correlations.

Lévy flights seem to model efficient search patterns in foraging animals [673], for example, the foraging behavior of the wandering albatross *Diomedea exulans* [831]. The explanatory power of Lévy flights is, however, discussed [219].

There is also the concept of Lévy walks that assign a time cost to moving a given distance (e.g., by directly drawing a step duration T followed by movement at a given speed for time T). A long step takes proportionally longer, while Lévy flights assume immediate arrival also at more distant positions. Hence, Lévy walks are the more realistic model for robotics, but the mathematical difference is minor.

The application of robots does not seem as straightforward as a simple, uncorrelated random walk with a constant step size. The long jumps occurring in Lévy flights would require considerable navigation skills in mobile robots, particularly in terms of moving in straight lines for extended periods while also avoiding obstacles. Multiple papers are using Lévy flights for search and foraging in swarm robotics [188, 261, 762].

Run-and-Tumble Motion

A commonly studied random motion in microbiology (e.g., for *E. coli*) is called run-and-tumble motion. In the run phase, the bacterium runs (moves in roughly a straight line) for a particular duration (often modeled as exponentially distributed). In the tumble phase, it tumbles (random movement on the spot to turn), randomly changing direction. Although it may seem similar to Lévy flights, they differ in the distribution of step durations.

There are also other types of random walks that we do not treat in detail here. One more example is the reinforced random walk, which incorporates memory [149, 174]. Each of the robot's visits to a site increases its attractiveness. This creates a feedback loop where frequently visited locations become ever more likely to be revisited. This leads to significantly different behavior compared to standard random walks.

5.2.5 *Flocking and General Collective Motion*

Flocking is a collective motion in which a group of agents or robots moves in a coherent, coordinated way. For swarms of robots, this corresponds to motion in formation, where they maintain a close proximity to one another while moving in the same direction and still avoid obstacles. There are many published models on flocking or, more generally, collective motion. We examine the three most prominent ones here. The Reynolds model is the contribution from computer science that was initially meant for the animation of birds. The Couzin model is a biological contribution that serves as a behavioral model for fish and other species. The Vicsek model is the contribution from physics that is supposed to be a minimal model of self-propelled particles. Note that implementing any of these models on mobile robots is generally challenging. These models typically require precise detection of neighbors, including their current heading and direction. This, however, is challenging for a robot to determine in most implementations of swarm robotics that use onboard sensors (see Sect. 5.3.7 for more details). Keysberg and Wakamiya [426] comprehensively compares flocking models in simulations, but with a focus on robotics. Their comparison also includes the Cucker–Smale model [168]. The model is based on a rigorous analysis of a continuous-time version of the Vicsek model, which incorporates a distance-dependent communication function into a system of differential equations. The model is framed in terms of graph Laplacians, making it analytically tractable. In their comparison, Keysberg and Wakamiya [426] use metrics, such as uniformity and polarization.

In summary, there is still room for innovation and novel approaches for flocking in swarm robotics. One could try to implement criticality or models that take into account more cognitive aspects of collective motion, as discussed at the end of this section.

Reynolds Model

In one of the most classical swarm papers, Reynolds [674], published three simple rules that allowed him to simulate a bird flock. The rules require the definition of a neighborhood, for example, by a maximal distance.

1. Alignment: adapt your direction to that of your neighbors.
2. Cohesion: stay close to your neighbors.
3. Separation: if you are too close to one of your neighbors, then avoid a collision.

The original concept of Reynolds [674] also required to ignore agents in the back of an agent.

The approach of Reynolds is metric because the neighborhood is defined on distances. Meanwhile, we know that birds probably use a topological approach, that is, they check, for example, their five closest peers and ignore distances when determining their neighborhood [52]. The accompanying classical theoretical paper for Reynold's work is that of Vicsek et al. [824]. They report a kinetic phase transition, which is interesting from a physics point of view. Another early theoretical work on flocking is that of Toner and Tu [786] and Savkin [699], which does control theory for flocking in autonomous mobile robots.

Flocking behaviors can easily be simulated to gather some data that may be of interest. In Fig. 5.4a you see such a measurement from a simulation based on Reynolds [674]. It shows the probability density of all headings in radians (from $-\pi$ to π) found in the swarm at a given time. The headings are actually relative to the current average heading of the swarm. This allows for averaging data over several independent simulation runs. Initially, this density is homogeneous because the headings are assigned uniformly at random. Over time, a peak around heading zero emerges clearly. This is the expected flocking behavior with all agents moving approximately in the same direction. In Fig. 5.4b, you see another measurement from the same simulation runs. This one shows the connectivity, that is, the number of neighbors, depending on the agent's heading. Initially, the connectivity is below two. That is an arbitrary value due to the chosen swarm density in this particular experiment. Over time, the connectivity increases independent of the agent's heading. Only towards the end of the measurement, a peak emerges around heading zero again. However, the difference between the average connectivity of an agent with heading π (i.e., 180°) and heading zero is small and even less than one. This data shows that flocking also comes with an aggregation behavior, probably due to the cohesion rule. At the same time, we can say that all agents in this flocking simulation are safe and are unlikely to lose connection. Even if an agent temporarily flies in the totally wrong direction of heading π, it still has about seven or eight neighbors that may help it to turn around following the alignment rule.

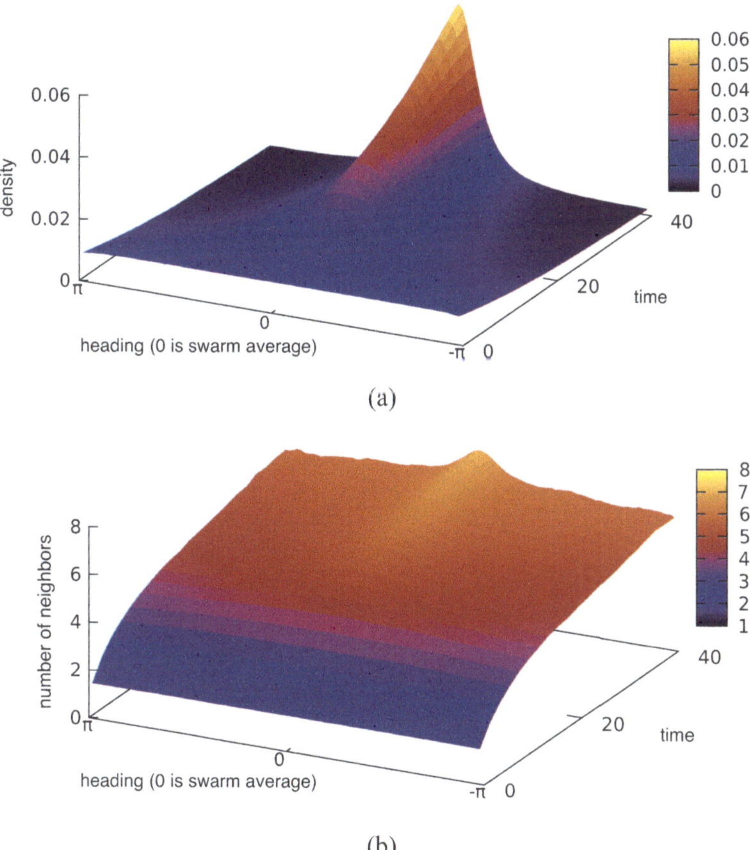

Fig. 5.4 Measurements of probability density of heading and connectivity in a simulated flocking swarm. **a** Heading. **b** Connectivity

Couzin Model

Couzin et al. [158] introduce a swarm model, meanwhile called "Couzin model," to simulate collective motion in animals, such as fish schools and bird flocks. The model reveals that small changes in local interaction rules can result in significant transitions between collective behavioral states, such as flocking, milling, and dynamic parallel group motion (see Fig. 5.5).

The Couzin model is an agent-based model focused on collective motion in animal groups. Each agent i is described by its position $\mathbf{r}_i \in \mathbb{R}^n$ in $n = 2$ or $n = 3$ dimensions, velocity $\mathbf{v}_i$, and orientation $\mathbf{u}_i = \frac{\mathbf{v}_i}{|\mathbf{v}_i|}$. Agents interact with their neighbors within three distinct interaction zones. Each agent perceives its neighbors within three circular or ring-shaped zones. (1) Zone of repulsion (r_r) to prevent collisions with agents that are too close. (2) Zone of orientation (r_o) to align an agent's direction with the average direction of its neighbors in this zone. (3) Zone of attraction (r_a) to move agents towards their neighbors for group cohesion. These zones are hierarchical: repulsion before orientation before attraction with $r_r < r_o < r_a$. In the following, we assume a notation of index sets R for each zone. R_r contains all agent indices within the repulsion zone, R_o contains all agent indices within the orientation zone, and R_a contains all agent indices within the attraction zone. If at least one neighbor is in the repulsion zone, we have

$$\mathbf{u}_i = -\sum_{j \in R_r} \frac{\mathbf{r}_j - \mathbf{r}_i}{|\mathbf{r}_j - \mathbf{r}_i|} . \tag{5.14}$$

The agent moves away from neighbors in the repulsion zone as the vector points away from the nearest neighbors. If there is no neighbor in the repulsion zone, but neighbors are in the orientation zone, we have

$$\mathbf{u}_i = \frac{1}{|R_o|} \sum_{j \in R_o} \mathbf{u}_j . \tag{5.15}$$

The agent aligns its direction with the average direction of all neighbors in the orientation zone. If no neighbors are in the repulsion or orientation zones, we have for the attraction zone

$$\mathbf{u}_i = \sum_{j \in R_a} \frac{\mathbf{r}_j - \mathbf{r}_i}{|\mathbf{r}_j - \mathbf{r}_i|} . \tag{5.16}$$

The agent moves towards neighbors in the attraction zone. The overall movement direction $\mathbf{u}_i$ is

$$\mathbf{u}_i = \mathbf{u}_i^{(r)} + \mathbf{u}_i^{(o)} + \mathbf{u}_i^{(a)} . \tag{5.17}$$

The vector is normalized to ensure unit length by $\mathbf{u}_i = \frac{\mathbf{u}_i}{|\mathbf{u}_i|}$. The agent has a maximal turning rate β for a given time step size Δt. So possibly the agent only turns towards $\mathbf{u}_i$ capped by maximal turning rate β. The agent then takes its position using the new direction

$$\mathbf{r}_i(t + \Delta t) = \mathbf{r}_i(t) + v\mathbf{u}_i(t)\Delta t , \tag{5.18}$$

for constant speed of the agent v and a discretized time step Δt.

As an extension of the model, agents may have a limited field of view, defined by an angle α. Only neighbors within this angle are considered. If the angle between an agent's heading $\mathbf{u}_i$ and the vector to a neighbor $\mathbf{r}_j - \mathbf{r}_i$ exceeds α, that neighbor is ignored:

$$\arccos\left(\frac{(\mathbf{r}_j - \mathbf{r}_i) \cdot \mathbf{u}_i}{|\mathbf{r}_j - \mathbf{r}_i||\mathbf{u}_i|}\right) \leq \alpha \tag{5.19}$$

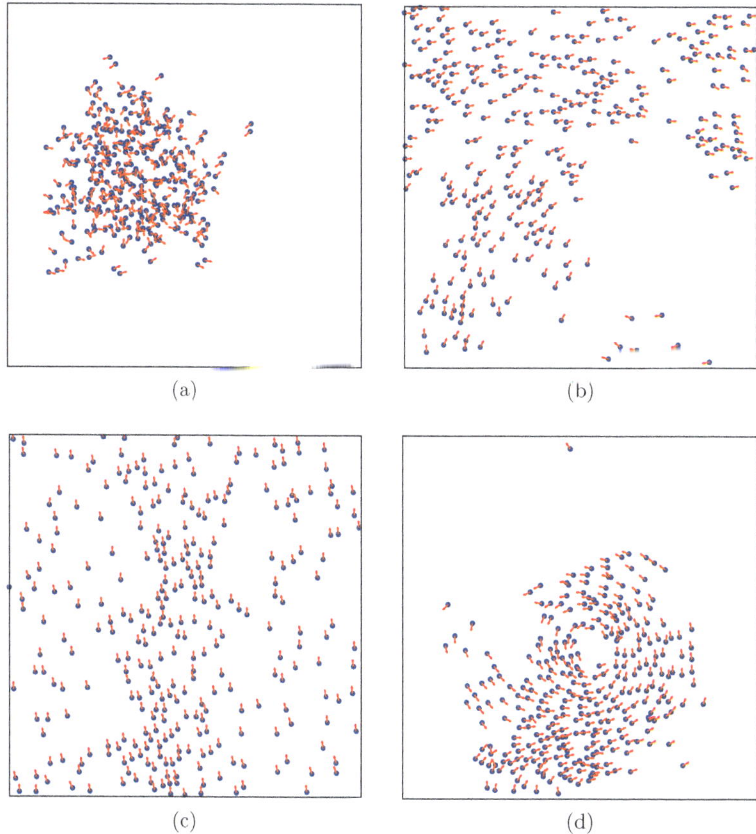

Fig. 5.5 Typical forms of collective motion as defined by the Couzin model: **a** swarm (cohesion but low level of polarization; $N = 300, r_r = 1, r_o = 1.1, r_a = 13, \beta = 5.7°, \alpha = 300°$), **b** dynamic parallel group (mobile and medium level of polarization; $N = 300, r_r = 1, r_o = 5, r_a = 10, \beta = 17.2°, \alpha = 300°$), **c** highly parallel group (mobile and high level of polarization; $N = 300, r_r = 1, r_o = 1.1, r_a = 13, \beta = 2.3°, \alpha = 250°$), **d** milling (also called torus, rotation around a common center; $N = 300, r_r = 1, r_o = 4.1, r_a = 35, \beta = 2.3°, \alpha = 250°$)

We can also add noise to the model by introducing a noise term $\boldsymbol{\eta}_i$: $\mathbf{u}_i = \mathbf{u}_i + \boldsymbol{\eta}_i$ for a random vector $\boldsymbol{\eta}_i$ that is sampled from a noise distribution. The Couzin model can describe, for example, flocking (coordinated, aligned, and directed motion), swarming (cohesive but disordered motion), and milling (circular collective motion).

Vicsek Model

Vicsek et al. [824] defined a model of what is called "active matter," which refers to self-propelled entities that exhibit collective behaviors. The model is focused on simplicity. The agents or particles are without volume. Each agent i has a position $\mathbf{r}_i$ and a heading θ_i. The update of an agent's position is defined by

$$\mathbf{r}_i(t + \Delta t) = \mathbf{r}_i(t) + v_0 \mathbf{v}_i(t) \Delta t \ , \tag{5.20}$$

for constant speed v_0 and time step size Δt. The update of the agent's heading is defined by

$$\theta_i(t + \Delta t) = \langle \theta_j(t) \rangle_{|r_i - r_j| < r} + \eta_i \ , \tag{5.21}$$

for the average heading of all particles j within radius r of agent i defined as $\langle \theta_j(t) \rangle_{|r_i - r_j| < r}$ (including agent i itself) and noise term η_i. We observe for low noise η (or high swarm density ρ) that agents move in a highly aligned, ordered state, exhibiting collective motion (see Fig. 5.6a). At high noise η (or low density ρ) agents move randomly without any system-wide alignment (see Fig. 5.6b). The main contribution of Vicsek et al. [824] is the finding of a phase transition between alignment and disordered swarm states depending on noise or density.

Criticality and Scale-free Correlations in Swarms

Cavagna et al. [130], led by Giorgio Parisi (Nobel Prize in Physics, 2021), tracked starling flocks and discovered a surprising scale-free pattern in the correlations of individual birds' motions across the entire group. Based on empirical data and

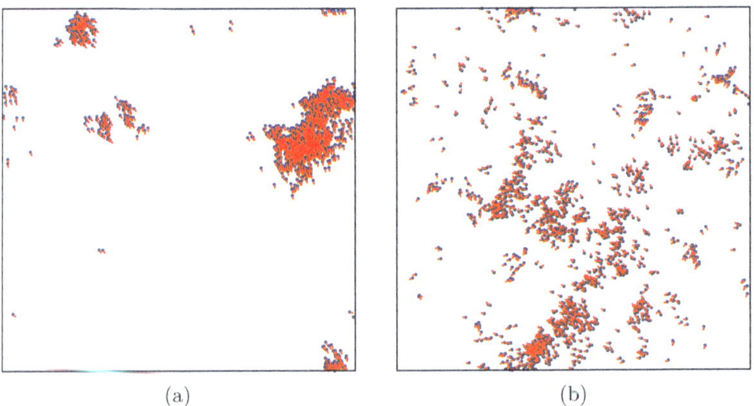

Fig. 5.6 Typical forms of collective motion as defined by the Vicsek model. **a** Ordered state ($\eta = 0.1$), **b** disordered state ($\eta = 0.5$). Noise is sampled from a uniform distribution $\eta_i \sim U(-\eta, \eta)$, $N = 1500$

supported by ideas from alignment-based spin models in statistical physics, they study correlation lengths. In this case, how far the effect of a single bird's movement can be felt across the swarm. They found that this correlation length grows in proportion to the flock's size rather than staying limited to a finite neighborhood. This signifies criticality: the system operates at a tipping point where a tiny deviation by one bird can propagate through the flock, enabling a rapid and cohesive group response. Mathematically, the correlation function does not drop off at a fixed, finite distance. Instead, correlations remain strong over the entire group. Even two birds at opposite ends of a large flock can be highly synchronized.

Lei et al. [471] experimentally investigate the criticality hypothesis. They utilize a swarm of 50 mobile robots (simple $80 \times 80\,\mathrm{mm}^2$ differential-drive robots equipped with a monocular camera) featuring Vicsek-like local alignment interactions. The authors investigate how various factors influence the system's capacity to respond to environmental stimuli. The experiments reveal that collective responsiveness is maximized when the swarm operates near the critical point, provided that criticality is induced by alignment parameters (i.e., alignment strength and interaction range). In contrast, criticality caused by noise, social attention, or action timing does not enhance functionality. The results support the criticality hypothesis.

In swarm robotics, this highlights a key principle: when robots are configured to operate near a critical threshold that balances coordination and variability, the swarm can achieve both robustness and sensitivity, enabling rapid, system-wide adaptation to changing conditions.

But we Need more Cognitive Models

Now, you have learned about the Reynolds, Couzin, and Vicsek models, which treat swarming agents as if they align according to straightforward, egocentric rules. The only problem is that some say, "All these models are wrong!" While that may be a rather extreme position, we should be ready to challenge these classical models of collective motion.[3]

The classical Reynolds, Couzin, and Vicsek models treat swarming agents as if they aligned according to straightforward, egocentric rules. Sayin et al. [702] suggest organisms might rely on cognitive architectures, such as ring attractors, using a world-centered perspective to encode neighbor positions. Order in animal groups may emerge from neural mechanisms tracking spatial cues, rather than from purely local, alignment-based interactions. Such findings encourage us to question the prevailing models that were motivated more by physics and mathematical convenience. These models also cannot explain the criticality observed in swarms, as, for example, the Couzin model is dampening any deviation. Instead, we observe in natural swarms fast, large-scale collective responses to actions of one individual, once these are

[3] One should always be motivated to question models because some are even more extreme as indicated by the quote often attributed to George E. P. Box: "All models are wrong, but some are useful."

beyond noise, by implementing selective attention. Each individual can precisely focus on specific neighbors or cues that matter most and react to those quickly. By focusing only on relevant signals, the swarm avoids being confused by random noise while responding rapidly to relevant external stimuli.

Another example is the preprint by Salahshour and Couzin [693]. Traditional models of collective motion assume that agents interact using simple rules based on an egocentric frame of reference. They align with neighbors based on relative positions and headings. However, recent neurobiological evidence suggests that animals, from insects to vertebrates, encode spatial information in an allocentric (world-centered) frame instead. Salahshour and Couzin [693] introduce a ring-attractor neural network model and demonstrate that coherent flocking can emerge naturally from neurobiological principles without requiring explicit alignment rules. This challenges the long-standing view that local alignment is necessary for collective motion. Instead, one can argue that animals achieve order by allocentric processing of spatial cues. This framework produces stable, long-range order that is largely independent of swarm density, calling for a paradigm shift in swarm robotics and collective behavior research.

5.2.6 Gossiping

In swarm robotics, we want to achieve maximal scalability. When our robots need to exchange information, they cannot do so through point-to-point communication (i.e., direct data transfer between two specific devices), which would result in $\mathcal{O}(N^2)$ scaling. Swarm robots are only allowed to communicate with their direct neighbors. If information needs to be distributed across large parts of the swarm, it must be disseminated. To spread information across the swarm, each robot passes the information locally to its neighbors, who in turn pass it to their neighbors, and so on. This is analogous to how heat spreads through a medium. The most typical way of implementing that is by gossiping.

Demers et al. [181] introduced the concept of gossiping protocols, which were initially referred to as epidemic protocols. These are simple, randomized algorithms (or sometimes even referred to as a design pattern) to spread information across a distributed system. Inspired by the spread of diseases and the rapid dissemination of rumors, they proposed an approach to distribute information with minimal assumptions about the network's reliability. Instead of coordinating systematically who tells whom, they use random contacts and probabilistic spreading. This is similar to how a virus spreads. They analyzed different strategies, comparing tradeoffs in convergence speed, network traffic, and robustness. The approach turns out to be scalable and robust.

Demers et al. [181] also explicitly mentioned the SIR model from mathematical epidemiology [420]:

$$\frac{ds}{dt} = -si \ , \tag{5.22}$$

$$\frac{di}{dt} = +si - \frac{1}{k}(1-s)i \ , \tag{5.23}$$

with susceptible s (robots who have not yet received the message), infected i (robots that actively spread the message). With a probability $1/k$, infected robots i "lose interest" and stop gossiping. This prevents unnecessary traffic once the update has been widely disseminated. The first equation models the spread: when an infected agent i encounters a susceptible one s, the message is passed, reducing s and increasing i.

Gossip protocols are known for their scalability and robustness in networks. Gossiping is resilient to robot failures as it is inherently redundant (i.e., robots are likely to receive information from several other robots). Usually, we assume dense gossiping where the majority or all robots in the swarm participate, which implies using broadcasting. Given our robots move, we speak of delay-tolerant gossiping. Network connectivity may rely on the mobility of robots, while the network may be disconnected for specific durations (due to the disconnection of several connected components). Robots opportunistically offload messages to other robots they encounter (opportunistic gossiping, cf. data ferrying). The convergence of gossiping in general is proportional to the network diameter (the longest shortest path between any two robots), that is, the time steps required to spread information across the whole network is directly proportional to the diameter. In a fully connected network, where each node can potentially communicate with any other node, the expected time to convergence is $\mathcal{O}(\log N)$ [97, 421].

Friedman et al. [260] apply gossiping to mobile ad-hoc networks (MANETs). Wireless networks have unreliable links, dynamic topologies, and constrained resources. The authors introduce three application scenarios: sparse, dense, and delay-tolerant. They also propose variants of gossiping algorithms, such as broadcast gossip and opportunistic gossip. Remaining challenges are resource management, convergence analysis, and robustness under dynamic network conditions.

5.2.7 The Anti-robot and "Less is More"

We quickly study two counterintuitive effects observed in robot swarms and related systems. Both effects have the potential to change how we engineer large-scale multi-robot systems. First, we introduce "anti-robots" that intentionally disrupt collective behavior that can paradoxically enhance system performance. For example, stubborn agents in collective decision-making can maintain diversity, improve adaptability, and prevent suboptimal outcomes by resisting consensus and introducing variability.

Second, the "less-is-more" effect highlights how robot systems can paradoxically perform better when individual robots operate less intensively. For example, reducing communication, sensing frequency, or driving speed can prevent instabilities, improve adaptability, and enhance overall system performance.

The Anti-robot

Is it a skill to be against virtually everything? Modern times seem to suggest that. In swarm robotics, the concept of the anti-robot or anti-agent exists. Most prominently in collective decision-making (probably because opinion dynamics inspire it) in the form of contrarians, but also zealots.

There are concepts of intentionally introducing anti-collaboration as conflicts into swarms in the form of anti-agents or anti-components [548, 549]. For example, in an aggregation task, one could introduce robots that do not aggregate but rather command clusters of aggregated robots to dissolve the cluster. Such an approach may be somewhat counterintuitive, but it can increase an algorithm's performance.

Merkle et al. [548] study clustering of objects by swarms. A regular agent picks up an item with probability $p_p = \frac{k_1}{k_1+f}$, for a density f of items in the surrounding and a threshold k_1. A higher local density f reduces the probability p_p. The probability that a laden agent drops an item is $p_d = \frac{f}{k_2+f}$ for a threshold k_2. A higher local density f increases the probability p_d. They test three kinds of anti-agents: (1) Reverse anti-agents: These agents invert the standard behavior by swapping the probabilities of picking up p_p and dropping items p_d, but they are largely ineffective at preventing clustering. (2) Random anti-agents: These agents pick up items deterministically but drop them randomly. They are disrupting clustering when present in sufficient numbers. (3) Deterministic anti-agents: These agents always pick up items in crowded areas and drop them in empty areas. They prevent clustering with a moderate presence while in some cases even enhancing clustering performance. Merkle et al. [548] report an experiment with $N = 80$ agents of which 30 are deterministic anti-agents. The anti-agents improve the clustering performance.

Prasetyo et al. [648] report a positive effect of including "stubborn" individuals in their approach to collective decision-making. A stubborn robot (also called zealot) has a fixed opinion as they remain committed to their initial choice throughout the process. The inclusion of a few stubborn agents ($0.2N$ and less) significantly enhances the swarm's adaptability. In a dynamic environment, stubborn robots help to keep exploring alternative options.

As discussed in Sect. 8.5.14, there is a related concept of contrarians in collective decision-making. According to Galam [264] "a contrarian is someone who deliberately decides to oppose the prevailing choice of others, whatever this choice is." Additionally, contrarians can have positive effects in collective decision-making, such as increasing adaptability, similar to stubborn agents, and preventing premature convergence on a suboptimal decision by maintaining diversity. A similar effect can be achieved through spontaneous switching or noise [321], and noise in communication is also known sometimes to contribute positively [893].

Related to the idea of anti-agents are also scenarios where two swarms operate independently but within a shared area [85].

"Less Is More" and "Slower Is Faster"

Gershenson and Helbing [279] discuss the "slower-is-faster" effect, where systems paradoxically perform better when individual components operate less efficiently. They illustrate this phenomenon across various domains, including pedestrian evacuation, vehicle traffic, social dynamics, and ecological systems. They argue that such systems often face instability or cascading failures when pushed beyond a critical threshold of efficiency. The study emphasizes the importance of understanding and managing nonlinear dynamics to optimize systemic performance across various contexts. They emphasize the need for future research to develop a unified mathematical framework that could comprehensively describe and predict the slower-is-faster effect across different systems.

Talamali et al. [770] follow the idea of "slower is faster" to develop and study robot swarms. They focus on how robot swarms can effectively adapt to dynamic environments in the best-of-n decision problem (see Chap. 8). They find that reducing communication range and limiting the number of connections between robots (i.e., node degree) improves the swarm's adaptability to environmental changes; hence, they reinterpret the motto as "less is more." This is in contrast to the standard assumption that more communication usually enhances performance. The robots integrate locally gathered environmental data with information exchanged through interactions with neighboring swarm members. Their behavior follows the voter model,, where each robot updates its opinion based on input from a randomly selected neighbor. To resolve conflicting information, they use a cross-inhibition mechanism: when two robots with differing opinions interact, they reset their opinions and temporarily poll others. Talamali et al. [770] support their claims by mathematical models, simulations, and experiments with Kilobot robots. They show that reducing communication links prevents a large uninformed majority from overpowering a well-informed minority.

In a similar paper, Aust et al. [40] investigate how slower sensing and opinion dissemination enhance the swarm's ability to reach consensus in collective perception. Using simulations with Kilobot robots in a dynamic collective perception scenario, they study how these effects depend on an interplay between communication range, sensing frequency, and environmental noise.

The "slower-is-faster" effect was also shown for other domains. For example, Pastor et al. [620] validate it in experiments for three systems: humans evacuating through a narrow door, sheep passing through a gate, and granular particles flowing through a vibrated hopper (funnel-shaped structure). The results show that excessive driving force leads to clogging and intermittent flow.

Another related paper follows the idea of improving system performance by "being less capable." Manrique et al. [512] investigate an abstract model inspired by the behavior of *Drosophila* larvae. The scenario is a navigation task that is solved by a trial-and-error behavior (klinotaxis: step-by-step navigation based on local sensory information and iterative course corrections). They find that intermediate agent capabilities optimize system performance. When agents are too capable, their actions become uncoordinated, resembling random noise, while too limited agents cause excessive synchronization, leading to overcorrections. Hence, the key lies in striking a balance between independence and coordination. What they interpret as an agent's capability is its memory of past decisions (a binary string representing past decisions on clockwise or counterclockwise movement). Having too large a memory increases combinatorial options and complicates agent-agent coordination. With insufficient memory, agents often repeat the same actions, resulting in over-rotation and a suboptimal, zigzag-like trajectory.

Phenomena like the "less-is-more" effect and the anti-robot concept indicate that engineering robot swarms can be counterintuitive at times. However, they also teach us that we can build multi-robot systems that perform high-level tasks collectively despite individual robots being limited.

5.3 Scenarios

We study a set of scenarios for swarm robotics that is roughly sorted by complexity. Note that the earliest papers on swarm robotics already mentioned a set of "basic interaction primitives" including aggregation, dispersion, and flocking [521].

5.3.1 Aggregation and Clustering

Aggregation is one of the most fundamental swarm behaviors frequently observed also in natural swarms (see Fig. 5.7). Not only that, but especially for natural swarms, it is crucial to stay together. Otherwise, the swarm would separate into several parts, decrease in size, and eventually its survival may be endangered. Aggregation behaviors are observed in many natural swarms, such as young honeybees [766], ladybugs [468] (see Fig. 5.7, top left), and monarch butterflies [856, see Fig. 5.7, lower left]. In aggregation, the task for the robots is to position themselves close to each other in one spot. This is achieved by minimizing the distances between robots. The actual position of the aggregation spot may be specified or unspecified. If it is unspecified, then the robot swarm self-organizes to find a consensus on its aggregation position. Hence, aggregation at an unspecified spot has an inherent collective decision-making component. The aggregation spot can be specified, for example, by saying the swarm should gather at the brightest or warmest spot. Then each robot, possibly in collaboration with others, has to find that position and stop there.

The naive control approach to aggregation in swarm robotics is to move in a random direction. Once you see another robot, you stop. Although initially robots may reduce the average distance between robots, the swarm system gets caught in a

Fig. 5.7 Aggregations of insects (top left photo of ladybugs by Katie Swanson (Nature ID blog) with permission, other photos are CC0)

local optimum immediately. Most clusters of robots are small, maybe two or three robots meet at most. Even in this simple problem of aggregation, we already face a challenging problem of the exploitation of local knowledge and the exploration of the environment. Robots should not greedily exploit their environment by staying stopped close to the first robot they meet. Instead, they should also explore their surroundings sometimes and check whether bigger clusters exist that they should join.

Also note that a generic optimal solution to this exploration–exploitation tradeoff does not exist because it depends on the respective scenario. How often an explorative behavior is proper depends on the swarm density (robots per area), the swarm size, and the robot's speed. In addition, the robots are usually not synchronized, certain behaviors may create deadlocks, and the robots can hardly know when the exploration phase in the swarm is over. Finally, in the case of aggregation at a specified spot, the environment may be dynamic. Hence, the robots should always stay explorative and send out scout robots that check whether a better aggregation spot has emerged at a different location.

The literature about aggregation in swarms is rich. The most popular algorithm is, as mentioned before, BEECLUST [710] originally published by Schmickl et al. [712] and Kernbach et al. [422]. For details about BEECLUST, see Chap. 9. BEECLUST was analyzed by Hereford [356] and extended with a fuzzy-based approach by Arvin et al. [32, 34]. The phenomenon of symmetry breaking in BEECLUST, when the swarm is presented two options (locations) of the same utility, was investigated by Hamann et al. [326, 330]. Symmetry breaking may seem like a simple problem. The swarm has to find consensus about options A and B. Both options have identical utility. A human being faced with such a problem randomly picks one. The swarm, however, needs to avoid splitting and has to collectively decide about which one of the two to choose based on local information only. This problem corresponds directly to the problem of having two clusters of robots to merge them. Who stays and who moves to the other cluster? This is an even more difficult question if the other cluster is out of the robots' perception range. This is again an exploitation–exploration tradeoff that cannot easily be solved. Depending on the environmental conditions and the robot's capabilities (e.g., speed and sensor range) in the given scenario, appropriate probabilities of leaving and joining a cluster need to be defined. Joining and staying at bigger clusters should be more desirable, hence, creating a positive feedback effect (i.e., big clusters tend to get bigger).

A multi-robot aggregation behavior is simple enough to allow for a mathematical formalism by control theory. We follow the presentation and notation style of Egerstedt [221]. Robot i has position x_i and the swarm's joint state is $x = [x_1^T, \dots, x_N^T]^T$. We define a cost via a measure of robot-robot distances by

$$\varepsilon(x) = \frac{1}{2} \sum_{i=1}^{N} \sum_{j \in \mathcal{N}_i} \|x_i - x_j\|^2 \ , \tag{5.24}$$

with the neighbors of robot i are $\mathcal{N}_i$ (depending on some sensor range) and the norm $\|\cdot\|$ can be Euclidean. Measuring the distance to all neighboring robots is not trivial, but we assume our robots can accomplish this task. We want to minimize $\varepsilon(x)$ and define a gradient descent flow by

$$\frac{\partial \varepsilon(x)}{\partial x_i} = \sum_{j \in \mathcal{N}_i} (x_i - x_j) \, , i \in \{1, \dots, N\} \ . \tag{5.25}$$

By taking the negative gradient flow, we get an equation of motion for each robot

$$\dot{x}_i = - \sum_{j \in \mathcal{N}_i} (x_i - x_j) \ . \tag{5.26}$$

This equation is also known as the consensus equation because, by moving around, we can say that the robots agree on where they want to meet today. Will this aggregate all robots in one cluster? The swarm dynamics crucially depend on the graph implied by the robots' neighborhoods $\mathcal{N}_i$. For the initial swarm state, is there a path from any robot i to any robot j (i.e., all robots are connected via a neighbor's neighbor...)? If not all robots are members of the same initial connected component, then they will not form only one cluster. Obviously, this approach is too greedy and lacks an exploration strategy.

For a theoretical computational approach to collective behaviors and aggregation, specifically, you can read the paper of Chazelle [137] and his other works. Popkin [645] starts from flocking and then goes into a rather philosophical but inspiring discussion of whether there is "the physics of life."

Na et al. [586] introduce a novel light-based artificial pheromone system that models realistic spatio-temporal pheromone dynamics, such as diffusion, evaporation, and advection, derived from fluid-flow equations. The system utilizes an LCD screen to project pheromone fields, while a camera-based tracking system enables robots to inject pheromones back into the environment, allowing for bidirectional communication. Experiments with Colias micro-robots (see Sect. 3.6.1) demonstrated how environmental factors and pheromone injection affect aggregation behaviors, showing that diffusion increased swarm cohesiveness. In contrast, pheromone injection countered the dispersive effects of diffusion and moving cues.

A pretty different work is that of Correll et al. [154]. They report a technological system to aggregate cattle via social control. The idea is to exploit the adaptive behavior of cows and control only a fraction of the herd. Some cows follow other cows; hence, not all need to be controlled directly. The cows carry the device and trigger an undirected stimulus when they enter undesired areas, hence, implementing a "virtual fence."

The aggregation problem of robots choosing in a greedy way with whom they want to join and form a cluster can also be understood in a more fundamental and theoretical way. Think of the robot swarm as a directed graph. Each node represents a robot, and a directed edge indicates which robot wants to join which other robot. Say, each robot chooses exactly one other robot to join with. That corresponds to one outgoing edge for each node. For simplicity, we ignore that robots can only perceive a sub-population of the swarm due to local perception. If we randomly choose one target node for each node, what is the expected size of the clusters formed? We can use methods from graph theory to answer that. Our problem is known as directed 1-forests, also referred to as functional graphs. They are directed graphs in which each node has outdegree exactly one. What we consider a cluster is known as a 1-tree (or unicyclic graph), which is a connected (sub-)graph containing exactly one cycle. Early methods were published in the context of directed pseudoforests [446]. We are more interested in the combinatorial aspects [244]. It is possible to estimate the expected number of emerging 1-trees and their size distribution. Once we relax the assumption that robots do not see all other robots and can only choose to join with neighbors, the methods from directed 1-forests tend to break.

5.3.2 Dispersion

Dispersion is the behavior directly opposite to aggregation. A robot is supposed to position itself as far as possible from every other robot, but while staying in contact. The distances between robots should be maximized, but the robots should stay within communication range. This is a proper behavior if large areas need

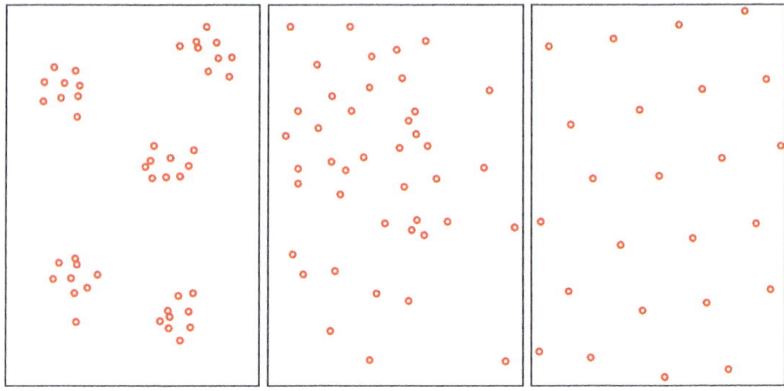

Fig. 5.8 In ecology, there are three types of population distributions: clumped, random, and regular. (GNU Free Documentation License)

to be monitored or if the swarm should minimize its robot density. The expected spatial distribution of robots is primarily uniform, meaning that the robots position themselves in a regular pattern. The task can, however, also be approximated by a random uniform distribution, or even clusters may be acceptable depending on the underlying task. This corresponds well to the three types of population distributions in ecology: clumped, random, and regular (see Fig. 5.8).

An implementation is reported by Payton et al. [623] in a popular paper on so-called pheromone robotics. The term "pheromone" only serves as a buzzword here and should not be interpreted as indicating an appropriately bio-inspired pheromone approach. Instead, the robots communicate by infrared messages and measure distances between robots by measurements of signal strength only. A similar approach is reported by Ludwig and Gini [502] but focuses on radio communication. The study is, however, only done in simulation.

Also McLurkin and Smith [535] study in robot experiments with 56 robots how a swarm of robots can disperse efficiently in terms of time and form a uniform distribution. They use a gradient-based multi-hop messaging protocol. Their algorithm alternates between two sub-routines, with one spreading the robots evenly and the other one directing robots towards unexplored areas.

Similar to the above example for aggregation, we can also follow Egerstedt [221] here. Dispersion can be treated using control theory. We do not consider any limitations due to swarm connectivity that could potentially compromise the effectiveness of this formalism. We require a prescribed inter-robot distance δ. As cost for all relevant robot pairs (i, j) we define

$$\varepsilon_{ij}(\|x_i - x_j\|) = \frac{1}{2}(\|x_i - x_j\| - \delta)^2 \quad . \tag{5.27}$$

We define again a gradient descent flow

$$\frac{\partial \varepsilon_{ij}(x)}{\partial x_i} = w_{ij}(\|x_i - x_j\|)(x_i - x_j) \ , \tag{5.28}$$

this time defined generically for any task-specific weight w_{ij}. Based on Eq. 5.27 we get as weight

$$w_{ij} = \frac{\|x_i - x_j\| - \delta}{\|x_i - x_j\|} \ , \tag{5.29}$$

and as the motion equation

$$\dot{x}_i = -\frac{\partial \varepsilon}{\partial x_i} = -\sum_{j \in \mathcal{N}_i} w_{ij}(\|x_i - x_j\|)(x_i - x_j) \ . \tag{5.30}$$

5.3.3 Pattern Formation, Object Clustering, Sorting and Self-assembly

Once we go from aggregation behaviors one step further in complexity, we talk about pattern formation, clustering and sorting of objects, and self-assembly of robots. Pattern formation means robots aggregate in defined shapes or shape their environment. In clustering, robots move objects around to aggregate the objects or to sort them according to features, such as color, size, or age. Self-assembly means robots physically connect to form more powerful super-robots.

Pattern Formation

Once we go beyond mere aggregation, robots may be asked to form certain patterns, to sort themselves, to self-assemble into a super-robot, or to sort objects. Self-organized pattern formation is also found in many natural systems [784], predominantly in biology, but even in self-organizing stone and salt patterns [424]. Many of these patterns can be explained by reaction–diffusion mechanisms and activator–inhibitor schemes [51]. Although Turing [795] published his fundamental paper about pattern formation already in 1952, as late as in the 1970s, biologists became aware of how biological pattern formation may be explained by self-organization [283]. Well-known phenomena such as animal coat markings [583, 584] and patterns in insect embryogenesis [402] were addressed. Another popular example are the pigmentation patterns of shells [539, 542]. More relevant for swarm robotics are, for example, the *Physarum* transport networks [391, 392] and pattern formation in simple, simulated agent swarms [329].

Besides the obvious colorful patterns, such as animal coat markings, there are also spatio-temporal patterns. The formation of a particular traffic flow in pedestrians, where people walk in one direction on the right half and in the opposite direction on the left half of the path, is also an example of self-organized pattern formation. So there is collective behavior in vertebrates and especially also in human beings [158]. Often, people ignore the paved system of walkways, take shortcuts, and create new trail systems [351, 352]. A special example of pattern formation is the trail networks formed by ants. It is well-known that ants use stigmergy to organize their foraging behavior, that is, they mark trails with pheromones. The navigation within these trail networks seems to be based on the specific geometry of the trail system. In particular, defined angles of trail bifurcations can be leveraged by the ants to know which direction leads to food and which to the nest [376]. This approach can also be used in swarm robotic systems for orientation in a trail network [325]. However, one requires a feasible emulation of pheromones, such as proposed by Mayet et al. [530]. Alternatively, one can use heat trails [689].

Roughly following the idea of trail networks, Nouyan et al. [599] implemented a robot swarm system that positions robots in a chain formations. These robot chains can, for example, be used in a foraging scenario to guide another robot along the chain to the food.

Sayama [701] presents a framework for what he refers to as morphogenetic robotic swarms. These are large-scale, self-organizing systems composed of simple, mobile robots with minimal sensing, computation, and communication capabilities. Building on the Swarm Chemistry model [700], the approach uses local kinetic interactions and stochastic differentiation to generate complex structures and dynamic behaviors. Robots can self-assemble and self-repair without the need for a central controller. The system demonstrates how sophisticated patterns can emerge from decentralized local rules.

Slavkov et al. [736] introduce embodied capabilities in mechanical systems that enable functional, adaptive behavior in the form of morphogenesis. Drawing on examples from soft robotics and biological systems, the authors argue that physical intelligence emerges from the interplay of material properties, structure, and environmental interactions, enabling systems to sense, respond, and adapt autonomously. They report experiments with a swarm of up to 300 Kilobots to demonstrate emergent shape formation. Each robot runs a gene regulatory network that produces Turing-like patterns (simulating biological reaction-diffusion dynamics) and uses these patterns to guide migration toward high activator concentration areas, forming protrusions (limb-like or finger-like extensions) in the swarm shape. The system required no global positioning or centralized control and demonstrated robustness to damage, as well as adaptability, including the ability to self-repair after cutting parts of the swarm or altering its initial shapes.

Divband Soorati et al. [193] explores a bio-inspired approach to robot self-assembly and pattern formation, utilizing principles from what they term photomorphogenesis. They combine the pattern formation scenario with collective decision-making, self-repair, and dynamic environments. The work is inspired by light-directed growth in plants. The robot swarm forms structures by aggregating in response to changes in light intensity, which serve as an environmental stimulus. The robots need to localize the arena half in which they are, and to detect dynamic changes in the environment. The robots collectively agree on a leader, a growth location, and indirectly their self-repair actions. This method enables a robot swarm of up to $N = 70$ robots to dynamically reconfigure, repair damaged components, and autonomously optimize resource allocation.

Centroidal Voronoi tessellations

Our swarm robots often need to spread out efficiently to disperse, monitor, or cover an area. Reactive behaviors, such as mere repulsion from neighbors, can achieve rough coverage (see Sect. 5.3.2), but more structured coordination is sometimes needed to ensure efficiency. The field of computational geometry offers methods that can be implemented in a distributed manner, with potential applications in swarm robotics. Two key methods are Lloyd's algorithm by Lloyd [497] and centroidal Voronoi tessellation [209]. Lloyd's algorithm (similar to the k-means clustering algorithm) iteratively assigns points to their nearest center. Then it updates each center to the centroid of its designated region, gradually converging to a close approximation of the centroidal Voronoi tessellation.

A Voronoi tessellation divides space so that each robot is responsible for the area closest to it. Such an approach creates a natural partition of the environment. A centroidal Voronoi tessellation (CVT) moves each robot toward the center of mass (called the centroid) of its assigned region. When all robots reach the centroids of their respective areas, the system forms a CVT. This configuration distributes the robots evenly across the space, minimizing overlap and gaps. CVTs can emerge from local interactions. Each robot needs to estimate its own region and adjust its position to align with its centroid. The behavior triggered by CVT can be seen as an extension of a basic collision avoidance approach.

Du et al. [209] present centroidal Voronoi tessellations: Voronoi diagrams where each generator coincides with the centroid of its region. They survey a broad set of applications, mathematical properties, and computational algorithms.

Durham et al. [214] introduce a decentralized algorithm for coverage control over discrete environments. Unlike Lloyd-type methods, which separately compute centroids and partitions and often get trapped in local optima, the proposed pairwise-optimal gossip algorithm allows asynchronous, pairwise communication between agents to optimize their joint coverage directly. The method provably converges to a pairwise-optimal partition, a strict subset of centroidal Voronoi partitions, leading to significantly improved coverage performance. Simulations confirm that the algorithm consistently outperforms Lloyd-type approaches.

Clustering

Clustering of objects was often investigated in simulation, but there are also many hardware implementations [91]. The classical termite simulation model of Resnick [672] resembles a simple behavior of termites that collect wood chips and cluster them. The two only rules are: if you do not yet have a wood chip, then pick one up randomly; if you have a wood chip already, place it next to a randomly chosen wood chip. As a refinement, one can introduce probabilities, such that it is less likely to pick up a wood chip positioned close to other wood chips and more likely to place a wood chip close to many other wood chips.

In object clustering, high robot density often leads to spatial interference, resulting in frequent collisions that reduce performance. Vardy [816] addresses the problem in planar construction, where robots gather and push objects to form a two-dimensional structure. The proposed solution subdivides the workspace into lanes around the goal region. Each robot determines its lane and follows a control law that keeps it within its lane while nudging objects inward. Simulation results show that while performance gains in construction are modest, lanes significantly increase the distance maintained between robots. Simple spatial partitioning can mitigate interference in swarm construction tasks.

Vardy [818] introduced another method to reduce spatial interference in collective construction tasks. The proposed approach constrains robot motion to a labyrinth-like guiding path [624] projected onto the environment. Robots follow this curve, deviating only to push objects or avoid collisions. Simulations show that the method scales well with increasing swarm size up to a point, distributing robots uniformly along the labyrinth. A validation study with physical robots demonstrates that such guiding structures can effectively reduce interference and support robust planar construction.

An interesting alternative work on object clustering, as mentioned in Sect. 5.2.7, is the approach of "anti-agents" by Scheidler et al. [704] and Merkle et al. [548]. The standard approach to object clustering is probabilistic with a probability P_{pick} to pick up an object and P_{drop} to drop it. An anti-agent is a robot that behaves differently from the majority of robots. Scheidler et al. [704] investigate different choices of anti-agents, such as random behavior, inverted neighborhood (i.e., inverted perception of the neighborhood), and reverse anti-agents. Reverse anti-agents invert the above probabilities to $1 - P_{\text{pick}}$ and $1 - P_{\text{drop}}$. Scheidler et al. [704] found a nonintuitive effect, namely that certain numbers of reverse anti-agents increase the observed clustering effect. This could lead to a concept of increasing the efficiency of swarm systems with an appropriate number of anti-agents.

Sorting

Once there are multiple types of objects that need to be clustered, we are entering the domain of sorting. For example, there is a multi-object sorting in "annular structures" by Wilson et al. [862]. Annular structures are defined as forming a central cluster of one class of objects, and surrounding it with annular bands of the other classes, each band containing objects of only one type. This research is inspired by ants (*Leptothorax albipennis*) that build a perimeter wall around their nest [254]. A related behavior is brood sorting in ants [253]: "Different brood stages are arranged in concentric rings in a single cluster centred around the eggs and micro-larvae."

A similar task is patch sorting [545]. Patch sorting is defined as "grouping two or more classes of objects so that each is both clustered and segregated, and ... each lies outside the boundary of the other" [546]. Melhuish et al. [545] implemented an appropriate behavior on six robots that successfully sorted 30 frisbees (three types) into three patches.

Alternatively to sorting objects, one can also sort robots directly. Say there are two types of robots in the swarm, red and blue robots, and there are two aggregation areas. The swarm is supposed to collectively decide which area belongs to the red and which area to the blue robots. Then they should aggregate there and hence sort themselves [190].

A particularly remarkable study of swarm robot sorting was done by Groß et al. [301] and Chen et al. [138]. They studied a segregation behavior in a swarm of robots based on the so-called Brazil nut effect [567]. The Brazil nut effect is the unfortunate phenomenon of self-organized sorting in breakfast cereals. The big Brazil nuts gather at the top, and all the small nuts end up at the bottom of the bucket. Obviously, a dedicated cereal eater does not want to eat big nuts only on Monday and exclusively tiny nuts on Friday. Self-organized sorting in nuts works only if energy is added to the system, for example, by vibrations during the transport of the cereals in a truck. All nuts jump up and down, but small nuts gather in the cavities below big nuts before they can fall again (see Fig. 5.9, bottom). Hence, the big nuts move up and the small nuts move down. Groß et al. [301] and Chen et al. [138] cleverly transferred this self-organizing, physical effect to the domain of swarm robotics. Although the robots are physically homogeneous, unlike the nuts, they are programmed to avoid collisions with different minimal distances. The robots have different collision avoidance radii, which correspond directly to nuts of different sizes. As an effect, once the robots are asked to aggregate at a specific spot (e.g., approaching a light, Chen et al. [138]), robots with small collision avoidance radius gather in the middle and robots with big collision avoidance radius gather at the boundaries (see top right photo in Fig. 5.9, green robots have a smaller collision avoidance radius than red robots).

Self-assembly

"Self-assembly can be defined as a process by which pre-existing discrete components organize into patterns or structures without human intervention" [299]. In contrast to the above pattern formation, the idea is to connect smaller autonomous parts physically to construct a bigger entity that has relevant physical or architectural function, such as "bivouacs, bridges, curtains, droplets, escape droplets, festoons, fills, flanges, ladders, ovens, plugs, pulling chains, queen clusters, rafts, swarms, thermoregulatory clusters, tunnels, and walls" [24]. Self-assembly is relatively common in insects [24].

Fig. 5.9 The Brazil nut effect as inspiration for robot segregation (photo of nuts by Melchoir, CC BY-SA 3.0; photo of robots by Roderich Groß with permission [138]; schema by Gsrdzl, CC BY-SA 3.0)

The s-bot of the swarm-bots project (see Sect. 3.6.1) was developed to allow for self-assembly [202]. Groß and Dorigo [299] use self-assembly with s-bots to form chains of robots that can pull a heavy object collectively (i.e., collective transport, see Sect. 5.3.5). The robots can physically connect with a gripper, coordinate their motion, and pull collectively. In similar studies, the authors show how groups of self-assembled robots can cross gaps and steep hills that they could not have crossed as single robots [601, 792].

The concept of self-assembly in robotics is taken to the next level in reconfigurable modular robotics. The idea is to design autonomous robot modules that can physically connect and self-assemble into super-robots (cf. the quote of Stanislaw Lem at the beginning of this chapter). Rubenstein and Shen [684] give an overview based on simulation of what is theoretically possible following this concept. The full power of this approach is available once the robot modules can autonomously reconfigure the shape online at runtime. Then the robot system has levels of robustness similar to swarm robots because it can self-repair [26, 27, 751]. Self-reconfigurable modular robotics with aspects of swarm robotics was investigated in the projects SYMBRION and REPLICATOR [327, 484, 714, 715]. The robots can dock to each other, and the docking allows them to connect physically, to share energy, and to establish a communication bus. The control of reconfigurable modular robots is considered to be challenging, especially if the self-assembled shapes and topologies of the super-bot are allowed to be adaptive. Not all shapes can be anticipated before, and the control needs to deal with them at runtime [328, 889].

The Kilobot (see Sect. 3.6.1) was used by Rubenstein et al. [686] to emulate self-assembly by merely positioning the robots next to each other (see Fig. 5.10). The robot swarms in these experiments are the biggest reported until now, with up to 1024 Kilobots. The task for the robot swarm is to position all robots properly to form predefined shapes. A challenge is to locate robots based on local information only. Rubenstein et al. [686] use four seed robots as anchor (see Fig. 5.10h) to form a gradient as coordinate system. The robots then move along the edge of the already formed aggregation in a line, which has a strong impact on the time overhead as waiting times scale linearly with the swarm size (see Fig. 5.10i). A problem that

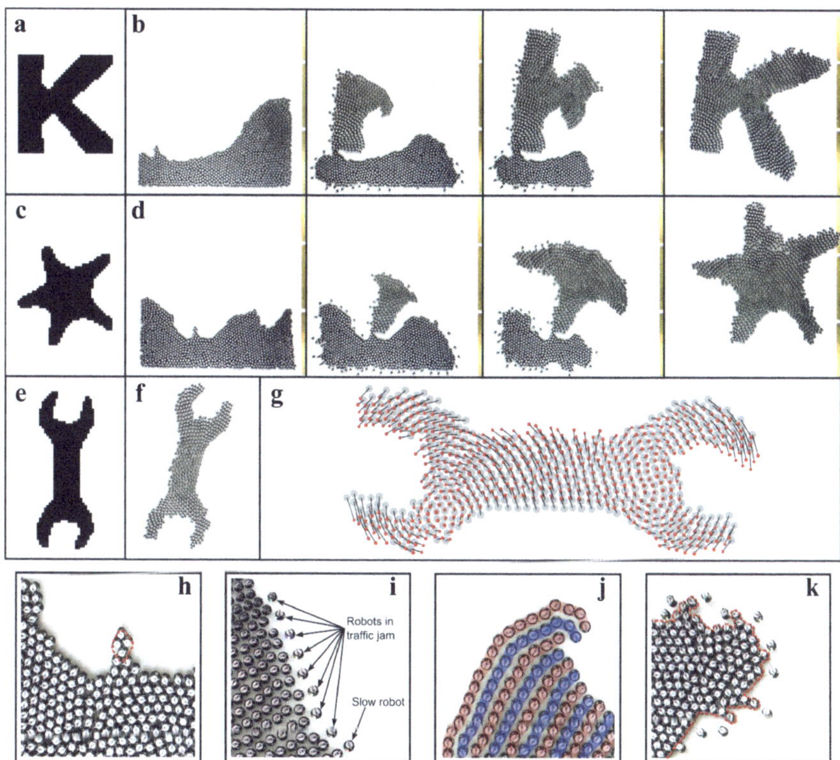

Fig. 5.10 Emulation of self-assembly with Kilobots [686] (printed with permission) **a**, **c**, **e** desired shape, known by each robot **b**, **d** self-assembly process from initial configuration to final configuration **f** final shape slightly bent due to errors in the process **g** accuracy of the final shape is measured by comparing actual to desired robot positions **h** four seed robots form an anchor **i** robots move along the edge in a line **j** robots in each of these rows share the same gradient value (**k**), erratic boundary due to moving robots pushing stopped robots

was addressed later by inverting the whole approach, that is, doing self-organized disassembly [275]. Similarly, the problem of error propagation using the gradient approach was investigated [276].

Another example of self-assembly using Kilobots is presented by Divband Soorati and Hamann [192]. They present an adaptive strategy to form only qualitatively predefined tree-structures. The trees are specified, for example, by their branching factor (i.e., bushiness of the tree). A related model that can be used in swarm robotics to control the self-assembly of growing tree structures is the so-called Vascular Morphogenesis Controller [891].

Sun et al. [759] propose a novel, fully distributed strategy for pattern formation. Instead of assigning goal positions, robots use a modified mean-shift algorithm to iteratively explore and occupy unoccupied high-density regions of a desired shape. A key feature of this strategy is its adaptability: robots actively vacate crowded areas to allow others to enter, which supports the formation of complex, nonconvex shapes and introduces much more robust behaviors compared to previous approaches as shapes can regenerate. The method shows good scalability and efficiency compared to state-of-the-art techniques, and remains robust under varying swarm sizes and robot failures.

5.3.4 *Collective Construction*

Using a robot swarm to construct buildings or any other collectively The concept of an architectural artifact is probably an intuitive idea. Construction is a task that seems easily paralellizable. In addition, collective construction may incorporate several other basic behaviors of the swarm robotics, such as collective decision-making (e.g., where to start construction) and task partitioning (e.g., "I place my block here and you there").

Collective construction is even more fascinating once we investigate how animals build their homes [336, 833]. In particular, we are interested in building behaviors of social insects, such as honeybees, wasps, ants, and termites. They operate based on local information and in a self-organized way. An essential work in this context is the papers by Theraulaz and Bonabeau [781, 782]. They model the building behavior as cellular automata, that is, the world is discretized as a lattice with cells (see Fig. 5.11). The animal looks only at its directly surrounding environment (e.g., where is building

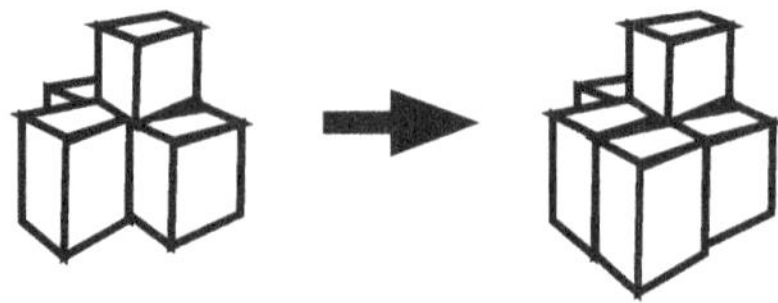

Fig. 5.11 An example rule of a cellular automaton in 3-d with a neighborhood of $3 \times 3 \times 2$ blocks

material and where are holes) and follows simple rules that map the state of the local environment to actions, such as "move forward" or "place material here." The major challenge for this cellular-automaton approach, but also for any kind of collective construction is to break down the global blueprint (i.e., how the "house" looks like eventually) to simple local rules (i.e., if your neighborhood looks like this, then place material here). This approach is inherently based on stigmergy, and it is unclear whether it is always possible to derive local rules for any given global blueprint. Even more so, it is unclear how that can be done efficiently.

The most popular paper on swarm construction was published by Werfel et al. [857]. They designed a dedicated robot platform that picks up flat blocks, puts them on its back, climbs stairs built of these blocks, and places the block at an appropriate position (see Fig. 5.12). Even for a single robot, the challenge is to ensure that it does not place building blocks at positions that may result in deadlocks. For example, a robot may not be able to climb down the stairs anymore to pick up the next block. Once we allow multiple robots to construct in parallel, additional deadlock situations may arise because robots block each other's way. Using building blocks also requires relatively high precision in all actions of the robot, ranging from picking up blocks to placing them, and not falling off the construction.

Another example of an approach based on building blocks is by Allwright et al. [14–16]. This is a modular, fully open-source platform designed for research in autonomous multi-robot construction. The robots are equipped with a forklift and move "smart blocks" around. These blocks are equipped with microcontrollers, LEDs, and near-field communication, enabling robots to interact with and program the environment. The robots can change the RGB LEDs of these smart blocks via radio to implement even more possibilities for stigmergy. For example, the colors can encode different levels of the construction. presents SRoCS (Swarm Robotics Construction System). It comprises mobile robots called BuilderBots and programmable

Fig. 5.12 Example of a robot swarm collectively constructing an architectural artifact [857] (printed with permission)

building elements known as Stigmergic Blocks. The robots autonomously build structures up to three blocks high using reactive control. They use onboard computer vision and sensors for navigation and manipulation. All design files, software, and assembly instructions are publicly available, encouraging replication and further development in research and education.

Augugliaro et al. [39] report the art installation "Flight Assembled Architectures," a swarm of quadrotors to construct architectural artifacts. As an example, they built a 6-m-tall tower out of foam modules. The system makes use of global information, a central element, and robots are controlled offboard. There is an overhead motion capture system, a 19-camera Vicon T-40 system, to provide high-accuracy position information. The information and the robots' destinations are processed with a complex trajectory planning and trajectory generation algorithm.

A quite different approach was proposed by Heinrich et al. [348] and Eschke et al. [235]. This concept of collective construction operates with continuous building material in the form of strips or strings that are braided together (see Fig. 5.13). Initially, it may seem an unnecessary limitation to working on braids, but the possibilities are actually almost infinite [822].

In Fig. 5.13a the braiding approach with mobile robots is shown. The applied robot is the Thymio II robot. Robots collectively braid, one strip each, by a simple line-following behavior combined with a collision avoidance behavior. The resulting braid is shown on the right-hand side in Fig. 5.13a. In Fig. 5.13b the inspiration of this work is shown, a traditional May dance as practiced, for example, in Germany and Austria.

Eschke et al. [235] explore self-organized adaptive path planning for multi-robot fiber deployment. They propose a swarm-based braiding system where mobile robots collaboratively manipulate continuous fiber materials. Using Thymio II robots, the system autonomously assigns robot groups, adapts to environmental cues, and maintains braiding patterns without requiring centralized control. The experiments demonstrate successful pattern formation, real-time adaptation, and scalability.

For a short overview of the literature on self-organized construction, see the review paper of Gerling and Mammen [278].

5.3.5 *Collective Transport*

Collective transport is a good example of a scenario with potential for an actual swarm effect. A few robots may not be able to move a heavy object even for one centimeter, but once a certain threshold in the swarm size is reached, they can move it for any distance. Such an extreme steep increase in swarm performance (super-linear performance increase, cf. Sect. 2.4.5) can be observed only in a few scenarios. This difference as a discrete jump in performance from zero to infinity exists basically due to the physical effects of stiction (friction of stationary objects) and kinetic friction (friction of moving objects). Otherwise, there should be a continuous increase in the covered distance of the transported object with increasing swarm size.

(a)

(b)

Fig. 5.13 Swarm construction by braiding continuous material [348]. **a** Braiding Thymio robots based on line following and resulting braid **b** Traditional May dance (photo by Martin Ladstätter, with permission)

There is a review paper by Tuci et al. [793] on collective object transport in multi-robot systems. The authors classify transport strategies into three categories: pushing-only, grasping, and caging. Each strategy entails unique coordination, control, and communication challenges. The review is focused on design tradeoffs, communication modes (direct vs. indirect), and system architectures (homogeneous vs. heterogeneous). It also highlights open challenges, such as designing functional heterogeneity that allows robots to allocate and re-allocate resources in a fully autonomous and adaptive manner. This would also increase fault tolerance.

There are several implementations that utilize robots. Kube and Bonabeau [447] investigate box-pushing inspired by ant behaviors. They discuss several issues, such as how to overcome stagnation during the pushing action and taxis-based transport, that is, pushing the box towards a light. Coordinating the actions of the robots based on local interactions and making use of appropriate sensors (e.g., touch, proximity, and light sensors).

A similar behavior but including self-assembly is the above-mentioned work of Groß and Dorigo [299]. Nouyan et al. [600] make use of the same behavior but in the context of a rather complex foraging scenario where the object needs to be collectively transported along a chain of robots pointing the way back home. Berman et al. [76] report an in-depth analysis of biological data of collectively transporting ants, its application to a robot control strategy, and its verification in simulation [864]. Habibi et al. [313] report a scalable distributed path planning algorithm for transporting a large object through an unknown environment. The area is sampled, and the distributed Bellman-Ford algorithm [250] is used to build a shortest-path tree. The robot experiment was done with eleven "r-one" robots (see Sect. 3.6.1) [537].

Exciting opportunities arise when we consider quadrotors as an option for collective transport. Mellinger et al. [547] report their approach to grasping and transporting building material in groups of quadrotors. Given that this is one of the earliest works focusing on flying robots for collective transport, this is certainly a big step forward. A drawback though an indoor-GPS system is used, and the robots receive control values from a central computer. An early study was published by Nardi and Holland [592], where they discuss a few of the challenges and propose an automated system identification approach to generate an appropriate model of micro aerial vehicles (MAV).

Another relevant work is that of Heuthe et al. [357]. They study embodied swarms that emulate a micro-robot system composed of up to 200 individually laser-controlled particles with a size of 6 μm. The primary motivation is to utilize this embodied system to simulate a swarm of tiny robots. The particles have neither sensors nor onboard computational units; however, they are tracked, sensor input is calculated offboard, their behavior for a given control algorithm is also calculated offboard, and the action is executed using the laser. The laser setup uses an acousto-optical deflector to precisely and simultaneously control the swarm. The laser induces localized heating on the carbon-capped hemisphere of each particle, enabling forward motion, rotation, or stationary positioning based on the laser's intensity and placement. The swarm's task is to collaboratively rotate and push a rod of size 100 μm to a target pose. Multi-Agent Reinforcement Learning (MARL) is employed to develop a robot controller that effectively solves the task. The paper also addresses a learning challenge. Attributing each robot's contribution in a swarm is challenging because the global performance of the swarm emerges from complex interactions between individual robots, making it difficult to isolate the impact of a single agent. Using a shared swarm reward can result in the "lazy agent problem," where some robots benefit from others' work without actively contributing. Counterfactual rewards are a reward assignment method that evaluates the contribution of each agent to the overall group performance by comparing the system's performance with and without that specific agent's contribution. The swarm successfully rotates and transports rod-shaped cargo objects to arbitrary target positions. Potential applications include automated assembly of mobile micromachines and programmable drug delivery systems.

A pretty surprising concept of collective transport is that proposed by Kemsaram et al. [416]. They present an acoustophoretic (i.e., manipulation of particles using sound waves) robotic system that uses phased ultrasonic transducers to enable both independent and cooperative mid-air contactless transport of objects. They use phased arrays of 8×8 ultrasonic transducer boards operating at 40 kHz, producing stable acoustic traps a few centimeters away from the array. In the experiments, they levitate and transport lightweight expanded polystyrene particles of 1 mm radius for single-robot transport and 2 mm radius for cooperative transport. Possible future applications of this approach include contactless material handling in cleanrooms or environments where contamination must be avoided, where swarms of such robots could scale manipulation to larger or multiple objects, increase robustness through redundancy, and flexibly adapt their formations to different tasks.

5.3.6 Collective Manipulation

The rather early study by Ijspeert et al. [373] from 2001 is an evergreen of swarm robotics because they obtained exciting results concerning swarm performance. The investigated task is "stick pulling." Robots are equipped with a gripper and are supposed to pull sticks out of the ground. The experiment is designed in a way, however, that a single robot cannot pull the stick completely out of its hole in the ground. Pulling a stick out, hence, requires two robots and a sequence of the following actions. First, a robot needs to find a stick. Second, the robot has to do the so-called grip1, which moves the stick out of the hole about half-way. Third, the robot has to wait in this position for help. Fourth, a second robot may come by (if the first robot does not just give up, releases the stick into the hole and moves on) and, fifth, does grip2 to remove the stick completely. The third step (waiting for help) comes with a crucial parameter: the waiting time. If the robot waits for too short a period, it is unlikely that another robot will come by in that short period of time. If the robot, however, waits for too long, the swarm may be quite inefficient, or even worse, we may run into a deadlock situation because all robots do grip1 and wait.

Another question is that about the optimal swarm size for a given environment and waiting time. Ijspeert et al. [373] observed a superlinear performance increase (cf. Sect. 2.4.5) going from swarm size $N = 2$ to $N = 6$ robots. For swarm sizes $7 \leq N \leq 11$, the performance increase is linear. For swarm size $N > 11$, the performance breaks down because the robots interfere with each other. Hence, we have a nice swarm effect in this scenario. A single robot receives zero performance, and for $2 \leq N \leq 6$ the performance increases superlinearly, that is, each swarm member gets more efficient with increasing swarm size. That is a property we usually do not observe in human groups.

5.3.7 *Implementation of Flocking and Collective Motion*

Flocking is the coordinated motion of bird flocks (see Fig. 5.14). At any time, all birds of the flock fly about in the same direction with the same speed.

Collective motion is a diverse field as reflected, for example, by the extensive review paper of Vicsek and Zafeiris [823]. A particular example is the paper by Yates et al. [886]. In this rather theoretical work, they investigate the collective motion in locusts. In their nymph state, locusts cannot yet fly. Instead, they show collective motion, which is called "marching bands" because they walk in long groups of extremely many animals [115]. Instead of measuring their behavior in the wild, biologists prefer to reduce the complexity and bring them into the lab. There, the locusts are allowed to move within a ring-shaped arena. Depending on the swarm size and swarm density, one observes a synchronization in their walking directions. From the two options of clockwise motion and counterclockwise motion, they choose by majority only one (cf. collective decision-making, see Chap. 8).

It may be surprising to realize that, despite early claims of successful implementations [521], achieving flocking behavior in real robots remains a challenging task. Probably most problematic is the alignment rule because it assumes that each robot knows the heading of all its neighbors. Working with sensors only and without communication, it is unclear how the heading of neighbors can be perceived. While the birds probably follow visual cues, swarm robots usually do not have a good vision

Fig. 5.14 Starlings as a natural example of flocking (Photo by Oronbb, CC BY-SA 3.0)

system. Even a single camera may not help because of the wide range that needs to be covered. Once we allow communication, it is still unclear what information should be transmitted. If the robot has no global information, it can only try to communicate some information about the relative angle between the two considered robots. If the robots have digital compasses, which arguably gives them access to global information, then a robot can simply transmit its absolute heading. This is the approach taken by Turgut et al. [794]. They use a digital compass and communicate the heading via a wireless communication module (IEEE 802.15.4/Zigbee compliant XBee wireless module). They report successful experiments with seven robots (they use the so-called Kobot). One of the earliest approaches to group formation in robotics goes back to Yamaguchi et al. [880]; however, the distributed control scheme is only validated in simulation.

Hauert et al. [343] investigate Reynolds-style flocking [674] in aerial swarms. The robots have GPS modules and communicate their heading information to neighboring robots by WiFi (ad-hoc mode). They are interested in two key physical constraints of fixed-wing flying robots: their limited communication range and restricted turning ability. They report a combination of simulation and outdoor experiments with up to 10 autonomous fixed-wing aircraft. They achieve stable, coherent flocking only if the communication range and turn rate are correctly balanced. Their study emphasizes that real-world platforms introduce nontrivial design tradeoffs not present in simulations of idealized agents.

A different approach is followed by Moeslinger et al. [569]. They propose a simplified flocking algorithm for simple swarm robots that does not need communication or global information. The algorithm relies on the localization of neighbors according to a sophisticated sector design. There are virtual sectors around each robot, and we assume that a robot can distinguish four directions to its neighbors. This can be implemented with at least four infrared sensors. For detecting neighbors in each sector, there are rules on whether the robot should turn or keep moving. For example, if there is a neighbor close by, the robot will turn away (separation). If there is a neighbor relatively far at the sides or in the back, the robot turns towards it (alignment and cohesion). Once the sectors are cleverly chosen and the different rules come with the correct priorities, flocking is achieved. Moeslinger et al. [568] validated that behavior in a tiny swarm of three e-puck robots.

Rasouli et al. [665] report a simulation study of a flocking-based aggregation behavior in swarm robotics using Reynolds' flocking rules combined with a K-means clustering algorithm. They validate performance under dynamic conditions, including robot failures and the presence of obstacles. Compared to other Reynolds-based methods, their algorithm performs better.

Mezey et al. [555] explore a purely vision-based approach to decentralized collective movement in ground robot swarms, inspired by animal behaviors and a previously published model [60]. Each robot perceives its surroundings through a visual projection field that encodes the relative angular positions of other robots as binary signals. The robot's movement is driven by simple attraction-repulsion forces that also adjust the robot's speed based on the perceived density of visual signals in different regions of the visual field. Using that vision-based model, terrestrial robots achieve collision-free, polarized movement solely through local visual information (see Fig. 5.15). The robots don't need to detect the headings of neighboring robots. They also don't communicate directly, and they don't require memory. Simulations and experiments demonstrate that the flocking behavior is robust to a limited field of view and confined spaces.

Boldrer et al. [89] present a fully distributed, communication-less flocking algorithm for swarms of aerial robots operating in cluttered, GPS-denied environments. Building on a Lloyd-based framework (each robot moves toward the utility-weighted centroid of its local spatial partition), they move each robot toward the centroid of a dynamically defined cell (convex weighted Voronoi diagram), shaped to ensure both safety and proximity to neighbors. The flocking behavior emerges as robots maintain bounded distances to selected swarm mates while avoiding collisions with obstacles and each other. The method tolerates sensing and tracking uncertainties and is validated in real-world experiments with UAVs navigating dense forests.

5.3.8 Foraging and Distributed Source Seeking

Foraging is a fundamental behavior in swarm robotics, even though robots, unlike biological organisms, do not consume food. Instead, foraging in robotics refers to a search and retrieval process, where robots locate and transport objects of interest. This concept is widely applicable to various real-world tasks, such as search and rescue, resource collection, and environmental monitoring. Foraging can be performed by a single robot or, more effectively, by a swarm of robots.

Foraging behaviors are omnipresent in the animal world because all animals need nutrition. We are interested in the foraging strategies of social insects. We skip a sophisticated review of the biological literature here. From the modeling perspective of swarm robotics, for example, the foraging model of Schmickl and Crailsheim [708] for honeybees is interesting, and so is also the relevance of noise in the decision-making process of foraging ants as reported by Meyer et al. [553]. The work by Jackson et al. [376] was mentioned above in Sect. 5.3.3 and it is relevant here because it explains how foraging ants navigate their trail network. For foraging in animals, it is essential to "minimize energy expenditure with respect to energy acquisition, maximizing their net rate of energy gain" [863]. This is often overlooked in robotics, as most robot systems rely on an energy reservoir rather than striving for long-term autonomy through energy harvesting. Another modeling aspect is that Lévy flights are often assumed as a motion model in animal foraging, as discussed above in Sect. 5.2.4. An example from swarm robotics research is the work by Nauta et al. [593] who improve search efficiency by Lévy flights. The proposed approach mitigates the effects of collisions in dense swarms, preserves heavy-tailed movement distributions, and dynamically adjusts Lévy parameters based on local estimates of target density.

In control theory terminology, there is a scenario known as distributed source seeking [490], which is related to foraging. Foraging is about finding and collecting

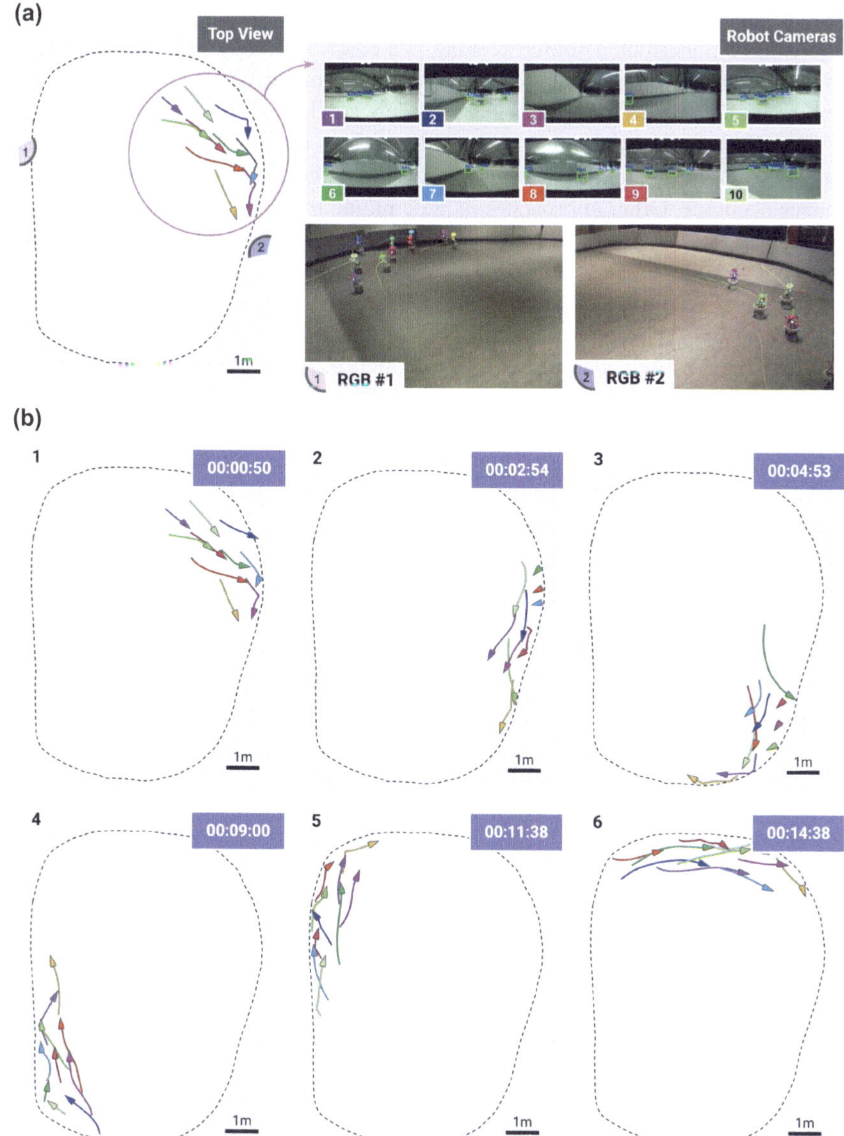

Fig. 5.15 Vision-based flocking in ground robot swarms. Experimental results from Mezey et al. [555], showing decentralized collective movement of ground robots using only local visual input. **a** Each robot perceives others via a camera (top right). Bounding box sizes are used to calculate range and bearing. These inputs are fed into an attraction-repulsion controller that regulates the heading and speed. **b** Snapshots from an experiment with $N = 10$ robots in a confined arena show the smooth polarized, collision-free collective motion over time without wall effects. Reproduced from Mezey et al. [555], licensed under CC BY 4.0, http://creativecommons.org/licenses/by/4.0/

discrete resources, whereas distributed source seeking involves tracking a continuous signal source. Distributed source seeking is a control strategy in which a group of autonomous agents collaborates to navigate an environment and locate the source of a signal (e.g., light, heat) using only local sensing and decentralized decision-making. Distributed source seeking methods include gradient estimation, where agents estimate and follow the signal gradient, often utilizing consensus-based techniques to enhance accuracy. Other approaches include plume tracking, which follows dispersion patterns (e.g., in gas leak detection), and coverage-based methods, where agents spread out to collectively explore and localize the source, as used in search and rescue. A related scenario is that of plume detection and odor source detection [252, 876]. For example, Chen et al. [140] present a low-cost robotic system for particle source localization, adapting an infotaxis algorithm [821] to airborne particulate matter by integrating multi-modal sensor data. The proposed method refines probability map fusion strategies, evaluates its performance in simulations and wind tunnel experiments, and demonstrates robust and efficient source localization under varying wind conditions. A related approach is that of Hayes et al. [346]. They present a distributed odor source localization approach using robots equipped with conducting polymer-based odor sensors. The robots collaboratively perform plume finding, plume traversal, and source declaration. The results demonstrate that simple local sensing tightly coupled with robot behavior is sufficient for effective odor localization. A later paper from the same lab compares casting and spiral-surge algorithms for odor source localization in laminar wind flow. Experiments in a wind tunnel show that spiral-surge is significantly faster and more robust than casting [498].

Interesting for modeling of foraging in swarm robotics is the model by Lerman and Galstyan [479]. They investigate the effect of interference in a foraging group of robots. Robots search objects, pick them up, and bring them to a specific target area. It is expected that there will be quite a lot of traffic in that target area, which may trigger interference between robots. For example, they find the typical swarm performance curves as described in Chap. 2. Hamann and Wörn [323] report a spatial model of a foraging behavior based on pheromones.

Probably one of the earliest works on foraging in robots was reported by Sugawara and Sno [757] and Sugawara et al. [758]. They report successful experiments with five robots and find that cooperation is practical.

Mayet et al. [530] report a method to emulate pheromones for a foraging behavior similar to ants based on ultraviolet light and glow paint (see Fig. 5.16). They use extended e-puck robots. Robots can "draw" trails on the ground by lighting their ultraviolet LED. The trail is then detected by vision. The approach also includes a light beacon to emulate a compass.

The above-mentioned approach by Nouyan et al. [600] is one of the most sophisticated experiments in swarm robotics. The robots form chains, search for an object, and once the object is found, they collectively transport it along the chain of robots back to their base station.

The probably most complex foraging behavior reported in swarm robotics is that of Dorigo et al. [196]. It was done within the Swarmanoid project that investigated heterogeneous swarm robots. The eye-bot (quadrotor), hand-bot (robot with grip-

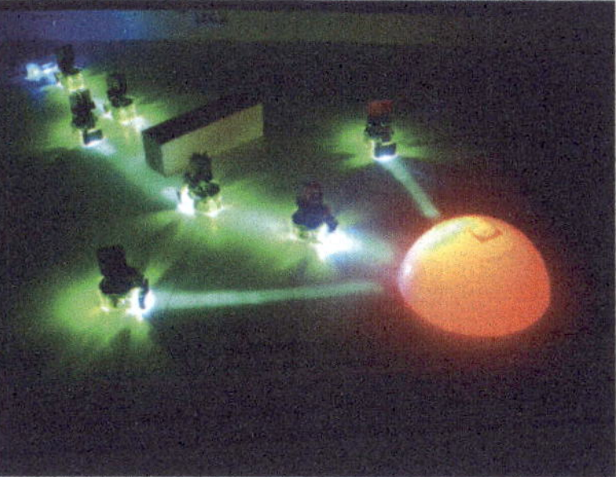

Fig. 5.16 Emulation of pheromones with ultraviolet light and glow paint by Mayet et al. [530]. Note that the right-hand photo is contrived

pers), and foot-bot (similar to the s-bot) cooperate to pick up a book from a shelf. First, they have to find it, which is done by the eye-bot; then they form a relay-station line of foot-bots, collectively transport a hand-bot to the shelf, and the hand-bot finally picks up the book.

Mezey et al. [554] introduce a mechanistic and spatially explicit agent-based model to explore how individual decision-making processes shape collective outcomes in foraging tasks. They find that social information influenced by physical and perceptual constraints, such as visual occlusion and limited fields of view, determines group performance. They study tradeoffs between individual exploration and social relocation, showing that real-world limitations often promote more efficient and adaptive group behaviors.

5.3.9 *Collective Tracking*

In swarm robotics, the task of tracking refers to a collective behavior in which a group of robots follows or localizes a moving target over time. In distributed source seeking and plume detection, the target is assumed to be static. In tracking, the target is in motion, may emit a detectable signal (e.g., sound, odor, radioactivity), and the goal is for the swarm to remain close to it, even as it moves through the environment.

Coquet et al. [152] propose an algorithm inspired by flocking behavior and Particle Swarm Optimization [418]. It is designed for tracking dynamic targets, such as chemical plumes. The algorithm is based on potential fields and naturally forms a stable configuration. The robots move collectively toward the target, avoiding collisions and adapting to changes in the target's movement. Simulation results suggest that it is suitable for real-world applications, such as underwater tracking.

Kwa et al. [456] investigate how swarm density affects the ability of a robot swarm to track a target collectively. The authors identify three distinct performance phases as density increases: (1) a low-density phase where agents fail to track the target, (2) a transition phase where tracking performance improves sharply, and

(3) a high-density phase where near-perfect tracking is achieved. A key contribution is the identification of a tradeoff between exploration and exploitation during the transition phase. Importantly, the authors argue that system size alone is insufficient for predicting performance due to finite-size effects, and they advocate for designing swarms that can adapt to varying densities (see Sect. 2.7.1).

5.3.10 Division of Labor and Task Coordination

The challenge of properly partitioning and allocating work in a swarm is essential, complex, and is an aspect of swarm robotics that comes with a lot of algorithmic features. Division of labor means the work is organized, not everyone is doing the same tasks, and consequently, the work needs to be usefully distributed among the workers. Task partitioning is the problem of defining smaller pieces of work (tasks). They should, of course, be self-contained, and each task should have similar requirements and a similar volume. Task allocation is then the problem of assigning tasks to the robots, hopefully in an efficient way. Both task partitioning and task allocation can be done offline (before deploying the system) and online (at runtime). An offline approach is more straightforward, but an online approach allows for adaptations to dynamic changes. If the task allocation is dynamic, then a strategy for task switching is required. One wants to avoid frequent switching because each switch may come with a certain overhead [331]. Related problems and skills include task specialization (where robots choose to be specialists for a specific task) and task sequencing (learning a particular order for executing a set of tasks).

Task Partitioning

Again, it is useful to look at biological role models, in particular social insects. Ratnieks and Anderson [666] define division of labor as

> the division of the workforce within the range of tasks performed in the colony, whereas task partitioning is the division of a discrete task among workers.

In their paper Ratnieks and Anderson [666] review task partitioning in insect societies with examples in ants, bees, wasps, and termites. They say benefits of task partitioning

> occur either through enhancement of individual performance or through enhancement of the overall system (e.g., where partitioning itself eliminates a constraint affecting task performance, such as when a forager can collect sufficient material for several builders).

Anderson et al. [23] study task partitioning in insect societies using the example of the bucket brigades.

An interesting model of task partitioning is reported by Karsai and Schmickl [406]. The question of how to obtain knowledge about the progress of a task based on local cues as described by Gordon [291] is answered here by the concept of a "common stomach." Karsai and Schmickl [406] investigate nest construction in social wasps. The common stomach is formed by a group of wasps that are generalists and can store water in their crops. The common stomach then serves not only as a storage of water but also as an information center about how well the task partitioning is currently being done. A set of positive and negative feedbacks is based on the common stomach and regulates the task partitioning process.

Pini et al. [637] present an approach for task partitioning that is based on methods from machine learning. They consider task partitioning, in particular, the question of deciding at runtime whether to utilize task partitioning at all. They apply well-known methods from the multi-armed bandit problem. The considered swarm has to transport objects from A to B and robots can choose to use a cache of objects C (see Fig. 5.17). Hence, a robot can do the full unpartitioned service of transporting objects from A to B, or a robot can decide to go for task partitioning and transport objects only from A to C or from C to B. The robots do not know the costs of either of these three options beforehand. Hence, they have to explore and estimate the costs online while working on the task. This corresponds directly to the typical challenge in the multi-armed bandit problem. A gambler can choose to play any of the bandits' arms, but initially doesn't know which one gives, on average, the biggest profit. The gambler, or in our problem, the robot, has to explore the different options (exploration), estimate the underlying probability density, and make a choice of which option to go for (exploitation). In addition, the costs or the payoff may be dynamic, that is, each option needs to be explored from time to time. Pini et al. [637] directly apply methods from machine learning to task partitioning in swarm robotics. There is no robot-to-robot communication, and they verify their approach in simulation.

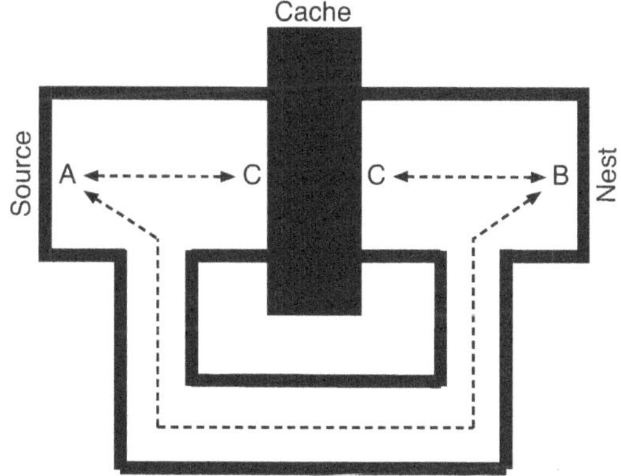

Fig. 5.17 Task partitioning scenario as described by Pini et al. [637]. Robots can decide against task partitioning and transport objects from A to B directly, or a robot can decide to go for task partitioning and transport objects only from A to C or from C to B

Task Allocation

Gordon [291] reports early studies on task allocation in social insects. She identifies both internal factors (e.g., body size) and external factors (e.g., the amount of already collected nectar, the number of workers engaged in the task) for task allocation. Necessary conditions for task allocation are "local cues that reflect the numbers of workers performing a task and the extend to which the task is accomplished" [291].

The well-known task allocation in honeybees determined by their physiological age is taken as inspiration by Zahadat and Schmickl [890], Zahadat et al. [888]. Their focus is on regulating robot's physiological age based on their local information about the current recruitment demand of their task. The approach is tested in an AUV swarm (autonomous underwater vehicles) with the task of monitoring at different depths.

Enough about the biological systems, now we turn to the multi-robot and robot swarm systems. Nair et al. [590] investigate task allocation in the RoboCup Rescue Simulation Domain. The approach is distributed and based on an auction system. All-to-all communication is required, and hence the system does not scale.

Lemaire et al. [474] discuss task allocation in a multi-UAV system (unmanned aerial vehicle system). The architecture is completely distributed, and tasks are also allocated in an auction process. Centralized planning is avoided, but the system is synchronized by token circulation, and all-to-all communication is required. The approach is verified in simulation only.

Kanakia [403] presents an approach to task allocation based on a response threshold approach. The focus is on allowing only local information and local communication. A theoretical result is a proof that the system converges. Kanakia [403] makes use of game theory, in particular, so-called global games: "Global games can be seen as a particular instance of equilibrium selection through perturbations." [123]. The idea of global games is to focus on "strategic environments with incomplete information." Global games raise an essential question of whether to commit or not. An interesting example is a bank run (see Fig. 5.18), the question of withdrawing your money from the bank in a crisis or not [285]. Kanakia [403] investigates scenarios similar to the stick-pulling task allocation problem (see Sect. 5.3.6), macroscopic modeling includes a master equation approach, and also verifies the results in physical experiments using the Droplet swarm robot (see Farrow et al. [239] and Sect. 3.6.1).

An approach to task allocation in sequentially interdependent tasks is presented by Brutschy et al. [112]. The robots measure the delay between having finished their task and having to wait for another subtask to finish. They use this local cue to decide about task switching. The approach relies on these measured delays and on a probabilistic task switching behavior. The results are verified in a simulation.

Khaluf et al. [428] present a specific approach to task allocation and task switching with soft deadlines in swarm robotics. A soft deadline is a deadline that should be met but if it is not met the quality of the result is only decreased in quality but has no catastrophic results. The approach is probabilistic and offline, that is, each robot is equipped with a so-called decision matrix that defines probabilities of switching between tasks at runtime. This matrix is not changed at runtime but pre-calculated and optimized offline. The system is still adaptive to changes in the swarm size, for

Fig. 5.18 Example of a bank run, group of people in front of the closed American Union Bank, New York City, April 26, 1932 (public domain)

example, in the case of robot failures. There is no robot-to-robot communication, and results are verified in simulation.

Task Specialization

Ferrante et al. [242] uses evolutionary algorithms in a robot swarm to evolve behaviors that exhibit specialization, specifically task partitioning. They investigate a scenario inspired by leaf-cutting ants that has potential for task partitioning. The ants can divide the work into two jobs: (1) dropping leaves from trees and (2) transporting them to the nest. Neither the basic behaviors of what to do when nor the actual task partitioning is predetermined. Still, Ferrante et al. [242] observed a self-organized, emergent task partitioning in the system.

Task Sequencing

Garattoni and Birattari [269] present a system that is capable of collectively learning and executing an unknown sequence of tasks. Individual robots operate reactively, but the swarm as a whole achieves a planning-like behavior through coordination and feedback. Robots form physical chains that act as both navigational guides and memory structures, encoding the task order. The authors explore how the swarm can handle different levels of complexity, including delayed feedback and longer sequences. This is an interesting contribution to swarm robotics as it is pointing to how reactive individual behavior can be used to implement deliberative control on the collective level. This is how collective behavior can give rise to emergent planning.

5.3.11 Shepherding

Shepherding behaviors are interesting for swarm robotics because they often require more than one shepherd robot, the behaviors of the shepherds are mutually dependent, and the multi-robot approach is not a mere parallelization of a task that could also be done by a single robot. Shepherding is also one of the rare cases of multi-swarm interaction, where two interacting swarms are involved (also see Sect. 6.10). However, our focus is on engineering the shepherd behavior while the sheep behavior is considered a given.

Shepherding can be done with a single shepherding agent (such as a dog or a shepherd) or a group of cooperating shepherds. The shepherding agent follows different behaviors. Two of the most important ones are (1) collecting: moving to the far sides of the flock and bringing the animals together, and (2) driving: pushing the herd forward (e.g., towards the target) from behind. Especially, robot shepherds may also require a searching or exploration behavior to find the sheep flock.

The difficulty of the shepherding task is mainly influenced by the cohesion of the sheep and the swarm size of the sheep. However, it is also influenced by the relative speed between the shepherd and the sheep (i.e., how much faster the shepherd is compared to the sheep), the number of shepherds, and the initial positioning of the sheep. The flock of sheep can vary in size, but what controls the difficulty of shepherding more directly is their cohesion. If they tend to stick together by default, the shepherd's job is easier. In contrast, if there is only low cohesion in the sheep, the shepherd needs to invest most of their time into collecting the flock. This is important to remember when benchmarking different algorithms, as task difficulty may vary depending on the sheep flock's cohesion. Additionally, larger sheep flocks with a fixed cohesion may be more challenging.

Insights from Biology

A recent study in biology, as cited by Jadhav et al. [378], investigates how directional information propagates through a flock of sheep ($N = 14$) herded by a trained border collie. Despite the dog driving the flock from behind, directional information primarily flows from the front to the rear of the flock. Sheep in front positions exhibit higher leadership influence, steering the flock's movement, while the dog adjusts its motion in response to the flock's collective behavior. The flock maintains high cohesion and polarization during herding events, optimizing both alignment and proximity. The dog employs zigzag movements and continuously adjusts its speed and direction in response to the flock's reactions.

Overview and Early Work

Long et al. [499] give a review of shepherding as a bio-inspired swarm robotics problem and technique. Basic skills required by a sheepdog are: outrunning, lifting, fetching, driving, shedding, penning, singling, and gathering/collecting. Challenges such as scalability, decentralized control, and dynamic adaptation are explored. They also discuss machine learning in advancing shepherding techniques.

There is an early work by Schultz et al. [720] using genetic algorithms to learn behaviors for shepherding. A similar approach is reported by Potter et al. [646], which focuses on a heterogeneous approach and the emergence of specialists. One step further, but following a similar line is the study of Gomes et al. [287]. They investigate a coevolutionary approach with a special technique of merging and splitting populations.

Algorithmic Models and Simulations

The paper by Strömbom et al. [756] is also significant. They study a single shepherd and how it can herd a group of sheep towards a target. The shepherding control algorithm has two core parts: collecting (gathering dispersed sheep) and driving (pushing the cohesive group towards the target). The shepherd keeps switching between these two modes. The sheep are attracted to their neighbors, repulsed by the shepherd, and avoid overcrowding. The shepherd executes the typical side-to-side or zigzag motion behind the flock. The model was validated post hoc using real-world data from GPS-tracked experiments, showing good agreement between model predictions and observed behavior. For larger groups, the shepherd occasionally fails as subgroups of sheep form, and the shepherd needs to focus on smaller subgroups sequentially.

Özdemir et al. [610] explore a minimalist approach to the shepherding problem using simulated robots that lack computation and memory capabilities. The shepherd robots, equipped with only a simple line-of-sight sensor providing two bits of information, directly map sensor inputs to motor outputs without performing arithmetic calculations or storing information. Simulations demonstrate that these simple robots can herd dynamic, sheep-like robots toward a goal by relying solely on reactive behaviors. Their approach is robust to sensory noise and can adapt to varying numbers of agents, showing that coordinated control can emerge from simple perception-action couplings.

Robot Shepherding in Hardware

Generally, hardware implementations of shepherding robot swarms seem relatively rare. This is probably due to the high requirements of having a flock of robot sheep and one or more shepherd robots. Using onboard sensing and computing to enable the shepherds to detect the poses of the sheep is also challenging. Interestingly, there is, for example, "SwagBot" that was developed at the Australian Centre for Field Robotics, University of Sydney, which can herd cattle.[4]

Hu et al. [364] present robot experiments with up to $N = 10$ real robots. They propose an occlusion-based coordination protocol for shepherding tasks, utilizing autonomous robotic dogs to efficiently guide a flock of sheep-like agents to a target location. The sheep flocking behavior is modeled using adaptive consensus algorithms and artificial potential fields, ensuring realistic movement and cohesion among agents. The robotic shepherds use an occlusion-based motion strategy, which means they position themselves in areas where the sheep obstruct the view of the target, thereby optimizing their herding efficiency. The proposed strategy reduces the number of robotic dogs required compared to traditional formation-based approaches. The approach is validated even with real-world experiments. They used $N = 10$ Mona robots [36], with three robots acting as shepherd dogs and seven robots acting as sheep. Initially, the flock of seven sheep robots was divided into two groups, each placed in opposite corners of the arena. The shepherd robots successfully herded the two groups towards the target area and merged them into one flock at the goal. The task was completed within 80 s, demonstrating the efficiency of the occlusion-based coordination protocol.

UAVs and Ethical Considerations

King et al. [436] outline a roadmap for using uncrewed aerial vehicles (UAVs) to herd animal groups in a bio-inspired and ethical manner. Drawing on natural herding strategies as used by sheepdogs, they propose a dual-UAV system: one UAV performs surveillance to monitor animal movement. At the same time, the second UAV actively herds the group. The paper emphasizes the need for species-specific understanding of animal responses, the development of behavioral models, and the integration of algorithms into real-world UAV systems. This approach has potential applications in wildlife management, conservation, and mitigating human–animal conflicts, while also raising critical ethical and technical challenges.

[4] https://spectrum.ieee.org/automaton/robotics/industrial-robots/swagbot-to-herd-cattle-on-australian-ranches.

Control-theoretic Perspectives

El-Fiqi et al. [228] study the relative speed between the sheepdog and sheep, the spatial configuration of the sheep at the start of the task, the number of sheepdogs, and the density of obstacles in the environment. They focus on the limitations and potential of reactive strategies.

Lama and di Bernardo [460] investigate a shepherding problem for a specific sheep motion model (stochastic motion of diffusion with drift, Brownian motion, modeled by Langevin equations, see Sect. 7.4.2) from the perspective of control theory. Shepherds must guide non-cohesive sheep to a desired goal region. Generally, the cohesiveness of the sheep serves as a parameter to vary the difficulty of shepherding (highly cohesive sheep simplify the shepherding task). However, their assumption of Brownian motion for sheep may simplify the task also partially because the mean displacement is in the order of $\sqrt{T}$ for T time steps. Lama and di Bernardo [460] identify a scaling law for the relationship between the number of shepherds N and sheep M. Under certain assumptions, the number of required shepherds N is in the order of $\sqrt{M}$.

Many Shepherds

Di Lorenzo et al. [187] consider a large-scale shepherding problem (many shepherds or "herders") for the above-mentioned specific sheep motion model (non-cohesive, Brownian motion) from the perspective of control theory. A large group of herders directs a large flock of sheep to a goal region. They introduce a macroscopic continuum model in the form of a partial differential equation (convection-diffusion equation, similar to a Fokker-Planck equation, see Sect. 7.4.2). With their approach, one can identify the minimum number of herders required. Validations through simulations and experiments with $N = 20$ robots demonstrate the feasibility and scalability of their control strategy.

Van Havermaet et al. [810] present a decentralized multi-robot shepherding algorithm designed to guide animal herds away from dynamic dangers. The proposed method uses local sensing and communication to maintain a "caging" formation around the herd. Robots detect threats near the perimeter and selectively approach the herd to steer it away from danger through induced aversive responses. Van Havermaet et al. [810] evaluate the method across various collective motion models and show that successful shepherding requires the number of robots to meet a model-dependent threshold. Two dynamic scenarios are explored: (1) avoiding spontaneously appearing dangerous zones and (2) confining the herd within a shrinking safe area. Simulation results demonstrate the robustness and effectiveness of the approach.

In a similar study, Van Havermaet et al. [811] investigate multi-robot shepherding to guide groups of animals along a predefined path. Results show that narrow paths and sharp turns pose the most significant challenge, while larger robot teams and moderate group velocities improve stability and containment.

These two studies examine shepherding as a mixed-species system and focus on danger-avoidance by using a protective cage around a herd.

Learning-based Approach

Napolitano et al. [591] introduce a decentralized reinforcement learning (RL) approach to address the multi-agent shepherding problem, which involves herding targets into a goal region using multiple autonomous agents. Unlike conventional methods, it eliminates the assumption of cohesive target behavior and minimizes reliance on heuristics. The proposed two-layer control architecture comprises a low-level controller for guiding individual targets and a high-level policy for determining which target each herder should pursue. Using deep Q-learning, the approach enables herders to autonomously learn cooperative strategies, improving efficiency without requiring explicit communication. The decentralized two-layer control architecture, where each herder independently queries the high-level policy to select a target, guiding it to the goal region using low-level control inputs. The method demonstrates scalability to larger systems by limiting each herder's interactions to a fixed subset of agents, offering a robust solution for complex, real-world multi-agent systems.

5.4 Catastrophic Collective Behavior

Natural and artificial collective behavior is not immune to failures and even catastrophic failures. One of the most prominent examples of a negative collective behavior is possibly circular milling as found in army ants (also see Sect. 5.2.5). It emerges "when moderate numbers are separated from a colony and restricted to a confined area, either in the laboratory or naturally in the field during exceptionally severe rainstorms" [157]. This phenomenon arises when a disturbance disrupts the pheromone-based communication, causing them to become isolated from the main column of the raiding swarm [716]. A circular pheromone trail is established, reinforced, and each individual follows it until exhaustion, and ultimately the animals die [717]. This is due to a lack of global information (circular pattern) as well as limited localization and memory ("O, I think, I have been here before."). Modern phylogenetic analysis shows that all army ants share a single common ancestor. The circular milling behavior has probably evolved only once. Molecular dating suggests their origin around 105 million years ago. More generally, this collective behavior is called "vortex" and it can also be modeled by the Vicsek model [156].

Delcourt et al. [180] go beyond ant mills, they review similar phenomena ranging from bacteria to vertebrates, including humans, such as caterpillar circles, bat doughnuts, amphibian vortices, duck swirls, and fish toruses as collective vortex behaviors. They explore the diversity, proximate mechanisms, and potential evolutionary causes of these circular movements. They consider both potential benefits, such as improved feeding efficiency and predator avoidance, as well as negative aspects, because vortices can become traps that lead to significant energy loss and wasted time, with a negative impact on reproduction and survival.

Bartashevich et al. [58] study transient milling in an approach focused on embodiment using an agent-based model with a limited field of vision, visual occlusions, and a collision avoidance mechanism. They find short (transient) occurrences of milling, but they recur frequently.

Liu et al. [496] investigates the behavior of dielectric colloidal particles (Quincke rollers) in a conducting fluid under a subcritical electric field. They find activity waves and self-organized freestanding vortices. The collective dynamics relies on hydrodynamic effects.

Human beings, as highly cognitive animals, rarely show collective behavior that could be described directly with similar methods as collective behavior in, for example, insects or birds. However, exceptions are lane formation [580], trail formation [351, 352, 369], and human behavior in panic situations [353].

Pedestrians show self-organizing lane formation in certain densities (persons per square meter) and bi-directional flow [776]. Crowds of pedestrians form separate lanes of uniform walking direction [353]. Obstacles and whether the pedestrians, for example, carry baggage influence this behavior significantly [731]. This can be modeled by social forces and potential fields [308, 389]. This form of collective behavior may help to increase the efficiency of the pedestrian flow. However, in too high densities and possibly combined with panic situations [581], the collective behavior of crowds can result in disasters. Examples are the 2015 stampede in Mecca, Saudi Arabia, during the annual Hajj pilgrimage and the 2010 Love Parade disaster in Duisburg, Germany.

Moussaïd et al. [582] explore crowd behavior during high-stress evacuations using immersive virtual environments. Under high-stress conditions, participants exhibited herding behavior and dangerous overcrowding driven by increased density, rather than an increased tendency to imitate others. Helbing and Mukerji [350] examine the 2010 Love Parade disaster in Duisburg, Germany. They view it as a systemic failure rather than the result of individual behavior or mass panic. They stress the role of crowd turbulence, inadequate flow management, and poorly designed infrastructure in creating dangerous conditions. Also Zhao et al. [896] investigate the 2010 Love Parade disaster using computer simulations and virtual reality. By replicating the event with a three-dimensional model calibrated with real data, they evaluate interventions such as removing police cordons, opening additional exits, and separating inflow from outflow. The results demonstrate that these strategies could significantly reduce congestion, improve crowd throughput, and prevent casualties.

Catastrophic collective behavior in swarm robotics research literature is rarely reported. One possible mention is the finding of two-phase performance as discussed in Sect. 2.5.3. Specific swarm systems can seemingly be put "on the edge" in terms of their performance. If the system has too high a swarm density, an additional robot or another disturbance could lead to a complete breakdown of performance due to an exhausted shared resource. This is comparable to the emergence of a spontaneous traffic jam on the highway without a construction site or an accident causing it.

Another example is the so-called lock-in state in a system of collective decision-making. Zakir et al. [894] report consensus lock-in, where strong social reinforcement causes the entire swarm to converge on an outdated or incorrect option. Once locked in, the swarm may ignore new environmental evidence, preventing adaptation when conditions change. This demonstrates the risk of overly relying on social information and shows the need for mechanisms, such as cross-inhibition, to maintain adaptivity in robot swarms.

5.5 Heterogeneous Swarms

In Chap. 1 we have defined robot swarms as (quasi-)homogeneous systems following Beni [72]. However, in recent works, the advantages of using heterogeneous robot swarms have been investigated. For example, Arkin and Egerstedt [29] discuss heterogeneity in robotics in general and point to a niche that seems not well covered yet and that they nickname "slowbots." Just as in the animal world, we see animals across all scales from super-fast to super-slow, we might want to think about a similar diversity in robotics. The above-mentioned project, Swarmanoid by Dorigo et al. [196], explicitly investigated heterogeneous hardware in swarm robotics with their eye-bots, hand-bots, and foot-bots. Especially the combination of flying robots and ground-based robots seems an advantageous approach, for example, by using the visual overview of a quadrotor to instruct wheeled robots on the ground. This way, still local information is used, and the system can scale up, but each robot has much more information available. Kengyel et al. [417] investigate the impact of software heterogeneity (i.e., the robot hardware is homogeneous but robots have different controllers) in an aggregation task inspired by a honeybee behavior. They use methods of evolutionary algorithms to combine appropriate numbers of predefined robot types. Prorok et al. [653] investigate the impact of diversity in a robot swarm on performance in a task allocation problem, where each task has different requirements for the robots working on it. They also define a diversity metric called "eigenspecies" to quantify differences in heterogeneity. In a much earlier work Balch [50] defined a measure of heterogeneity called "hierarchic social entropy." It takes into account what fraction of the swarm is of the same kind and the differences between the types. Another measure of heterogeneity by Twu et al. [797] stresses that heterogeneity basically depends on two components: complexity and disparity. Complexity describes the overall number of differences. For example, if not two robots are of the same kind, then the complexity is high. Disparity describes how much the overall difference between robots is maximized. For example, if there are only two groups of robots but both are at one opposing end of the "diversity scale," then the disparity is high. For further reading, see ([212, 337, 507, 527]).

5.6 Further Reading

The review papers by Schranz et al. [718], Brambilla et al. [102], and Bayindir [62] go through different scenarios of swarm robotics. The recent book by Schranz et al. [719] give a more applied and fast introduction to a few selected scenarios and methods. Floreano and Mattiussi [247] give a short overview of collective systems, including reconfigurable robotics and co-evolutionary systems. Eftimie [220] surveys collective motion from the perspective of mathematical biology, focusing on hyperbolic (directed movement with finite speeds) and kinetic models (tracking agent distributions in space and velocity). Covering local and nonlocal interactions, such as alignment and attraction, the paper demonstrates how simple rules can generate complex spatiotemporal patterns, including pulses, ripples, and zigzags.

5.7 Exercises

Exercise 5.1 (*Random motion*) Implement three types of random motion: uncorrelated random walk, correlated random walk, and Lévy flight. Use a bounded environment and compare the strategies based on area coverage for a defined period of time. Evaluate your method in terms of the following aspects:

- Coverage efficiency: How fast and evenly is the space covered?
- Redundancy: Do robots revisit areas frequently?
- Reliability: Does the robot get stuck sometimes or explore too erratically?

You can visualize your results using heatmaps that show the explored space across multiple independent runs. Also, plot a graph of the average percentage of area covered for many independent runs over time.

Exercise 5.2 (*Synchronization of a Swarm*) We create a simple model of synchronization in a firefly population. The population is scattered randomly over a square of 1 by 1 (uniform distribution). We assume that fireflies are stationary and can only perceive local neighbors in their vicinity. We say two fireflies are within the vicinity of each other if the distance between them is smaller than r. Hence, there is a virtual disc centered at each firefly's position, and every other firefly sitting on that disc is a neighbor. The fireflies flash in cycles. We define the cycle length by $L = 50$ time steps. The firefly flashes for $L/2$ steps followed by $L/2$ steps of not flashing. This holds except for those cases when the firefly tries to correct its cycle to synchronize. In the time step after it has started to flash, it checks its neighbors and tests whether the majority of them are actually already flashing. If so, the firefly corrects its clock by adding 1, that is, it is decreasing the current flashing cycle from $L/2$ to $L/2 - 1$ steps and will consequently flash 1 step earlier next time.

(a) Implement the model for swarm size $N = 150$ and cycle length $L = 50$. Calculate the average number of neighbors per firefly for vicinity distances $r \in \{0.05, 0.1, 0.5, 1.4\}$. Plot the number of currently flashing fireflies over time for vicinity distances $r \in \{0.05, 0.1, 0.5, 1.4\}$ for 5000 time steps each. When plotting the number of currently flashing flies, make sure you plot the full interval of $[0, 150]$ for the vertical axis.
(b) Extend your model to determine the minimum and maximum number of concurrently flashing fireflies during the very last cycle (last $L = 50$ time steps starting from $t = 4950$). By subtracting the minimum from the maximum, you get twice the amplitude of the flash cycle. Average the measured amplitudes over 50 samples each (50 independent simulation runs with 5000 time steps each) and plot them over vicinities $r \in [0.025, 1.4]$ in steps of 0.025. What seems a good choice for the vicinity and the swarm density?

Exercise 5.3 (*Crowd dynamics on wobbly bridges*) We study a bit more the biomechanically inspired model of crowd dynamics of the Millennium Bridge in London [68, 755] as discussed in Sect. 5.2.2. We need to implement a solver for Eqs. 5.4 and 5.5. Feel free to discretize these differential equations as difference equations in the following way

$$
\begin{aligned}
\frac{\Delta x_i}{\Delta t} &= z_i \ , \\
\frac{\Delta z_i}{\Delta t} &= \lambda_i \left(z_i^2 + x_i^2 - a^2\right) z_i - \omega_i^2 x_i - \Delta u \ , \\
\frac{\Delta y}{\Delta t} &= u \ , \\
\frac{\Delta u}{\Delta t} &= -2hu - \Omega^2 y - r \sum_{i=1}^{N} \Delta z_i \ .
\end{aligned}
\tag{5.31}
$$

Choose an appropriate time step (order of $\Delta t = 0.1$ or smaller) or a more sophisticated numerical solver. In our model, we set the parameters $a = 1.0$, $h = 0.05$, $r = 0.0023$ and $\Omega = 1.2$. For all agents, we set $\omega_i = 1.097$ and $\lambda_i = 0.5$.

(a) Implement the Van der Pol model for the Millennium Bridge oscillations using a swarm of $N = 200$ agents. Initialize the lateral gait displacement x_i and $\frac{dx_i}{dt}$ of each agent uniformly at random $x_i \in [-0.4, 0.4]$. Simulate the system for 500 time steps ($500\Delta t = 500 \cdot 0.1 = 50$). Plot $\frac{dx_i(t)}{dt}$ (in the case of the discretization $z_i(t)$) for all agents i over time in a useful way. Adjust the parameters so that the system converges to oscillations with a constant amplitude (e.g., by adjusting coupling strength r). Do you observe a synchronization of the pedestrians (oscillators)?

(b) Next, we aim to develop a data-driven macroscopic model. We are interested in only two classes of pedestrians: those currently accelerating to the left and those currently accelerating to the right. From the perspective of a pedestrian (i.e., microscopic level), they can be in either of two states: left or right. After having accelerated to the left for a while (i.e., dwelling in state *left*), the pedestrian will experience a transition to state *right*. On a macroscopic perspective, this can be understood as a population model. A certain fraction $k \in [0, 1]$ of all pedestrians is in state *left* and a fraction of $1 - k$ pedestrians is in state *right*. We are interested in the macroscopic transition rates, that is, for a given system state k, how many pedestrians will, on average, experience a transition to the other state. We want to measure a probability $P(k)$ of making a transition in system state k. We focus on the ratio of pedestrians $k \subset [0, 1]$ that accelerate to the right ($\frac{dx_i}{dt} > 0$ or $z_i > 0$), and we choose to check for $\frac{dx_i}{dt} > 0$ only every 65 time steps ($65\Delta t = 65 \cdot 0.1 = 6.5$) that is roughly the oscillation period of the converged system. If a considered pedestrian has made a transition (*left*→*right* or vice versa), you should count that for the previously observed system state k. Do as many independent repetitions as you can afford of 500 time steps ($500\Delta t = 500 \cdot 0.1 = 50$) each. Plot your measured transition probability $P(k)$. Try to interpret the plot. How can you use it to predict stable system states?

(c) We want to modify the model to remove the global coupling of pedestrians. We assume that the bridge is closed to southbound traffic and has been converted into a one-way bridge. People are allowed to enter the bridge in groups of $g = 5$ people at a time. There are still $N = 200$ people on the bridge at any time, but each pedestrian has a small group $\mathcal{G}_i$ of people they subconsciously fall in step with due to visual cues. This is modeled by introducing an additional term to the system of differential equations:

$$\frac{d^2x_i}{dt^2} = \lambda_i \left(\left(\frac{dx_i}{dt} \right)^2 + x_i^2 - a^2 \right) \frac{dx_i}{dt} - \omega_i^2 x_i - \frac{d^2y}{dt^2} - c_i \sum_{j \in \mathcal{G}_i} \frac{d^2x_j}{dt^2} , \tag{5.32}$$

$$\frac{d^2y}{dt^2} = -2h\frac{dy}{dt} - \Omega^2 y - r \sum_{i=1}^{N} \frac{d^2x_i}{dt^2} . \tag{5.33}$$

The parameter $c_i = 0.00125$ determines the strength of the visual coupling between neighboring pedestrians. All other parameters remain as above. Implement the modified model with neighborhoods. The local neighborhood (size $|\mathcal{G}_i| = 5$) stays consistent during the walk over the bridge. Add a local coupling between neighboring agents, representing the psychological aspect of locking step with neighboring walkers. Maybe adjust the parameters so that the system converges to oscillations with a constant amplitude. Do you still observe synchronization among pedestrians?

Exercise 5.4 (*Estimating swarm size*) Implement the stochastic experiment as described by Wang and Rubenstein [846]. Generate N random numbers r_i uniformly distributed $r_i \in [0, 1]$. Determine the maximum random number $r_{\max} = \max_i r_i$. Calculate the estimated swarm size $N \approx \frac{r_{\max}}{1-r_{\max}}$. Do M independent stochastic experiments to calculate the mean of maxima $\bar{r}_{\max}$. Do experiments and analyze the data to determine the dependency of swarm size estimation accuracy on the number of trials M and the swarm size N.

Exercise 5.5 (*Phase transition in Vicsek model*) Implement the Vicsek model as defined in Sect. 5.2.5. For a fixed area and fixed swarm size N (i.e., constant swarm density ρ), we want to vary noise η and measure how much aligned the swarm gets. Following Vicsek et al. [824], we define the average normalized velocity as an order parameter by

$$v_a = \frac{1}{Nv} \left| \sum_{i=1}^{N} v_i \right| . \tag{5.34}$$

Choose a proper interval and step size for varying noise intensity η. Run a high number of statistically independent repetitions for each value of η and measure v_a at the end of each run. Plot v_a as a function of noise η. Note that the phase transition found by Vicsek et al. [824] is a continuous (second-order) phase transition. The transition is characterized by a smooth change in the global order parameter v_a. Still, there is a critical noise intensity η_c that separates an ordered phase ($\eta < \eta_c$) from a disordered phase ($\eta > \eta_c$). Try to improve your plot by pushing the boundaries of how many particles N you can simulate (preferably keep swarm density ρ constant). The more particles you can simulate, the clearer the phase transition will be visible in your plot.

Exercise 5.6 (*Simple flocking*) Implement a Reynolds-style flocking algorithm. Robots should follow the three rules: alignment, cohesion, and separation. Simulate at least 20 robots in a 2D plane with periodic boundary conditions (i.e., toroidal space). Optional: add slight heading noise. Reproduce plots similar to Fig. 5.4: Distribution of headings over time (relative to swarm average) and average number of neighbors over time.

Exercise 5.7 (*Modeling simple aggregation with directed 1-forests*) Implement a graph simulator to model directed 1-forests. A 1-forest is a directed graph in which each node has exactly one outgoing edge. The connected components of such a graph are 1-trees (unicyclic graphs), each containing exactly one cycle.

(a) Choose several nodes $N \in \{10, 50, 100, 500, 1000\}$. For each node i, pick a target node $j \neq i$ uniformly at random and create a directed edge from i to j. We do not allow self-loops (i.e., no robot chooses to join itself). Store the resulting graph in a suitable data structure. Identify 1-trees and their sizes. Each connected component of a 1-forest contains exactly one cycle and any number of trees "rooted" at that cycle. We call these components 1-trees. There are various algorithms available to detect cycles. One option is to use a depth-first search for cycle finding. For each node, follow the chain of outgoing edges. If you encounter a node you have visited before during this chain, you have found a cycle. Once a cycle is found, mark all nodes that lead into the cycle and all nodes in the cycle as members of this 1-tree. For each 1-tree (connected component), count the total number of nodes within it and store that number. This gives you the total number of 1-trees and a list or histogram of the 1-tree sizes. Do repeated independent trials to gather reliable statistics.

(b) Plot a histogram of 1-tree sizes for each $N \in \{10, 50, 100, 500, 1000\}$ and plot a histogram of 1-trees numbers.

(c) What changes if self-loops are allowed? What changes if you introduce a probability $0 \leq P < 1$ of a node having an edge? What would be ways to model neighborhoods of robots, that is, edges can only connect to neighbors?

Exercise 5.8 (*Aggregation behavior*) Initialize a robot arena with 20 randomly distributed robots. All robots should be operated by instances of the same controller (i.e., copies of the same controller).

(a) Program a behavior that makes robots stop if they get close to another robot. For that, your simulated robots require sensors that detect nearby robots. Possibly, you can also fake such a sensor based on distances between robots to keep the simulation simple.

(b) Extend your program to limit the time a robot stays stopped according to a defined waiting time. When a robot wakes up and moves again, it might immediately stop again, depending on your implementation. However, the idea is that the robot leaves at least small clusters before stopping again. Think of a strategy to implement this behavior and change your program accordingly.

(c) Tweak your implemented behavior and parameters such that your robot swarm eventually aggregates in one big cluster with robots leaving only from time to time but rejoining it soon again.

Exercise 5.9 (*Dispersion*) Design and simulate a simple dispersion behavior for a swarm of 20 to 30 robots in a 2D environment with unknown obstacles. Each robot uses local sensors (range and bearing to nearby robots and obstacles) to spread out and maintain connectivity with at least one neighboring robot. Alternatively, it can use only a "ping" from neighbors without actual range and directional information. Implement a force-based model: robots are repelled by nearby robots and obstacles. Optional: Introduce a rule for robots to move randomly if they become stuck. Observe and evaluate how evenly the swarm spreads over time. Study the resulting final spatial distributions (e.g., Voronoi plot, histogram of inter-robot distances).

Exercise 5.10 (*Sparse-Target Swarm Exploration Using Lévy Flights*) You are deploying a swarm of autonomous ground robots into a large, unknown environment. The mission is to locate sparse and unpredictably placed items. These times do not emit a detectable gradient and can only be detected upon proximity. Design and simulate a distributed exploration algorithm for the swarm based on Lévy walks that: (1) efficiently covers the environment, (2) avoids redundant exploration, (3) maximizes detection probability of sparse sources, and (4) includes coordination mechanisms (optional) to improve performance.

(a) Robots move according to a Lévy walk: At each decision point, sample a step size s and turning angle θ. Sample step size s from a power-law distribution as in Eq. 5.13. You may choose to require a minimal step size. The turning angle θ is sampled from a uniform distribution over $[0, 2\pi)$. Optionally, you can use correlated turning to model inertia (e.g., Gaussian bias around current heading).
(b) Implement a coordination mechanism between the robots. An option is to implement an implicit coordination. Robots avoid each other through proximity sensing and change direction if another robot is too close. Another idea is to adapt the step size. You can change the exponent α of your power-law distribution over time, for example, based on the success rate. For example, shorten the step size if a robot is informed by its neighbors that many sources have been detected early.
(c) Measure the coverage efficiency. For example, measure the average time until the first source is detected depending on swarm size.

Exercise 5.11 (*Task allocation with dynamic switching*) Simulate a group of robots faced with two types of tasks: collecting and transporting red and blue tokens. Tasks appear randomly at different red and blue stations. Robots must decide which task to perform based on local sensing. Robots pick up tokens and take them to the base station. Simulate 10 to 20 robots in a 2D environment. Each station can broadcast a local demand signal (e.g., low-frequency beacon increases if unserved). You can complicate this task by removing that signal and instead having the robots detect supply/demand based on other cues (e.g., the number of tokens provided to be picked up at a station or the number of robots carrying tokens of either color). Robots use threshold-based decisions: if perceived demand exceeds the internal threshold, the robot commits to that task. Robots periodically reconsider and can switch tasks. Analyze the time-series plot of the number of robots per task over time. Study whether your system balances allocation and adapts to changes in task distribution.

Exercise 5.12 (*Design outdoor robots*)

(a) Design your own swarm robot for operations on water. Consider different options for locomotion (e.g., thrusters, sails). What are the minimal feasible sizes?
(b) Design your own swarm robot for operation in the air. Consider different options of multi-rotors, gliders, and blimps.
(c) Try to face the challenge and design your own swarm robot for operation in off-road conditions. Choose whether you want to go for a wheeled, legged, or tracked option. What sensors do you need to ensure your robot doesn't make stupid moves and gets stuck frequently?

Chapter 6
Modern Approaches to Robotics for Swarms

They were many and they were one.

—Adrian Tchaikovsky, Children of Time

The swarm actually solved the problem they had set for it.

—Michael Crichton, Prey

HAL is [...] the almost human computer simulacrum who gains consciousness and agency [...] while the humans around him lose both.

—Robert Kolker, A Cinema of Loneliness

Abstract We review recent advances in swarm robotics, highlighting human– and animal–swarm interaction, decentralized programming, robustness, field applications, and emerging directions. This chapter reviews recent advances in swarm robotics, focusing on human–swarm and animal–swarm interaction, decentralized programming frameworks, and strategies for robustness. Frameworks such as ROS 2 and Buzz enable scalable programming, while n-version programming, fault detection, and blockchain improve reliability, fault tolerance, and security. Distributed SLAM (or C-SLAM) supports collaborative mapping, and new interfaces use swarms as displays or integrate augmented reality to enhance sensing. The move toward field robotics introduces swarms to unpredictable environments on land, in the air, and on water, presenting new challenges for robustness and autonomy. Future directions include morphological computation, where robot design influences behavior, and Green Robotics, which prioritizes energy harvesting for sustainability.

H. Hamann, *Swarm Robotics*,
https://doi.org/10.1007/978-3-032-10584-4_6

6.1 Introduction

6.1.1 Learning Objectives

After reading this chapter, you should be able to:

- Understand human–swarm interaction, including semi-autonomy, control modalities, and cognitive load.
- Understand important software frameworks and programming paradigms, including ROS 2, Buzz, shared-memory architectures (e.g., SwarmMesh), and onboard computation for scalable, decentralized behaviors.
- Comprehend distributed localization and mapping, especially the principles and challenges of multi-robot SLAM.
- Design for robustness, fault management, and security through redundancy, fault detection, and blockchain-based resilience against failures or adversarial actions.
- Explore swarm robots as interfaces and the role of augmented reality in enhancing sensing and actuation.
- Recognize the shift toward field robotics and the challenges of deploying swarms in real-world waterborne, airborne, and groundborne applications.
- Investigate computation-free paradigms such as morphological computation, and understand the principles of Green Robotics with energy harvesting and sustainable long-term operation.
- Analyze the complexities of coordinating multiple swarms with distinct objectives or owners in shared environments.

6.1.2 Guiding Questions

- What are the key challenges in human–swarm interaction, including semi-autonomy, cognitive load, and control granularity?
- How do frameworks like ROS 2 and languages such as Buzz enable scalable, decentralized swarm behaviors, and what is the role of advanced architectures (e.g., SwarmMesh)?
- What are the principles and challenges of distributed SLAM, and how can multiple robots collaboratively map unknown environments under limited communication and resources?
- Beyond hardware redundancy, what defines true robustness in swarm robotics, and which techniques (e.g., n-version programming, model-based or peer-based fault detection, fault tolerance) are most effective?
- How can blockchain enhance cybersecurity, coordination, and accountability in decentralized swarms, and what challenges arise regarding resources, scalability, and network dynamics?

- In what ways can swarms act as human–robot interfaces, and how can augmented reality extend the sensing and actuation of simple robots?
- What motivates the deployment of swarms in real-world environments, and what unique challenges arise in waterborne, airborne, and groundborne field robotics?
- How can complex swarm behaviors emerge without explicit algorithms, through morphological computation based on robot shape, materials, or physical interactions?
- What are the principles of Green Robotics for swarm systems, and how can strategies like energy harvesting or robot repurposing support sustainability?
- What new challenges emerge in coordinating multiple distinct swarms with different goals or owners, and how does this reshape interaction and resource management?

6.2 Human–Swarm Interaction and Animal–Swarm Interaction

For many real-world applications, it is rather the roboticist's optimistic imagination that their robots could operate in a fully automated environment without any contact with humans or other animals. Instead, it is most likely that, in the future, robots will interact with humans in many application scenarios. Although we often focus solely on achieving (semi-)autonomy during the robot's mission itself, it is equally important to recognize that human involvement is typically also required during the deployment and recovery phases. On the one hand, robots are increasingly entering spaces inhabited by humans due to technological advances in robotics that extend beyond vacuum cleaning robots, lawn mowing robots, and pool cleaning robots. Once robots can navigate unstructured and dynamic environments, they will also coexist in spaces populated by humans. Robots may assist humans in surgeries, in physically demanding tasks, or in hazardous environments (e.g., bridge inspections, sewer repairs). On the other hand, automation may reach its limits in many domains, necessitating the need for robots to collaborate with humans to achieve optimal performance. Human–robot collaboration may often optimize workflows, reduce downtime, and increase productivity. Even in fully automated factories, the human may remain in the loop [78, 796].

Robots may also share spaces with animals, for example, for environmental monitoring and conservation, or to improve animal welfare. In a bio-hybrid system (also known as mixed society), we integrate living organisms (e.g., animals, natural plants) with technological components, creating a functional unit where biological and technological elements interact, cooperate, and enhance each other's capabilities [680]. The idea is to leverage the best of both worlds to achieve goals that neither could accomplish independently. Example implementations of bio-hybrid systems combining mobile robots with living organisms are cockroaches [121, 316, 727], chicken [297], rats [730], fish [94, 132, 462], and natural plants [118, 333, 839]. Bio-hybrid systems may be able to stabilize ecosystems, improve species resilience, and address global environmental challenges in the future [374].

6.2.1 Human–Swarm Interaction

What is generally known as human–robot interaction (with origins in human–computer interaction) is called human–swarm interaction (HSI) in the context of swarm robotics. This naming convention is likely intended to emphasize that it presents a distinct challenge to have a human operator manage a swarm of robots as opposed to just one. The field of HSI is experiencing significant growth and garnering increasing attention. From the perspective of robotics, allowing for human–robot interaction represents a qualitative shift in concepts. Typically, the primary focus of robotics is automation and autonomy. Introducing a human operator means we need a concept of semi-autonomy, that is, a robot that is neither fully autonomous nor fully radio-controlled by a human. Many questions arise immediately. Should human operators control all robots, a few, or just one? Is the operator controlling the whole swarm or only guiding it? How many robots can a human possibly handle? Should the operator be able to switch control from one robot to another? How should such handovers be organized to ensure safety and security? Should we consider scenarios with several human operators controlling a swarm? Figure 6.1 provides an overview of relevant aspects in human–swarm interaction and Fig. 6.2 shows an example implementation with TurtleBots in the lab.

The more common interface aspect is the question of how to create an interface that allows, for example, a human supervisor to efficiently interact with the robot swarm (another flavor of the human factor, cf. Sect. 1.6).

An overview is given in a survey paper by Kolling et al. [440]. They discuss essential aspects, such as remote interaction with the swarm, proximal interaction (supervisor is with the swarm, and the robots can perceive the supervisor), visualization, control, levels of automation, and organization in sub-swarms. Naturally, also for HSI, system size scalability is a challenge for both $M = 1$ human (partially) controlling $N \gg 1$ robots and for $M \gg 1$ humans (partially) controlling $N \gg 1$ robots.

The future implications of human–robot collaboration in various industries will require human-centric approaches in designing robotic systems [861]. The integration of advanced robotics is transforming workplaces by enabling closer human–robot collaboration [678]. This may lead to enhanced capabilities and increased productivity, especially in large-scale systems with numerous coordinated robots. Still, there are also challenges, such as the psychosocial impacts and diminished worker autonomy. Human-centric robotic integration should involve workers as co-creators of technology and supporting ethical design practices.

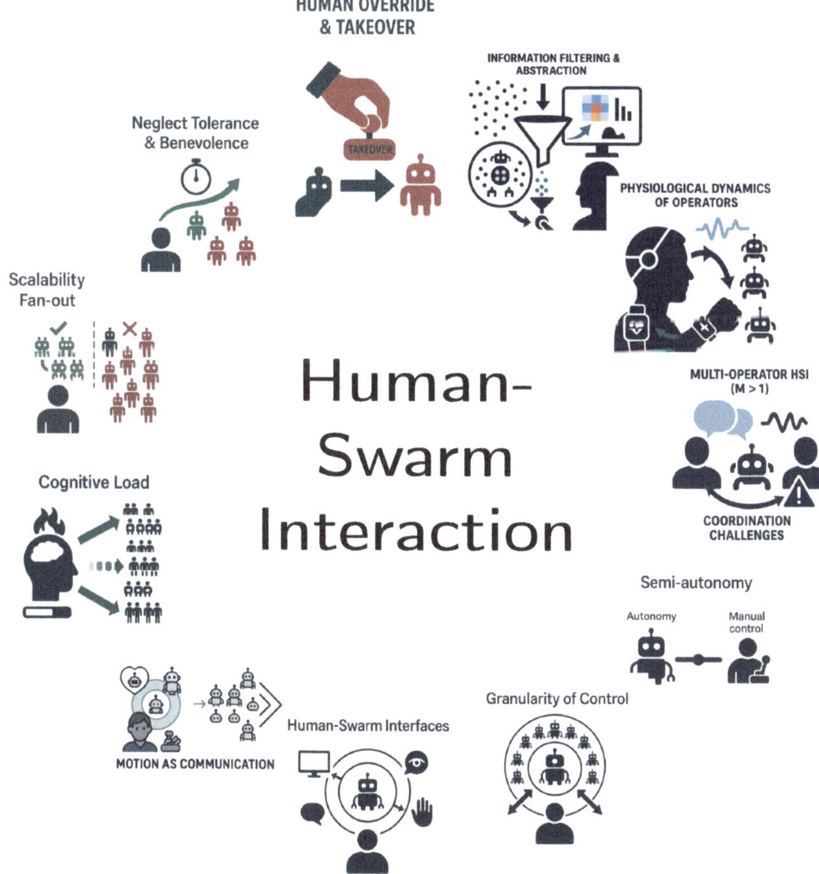

Fig. 6.1 Eleven important aspects of human–swarm interaction

Human–Swarm Interface and Modalities of Interfaces

The human operator controls and monitors the swarm using an interface (e.g., remote or proximal interaction). We face the challenge of mediating the distributed nature of swarms with centralized human control and information collection. While individual robots may enable global communication with the operator, or commands can be broadcast to the entire swarm, a gateway robot often facilitates communication to the swarm. Key practical challenges include managing latency, bandwidth, and asynchrony in these interactions [440].

A centralized user interface provides the operator with swarm state information. This interface may have limitations in communication, and it needs to inform the operator about the state of each robot or a form of holistic swarm-level state [440].

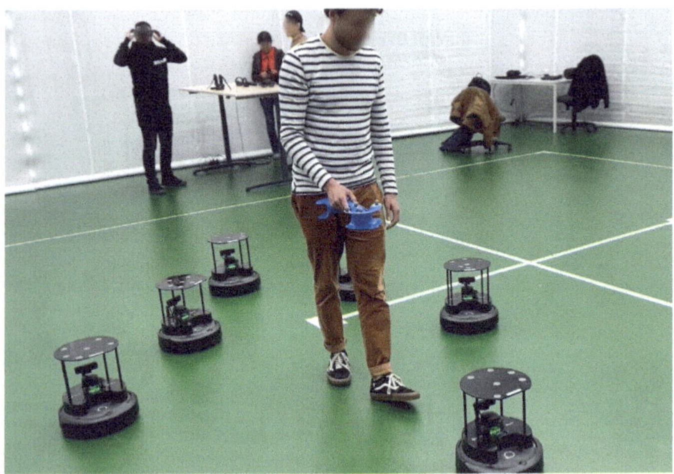

Fig. 6.2 Implementation of human–swarm interaction in the lab with six TurtleBot 4 robots (work by Julian Kaduk, Till Aust, Jonas Kuckling, and Genki Miyauchi within the EU-funded project *ChronoPilot*)

When the operator sends a command, it must be communicated to a specific robot, to all robots, or to a group of robots. If there is no global bidirectional link to each robot, this message must be routed through the swarm using multi-hop messaging. A challenge can be that the swarm breaks into several connected components, such that there is not always a communication path from any robot to any other robot. For a flocking behavior, for example, it may be an option to control only one robot as the leader of the swarm. This simplifies the communication requirements as we only need a global link to one robot. However, to ensure the robustness of the approach, we would need a method to deal with failures of the leader robot. Additionally, there is the concern that an adversary could identify the leader robot through observation. Zheng et al. [897] propose a method to shape the leader's motion pattern to hide its identity. This raises the challenge of ensuring that a human operator can still focus and follow the leader robot, if required, in a human–swarm interaction scenario.

When designing the actual interface, one must decide on the key modality. Options include visual, auditory/speech, or tactile/touch, with technological options such as display screens, virtual reality, augmented reality, gesture detection, and wearables. Some of these modalities also relate to a defined direction of communication: human to swarm (e.g., gestures, voice commands), swarm to human (e.g., lights, augmented reality), or bidirectional (e.g., human operators adjusting commands based on real-time swarm feedback). Other, less obvious modalities include cue-based communication. For example, the robot swarm signals its intentions or states through motion-based cues (e.g., collective motion) or spatial cues (e.g., the aggregation of robots). Faster response times can be achieved with a multi-modal approach [312].

Fang et al. [238] propose a swarm control method using eye-tracking technologies. Based on eye gazing and eye tracking, the operator's eye movements are used to influence robot behavior. The approach is validated through simulations and experiments using augmented reality (Microsoft HoloLens 2) and TurtleBot3 Burger robots, demonstrating its feasibility.

Ferro et al. [243] explore a haptic shared control teleoperation system designed for the precise navigation of multiple untethered magnetic microrobots in a cluttered 3D environment. Using a combination of autonomous control algorithms and haptic feedback provided through a kinesthetic interface, the system enables a human operator to guide microrobots effectively. The study demonstrates the system's scalability for controlling larger microrobot formations.

There are several ways of proximal interaction, such as using gestures to control a robot swarm [572, 589]. Kim et al. [435] study how non-expert users intuitively interact with tabletop robot swarms using gestures, touch, and verbal commands. Experiments involving up to $N = 20$ robots demonstrated that group size and proximity have a significant influence on interaction styles. Gestures were the most common control method, with larger groups prompting two-handed gestures, while smaller groups favored single-hand controls. Proximity affected precision and modality choice. Kim et al. [435] emphasize the need for adaptive interaction vocabularies that adjust to swarm size. Pourmehr et al. [647] use face engagement and voice commands to control groups of robots. Maria et al. [515] try to go beyond limitations of gestures by adding "roleplaying metaphors as a scaffold for designing richer interactions."

A specific way of communicating from swarm to human is by motion-based cues. An example for expressive motion to convey information in swarms is the approach by Santos and Egerstedt [695]. They study how a swarm of robots can express fundamental emotions, such as happiness, anger, or sadness, through collective motion. Inspired by social psychology, they designed swarm behaviors to embody specific emotions and tested their effectiveness in a user study. Results showed that participants could reliably identify the intended emotions for most behaviors. Santos and Egerstedt [695] implemented these behaviors on real robots, demonstrating the feasibility of their approach.

Kaduk et al. [398] study emotionally expressive behaviors in a leader–follower robot scenario. In contrast to Santos and Egerstedt [695], Kaduk et al. [398] focused on the interactional dynamics. Using a follower robot that displayed behaviors corresponding to anger, happiness, sadness, and a neutral baseline, the study assessed how these emotional motions influenced the attention, stress, and excitement of human participants controlling the leader robot. Expressive behaviors shifted attention toward the follower.

Cognitive Load and Scalability

The task of a human operator interacting with a robot swarm entails a specific cognitive load. Cognitive load is a measure of the cognitive resources used, often also applied as a reference for task difficulty. It describes the mental effort required to process information in working memory, which is a limited-capacity system. It is clear that with increasing swarm size N, any human operator reaches their personal limit of how many robots they can manage. The HSI system is, hence, constrained by the human operator's cognitive capabilities, decision-making speed, fatigue, situational awareness, and performance under stressful conditions [368]. The system performance for increasing swarm size N may saturate or even drop when human operators are overwhelmed by the task [368, 486]. This is also similar to the swarm performance of fully autonomous robot swarms without a human operator (see Chap. 2). The cognitive complexity of dealing with a swarm of a given size, as well as the scalability of HSI, has been discussed frequently in the literature [312, 440, 626, 647]. It is also known that scaling up in swarm size can have a psychological effect on the operator when in proximity. For example, even without active interaction between a human and the robot swarm, the mere presence of multiple robots affects the human being in proximity [642]. Similarly, Kaduk et al. [396] study the influence of swarm size N and robot speed on an operator in proximity but in an active supervision task involving up to $N = 15$ robots. They also introduce the concept of subjective time perception to human–swarm interaction. They find that larger swarms increase cognitive load, leading to perceived compression of time and altered flow states.

Olsen Jr and Wood [605] define the fan-out measure, a metric for HSI to quantify the swarm size N that $M = 1$ human operator can effectively manage. The underlying idea is that in an efficient HSI approach, robots should have high neglect tolerance, meaning they can function effectively without supervision for an extended period (i.e., high autonomy). The concept of neglect tolerance measures how well an autonomous system maintains its performance when the human operator is not paying attention, hence, reflecting how slowly performance degrades during neglect [161]. Also, we require short interaction times, that is, the duration an operator spends actively supervising a robot (i.e., efficient interface). To compare competing HSI approaches, Olsen Jr and Wood [605] propose to measure and calculate the swarm size $N_{\text{fan-out}}$ that a human can control, defined as

$$N_{\text{fan-out}} = \frac{T_{\text{activity}}}{T_{\text{interact}}} , \tag{6.1}$$

for activity time T_{activity} and interaction time T_{interact}. Activity time T_{activity} is measured as the time between the operator entering a command to a robot until that robot stops doing valuable work. Olsen Jr and Wood [605] argue that interaction time T_{interact} cannot be measured as it would require reading the human's mind. In an empirical approach, the scenario should therefore be chosen such that interaction time T_{interact} is constant.

Lewis et al. [485] discuss a formal approach to scaling properties of task difficulty for the human operator of the swarm with increasing swarm size N in Big $\mathcal{O}$ notation. Verifying actions of robots is $\mathcal{O}(N)$ and assigning N robots to K tasks is $\mathcal{O}(KN)$ [277]. Also, Humann and Pollard [368] explicitly discuss scalability in HSI. They summarize empirical studies and simulations, finding that a single operator can typically manage $N \in \{4, 5, \dots, 8\}$ robots, with scalability constrained by human cognitive and physical limitations. Humann and Pollard [368] find that

> [t]o remain insensitive to scale, control methods do not allow users to address robots individually. Instead, control can take the form of simulated leaders or predators that influence the swarm [...]

Pourmehr et al. [647] propose grouping and gesture control for HSI. The somewhat counterintuitive idea of "predators" by Pendleton and Goodrich [626] describes human-controlled robots that "repel swarm members within their radius of influence rather than attracting them."

Paas et al. [613] study the impact of control strategies and interfaces on the human operator's cognitive load. An intense study with 40 participants and $N = 6$ micro-UAVS (Crazyflies) compared a tablet-based interface and a tangible robot interface (six Zooid robots) using individual waypoint control and group self deployment. The tangible interface led to higher involvement but mixed outcomes on cognitive load. Group control generally reduced the workload compared to individual control.

Kaduk et al. [397] introduce active (robots that exhibit movement and behavior) and passive robots (robots that remain stationary). Maintaining a constant total number $N_{\text{const}} = N_a + N_p$ of active (N_a) and passive (N_p) robots allows the study to isolate the effects of robot activity by ensuring stable visual complexity, controlled task difficulty, motion salience, and attention allocation. A possibility to design HSI systems is to adaptively activate or deactivate robots in the swarm in response to human needs. They study how varying the number of active robots in an $N = 15$ robot swarm affects human cognitive and perceptual processes. They find that larger active sub-swarms accelerate time perception and increase flow experience, but also heighten perceived task difficulty and emotional arousal.

Granularity of Control, the Semi-autonomy Challenge, and Human Override

The granularity of control in HSI defines how many details of the robot swarm a human operator can influence. It defines whether the operator can control individual robots, groups of robots, or the entire swarm. From the robot's perspective, it means that robots need to switch between different modes of autonomy: from full autonomy to a remote-controlled state, or forms of semi-autonomy when the operator only defines goals or constraints. The granularity of control needs to be carefully selected, as the operator may be overwhelmed by excessive micromanagement of individual robots or feel left out if the swarm operates autonomously for most of the time. Additionally, the distribution of responsibilities between the human operator and the robots must be clearly defined. Harriott et al. [338] define HSI metrics, for example, the critical distinction between micro-level movements and macro-level movements. Inspired by Bjurling et al. [82], we can introduce several levels of control: individual robot, subswarm or group, swarm, $M = 1$ operator and swarm, and $M > 1$ operator and swarm. In a micro-level control approach, the operator directly controls individual robots within the swarm; however, this approach cannot possibly scale. In macro-level control, the operator provides high-level commands to the swarm as a whole (e.g., "move to this region" or "form a line"). In that case, the robots execute many actions autonomously, and the operator may feel "out-of-the-loop," which could cause boredom and reduce the operator's attention (see Sect. 6.2.1). In a mixed control approach, the operator can switch dynamically between macro- and micro-level controls. These dynamic handovers (from macro to micro or within micro from robot to robot) require well-defined switches in autonomy levels as well as efficient and intuitive interfaces.

An important feature is the human override. It refers to the ability of a human operator to take direct control of a robot's functions, overriding its autonomous systems. Human commands supersede the robot's decision-making algorithms. Human override is crucial in safety-critical applications where autonomous actions may lead to undesirable or dangerous outcomes.

Another aspect is to organize a defined transition procedure for both the human operator and the robot. Humans need to be informed at what exact moment they are supposed to take control of a robot, a group, or a swarm. Each robot needs to be notified as well when it is supposed to go back to full autonomy mode. Challenges similar to those in autonomous driving arise. For example, "what is the appropriate lead time for a safe transition from an automatic mode to a human takeover?" [38].

Physiological Dynamics of Human Operator

Measuring physiological features of the human operator (e.g., heart rate variability, skin conductance, skin temperature, eye tracking, and electroencephalography) can help gain insights into the effects of the swarm on the human. It can also be used to improve the HSI system. Analyzing physiological data can help to control better the cognitive load (neither overwhelmed nor underutilized) and improve decision-making efficiency. Monitoring physiological responses in real time can provide information to adapt the HSI system online. Given that there are significant differences between individual human operators, physiological data can help calibrate and adapt the system to these individual differences.

The above-mentioned paper by Podevijn et al. [642] investigates the effects of swarm size N on the psychophysiological state of human operators in a passive interaction context. Physiological measures, such as heart rate and skin conductance, in combination with self-reported metrics, indicate that larger robot swarms provoke stronger psychophysiological responses (e.g., increased arousal levels).

Villani et al. [827] propose using a wrist-worn device, such as a smartwatch, to let the human operator intuitively control robot velocities through wrist movements. Heart rate variability is monitored to estimate mental fatigue. When the user's cognitive load becomes excessive, the system adapts the robots' behavior to simplify interaction. Virtual reality tests show that this approach reduces mental fatigue and workload.

Distefano et al. [191] study physiological data of the human operator, such as electroencephalography (EEG) and eye tracking. They also study how swarm feedback of simulated robots affects human cognitive factors, such as cognitive load and mental engagement. The human operator has a supervisory role but must rely on swarm feedback to assess the feasibility of actions, even though they may not have full awareness of the environment. They study different feedback levels (varying amounts of information provided to the user) and different compliance levels (randomly denying the user's actions 50% to 90% of the time). The results reveal individual differences among operators that influence interaction outcomes, with expert users more effectively utilizing feedback compared to novices.

Mondada et al. [577] propose to use EEG signals, specifically preparing the sci-fi dream of a brain-controlled robot swarm. They exploit the steady-state visually evoked potential (SSVEP), a consistent neural response triggered by a visual stimulus blinking at a specific frequency. Each robot has LEDs blinking at a unique frequency. When a human operator looks at a particular robot, their SSVEP response can be extracted from EEG signals to identify and select that robot. No deliberate action is required from the user.

Zhu et al. [902] propose a research platform for HSI that supports diverse interaction modalities, including brain–computer interfaces (BCI), electromyographic (EMG) interfaces, eye tracking, and multitouch screens for real-time visualization. The system integrates flexible hardware and software, allowing researchers to test and develop HSI methods.

Filtering Information for the Human Operator

For HSI, we need to communicate the current state of the robot swarm to the human operator. This is potentially a two-step process of, first, estimating the swarm state (if required, entirety of N robot poses and their sensor input) and, second, processing that information such that it can be communicated to and understood by the operator.

There is a lot of research on swarm state estimation [86, 528], for example, based on ultra-wide band [877, 878], LiDAR-inertial measurements [900, 901], and acoustic [431]. The entire swarm state may be highly complex and impossible for the human operator to process in real time. We need ways to channel and filter the large amount of information into the human operator's mind.

An option, of course, is to restrict the operator's actions to macro-level control, providing only high-level commands. This addresses the operator's active part of their job. However, we also need to consider the passive aspect, namely, how to display and convey information to the operator. The entire swarm state can be broken down by aggregation and abstraction of information. Robots can be grouped in logical clusters, and each cluster can be summarized by specific metrics and statistical summaries (e.g., average sensor reading). One can also only display swarm densities or other forms of heat maps. An alternative approach can be event-based. Saliency-driven filtering can select only the most relevant data. For example, outlier detection could help channel the operator's attention to anomalies (e.g., broken robots). The operator could be notified when a subgroup of robots has achieved a subgoal (e.g., arriving at a waypoint). These options aim to reduce cognitive load while maintaining situational awareness.

Divband Soorati et al. [194] propose a scalable interface, where operators give high-level commands instead of micromanaging individual robots. A simulation study with 100 participants found heat maps to be more effective for swarms of size $N = 15$, especially in time-critical scenarios. The interface enables operators to dynamically influence swarm behavior through belief maps, eliminating the need for constant direct communication.

Kapellmann-Zafra et al. [405] study HSI when information for the human operator is heavily constrained, resulting in limited situational awareness. They focus on distributed control with the operator interacting with the swarm using only local information from the ego-perspective of a robot. They also consider the "neglect" of robots for periods when the swarm's autonomous dynamics are allowed to progress undisturbed. This refers to the above-introduced concept of neglect tolerance [161]. Nagavalli et al. [588] introduce "neglect benevolence" that extends this idea by pointing to situations where the system's performance may actually improve when the operator is not actively supervising it (for a limited period of time). A key finding by Kapellmann-Zafra et al. [405] is the importance of training and the concept of neglect benevolence, where allowing the swarm to self-organize before intervention increased HSI efficiency.

$M > 1$ Human Operators

Patel and Pinciroli [621] explain that a group of human operators ($M > 1$) introduces additional challenges among the human group, such as ineffective group dynamics, unbalanced workload, and inhomogeneous awareness. They point to the out-of-the-loop performance problem [231] that is indicated by reduced situational awareness, degraded decision-making ability, slower response times, skill degradation, and over-reliance on automation. It is "caused by a lack of engagement in the task, awareness of its state, and trust in the system and other operators" [621]. Endsley and Kiris [231] see granularity of control (see Sect. 6.2.1) and the degrees of freedom afforded to human operators as key for the out-of-the-loop problem. Micro-level control increases interaction and trust, but at the cost of high cognitive load. Macro-level control reduces cognitive load but at the danger of boredom and reduced situational awareness.

Patel et al. [622] focus on transparency, inter-human communication, and information loss in a collective transport task using a web-based simulator. User studies with 48 participants reveal that combining local and global information (mixed transparency) and supporting both direct and indirect communication (mixed communication) lead to better performance, trust, and awareness. The results also show that multiple operators are needed to handle the cognitive load of large swarms, and that uneven information loss across operators degrades coordination and trust.

Miyauchi et al. [563] introduce the concept of two HSI systems interacting. In total, there are $M = 2$ operators and N robots. The two operators A and B operate a group of robots each by sharing the robots $N = N_a + N_b$. N_a robots are assigned to operator A at a given time and N_b robots to operator B. An operator can request additional robots from the other operator to adapt to task requirements. The study compares two types of communication: direct verbal communication and indirect message-based interaction. Their findings emphasize the importance of well-designed communication interfaces in multi-human ($M > 1$) HSI setups.

6.2.2 *Animal–Swarm Interaction: Mixed Societies, Bio-Hybrid Systems, and Cyborgs*

Mixed societies and bio-hybrid systems in the context of swarm robotics are systems of interacting robots and living organisms. While this concept may initially appear daunting and evoke comparisons to science fiction, such as cyborgs in fiction, these systems have the potential to offer significant practical benefits. For example, integrating robots to control insect swarms could create new opportunities for pest management. Controlling an insect swarm opens up completely new options for pest control. Also, a symbiosis between robots and organisms is possible to leverage the strengths of both robots and living organisms within a single integrated system.

For an overview, Papadopoulou et al. [617] provide a survey of many bio-hybrid systems, including robots interacting with various animals such as lizards, birds, bees, fish, crabs, cockroaches, and rodents. These examples demonstrate how robotic agents can be designed to integrate into animal groups, facilitating the study of natural collective behaviors and providing insights applicable to the design of artificial swarms.

The probably most prominent example is the work by Caprari et al. [121] and Halloy et al. [316] on a mixed society of cockroaches and robots. In the project called "Leurre" they developed a small, simple mobile robot called "Insbot" that was able to usefully interact with cockroaches (see Fig. 6.3). Although one may think that it must be hard for such a robot to be accepted by the cockroaches as a peer, Halloy et al. [316] succeeded in doing so. It seems to require not much, as the robot was just clad in a sort of blotting paper that was soaked with cockroach odors.

Fig. 6.3 A bio-hybrid system of cockroaches and an "Insbot" robot [727]

The task for the robots within this mixed society was to influence the cockroaches socially. The society was put in an arena with two shelters of different sizes. Under normal circumstances, the cockroaches would take shelter that allows all cockroaches to just fit beneath it. However, the robots were programmed to try to convince the cockroaches that aggregation beneath the other shelter is more desirable, and they succeeded. A similar study was done by da Silva Guerra et al. [170] with robots and crickets.

There is a growing field of study examining cyborgs, which is short for cybernetic organisms. They have organic and biomechatronic body parts [148]. A particularly relevant concept for swarm robotics is the use of living insects augmented with electronic components to enable remote or autonomous control [698]. Recently, this has been pushed toward swarms of cyborgs. Bai et al. [45] present a novel approach to swarm navigation using cyborg insects, which are real insects equipped with electronic backpacks for remote control. The authors develop and experimentally validate a strategy to guide a swarm of cyborg cockroaches through unknown terrain. The algorithm alternates between two rules: a free motion mode when local density is high, and a move-toward-crowd rule when agents are isolated. This strategy leverages the insects' natural locomotion abilities, reduces electrical stimulation, and avoids habituation effects. Experiments with $N = 20$ cyborgs demonstrate successful navigation. Note that there is a commercial initiative in this direction. *SWARM Biotactics* is a company founded in 2024 that develops fully controllable cyborg systems based on living insects equipped with sensor backpacks and neural interfaces. Their goal is to scale such cyborg swarms from laboratory prototypes to operational deployments for defense, disaster response, and security applications.[1]

In the "Chicken Robot" project Gribovskiy et al. [297] developed a robot that can interact with young chickens (see Fig. 6.4). The robot makes use of the relatively popular, so-called filial imprinting: "shortly after hatching, we present the moving and vocalizing robot to the chicks, and they learn that the robot is the mother" [297].

Another research project that was fully focused on bio-hybrid systems was ASSISI$|_{\text{bf}}$ (see Fig. 6.5). The researchers investigated two different setups, one with multiple stationary robots and honeybees [516] and another one with a mobile robot and zebra fish [94]. The idea is to establish a form of communication between the robots and the animals (hence the project's name). Also, the robots are supposed to influence the animals in possibly user-defined ways.

In the project *flora robotica*, a bio-hybrid system of robots and natural plants is formed [245, 333, 334]. The robots are stationary and control the directional growth of the plants by the blue light of bright LEDs (see Fig. 6.6). The idea is to use this concept in applications of architecture. For example, the bio-hybrid *flora robotica* system could autonomously grow green walls, roofs, or maybe in the future even houses.

[1] https://www.swarm-biotactics.com.

Fig. 6.4 "Chicken Robot" project Gribovskiy et al. [297] (photo by José Halloy, with permission)

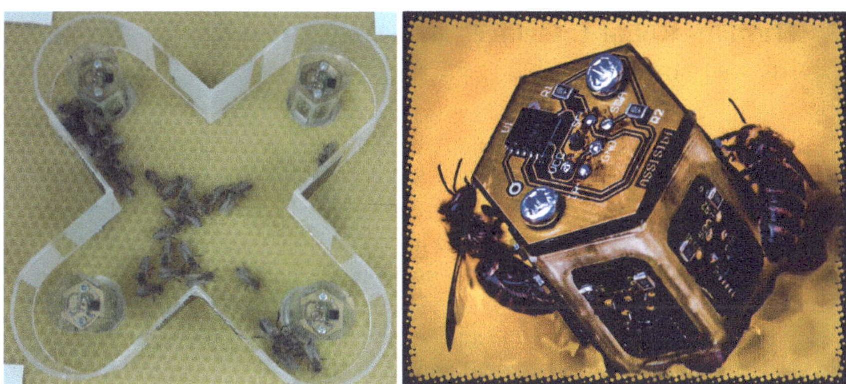

Fig. 6.5 A bio-hybrid system of interacting robots and honeybees in the project $\text{ASSISI}|_{\text{bf}}$. FER developed the robots, University of Zagreb (photos by Thomas Schmickl, Artificial Life Lab, Karl-Franzens University of Graz)

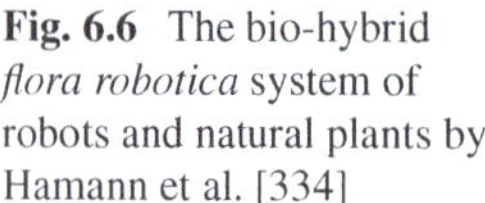

Fig. 6.6 The bio-hybrid *flora robotica* system of robots and natural plants by Hamann et al. [334]

6.3 Human–Robot Interfaces and Augmented Reality for Swarm Robots

Different from human–swarm interaction (see Sect. 6.2.1), systems of swarm robotics can also be used as an interface themselves. Similar to ideas of haptic displays [845] or programmable matter [134], swarm robots can be used to provide physical feedback to a human user, thereby implementing an interface. Robots can position themselves to communicate information or can provide haptic feedback to humans.

A related concept is to enrich the environment and the sensing capabilities of simple swarm robots to increase the number of opportunities in experiment design. This is essentially the concept of applying methods of augmented reality [843] to robots, rather than humans.

6.3.1 Human–Robot Interfaces: Swarm Robots as Display and Input Device

An idea for swarm robotics that relates to the concept of "robotic material" [533] is "ubiquitous robotic interfaces" [433]. The robot swarm displays information to users (e.g., collective motion to point to objects or to operate as a timer) but can also manipulate (e.g., pushing objects around) and sense (e.g., their location and surrounding objects). A new aspect is user interaction that allows the user to move around a robot and use it as an input device. Kim and Follmer [433] showcase their approach with a swarm of ten small differential-drive robots (diameter 26 mm). The swarm is capable of exhibiting standard behaviors, including aggregation, dispersion, circling, random motion, and flocking. Localization is implemented as a central system based on a DLP projector that produces a gray code pattern. Robot swarms as a robotic interface may have a future, but they need to be decentralized, probably even smaller, and reliable to be of use.

Kim and Follmer [434] introduce a haptic display system that uses a robot swarm to deliver tactile feedback through coordinated movement and force patterns. They explore the potential of swarm robotics for applications, such as haptic notifications, directional cues, and remote social touch. The authors present a design space for haptic interactions, investigating parameters and features, such as force type, frequency, and robot coordination. In a user study, they evaluate human perception of haptic stimuli.

A related effort is *MOSAIX*, a platform developed by Alhafnawi et al. [9]. The system consists of up to 100 small, brick-like robotic "tiles" that autonomously move and interact to support social tasks, such as brainstorming, opinion-mixing, and idea visualization in public spaces. In a real-world deployment, Alhafnawi et al. [10] utilized 63 MOSAIX tiles at a science museum, where approximately 300 visitors treated the tiles like smart sticky notes. Participants shared their ideas, and the swarm aggregated and displayed them in evolving visual patterns.

6.3.2 Augmented Reality for Swarm Robots

Augmented Reality (AR) is often used together with (single) robots [764]. A more radical concept is to use AR for robot swarms, that is, augmenting the capabilities of the robots, not humans. Swarm robotic platforms are typically designed to be as inexpensive and straightforward as possible, allowing for scaling experiments to many individuals. This simplicity, however, imposes substantial limitations on sensing and actuation. For example, Kilobots [685] can only perceive ambient light and exchange short-range infrared messages, but they cannot detect obstacles, gradients, resources, or chemical-like signals, such as pheromones. These restrictions complicate the study of collective behaviors in complex or realistic environments in the lab using overly simple swarm robots. Augmented reality systems, such as *Kilogrid*, *ARK*, and *LARS*, extend the limited sensing and actuation of simple swarm robots by embedding them in enriched virtual environments. In the following, we present these platforms to illustrate how different technical approaches based on distributed infrared communication, centralized camera tracking, and projected light fields enable the study of increasingly sophisticated swarm behaviors.

AR systems address this limitation by artificially enriching the robots' environment with virtual features. The robots continue to operate physically, but their sensors are supplemented with additional information delivered by the AR infrastructure. As a result, robots may "sense" virtual obstacles, "follow" virtual gradients, or "deposit" virtual pheromones, even though these features do not exist physically. This approach enables researchers to design and test experiments that go well beyond the hardware's native capabilities, making it possible to study more sophisticated scenarios, such as collective navigation, foraging, or more involved variants of collectivity decision-making, while preserving the advantages of using real robots with their inherent noise and embodiment instead of using simulators.

Valentini et al. [807] introduce the *Kilogrid*, an open-source augmented reality environment designed to extend the experimental capabilities of Kilobot swarms [685]. The Kilogrid consists of a grid of communication modules that interact locally with Kilobots via infrared, allowing them to sense and act upon virtual features, such as obstacles, resources, pheromone fields, or gradients that would otherwise be impossible to represent physically. This distributed design enables scalable and situated interactions, real-time data logging, and automated experiment control. The authors demonstrate the system through several case studies (e.g., site selection, obstacle avoidance, virtual pheromone-based foraging), showcasing the capabilities of the Kilogrid.

Kedia and Rao [411] propose *GenGrid*, an open-source, low-cost experimental platform for swarm robotics that enables stigmergic communication through light signals and magnetic sensing. Robots can interact bidirectionally with the grid cells to simulate pheromone deposition, evaporation, and environmental cues. These ways of experimentation, such as collective transport and shepherding, are supported. In contrast, the Kilogrid by Valentini et al. [807] was explicitly designed for Kilobots and relies on infrared-based communication to provide virtual features and centralized data logging. GenGrid emphasizes general compatibility, modular scalability, and direct support for bio-inspired pheromone communication.

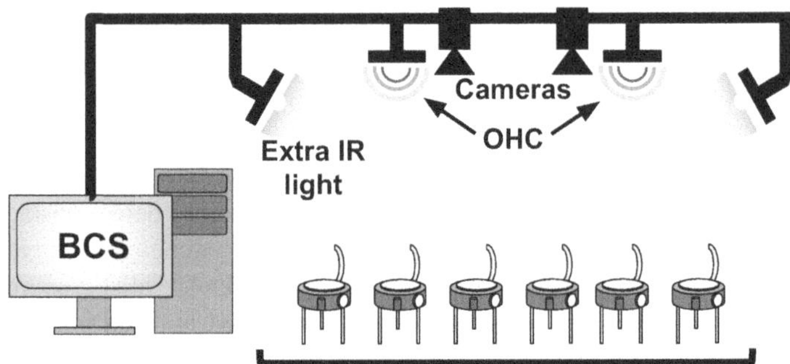

Fig. 6.7 The Augmented Reality for Kilobots (ARK) system provides a flexible and scalable augmented reality environment for swarm robotics. Its architecture consists of three main components: the Base Control Software (BCS), an open-source tool to manage and edit all system components; communication handled mainly through a modified Overhead Controller (OHC) with infrared transmitters for individual robot messaging; and tracking via several overhead cameras that continuously localize the Kilobots [635]

Reina et al. [670] introduce the *ARK* system that enhances the capabilities of Kilobots by embedding them in a virtual environment. Kilobots were widely used for swarm robotics research, but their sensing and actuation are minimal. The ARK system overcomes this by using overhead cameras for tracking, infrared communication for individual messaging, and a central control station to simulate virtual environments. This enables each robot to access location- and state-specific information, such as gradients, reference directions, or virtual pheromones, and to modify the environment accordingly. The authors demonstrate ARK in three experiments: automatic ID assignment, automatic robot positioning, and a foraging task in which $N = 50$ Kilobots interact with a virtual resource field. Compared to similar systems, such as Kilogrid, ARK is more cost-effective and flexible, allowing for large-scale experiments involving hundreds of robots (Fig. 6.7).

Raoufi et al. [664] introduce *LARS* (light augmented reality system), an open-source and cost-effective platform that uses light projections to enrich swarm robotics experiments. Instead of altering robot hardware, LARS augments the environment by projecting virtual objects, noise, or pheromone-like cues directly into the physical arena. Robots can sense and react to these projected patterns, enabling indirect communication through stigmergy as well as human–swarm interaction. The system supports different platforms (e.g., Kilobots, Thymio), provides real-time tracking

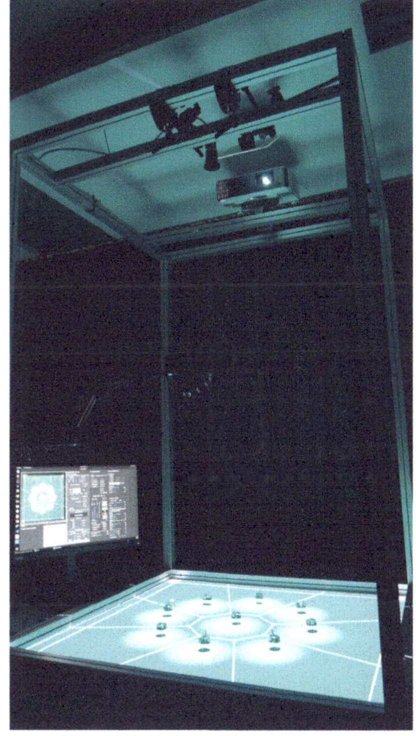

Fig. 6.8 The Light Augmented Reality System (LARS) setup. A projector adds virtual features to the arena, such as gradients, resources, or pheromone-like trails, while overhead cameras provide real-time tracking of the robots. By augmenting the environment rather than the robot hardware, LARS enables flexible swarm robotics experiments with realistic robot behavior in controllable virtual conditions (photo credit: Mohsen Raoufi, Humboldt University of Berlin)

without markers, and operates efficiently at high frame rates on a standard PC. By combining a realistic robot behavior with a controllable augmented environment, LARS helps bridge the gap between simulation and reality, making it a powerful tool for both research and education in swarm robotics (see Fig. 6.8).

6.4 Software Frameworks, Programming Approaches, and Computational Onboard Power

Designing and developing systems of swarm robotics is still primarily driven by research and rarely applied in an industrial environment. To simplify and streamline, for example, the software development for swarm robotics, it would be instrumental to have supporting tools. They are rare, but some exist. We discuss several approaches to provide special forms of infrastructure and tooling to robot swarms. Each of the following approaches is unique but has the potential to represent a category of efforts aimed at simplifying the programming of swarm robots and enabling them to be more efficient.

6.4.1 ROS 2: Software Architecture and Middleware for Mobile Robots

The development of mobile robotics and swarm robotics has historically been driven by individual research efforts, where researchers published their findings without a standardized approach to sharing and reusing software. This lack of interoperability often forces new researchers to reinvent the wheel, leading to inefficiencies in code development and hindering long-term progress. It became clear that a typical software architecture was needed to facilitate collaboration, code reuse, and scalability across different robotic platforms.

The Robot Operating System (ROS) emerged as a solution, providing a flexible, open-source middleware that enables modular robot development [657]. ROS was developed starting from 2007 by *Willow Garage* and released in 2010. ROS allows seamless communication between robotic components through a message-passing system; supports a wide range of debugging and visualization tools; and provides prebuilt capabilities, such as sensor drivers, motion planning algorithms, and control interfaces. Its growing global community of researchers, developers, and industry partners has transformed ROS into the de facto standard for robotic software development.

ROS simplifies the integration of sensors, actuators, and control systems, enabling developers to rapidly prototype and deploy robotic applications without being locked into a specific vendor. This has led to widespread adoption in academic research, industrial automation, and autonomous vehicles. ROS also supports hardware-agnostic development, allowing it to be used across diverse robotic platforms, from drones to industrial manipulators.

Recognizing the need for real-time performance, security, and multi-platform support, the ROS community introduced in 2017 the first distribution of ROS 2, a complete overhaul designed to support embedded systems, decentralized architectures, and swarm robotics [505]. Unlike its predecessor, ROS 2 replaces the centralized *roscore* architecture with the Data Distribution Service, a decentralized communication protocol that eliminates single points of failure. This shift makes ROS 2 particularly relevant for swarm robotics, where autonomous agents must operate without relying on centralized coordination. Additionally, ROS 2 expands compatibility beyond Linux to include Windows and macOS (however, support for these is still lacking). As interoperability and scalability become increasingly critical factors in robotics, ROS 2 could have some potential to support developers with a robust ecosystem. However, there are doubts that ROS 2 in its current form can actually scale to a relevant swarm size. For example, De Marchi and Bombieri [175] report in experiments with 80 ROS 2 nodes distributed across multiple devices, the default Standard Copy communication caused poor scalability. They propose ROS4K, which addresses this by automatically selecting optimized protocols for each node-to-node link. Relying on Wi-Fi in swarm robot experiments running ROS 2 is, however, generally not advised (for example, see Sect. 2.7.4).

ROS is used in many studies related to swarm robotics. For example, Walter et al. [844] use ROS to control autonomous firefighting multirotor UAVs. Meyer et al. [551] use ROS to control their underwater robot. In a first paper making use of ROS 2 in swarm robotics, Barciś et al. [56] used ROS 2 to implement the Swarmalator on a robot swarm (also see Sect. 5.2.2).

Kaiser et al. [400] introduce ROS2SWARM, an ROS 2 package designed for swarm robotics, providing a modular and reusable framework for implementing decentralized swarm behaviors such as aggregation, dispersion, and collective decision-making. The package supports multiple robot platforms (e.g., TurtleBot3 and Jackal UGV), ensuring hardware independence while facilitating the rapid prototyping of swarm behavior. Experiments demonstrate that the approach is practical in both simulation and real-world deployments.

Pichierri et al. [629] propose a modular framework based on ROS 2 for controlling and simulating swarms of Crazyflie nano-quadrotors, enabling distributed optimization, formation control, trajectory tracking, and pickup-and-delivery tasks.

Castillo-Sánchez et al. [129] report critical results in an experimental evaluation of the wireless communication performance of ROS 2. They analyze its latency, packet loss, and network congestion under different configurations. They found significant performance degradation as the number of robots increases due to inefficiencies in default ROS 2 settings for wireless swarm communication. They propose optimizations, such as multicast communication tuning and high-bitrate coding.

6.4.2 Buzz: Dedicated Programming Language for Swarms

Programming robot swarms is challenging, for example, because of the micro–macro problem (see Sect. 4.4). For swarm robotics, it is essential to simplify software development, especially addressing challenges that arise from the decentralized approach. For the broader adoption of swarm robotics in industry and other fields of application, it is necessary to facilitate ease of implementation, also for developers who lack experience in decentralized systems. A flexible, scalable, and expressive programming language tailored specifically for swarm robotics could be helpful.

Pinciroli and Beltrame [632] introduce Buzz, a domain-specific programming language designed for self-organizing heterogeneous robot swarms, offering both individual robot control and swarm-level coordination.[2] Buzz features a compositional programming approach, allowing developers to combine robot-wise and swarm-wise primitives to express complex behaviors. The programming language supports heterogeneous robot swarms. The run-time platform is designed for scalability and robustness, ensuring efficient execution even in large swarms with unreliable communication. Buzz integrates situated communication and virtual stigmergy for decentralized information sharing and coordination. Pinciroli and Beltrame [632] evaluate Buzz using swarm simulations and demonstrate that Buzz is effective in coordinating the motion of robots, in decision-making, and in swarm partitioning. The programming language is extensible, allowing developers to add new primitives and integrate with existing frameworks, such as ROS. Compared to other swarm programming approaches, Buzz strikes a balance between high-level swarm abstraction and low-level robot control, offering a flexible yet predictable programming model. Buzz has a C interface, which also helps integration with ROS. Buzz is a powerful tool for swarm robotics, facilitating code reuse, standardization, and the real-world deployment of large-scale, multi-robot systems. The study on shape formation by Li et al. [487] is an example paper that uses Buzz.

Related approaches. A related effort is the application of software engineering to swarm robotics. For example, Merlo et al. [550] study automated model extraction and verification of swarm robotics software using Buzz. Given the difficulty of programming swarms, ensuring the correctness and safety of swarm robotics implementations is essential.

A similar approach is MacroSwarm of Aguzzi et al. [6]. MacroSwarm is a field-based compositional framework for swarm programming that uses aggregate computing to define reusable and composable functional blocks for swarm behaviors. Their approach is influenced by "aggregate computing" [829]. The framework enables modular, scalable, and decentralized coordination by mapping sensing fields to actuation fields. Simulations demonstrate its effectiveness in dynamic environments, showcasing robustness to adversarial conditions and real-world applicability.

---

[2] See https://github.com/buzz-lang/Buzz.

Bachrach et al. [44] introduce an approach to programming robot swarms using the Proto language, which models a swarm as a continuous amorphous medium instead of programming individual robots. By specifying swarm behavior at the level of global spatial fields and compiling it down to local robot actions, Proto enables scalable, robust, and composable swarm programs. The authors extend Proto to mobile robots. They validate their approach both in simulations and on physical robots.

Bozhinoski and Birattari [98] propose the automatic design of robot swarm control software, starting from a formal mission specification. The authors introduce the Swarm Mission Language, a domain-specific language that allows precise, standardized description of swarm missions, which are then automatically transformed into optimization problems. Using model-driven engineering techniques and an optimization framework, the approach enables the systematic generation and deployment of control software onto real robots. Demonstrations on multiple missions, both in simulations and on real robots, show the potential of this method to create reliable and maintainable swarm behaviors.

6.4.3 Shared Memory for Swarms

Majcherczyk and Pinciroli [509] introduce the approach of "SwarmMesh" for data storage and management in robot swarms. In swarm robotics, we usually rely on information sharing by local communication and gossiping (see Sect. 5.2.6). When a robot receives a message, it may immediately utilize that information (reactive behavior). Still, there is no standard approach on whether swarm robots should store that information and in what form. SwarmMesh enables swarm robots to act both as a dynamic distributed storage system and as active participants in using the stored data. This innovation reimagines swarm programming, enabling the swarm to be programmed as a shared-memory machine. A shared-memory machine is a typical computational architecture where multiple processors can directly access a shared-memory space. This simplifies programming because processors can communicate implicitly by reading and writing shared variables. Instead, in a fully distributed memory system, each processor has a separate memory space, requiring explicit communication to exchange data. Programming in this paradigm demands carefully designed communication patterns, as data consistency and synchronization must be managed explicitly across distributed nodes. Instead of each robot handling data routing and synchronization manually, SwarmMesh creates an illusion of a global shared-memory space. This significantly simplifies programming logic and coordination.

SwarmMesh prioritizes storage decisions based on features such as data type or spatial relevance. The system self-organizes in response to the challenges of robot mobility and communication constraints, ensuring that critical data remains accessible even as individual robots move or disconnect from the network. SwarmMesh also incorporates mechanisms for efficient data querying and robust fault tolerance, ensuring low communication overhead and resilience to individual robot failures.

Data is managed using a dynamic hashing approach with key-value pairs. Hashing is a technique that maps data to a fixed-size key using a mathematical function, enabling efficient storage and retrieval in data structures. Dynamic hashing considers real-time conditions, such as memory availability, data distribution, or network topology. Hashing in SwarmMesh is performed both categorically, where data is prioritized by type, and spatially, where data is prioritized by its spatial location. The system determines which robot to store data based on node eligibility, considering both the availability of memory and the number of neighbors. If there are better places to store the data than locally, the data is actively rerouted to more suitable robots. Data resilience is achieved through a main-replica model (master–slave replication model). A main robot holding the data is active while a replica robot holding a copy is inactive until it is required to replace the formerly active robot. A heartbeat signal is sent between primary and replica to ensure synchronization. Efficient querying (data access) is achieved by flooding with hop gradients to provide low latency.

A similar approach is presented in RASS by Arseneault et al. [30]. They study risk-aware data storage and routing in mission-critical scenarios, focusing on data safety, integrity, and reliability in hazardous environments, while enabling data to reach a centralized base station. RASS utilizes a hop-count-based routing table with risk-avoidance mechanisms. It avoids failure points by actively routing data away from high-risk nodes.

6.4.4 *Swarm Robots with Heavy Onboard Compute*

Several tasks of a mobile robot may require a significant amount of computational power. A typical case is the processing of high-resolution image data. Also, SLAM can have high requirements [46]. More interesting seem applications of evolutionary swarm robotics [787], especially in its embodied form [105, 849]. A more speculative idea could be a modern approach to world models [311], where a robot basically builds its own simulation to improve controllers. While a more common idea in swarm robotics is to keep robots simple, there are also ideas to equip them with a range of capabilities. The idea of giving swarm robots considerable computational power is rare. However, Jones et al. [394] introduce the Xpuck, a high-performance swarm robotics platform that enhances standard e-puck robots with powerful mobile computing hardware. Each Xpuck integrates a system-on-chip to provide substantial onboard computational power. The Xpuck swarm collectively delivers over two teraflops of computing performance, supporting tasks such as onboard simulation, evolutionary robotics, image processing, and real-time decision-making. The system features a fast, GPU-accelerated robot simulator for in-swarm evolution, as well as a robust infrastructure for synchronized experiments.

6.5 Swarm-SLAM, C-SLAM, and Distributed SLAM

Simultaneous Localization and Mapping (SLAM) is a technique robots use to build a map of an unknown environment while simultaneously estimating their location. When SLAM becomes distributed, multiple robots independently build maps and estimate their positions, and collaboratively merge this information into a standard map.

In multi-robot systems, the term C-SLAM (collaborative or cooperative SLAM) is often used, and less frequently Distributed SLAM. C-SLAM refers to approaches in which robots share information to jointly build and refine a map. The focus is on the collaborative function of the robots. Distributed SLAM refers mainly to the computational organization of the system, where multiple agents solve the SLAM problem without relying on a central node. Focus is on how the solution is structured, with peer-to-peer communication, distributed optimization, and resilience against network failures or partitions. Swarm-SLAM refers to mapping by a large number of simple robots with a focus on scalability and distributed coordination mechanisms.

Following Kegeleirs et al. [415], the decentralized nature of robot swarms poses several unique challenges. Unlike centralized multi-robot systems, swarms have to manage information gathering, sharing, and map construction without global supervision. This complicates map merging. Current challenges include developing scalable exploration schemes that leverage swarm behaviors, efficient decentralized data sharing methods to maintain fault tolerance (see Sect. 6.4.3), and techniques for creating meaningful maps (topological or semantic) despite limited sensing capabilities and communication constraints. Achieving fully decentralized map merging and overcoming constraints in scalability and fault tolerance are open research questions crucial for advancing Swarm-SLAM techniques.

Lajoie and Beltrame [459] present their take to Swarm-SLAM specifically designed for swarm robotics. Swarm-SLAM emphasizes scalability, sparsity, and decentralized control, enabling robots to collaboratively build maps in environments without external positioning collaboratively, even when communication is sporadic. They propose a novel method for prioritizing inter-robot loop closures based on maximizing algebraic connectivity, which significantly reduces communication overhead and accelerates convergence. Extensive evaluations using datasets and real-world experiments with a Boston Dynamics Spot, Agilex Scout, and Agilex Scout Mini demonstrate that their approach is effective and efficient.

The extent to which an SLAM approach can be applied with minimal hardware requirements is demonstrated in the following paper. Niculescu et al. [595] introduce NanoSLAM, a lightweight SLAM framework designed specifically for simple robots, enabling fully onboard operation within tight resource constraints. They use low-power depth sensors and optimized parallel processing on a tiny drone weighing just 44 g. NanoSLAM achieves accurate real-time mapping with a power consumption of only 87.9 mW. Experiments demonstrate centimeter-level mapping accuracy and robustness in real-world environments.

6.6 True Robustness, Fault Management, and Security

Robot swarms are already robust because of the high degree of redundancy. However, on closer look, redundancy is relatively only a potential for robustness. Robustness requires more than just replacing robots that fail. What happens, for example, if a robot does not just break down but actually causes problems because it is still partially operational? Such a robot could cause errors and damage because other, fully functional robots may react in funny ways to its faulty behavior. In a hypothetical scenario, a defective robot could deposit a large amount of pheromone at one spot, attracting many robots to an irrelevant position. Hence, it is necessary to detect errors and to ignore faulty robots. Concerning security, we can also imagine a situation in which a saboteur has a robot infiltrate a swarm. The robot may manage to be accepted as a peer and may start to behave strangely.

6.6.1 True Robustness and Software Redundancy

Swarm robotic systems are designed to be robust and fault-tolerant. They are supposed to keep working despite individual robot failures. One of the key principles supporting robustness in swarm robotics is hardware redundancy. Swarms of robots typically consist of many identical robots, ensuring that if some fail, others can take over their tasks. This high redundancy and flexibility, where any robot can complete any task, contribute to resilience in collective behavior. However, hardware redundancy alone is not sufficient. Software redundancy should play an equally critical role in ensuring system robustness.

A common approach in swarm robotics is to have all robots run the same program, ensuring uniform behavior. However, this introduces a critical vulnerability: if a single crucial software bug is present, it affects the entire swarm, potentially leading to system-wide failure. Also, swarm robustness must account for faulty robots that remain operational but transmit incorrect data, which could mislead the entire system. A fundamental question arises: Can the swarm detect and isolate malfunctioning robots to prevent them from disrupting the system?

There is a whole field called fault management (also related to failure and error handling, predictive maintenance, and system hardening, among others), which attempts to find answers on how engineered systems can deal with faults [304, 830]. We will return to this topic below. One aspect of fault management is fault recovery, for example, based on rollback mechanisms (e.g., redundant software versions and rollback capabilities) as seen in spacecraft or aviation systems.

The aviation industry indeed provides valuable insights into robustness and fault tolerance, as it has developed stringent design assurance levels (DAL) to prevent catastrophic failures. In aviation, safety-critical systems, such as flight control computers, require a failure probability of less than 10^{-9} per flight hour. These strict standards underscore the importance of fault detection, redundancy, and fail-safe mechanisms—principles that can be applied to swarm robotics. When a single monolithic system cannot be made sufficiently fault-tolerant, an alternative approach is dissimilar redundancy. Instead of relying on one highly reliable component, multiple independently developed systems are combined to improve robustness.

For example, the Airbus A320 flight control system uses five dissimilar computers running four different software packages (with variations in operating systems and control algorithms). This ensures that even if one implementation fails due to a bug, others remain operational, reducing the risk of catastrophic failure. Applying this concept to swarm robotics could significantly enhance fault tolerance, preventing single points of failure in swarm control software.

To ensure fault-tolerant decision-making, robotic systems can employ voting mechanisms where multiple controllers or robots contribute to the overall decision. We have N voters with voting weights v_i. The voters are voting for options $x_1, x_2, \ldots, x_n \in \{0, 1, \ldots, m\}$. The vote tally for option y is $w_y = \sum_{i \in \{i|x_i=y\}} v_i$. Say option y has the highest vote tally.

Several voting strategies can be used, each offering different levels of resilience. (1) Consensus voting: The system accepts an option only if all votes agree ($w_y = \sum_i v_i$). (2) Majority voting: The system selects an option if it receives more than 50% of the weighted votes ($w_y > \frac{1}{2} \sum_i v_i$). (3) Byzantine voting: The system selects an option if it receives more than $\frac{2}{3}$ of the total weighted votes ($w_y > \frac{2}{3} \sum_i v_i$), ensuring robustness even when up to $\frac{1}{3}$ of the nodes are faulty (a common assumption in Byzantine fault tolerance). Voting-based decision-making is beneficial in swarm robotics, as it filters out incorrect information from faulty robots and ensures that the swarm continues to function correctly.

Example: *n*-Version Programming

To address software reliability, a software engineering method known as n-version programming was introduced by Chen and Avizienis [139]. This technique involves developing multiple independent implementations of the same functionality to reduce the likelihood of common bugs affecting the entire system.

In n-version programming, n independent teams develop according to the exact same software specification using different programming languages, algorithms, and development methodologies. A multi-version execution environment evaluates the outputs and selects the most reliable one based on voting mechanisms. The assumption is that software errors in independently developed versions will be statistically independent, reducing the risk of all versions failing simultaneously. However, a key challenge remains: Are software bugs truly independent, or do different teams make similar mistakes? While n-version programming increases fault tolerance, it is not a perfect solution and requires careful implementation. n-version programming can be applied to swarm robotics in two main ways.

(A) Microscopic level: Individual robots run multiple versions internally. Each robot runs all n controllers and hosts its own n-version execution environment. The robot evaluates different control outputs and selects the most reliable action, potentially using voting-based schemes.

(B) Macroscopic level: The swarm is divided into n sub-populations, each running a different controller. Each subgroup of robots runs one of the n different controllers, creating a heterogeneous swarm with multiple control strategies operating simultaneously. There are two sub-options: (B1) No explicit voting: The system relies on emergent behavior and hopes that diversity naturally leads to robustness. (B2) Explicit voting via communication: Robots share control decisions and vote on which actions should be taken. The macroscopic approach (B) introduces software diversity at the swarm level, potentially making it more resilient to software failures. However, coordinating different sub-populations may require additional mechanisms to prevent behavioral inconsistencies.

Ensuring robustness and fault tolerance in swarm robotics requires a combination of hardware and software redundancy, adaptive control mechanisms, and fault-tolerant decision-making strategies. While hardware redundancy (where multiple robots perform the same task) provides a baseline level of resilience, software redundancy (such as n-version programming and voting schemes) is crucial in preventing software bugs from compromising the entire system. Inspired by aviation safety practices, dissimilar redundancy and Byzantine fault tolerance may prove to be more relevant in future swarm robotics applications.

6.6.2 Fault Management and Change Detection

As briefly mentioned above, in fault management, one seeks methods to address faults [304, 830], such as the following. Fault detection identifies anomalies through sensor monitoring, anomaly detection, and self-diagnosis, such as detecting sensor drift in a robot. Once a fault is detected, fault diagnosis determines its root cause, distinguishing between hardware and software failures. To minimize the impact, fault mitigation employs redundancy and fault-tolerant control, allowing systems to switch to backup components when needed. Fault correction actively restores functionality through self-healing algorithms and recalibration. Fault prevention focuses on robust design and predictive maintenance to prevent failures. Fault recovery ensures that systems can return to a stable state after failure, using rollback mechanisms or backup software, as seen in spacecraft or aviation systems. We discuss change detection, fault detection, fault mitigation, and fault tolerance in more detail for robot swarms.

Change Detection

Change detection is the art of identifying any statistically significant deviation from the expected behavior of a system. The goal is to detect that something has changed, without necessarily identifying the cause.

Wahby et al. [840] present an extension of the BEECLUST algorithm that enables robot swarms to adapt their aggregation behavior to both dynamic environmental conditions (e.g., changing light distributions) and changing swarm densities. Robots estimate local light intensity and swarm density by measuring the time between encounters with other robots and share these parameters via local communication. These shared parameters are used to detect global changes and adjust individual behaviors collectively accordingly. The robots adapt to changes, for example, in swarm size by adjusting parameters in their control algorithms (e.g., waiting longer for other robots to arrive in situations with reduced swarm density). The approach is validated in physical robot experiments with dynamic changes in light and swarm size.

Pfister and Hamann [628] present an approach for collective decision-making in dynamic environments by combining the idea of Bayesian robots [218] with statistical change detection. The change detection here focuses on a distributed environmental feature that the robots need to measure collectively, and the robot swarm also needs to detect a global change in that environmental feature collectively. The authors introduce an enhanced algorithm that incorporates the PELT (Pruned Exact Linear Time) method [432] to detect environmental changes online. Robots continuously estimate environmental features and revise their beliefs if a change is detected, using local communication and feedback mechanisms (reset of belief, recruitment, and transformation). Simulation results demonstrate that this method improves both accuracy and speed of consensus formation across a range of task difficulties, outperforming previous methods and showing robustness even in dynamic environments.

Fault Detection

Fault detection is a specific type of change detection that focuses on identifying abnormal changes resulting from faults or errors. An essential first step in fault detection is to answer the question: how to define a fault? Generally, there are five main approaches: model-based, data-driven, threshold-based, task-based, and peer comparison. For the model-based approach, one needs to define an analytical or physics-based model. The actual system is compared to the model's predictions. Deviation from that known model of normal behavior is interpreted as a fault. For the data-driven approach, the model is a classifier or anomaly detector that was trained on historical or simulation data that was labeled as "normal." Statistically significant deviation from the learned patterns is interpreted as a fault. For the threshold-based approach, we define upper or lower bounds on metrics, such as speed, energy use, or sensor readings. A fault is a violation of these predefined limits. For the task-based definition, we monitor mission goals (e.g., object retrieval, navigation success). If performance drops, we infer a fault. The peer comparison approach is arguably the most interesting one in our context, as it leverages the multi-robot scenario. We make the robots monitor their neighbors. If a robot behaves very differently, this may indicate a fault [773]. This is also interesting because we are not required to define a norm, and the system responds online based on what appears normal at that moment. The robots collectively agree on outliers, and this significant deviation from peer behavior is used as a basis for fault detection. This approach, of course, implies that the majority is correct, as there is no other definition of "normal" than being a member of the majority.

Christensen et al. [142] address the problem of detecting faulty behavior by making all robots synchronize following the firefly algorithm (see Sect. 5.2.2). Robots are supposed to flash in synchrony. In their "fault detection scheme, the periodic flashes function as a heartbeat mechanism" [142]. A faulty robot can be detected if it does not synchronize properly. They show in robot experiments that "robots can detect and respond to faults by detecting non-flashing robots" [142]. The swarm stays robust even if there are multiple faults, and the robot swarm has self-repair capabilities to survive a relatively high failure rate.

A more sophisticated approach is inspired by the immune system [124]. Matzinger [529] proposes a "Danger Model" that challenges the classical self–nonself paradigm of immunology by proposing that the immune system is primarily activated not by foreignness but by signals from damaged tissues. Traditional models suggested that immune responses arise from detecting "nonself" elements. Still, these fail to explain tolerance to changing tissues, fetuses, or tumors, as well as the need for adjuvants in vaccines. The Danger Model instead proposes that antigen-presenting cells become activated by alarm signals released from injured cells, triggering immune responses regardless of whether the source is self or nonself. A complex and successful system like the immune system, of course, invites being used as inspiration also for fault detection in swarm robotics. Tarapore et al. [773] follow this path but in a complementary way to Matzinger [529] by basing their work on the so-called cross-regulation model [772]. The cross-regulation model is a theory in immunology that explains immune tolerance and activation through the interactions and balance between different populations of T cells. The swarm has to learn online what normal and abnormal behavior is. Robots use feature vectors to describe observed behaviors and share them with their neighbors. Finally, robots have a vote to consolidate their models and to determine faulty robots.

Millard [557] introduces a model-based method for exogenous fault detection (detection of faults in one robot by other robots, based on external observation of its behavior). Each robot runs simulations of its neighbors' expected behavior and compares these predictions to actual observations. Significant discrepancies are flagged as potential faults. This method addresses the limitations of endogenous fault detection, which may fail if a robot is unaware of its malfunction or is unable to communicate. He proposes two architectures: one that predicts future behavior and one that reconstructs past behavior, and demonstrates their effectiveness using experiments with e-puck robots.

Daigle et al. [171] present a distributed, model-based diagnosis framework for formations of mobile robots. The authors propose a qualitative fault-diagnosis approach that efficiently isolates faults in individual robots within a formation. This method enables each robot to diagnose faults locally using its own measurements and those selectively communicated to it. The system can detect and isolate process, sensor, and actuator faults. The approach was successfully tested in an experimental setup with four robots.

Also see other relevant papers on fault detection for robot swarms [125, 463, 467].

Fault Mitigation

As swarm robotics moves toward real-world deployment, ensuring robustness against failures is critical. Lee and Hauert [466] present a data-driven approach to identifying fault mitigation strategies in swarms. They assume robots can self-detect failures and execute predefined corrective actions. The method is evaluated in intralogistics scenarios, where robots transport boxes either individually or collectively. Through extensive simulations, Lee and Hauert [466] find that the effectiveness of mitigation strategies depends on both the fault type and task context. Specific faults do not benefit from mitigation, and interventions can sometimes degrade overall swarm performance.

Fault Tolerance

An influential paper on fault tolerance in swarm robotics is that of Winfield and Nembrini [866]. They challenged the often-assumed robustness of swarms by applying Failure Mode and Effect Analysis (FMEA) to a case study swarm that performed object encapsulation using wireless connectivity. The authors define a range of fault types (e.g., motor, communication, and sensor failures) and evaluate their effects on swarm behaviors (e.g., aggregation and taxis). They show that partial failures (e.g., motor failures that immobilize a robot) can be more disruptive than complete robot loss, as they can physically anchor and obstruct swarm motion. Winfield and Nembrini [866] also propose reliability modeling frameworks tailored to swarms, including multi-state and load-sharing models. The authors conclude that while robot swarms do exhibit fault tolerance through redundancy and distributed control, explicit design for dependability is still needed, especially for real-world, safety-critical applications.

Bjerknes and Winfield [80] focus on an emergent beacon-taxis behavior. They model reliability to analyze how different robot failure modes affect the overall swarm function. They introduce a k-out-of-N reliability model, showing that swarm performance collapses when the number of functional robots drops below a critical threshold. The concept of swarm self-repair is introduced to model how long it takes a swarm to recover from failed robots. They find that beyond a specific size, increasing the number of robots actually reduces overall reliability, because the failure rate eventually exceeds the swarm's capacity to self-repair.

O'Keeffe [602] introduces a predictive approach to fault tolerance in robot swarms by focusing on faults that develop gradually over time, rather than sudden failures.

Using the concept of predictive maintenance, robots detect early signs of hardware degradation (e.g., in motors and sensors) and autonomously return to a base station for servicing before failure occurs. O'Keeffe [602] compares this proactive method to traditional reactive strategies, which typically isolate or abandon failed robots. The study evaluates simulated swarms of 5, 10, and 20 robots using ROS 2 and the Gazebo simulator. Experimental results show that predictive maintenance not only sustains swarm performance but also reduces the cascading disruptions caused by failed robots in constrained environments. The underlying idea is not to accept individual robot failure as inevitable (a common doctrine in swarm robotics), but rather to utilize predictive mechanisms to enhance system autonomy and resilience.

6.6.3 Cybersecurity and Blockchains

With more and more robot swarms in application, the question of cybersecurity becomes more prevalent. Different scenarios become relevant, starting with increasing robustness against malfunctioning robots. Also, scenarios are assumed that include intruding robots that may attempt to be accepted as a swarm member and then sabotage the swarm by false messaging. Alternatively, only the communication between robots might be hacked to reduce the swarm's efficiency or shut it off. Given that there will be more robot swarms, for example, operating in the air, security is quickly becoming a safety issue.

Higgins et al. [358] identify the unique security challenges posed by swarm robotics, such as limited computational resources, the absence of centralized control, and the potential for emergent behaviors to be manipulated by adversaries. They emphasize that swarm robotics differs significantly from related technologies (e.g., MANETs, sensor networks, software agents) and hence requires tailored security mechanisms for communication, identity, and intrusion detection.

A prominent technology in this context is that of blockchains. Generally, blockchains as a method seem appropriate to secure a decentralized system. Blockchains were initially developed for financial applications. They offer decentralized trust, transparency, and robustness, eliminating the need for a central authority. However, they may come with challenging requirements, such as those related to memory. One of the first relevant papers started somewhat speculatively, Castelló Ferrer [128] discusses options of applying a blockchain approach to swarm robotics. The decentralized approach based on peer-to-peer networks can be transferred to swarms, and, for example, intruders may be detected by analyzing the blockchain. However, back then, it was still unclear how the resource-intensive requirements (computational power, memory, communication bandwidth) could be satisfied with typical robot–swarm hardware or whether these requirements could be reduced.

One of the first papers to demonstrate the potential added value of Blockchain technology for robot swarms was published by Strobel et al. [752]. They present the first proof of concept for using blockchain-based smart contracts (self-executing programs on a blockchain that enforce predefined rules) to enhance security and coordination in swarm robotics, particularly in the presence of Byzantine robots—those that exhibit faulty or malicious behavior. The authors propose a decentralized decision-making framework based on the Ethereum blockchain. In this approach, robots record their votes and opinions on a shared ledger, allowing for the detection and exclusion of malicious agents. Compared to traditional consensus methods, the blockchain-based system shows greater robustness in the presence of adversarial agents.

A recent survey and perspective paper provides a good overview of methods and possible perspectives for the blockchain approach in swarm robotics. Dorigo et al. [205] discuss how blockchain technology can enhance the capabilities and security of mobile multi-robot systems. The authors propose using blockchain to manage robot identities, coordinate actions securely, and detect malfunctioning (Byzantine) robots, and enable robots to establish an autonomous robot–robot economy. The authors also acknowledge technical challenges. For example, swarm robots have limited computational resources, and there are privacy concerns. Scalability is also a potential challenge for the blockchain approach. Another issue is how to make the swarm robust against intermittent connectivity. Say, the robot swarm was split into two connected components. The blockchain will then evolve into two different branches. How can robots efficiently merge blockchains after having reunited these two components again?

The paper identifies several key opportunities that blockchain provides. (1) Self-Governance: Blockchains allow robots to organize and manage themselves autonomously. Robots could collectively make decisions on tasks and whether robots should be excluded or added to the swarm. This is an opportunity to prepare robot swarms for coordination in dynamic and open environments without central control.

(2) Compliance and Accountability: By securely recording robot actions and interactions, blockchains can detect and isolate malfunctioning (Byzantine) robots. This may enforce compliance with legal and safety standards and provide tamper-proof records for post-mission accountability and auditing.

(3) Robot Economy: Blockchains can enable economic interactions between robots and humans. This could be used to establish a robots-as-a-service business model. Robots can autonomously exchange goods and services among themselves or with humans, using cryptocurrency tokens.

(4) Data Consistency: Blockchains may provide reliable decentralized data sharing and synchronization among robots. This may help to establish a consistent collective database, prevent double-counting, and reduce vulnerability to Sybil attacks (many robots with fake identities).

(5) Decentralized Supervision: Smart contracts on the blockchain can act as a decentralized supervisor and help to implement high-level collective decision-making.

Dorigo et al. [205] also identify several key challenges in using blockchain technology for robot swarms: (1) Limitations in computation, storage, and communication: Swarm robots typically have limited computing power, storage, and communication capabilities compared to systems that blockchains were initially designed for. Computationally intensive consensus protocols (e.g., proof-of-work) may not be feasible. We may need more efficient consensus mechanisms instead.

(2) Blockchain architecture selection: The choice of the blockchain architecture significantly impacts scalability, autonomy, and robustness. The paper discusses three architectures: internal blockchain maintained by robots, external blockchain, and hybrid sidechain solutions. Each architecture has tradeoffs in terms of reliability, connectivity, independence, and security.

(3) Authentication of real-world data (oracle problem): Blockchains ensure data integrity, but cannot guarantee the correctness of real-world data fed into the system. Robots must rely on trustworthy oracles to integrate real-world sensing information reliably. Such an oracle could be either centralized (straightforward but vulnerable) or decentralized (secure but complex).

(4) Transparency, confidentiality, and privacy: Blockchains are inherently transparent, which can potentially compromise sensitive information. Transparency is required for security and accountability. Privacy is required to protect sensitive robot data and system integrity. Achieving a balance between transparency and privacy remains a critical challenge.

(5) Network dynamics and partitions: Mobile robots face intermittent connectivity and network partitioning due to mobility and environmental conditions as discussed above. These network instabilities can disrupt the synchronization of blockchains, complicating the maintenance of a coherent chain.

Toychain, developed by Pacheco et al. [614], is a lightweight, customizable blockchain framework written in Python for research in swarm robotics. Unlike conventional blockchains, Toychain is designed to run efficiently in simulation environments and on physical robots with limited resources. It supports smart contracts, modular consensus protocols (e.g., proof-of-work, proof-of-authority), and synchronization with simulation clocks. Toychain facilitates the exploration of blockchain-based coordination, decision-making, and security mechanisms in swarms, eliminating the overhead and complexity associated with full-scale blockchain systems. The platform also supports custom consensus plug-ins, allowing users to prototype and test alternative mechanisms for trust and collaboration.

Blockchain may offer a novel infrastructure for trustworthy swarm behavior, but its implementation must be adapted to the constraints and dynamics of robotic systems. An alternative to blockchains could be any standard technique for encrypting and signing messages. The added value of blockchains compared to their required resources is up for discussion.

There are only a few other relevant works that focus on cybersecurity for swarm robotics. We deviate here a bit and also discuss centralized approaches for inspiration.

Wardega et al. [847] explore plan-deviation attacks in centralized multi-robot systems, where compromised robots deviate from assigned paths and attempt to conceal their movements. The multi-robot system mitigates these attacks by limiting the information robots receive about plans, and robots also report sightings of each other (cf. Tarapore et al. [772]). Experiments demonstrate that the approach effectively prevents stealthy attacks and significantly reduces the success of less cautious attackers.

Vijay et al. [826] propose an anchor-free integrity monitoring framework for distributed multi-robot systems, designed to detect and isolate faults and cyberattacks, such as GNSS spoofing, using inter-robot range measurements. They developed a distributed algorithm that enables robots to identify compromised agents and collaboratively reconstruct localization errors. The effectiveness and the robustness of the method were demonstrated through simulations and mixed-reality experiments using both real and simulated UAVs.

In summary, blockchain offers intriguing tools for decentralized trust and accountability in swarms; however, the associated overhead may exceed the capabilities of some swarm robots. Future work must continue to clarify whether lightweight blockchain variants offer a practical path for secure swarm coordination.

6.7 Field Robotics: Kicking the Robots Out of the Lab

A large majority of published work on swarm robotics research is based on lab experiments. For a long time, the standard approach was to design and study fundamental algorithms, with the lab experiment serving as the proof of concept in a final step. This "one-take" swarm robotics research is complex in its required methods and skills, spanning from theory to real-world testing in a wide arc: algorithm design, simulation testing, refinement, and multi-robot hardware deployment. This is scientifically powerful, as it demonstrates not only theoretical validity but also practical viability. However, we should more often try to push for projects that deploy swarm robotic systems in real-world, uncontrolled environments. This "kick out of the lab" would further mature the field, particularly in three aspects. (1) Realism: Taking robots into the field exposes them to the complexity, unpredictability, and noise of the real world for more accurate testing of robustness. (2) Validation: Field deployment is a complete benchmark for a system's reliability under real physical constraints. (3) Relevance: Applications of swarm robotics require operation in realistic, unstructured, dynamic environments.

Entering the field presents three significant challenges: (1) Engineering Robustness: Real-world deployments require robust hardware and software under potentially extreme conditions. (2) Autonomy: Without external infrastructure (e.g., motion capture), swarms must rely on onboard sensing, decentralized control, and local communication in harsh conditions. (3) Logistics and Scale: Swarm size comes at significant costs, and the duration of experiments is crucially limited due to practical constraints.

Field robotics comes with significant costs, but testing swarm algorithms under real-world conditions is essential to challenge the assumptions made in models and simulations. It is not only a milestone in terms of technological readiness, but also a methodological necessity to ensure that swarm robotics principles hold beyond controlled environments and translate into effective, real-world systems.

Duarte et al. [210] discussed this important issue back in 2016 (also see Duarte et al. [211]):

> Despite the potential of swarm robotics for real-world tasks, no demonstrations of swarm behaviors outside of controlled laboratory conditions have been performed so far [...] all real-robot experiments have, in fact, been performed in controlled laboratory environments.

Also Kegeleirs and Birattari [413] mention a "deployment gap" that questions whether "findings obtained on current research platforms are transferable to the advanced robots required for real-world applications." Duarte et al. [210] themselves reported a swarm of ten aquatic surface robots in the wild. The robot is a differential drive mono-hull boat. The onboard control runs on a Raspberry Pi 2. The robots have IEEE 802.11g Wi-Fi, GPS, and a compass. Reported experiments include homing (navigation to waypoints), dispersion, aggregation, and area monitoring.

From single-robot robotics we know that navigating robots in different media comes with unique challenges. The simpler media are air and water. Maybe we can also add on-road navigation. However, off-road is challenging even when we have to control only one robot. This is also reflected in the literature. We find robot swarms in the form of flying multi-rotors in the air, as well as robot swarms in and on water. Robot swarms that navigate on the ground in off-road conditions are rare.

6.7.1 Waterborne

It is not a mere coincidence that also the second outdoor swarm robot experiment is done in water (cf. the quote of Frank Schätzing at the beginning of this chapter; maybe we still underestimate the importance of our oceans). Within the European-funded project "subCULTron"[3] researchers developed an underwater swarm (culture) of robots (autonomous underwater vehicles, AUV). The final demonstration was done in the lagoon of Venice with a heterogeneous robot swarm of three types of robots and many dozens of robots in total. The main task is monitoring. For some impressions, see Fig. 6.9.

While submerged robots need to deal with currents and the challenge of communication, uncrewed surface vehicles (USV) deal with even more diverse dynamic water and weather conditions (e.g., waves, wind); however, communication is much easier. Obstacle avoidance is also more challenging because other vessels, swimmers, animals, and floating debris need to be detected and avoided.

Gershfeld et al. [280] introduce the Proposal-Based Adaptive Channel Search (PBACS) algorithm for rapid identification of navigable channels in bodies of water using USV. The PBACS approach utilizes adaptive sampling, decentralized estimation, and task allocation strategies to outperform traditional methods, such as

[3] http://www.subcultron.eu/.

Fig. 6.9 Work within the project "subCULTron" at the lagoon of Venice. The robots were developed by FER, University of Zagreb and SSSA Scuola Superiore Sant'Anna (photos by Thomas Schmickl, Artificial Life Lab, Karl-Franzens University of Graz)

lawnmower surveying and Markov Decision Process planning, particularly in multi-vehicle scenarios. Simulation and field trials demonstrate that PBACS is an efficient technology.

Paine and Benjamin [615] study collective decision-making in combination with multi-objective optimization for individual robot behaviors. This approach enables scalable and explainable coordination in multi-robot systems by reducing communication costs and scaling with the number of options rather than the number of robots. They report a 2-hour field experiment involving $N = 8$ uncrewed surface vessels in an explore-exploit-migrate scenario.

A strength of these approaches is the MOOS-IvP software suite developed by Benjamin et al. [73]. It is an open-source platform for autonomy in uncrewed marine vehicles, designed to support adaptive and collaborative operations in complex environments. MOOS provides middleware capabilities for communication and data handling, while the IvP Helm offers multi-objective optimization for behavior-based control. The system enables platform-independent operation and facilitates the integration of heterogeneous vehicles into a large-scale team for long-endurance missions.

Research on teams or swarms of autonomous sailboats is still in its early stages and remains relatively scarce. When multiple sailboats gather, it is typically in the context of a regatta, that is, a competitive race rather than a cooperative mission. From a mobile robotics perspective, sailing is exciting because wind conditions fundamentally constrain navigation. As a result, the shortest or fastest route between two waypoints is often not a straight line, but a dynamically changing trajectory that must account for wind direction, strength, and the vessel's sailing capabilities. This makes sailing a challenging testbed for adaptive, environment-aware swarm behaviors. Kedia et al. [412] present the development of Onyx, a custom-built autonomous sailboat designed for long-term operation in robotic swarms (see Fig. 6.10). Unlike conventional battery-powered robots, sailboats harness wind energy for propulsion, enabling extended missions with reduced human intervention. The authors demonstrate robust autonomous sailing and control strategies that emulate the skills of experienced human sailors, including tack planning and adaptive sail trimming. The envisioned swarm will utilize heterogeneous sensor payloads and energy-aware behaviors to explore and monitor aquatic environments collectively.

Zoss et al. [904] present the design, implementation, and testing of a large-scale distributed system of autonomous, self-propelled buoys for environmental monitoring of waterbodies. The system, called "Bunch of Buoys" (BoB), demonstrates scalability and robustness, successfully deploying up to 50 units. Experiments validate collective behaviors, such as flocking, navigation, and dynamic area coverage. Potential applications include water quality monitoring, detecting algae blooms, adaptive sampling of dynamic phenomena, and coastal surveillance.

Fig. 6.10 Autonomous sailboat intended for teams of boats [412]

6.7.2 *Airborne*

Drones have been popular research tools for many years, and the results from individual drones are awe-inspiring. For example, there have been fantastic results in using reinforcement learning to generate controllers for flying drones. Kaufmann et al. [409] introduce *Swift*, an autonomous quadrotor system that uses onboard cameras and inertial sensors to race at human world-champion level. By combining deep reinforcement learning in simulation with real-world data for perception and dynamics modeling, *Swift* outperformed expert human pilots in head-to-head races, marking a milestone in real-world AI and robotics performance.

Airborne robot swarms have also been around in research and applications for many years. It is important to note, however, that the popular coordinated aerial drone displays (drone light shows using many quadrotors) typically do not follow swarm strategies [841]. Instead, they are centrally controlled, and each drone follows a preprogrammed trajectory using GPS and an IMU (possibly also barometers and onboard cameras). Typically, these drones neither sense each other nor communicate with one another.

The advantages of flying robot swarms are as follows. There is the freedom of movement in three dimensions. Unlike ground and water surface swarms, drones can operate in three-dimensional space, providing greater flexibility in navigation and coordination. Drones are usually relatively easy to deploy, depending, of course, on the size of the swarm. They are fast in covering a wide area. They can access difficult or dangerous environments (e.g., disaster zones, tall structures) without requiring other infrastructure beyond a landing pad (even if improvised).

The disadvantages are primarily energy, safety, and regulations. A challenge for airborne swarms is their limitation in energy. Flight is energy-intensive, leading to limited mission durations and payload capacities. This is especially a problem for large swarms. If the time to deploy the swarm increases linearly with size, then preparing the swarm for flight will take significantly longer than its mission time. This raises questions about scalability (see Chap. 2). There are safety concerns. There are flying drones with weights above 20 kg that can potentially kill a person when falling. Collisions are dangerous and require strict safety margins, as well as more advanced conflict avoidance measures. A challenge is also that airspace use is heavily regulated, which limits deployment flexibility, especially in urban or shared-use airspace. This is especially complicated in the academic sector when executing experiments. The communication between drones can be limited by line of sight and can be affected by altitude, range, or interference. A challenge of flying multiple drones in a limited space is also the turbulence caused by the rotors. Bauersfeld et al. [61] present a lightweight, unified model that describes the downward airflow produced by hovering quadrotors as a turbulent jet. By combining classical fluid dynamics theory with 16 h of experimental flight data from six drones, they derive a scalable model that can be used in multi-drone control scenarios.

Coppola et al. [151] provide a comprehensive review of the technical and conceptual challenges in developing real-world swarms of Micro Air Vehicles (MAVs). It emphasizes the strong interdependence between local capabilities (e.g., sensing, processing, actuation) and global swarm behaviors. They argue that system design must carefully balance flight time, sensor weight, processing power, and dynamics to achieve optimal performance. They propose a layered approach to swarm design, spanning from MAV hardware to local control, intra-swarm sensing and communication, and collective behavior. They underline that robust, scalable, and flexible swarm behavior requires robustness at the individual level, especially in sensing and collision avoidance.

Baca et al. [42] introduce a modular, open-source control and estimation framework for multirotor drones. Designed for both simulation and real-world deployment, the system supports complex missions in GNSS-enabled and GNSS-denied environments (GNSS: Global Navigation Satellite System), offering features such as multi-frame localization, sensor fusion, robust feedback control, and trajectory generation. The system includes two feedback controllers: one for agile maneuvers and one for robust performance in noisy conditions (based on model predictive control). A unique feature is its bank-of-filters estimator, which allows switching between multiple localization hypotheses from different sensor sources (e.g., GPS, visual odometry, LiDAR, optical flow). The platform was extensively tested in real-world missions and competitions, including the DARPA Subterranean Challenge.

A significant contribution is that by Balázs et al. [49] who present a decentralized algorithm for managing high-density aerial traffic among autonomous flying drones. Their algorithm combines a velocity-based conflict avoidance strategy with a sense-and-avoid mechanism, enabling drones to navigate safely and efficiently toward individual goals. It is scalable, handles heterogeneous agents and priority hierarchies, and operates without centralized control or high-bandwidth communication. Simulations with up to 5×10^3 robots and real-world experiments with 100 drones demonstrate the system's robustness, efficiency, and potential for real-world unmanned aerial traffic management applications. The approach also extends to multi-layered 3D airspaces, enhancing throughput.

Another example is the "Swarm Robotics for Agricultural Applications" (SAGA) project by Trianni et al. [791]. Also, this project refers to the get-out-of-the-lab meme: "swarm robotics research is still confined to the lab, and no application in the field is currently available." Also, SAGA does not investigate robots on land but up in the air. A group of quadrotors is supposed to help in controlling weeds, such as detecting weeds and mapping infested areas. The quadrotors fly at relatively low altitude to inspect plants on the ground visually. The robots use GPS and onboard vision. A primary challenge is to perform efficient task allocation online.

6.7.3 *Groundborne*

Typical applications of swarm robotics in the field are in water and in the air, and not on land, probably because reliable robot control in water/air is easier than on land. Wheeled or legged robots can get stuck, fall over, and damage themselves or obstacles. Groundborne swarm robotics faces the same challenges as terrestrial robotics in general. To create a reliable robot, one usually needs a lot of sensors (cameras, laser scanners, radar) and computing power. Complex sensors are too big, too heavy, and too expensive for swarm robotics. Basically, it means that we are still not yet able to build any robot that gets even remotely close to the capabilities of an insect. However, if insects can control their legs and navigate robustly, then engineers will also be able to achieve the same on similar scales one day. We should accept the challenge.

Ground robots have been utilized in swarm robotics research from the outset, in numerous foundational experiments. They are typically used in indoor laboratories with flat surfaces and controlled lighting, making them ideal for systematic experimentation and repeatability. Popular platforms include e-pucks, Kilobots, Thymio II (see Sect. 3.6.1), and differential-drive robots based on the ROS ecosystem (see Sect. 6.4.1).

Besides simpler tests of multi-robot systems, there are basically no out-of-lab terrestrial swarm robot projects that have successfully deployed many ground robots in a challenging outdoor environment. Tarapore et al. [774] propose the sparse swarm concept, where robots cannot afford to get into proximity with their swarm members (they give "10 body lengths apart" as an example). They also provide an example of a forest application scenario for terrestrial robots, along with instance robot designs that may address the challenges of traversing rugged terrain, perceiving it, and facilitating long-range inter-robot communication.

Experiments in the lab have proven the principles, but the following years must show field-robust, scalable outdoor approaches. For success, it will be essential to integrate and standardize methods in ready-to-use software packages (similar to ROS 2, see Sect. 6.4.1) and long-horizon field experiments. Robot designs, hands-on experiences, and data should be shared openly to iterate faster across research teams.

6.8 Swarm Behaviors Without Computation

6.8.1 Swarm Robots Without Computation

In swarm robotics, we often follow a minimal-design strategy by restricting ourselves to simple approaches. This applies to equipping the robots with simple sensors, limiting the amount of communication, and reducing memory requirements. A new idea was introduced by Gauci et al. [273] when they allowed for only one binary sensor (indicating whether another robot is directly in its line of sight) with an extended sensing range while minimizing the amount of information that a robot needs to process (only a single bit of information). Gauci et al. [273] investigated reliable aggregation. They used methods of evolutionary computation to evolve two types of controllers: a memoryless (reactive) and a memory-based one. The controllers consist of only four parameters that determine the two wheel speeds for the differential drive, depending on whether the sensor outputs a one or a zero. Both controllers are shown to achieve effective aggregation. The approach is validated in both large-scale simulations and physical experiments with 20 e-puck robots.

In a follow-up paper, Gauci et al. [274] present a minimalist approach to object clustering with a swarm of robots that lack both memory and arithmetic capabilities (i.e., no computation). Each robot can only detect whether it is facing an object, another robot, or nothing, with no information about distance. Controllers were evolved in simulation and shown to reliably produce emergent object clustering behaviors, even when robots cannot distinguish between objects and other robots.

In yet another paper, Özdemir et al. [610] demonstrate that robots can also achieve robot shepherding without requiring computation. Using only two bits of sensory input (about the first object seen: sheep, shepherd, goal, or none) and a fixed mapping to motor commands, robots guide a group of sheep to a goal.

They also showed similar results, for example, in collective decision-making [611] and moving mixtures of particles [207]. This work indicates that a swarm of robots can be minimal in its capabilities and still collaboratively achieve relatively complex tasks. In realistic applications, the robot controllers will likely be more complicated than the ones reported here, but this clearly illustrates a key idea of swarm robotics: designing complex systems by engineering simple components.

6.8.2 *Programming Swarm Robots by Shaping Them*

Usually, and also for almost all content in this book, we program our robots by designing and implementing a controller in software. The robot's shape and how it is equipped with sensors only constrain that approach, but do not determine it. In evolutionary robotics and related works in artificial life, for example, there are approaches to co-evolve a robot's morphology in sync with its software. Given that we prefer simple robot hardware in swarm robotics, it is within reach to simplify the robot to an extent that most of its behavior is determined by its shape and material properties.

Ben Zion et al. [71] demonstrate how robot morphology can be used as a programming tool for swarm behavior. They specialize in the shape and flexibility of the robot bodies. They use the Kilobot [685] as a basis but enhance it with 3D-printed exoskeletons that change how they respond to physical forces and robot–robot collisions. The authors introduce two exoskeleton designs: (a) *frontiers* that push into obstacles and (b) *aligners* that slide along obstacles. The Kilobots still run a regular and identical software controller, but the two exoskeletons generate distinct emergent behaviors. This specifically designed physical interaction between robots is what the authors call morphological computation. The study also introduces a decentralized reinforcement learning algorithm that allows the swarm to adapt its behavior using only local sensing and communication. This work of Zion et al. [71] shows how morphological computation enables programming swarm robots not just by code, but by their physical shape. These physical differences affect key swarm tasks and have applications in collective transport and phototaxis. Small changes in design can replace or augment complex control algorithms.

Casiulis et al. [126] introduce a geometric design rule for programming collective behavior in robot swarms by shaping their bodies. These are simple robots without sensors. They have two counter-rotating vibration motors and a battery. A 3D-printed circular chassis (radius of about 3 cm) has a pair of soft legs and a stiff leg. The authors show that two key properties of each robot can be used to predict and control pairwise cohesion and emergent swarm behavior. They work on the morphological curvature and what they call the robot's *curvity*, which describes how the robot's heading responds to external forces. They find a simple criterion about *curvity* and radius. Using it, the authors can design the robots to attract each other and to form stable clusters. Experiments with passive vibration-driven robots and simulations confirm that the relation between radius and *curvity* governs transitions between swarm behaviors, such as disordered, clustered, and flocking phases.

A related work by McEvoy and Correll [533] addresses similar ideas, although it does not focus on mobile robots. They introduce the concept of what they call robotic materials. These are composite systems tightly integrating sensing, actuation, computation, and communication. Inspired by biology (e.g., skin, muscle), these materials enable decentralized, shape-responsive behavior. The authors discuss examples, such as artificial skins, shape-morphing beams, and tactile surfaces, where distributed microcontrollers embedded in the material respond locally to stimuli, enabling collective and adaptive responses. Although this work is not on swarm robots as such, it is foundational for programming collectives via morphology, and shows how shape-changing, self-sensing materials can embed control into the body itself. Hence, they define an approach to programming by designing shapes. For swarm robotics, this suggests that morphological programming could also help in self-assembly and other ways of coupling the robots tightly.

6.9 Green Robotics

With the rise of mobile robotics and as we will produce more and more robots, the question of how sustainable their production, maintenance, and recycling is will become more critical. The idea of green robotics [814] is to address such challenges to improve sustainability in robotics. An interesting aspect is also the concept of repurposing robots to increase their useful life and to reduce waste. Another challenge is energy, as mobile robots tend to be rather energy-hungry, especially in terms of their locomotion [116, 407]. A potential solution is, for example, energy harvesting [66]. The most feasible options seem to be: solar, wind (e.g., sailing or soaring), and possibly regenerative braking. Take the NASA Mars rovers, for example. At the same time, specific designs rely on nuclear energy, such as the Mars rover *Curiosity* [661] that runs since more than 13 years, NASA's Mars Exploration Rover *Opportunity* (see Fig. 6.11) ran for about 14 years on solar with 140 Watts peak power [163, 745]. An option for drastically reducing energy requirements is sailboats: autonomous vessels that use wind energy for propulsion and solar panels or miniature wind turbines for their electronics [166].

McGloin et al. [534] discuss the concept of robot repurposing as a sustainable approach to reduce electronic waste from robotic systems. Instead of recycling or disposing of robots after their primary life, repurposing includes altering the robot's skills (i.e., tasks it can perform) and also its applications. This way, one can extend the robot's useful lifespan. Repurposing follows the idea of a circular economy by retaining greater utility and reducing environmental impact compared to traditional recycling. Challenges include technical feasibility, economic viability, and attitudes within the robotics industry.

A potentially powerful technical effort is enabling mobile robots to harvest their own energy. From the robotics perspective, this has the potential to allow for long or even infinite energy autonomy. From the perspective of Green Robotics, it means that we can save on energy and materials for the charging station. With energy harvesting in place, the whole robot swarm can be managed with more focus on the individual and collective energy budget. Depending on the task and task deadlines, a lot of aspects of task allocation can be made energy-aware to manage the available energy budget intelligently.

Mokhtari et al. [571] propose a novel energy-aware scheduling method for collaborative robot swarms with energy-harvesting capabilities optimized through a variant of Particle Swarm Optimization [418]. Robot tasks are dynamically scheduled to

Fig. 6.11 NASA's Mars Exploration Rover *Opportunity* with solar panels (artist's portrayal, public domain)

manage energy resources efficiently, predict energy consumption, and minimize task execution time. The approach was successfully tested in simulations of a warehouse scenario using TurtleBot3 robots.

The autonomous sailboat mentioned above is an example system, where the whole design of the robot is governed by the energy-harvesting principle [412]. Sun et al. [760] provide a comprehensive review of autonomous sailing robots from an energy perspective, motivated by their potential for long-term maritime missions. They discuss strategies for actuation, energy harvesting (solar, wind, wave), and energy management to maximize operational duration and efficiency. The authors demonstrate how integrating energy harvesting and intelligent energy management can significantly enhance the endurance of sailing robots, and, consequently, their sustainability in autonomous marine exploration.

6.10 Multi-swarm Coordination

The effort of swarm robotics is to coordinate many mobile robots instead of only one, as in traditional robotics. We push it to the next level once we allow several robot swarms to operate independently in one space. The idea is that the robots do not need to be different in hardware and software, but could be owned by different parties. In contrast to the heterogeneous swarm idea (see Sect. 5.5), these are two or more robot swarms that follow different possibly even (partially) conflicting objectives. However, we also do not exclude the possibility that they may still collaborate. An example could be two robot swarms owned by different companies operating in the same area simultaneously.

Introducing another layer in the hierarchy, from single robot to swarm and now multi-swarm, seems consequential and follows a common divide-and-conquer strategy. It seems reasonable in the context of human–swarm interaction, for example, when there are too many robots to be controlled by one operator (see Sect. 6.2). Given the already increased complexity of running real-world robot swarm experiments, it is expected that there is not much literature on two or more interacting robot swarms. However, there are a few papers on simulated multi-swarm systems.

Miyauchi et al. [563] introduce a multi-swarm concept within the domain of human–swarm interaction. They investigate how multiple human operators can dynamically share control of two robot swarms. Here, the two robot swarms do not differ in hardware or software but only in their dynamic assignment to either of the two operators. Hence, one could argue that these are rather two robot groups belonging to the same swarm. We grant the fact that the two groups operate temporarily in spatial distance and could be seen as two independent swarms. The study focuses on the effects of direct (verbal) and indirect (interface-based) communication between the human operators. In a user study, participants controlled separate swarms but could exchange robots to adjust group size dynamically based on task demands. Results showed that sharing robots between the two operators reduced the total energy consumed by the swarm, allowed teams greater flexibility, and enabled more independent work. However, the authors did not find a significant increase in task performance scores. Direct communication slightly improved performance compared to indirect communication. Probably expected, clear human-to-human interactions are helpful in collaborative swarm operations.

Bodi et al. [85] investigate the robustness and interaction of robot swarms using the BEECLUST algorithm (see Sect. 5.3.1). Two distinct robot swarms perform opposite tasks as one is seeking bright spots, while the other is seeking shadows. The study is based on simulations. In a somewhat counterintuitive result, the authors found that small robot swarms benefit from interactions with other swarms. Although one might expect that two swarms with different or even opposite tasks would influence each other negatively, allowing swarms to share a space can be a design feature for small swarms. Medium-sized swarms remain unaffected, and large swarms show adverse effects due to too high robot densities (cf. Chap. 2).

Szwaykowska et al. [768] analyze the collective behavior of two interacting swarms of self-propelled agents. They study two populations with different dynamical properties (e.g., acceleration limits) using a delay-coupled mean-field model. The authors identify three emergent swarm behaviors: ring formations, collective rotation, and straight-line translation. They show how population segregation arises naturally due to heterogeneity. Using methods of nonlinear dynamics and numerical simulations, they identify bifurcations and derive conditions under which these behaviors emerge. Delay, coupling strength, and heterogeneity influence pattern formation and transitions between motion types.

Sartoretti and Hongler [696] present analytical results on the collective behavior of one-dimensional swarms composed of Brownian agents interacting through rank-based imitation dynamics. They study the interaction of two initially separate swarms with differing average velocities and analyze the outcomes of their, what they call, collisions. Depending on parameter values, the swarms either merge into a single cohesive group or return to independent motion. These results provide insight into flocking, cohesion, and phase transitions in systems of locally interacting stochastic agents.

These examples show that multi-swarm systems introduce new forms of interaction, from cooperation and resource sharing to competition and conflict. While still largely unexplored, this opens up a promising perspective for collective behavior at an even higher organizational scale.

6.11 Further Reading

Swarm robotics research is rapidly evolving, and many of the most innovative ideas remain at the forefront of the field. The best way to stay up to date is to follow the leading international conferences, where the latest results are presented and discussed. In particular, the proceedings of ICRA (International Conference on Robotics and Automation), IROS (International Conference on Intelligent Robots and Systems), DARS (Distributed Autonomous Robotic Systems), ANTS (International Conference on Swarm Intelligence), MRS (International Symposium on Multi-Robot and Multi-Agent Systems), and AAMAS (Autonomous Agents and Multiagent Systems) provide an excellent entry point for the latest developments in algorithms, systems, and applications of swarm robotics. For another look at applications, Kegeleirs and Birattari [413] review the current limitations and enablers of swarm robotics, offering a clear perspective on the challenges that still hinder real-world applications. For more details on blockchains, refer to Pacheco et al. [614] who introduce *Toychain*, a lightweight blockchain framework tailored for swarm robotics, enabling research on trust, coordination, and security under resource constraints.

6.12 Exercises

Exercise 6.1 (*Human-Guided Flocking*) Simulate flocking behavior in a robot swarm and investigate whether a human-controlled leader can steer the entire swarm. Analyze the conditions under which a leader's influence succeeds or fails.

(a) Program a simulation of flocking with live visualization using your favorite method for about $N = 20$ robots. Utilize a lightweight agent-based simulation framework that can simulate a robot swarm with minimal computational demand. Place the swarm either in a bounded space, ensuring that wall effects are not catastrophic, or place the swarm on a torus.
(b) Choose one agent to become the human-controlled robot. Develop a minimal interface to enable user control of this robot. Use the arrow keys for the input of a 2D velocity vector.
(c) Conduct a series of experiments to test the influence of the controlled robot. Vary parameters (e.g., swarm alignment strength, cohesion range, robot speed) to find under which conditions the human-controlled robot successfully steers the swarm to a target area.
(d) Allow the user to switch between robots. Does that help to get better control of the swarm?

Exercise 6.2 (*Human–Swarm Interaction: Analyzing Data*) Imagine we have a swarm of $N = 20$ robots that explore a space. Each of them sends a data stream to the human operator. The data is noisy, and only with clever analysis of several streams at a time, combined with robot navigation, can the human operator make sense of the data.

Generate $N = 20$ noisy data streams representing sensor readings of a robot swarm during exploration. The challenge for a human operator is that each stream looks random, while only aggregated statistics reveal a hidden event.

(a) Write a generator that produces $N = 20$ time series of Gaussian noise with length $T = 3000$. Insert a short event window (e.g., 60 s) where each stream is shifted by a tiny standard signal that is barely visible in any individual stream. Save the data.
(b) Plot the 20 time series in small multiples and visually inspect them. Can you detect the event from individual streams? Plot the mean signal across all streams. Do you now see the hidden event?
(c) Implement a statistical detection method. A possible option is a cumulative sum (CUSUM) or exponentially weighted moving average (EWMA) test on the group mean. Evaluate your detectors by comparing their output against the known event window. How early and how reliably can the event be detected? Experiment with changing the number of robots N or the signal amplitude to see how swarm size and signal strength affect detectability.
(d) Add explicit representation of space. Each robot moves around in a (correlated) random walk. Signals now only show a common effect if the robots were in a certain sub-space of the arena at the same time. Change your data generator

accordingly. Extend your approach to a two-phase system: First, robots move around and notify the human operator if they, by chance, share the same subspace at the right moment and notice a signal shift together. The human operator should decide whether they want to send all or many robots to that spot. With more robots, the data can be analyzed more easily. Experiment with this idea and try to simulate a scenario where the human operator's right decisions enable the robots to measure and identify the correct feature in the signal accurately.

Exercise 6.3 (*Scalability in ROS 2*) You will study how ROS 2 communication scales as the number of robots (nodes) increases and which middleware/Quality-of-Service (QoS) settings are most suitable for swarms. Use lightweight publisher/subscriber nodes that mimic swarm traffic (e.g., periodic state broadcasts). Run everything on a single machine first (multiple processes) and, if available, repeat key trials across two machines on a network.

(a) Minimal swarm traffic generator. Implement a `talker` and `listener` (C++ or Python) that publish/subscribe a custom message with fields: `stamp (builtin_interfaces/Time)`, `seq`, and `payload`. Make the rate (5, 10, 30, 60 Hz), payload size (128 B and 4096 B), and topic count N configurable via parameters or Command Line Interface (CLI). Provide a launch file to start N publishers and N subscribers (1-to-1 or many-to-many).

(b) Baselines and metrics. Use the baseline QoS on *both* publisher and subscriber: CycloneDDS. Measure:

- end-to-end latency (receive time minus send time carried in the message), latency jitter (std. dev.), and loss rate (sequence gaps);
- discovery time to steady state: elapsed time from process start until each publisher has ≥ 1 matched subscription and vice versa;
- per-process CPU and memory, total network throughput, and DDS participants/endpoints count.

Log results to CSV from callbacks; collect system stats with tools such as `pidstat`, `ps`, and `ifstat`. For multi-host trials, synchronize clocks, otherwise treat latency comparatively and measure inter-arrival jitter.

(c) Scaling sweep (N). With the baseline settings, sweep $N \in \{10, 25, 50, 80, 120, 160, 200\}$. For each N and each message size/rate, record all metrics. Plot latency, jitter, loss, discovery time, CPU, and throughput versus N. Identify the "knee point" where performance degrades sharply.

(d) QoS tuning. Repeat runs for `reliable` vs. `best_effort`, `KEEP_LAST` $\in$ {10, 100}, and `history` $\in$ {`KEEP_LAST`, `KEEP_ALL`}. Discuss tradeoffs (reliability vs. latency vs. CPU) and show whether `reliable` exacerbates congestion at higher N.

(e) ROS middleware interface (RMW) comparison and intra-process/zero-copy. Compare two RMWs (e.g., CycloneDDS vs. Fast DDS). Document versions. For each, toggle intra-process communication on/off. If your language/RMW supports loaned/zero-copy messages, test them using a fixed-size payload type. Report differences in discovery and steady-state performance.
(f) Network constraints and Wi-Fi caveat. If you have only one machine, emulate constrained networks with `tc/netem` (e.g., 20 ms RTT, 10 Mbit/s, 5% loss). If possible, repeat one condition on wired Ethernet vs. Wi-Fi. Record how latency, loss, and discovery behave and summarize the risks of wireless for large swarms.

Exercise 6.4 (*Evolving n-Version Controllers*) Explore how repeated runs of an evolutionary algorithm, seeded with different random initializations, can generate multiple diverse yet functionally correct controllers for a specified swarm behavior. Evaluate the resulting system from an n-version programming perspective in terms of fault tolerance, behavioral variance, and agreement.

(a) Choose a well-specified swarm task (e.g., aggregation, foraging, object pushing, area coverage). Clearly define success criteria for the behavior (e.g., aggregation: all robots within radius r within time T).
(b) Use an evolutionary algorithm (e.g., NEAT) and an open-source software framework combined with a simple lightweight robot simulation. Use the above-defined success criteria to formulate a fitness function. Ensure that your evolutionary algorithm produces diverse results, meaning different outcomes each time it is run. Do statistically independent evolutionary runs. Use the same task specification and fitness function for all runs. Select a set of $n = 3$ or more different but solving robot behaviors.
(c) Implement the n-version execution—variant 1: Each robot runs only one of the evolved controllers. Robots are randomly uniformly assigned one of the n controllers. Implement the n-version execution—variant 2: Each robot runs all n evolved controllers concurrently. Implement a voting mechanism (e.g., majority) when robots must agree on a discrete choice (e.g., direction).
(d) Evaluate robustness and agreement of both n-version approaches. Are the n-version swarms better than any homogeneous swarm running only one of the evolved controllers? Is the n-version variant 1 or 2 better?

Exercise 6.5 (*Controlling an Autonomous Sailboat Swarm*) Before starting to simulate the autonomous sailboat, study the fundamental principles of sailing. Fundamental concepts include points of sail (e.g., close-hauled, beam reach, broad reach, running), apparent versus true wind, tacking and jibing, sail trimming and rudder steering, as well as no-go zones and limitations in upwind sailing.

(a) Why can't sailboats sail directly into the wind? What are the typical angles for efficient upwind and downwind sailing? How do real sailors position their boats relative to others in a regatta? What sensors and actuators would an autonomous sailboat need?

(b) Single boat simulation: Program a simple simulation that allows you to control and simulate one autonomous sailboat. The input is wind direction (global) and a goal waypoint. The boat model includes position, heading, and speed. The boat's controls include rudder angle (for turning) and sail angle (which can be assumed to be actively driven into position). Implement the logic for tacking when sailing against the wind. Write a program that generates a path, that is, waypoints that account for wind and constraints for tacking to reach the goal point.
(c) Swarm of autonomous sailboats: Extend your simulation to control a swarm of $N = 5$ autonomous sailboats. Plan and visualize coordinated movement toward a shared goal (e.g., upwind target). Think of a way to maintain formation. Test different formations: line-abreast, V-formation, and staggered. Avoid wind shadowing ("dirty air") between boats. Explore how formation geometry could affect collective progress. How would boats influence each other's wind (qualitatively or via a simple occlusion model)?
(d) You can keep going further. What about adding wind variability (gusts, shifts)? What if the boats explicitly communicate with each other to inform others about their position in the formation or their subsequent actions? Add obstacle avoidance and include obstacles.

Chapter 7
Modeling Swarm Systems and Formal Design Methods

That relatively simple algorithm is how my model works, and it manifests itself as complex hunting behavior when scaled up to a swarm of stigmergic agents.

—Daniel Suarez, Kill Decision

Leon and I could try to model a swarm of them electronically. We could give them various characteristics and see how long it takes for them to start acting like a brain.

—Frank Schätzing, The Swarm

Abstract We review modeling and design methods, from mathematical abstractions to data-driven and search-based techniques. Modeling in swarm robotics focuses on reducing dimensionality to abstract and formalize inherently complex systems. A central challenge is local sampling, as robots perceive only limited and unreliable information. Modeling approaches include rate equations, spatial models using ODEs and PDEs (e.g., Langevin, Fokker–Planck), and network-based methods, such as random graphs and adaptive networks. Formal design techniques, including multi-scale modeling and global-to-local programming, address the micro-macro problem of linking local rules to global behavior. Automatic design methods range from data-driven approaches (reinforcement learning, deep learning, imitation learning, and inverse reinforcement learning, as well as large language models) to search-driven optimization (evolutionary algorithms, black-box methods). A persistent challenge across all approaches is the *reality gap*, the discrepancy between simulated and real-world performance.

H. Hamann, *Swarm Robotics*,
https://doi.org/10.1007/978-3-032-10584-4_7

7.1 Introduction

7.1.1 Learning Objectives

After reading this chapter, you should be able to:

- Understand the motivations of modeling in swarm robotics, particularly dimension reduction through abstraction, simplification, and formalization of complex systems.
- Recognize the challenge of local sampling, where robots rely on limited, often unreliable local perception not representative of the whole swarm.
- Identify and compare common modeling techniques, including rate equations, spatial models (ODEs, PDEs such as Langevin and Fokker–Planck), and network models (random graphs, adaptive networks).
- Describe formal design methods addressing the micro–macro problem, such as multi-scale modeling, verification, and global-to-local programming.
- Explain automatic design methods, from data-driven approaches (reinforcement learning, deep learning, imitation learning, and inverse reinforcement learning, large language models) to search-based methods (evolutionary algorithms, black-box optimization), and understand the persistent challenge of the reality gap.

7.1.2 Guiding Questions

- What is the primary motivation for modeling swarm robotics, and how do abstraction, simplification, and formalization help in understanding complex, high-dimensional systems?
- What is the "local sampling" challenge, and how does limited local perception affect the design of robust swarm behaviors?
- How do rate equations, spatial models (Langevin, Fokker–Planck), and network models (random, adaptive graphs) differ, and in which scenarios are they most useful?
- How do the Langevin and Fokker–Planck equations jointly link microscopic trajectories with macroscopic densities, and why is this micro–macro connection important?
- What are the core principles of formal design methods, such as multi-scale modeling and global-to-local programming, and how do they bridge local rules and global behaviors?
- What are the two main classes of automatic design methods (data-driven and search-driven), and what are examples of techniques in each?
- How do Reinforcement Learning and Deep Reinforcement Learning work in swarm robotics, and what challenges are addressed by centralized training with decentralized execution?

- How do Imitation Learning and Inverse Reinforcement Learning differ in using expert demonstrations to learn swarm behaviors?
- How are Large Language Models beginning to be applied in swarm robotics, for tasks such as code generation, planning, or human-in-the-loop coordination?
- What is the reality gap, why does it hinder the transfer from simulation to real swarms, and what strategies exist to mitigate it?

7.2 Introduction to Modeling

7.2.1 What Is Modeling?

Modeling is an essential part of the scientific approach in general. Modeling of systems usually has three objectives: abstraction, simplification, and formalization. Abstraction means to omit parts of the modeled system intentionally. Simplification means to transfer the modeled system into a simpler system that is easier to understand. Formalization means to describe the modeled system by a formal system that allows for deriving new insights by means of logic and mathematics. The option of getting new insights about an investigated system by means of formal derivation is essential because this is how arguably half of scientific progress is made (cf. history of science, such as Paul Dirac's prediction of neutrinos, etc.). The other half is made by a purely creative approach of gaining a deep (possibly informal) understanding of the system first (cf. history of science, such as Werner Heisenberg's creation of matrix mechanics, etc.).

Why is it important for swarm robotics to model swarm systems? The problem is that swarm robotics systems consist of many entities and hence are of high dimension. For example, consider a swarm of size N in two-dimensional space. Then each agent $i \in \{1, 2, \ldots, N\}$ has a position $\mathbf{x}_i \in \mathbb{R}^2$ with $\mathbf{x}_i = (x_i^1, x_i^2)$, a velocity $\mathbf{v}_i \in \mathbb{R}^2$ with $\mathbf{v}_i = (v_i^1, v_i^2)$ which includes the agents direction, and a discrete state $s_i \in \mathbb{N}_0$. Now we can define any of the system's configurations as

$$\gamma = (\mathbf{x}_1, \mathbf{x}_2, \ldots, \mathbf{x}_N, \mathbf{v}_1, \mathbf{v}_2, \ldots, \mathbf{v}_N, s_1, s_2, \ldots, s_N), \tag{7.1}$$

which also defines a configuration space Γ with $\gamma \in \Gamma$. This space has dimension $\dim(\Gamma) = 2N + 2N + N = 5N$. For a big robot swarm of size $N = 1000$ we get a huge 5000-dimensional space. There is no good way of trying to understand how such a huge system works except for running simulations. The simulation actually keeps track of all $5N$ dimensions, updates each variable in each time step (in a simulation, we have discrete time), and maybe even visualizes the system. That way, we can observe sequences of configurations (or trajectories) $\gamma_t, \gamma_{t+1}, \gamma_{t+2}, \ldots$ However, we can only observe one sequence at a time for a specific initialization γ_0. What we ideally need instead is to understand how the system operates in general for any setup. Therefore, it is obvious that we need to reduce the high-dimensionality of the

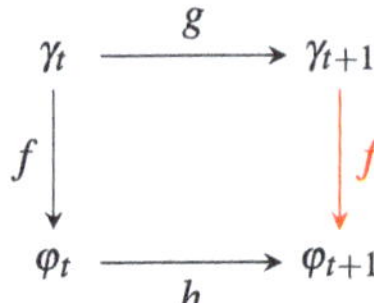

Fig. 7.1 Modeling represented as a series of mappings. The requirement given in Eq. (7.3) should hold

system by a modeling approach, that is, by abstraction and simplification. We need to omit certain parts of the system such that we obtain a different definition of a configuration, say φ, with a smaller configuration space $\varphi \in \Phi$. We would like to have $\dim(\Phi) \ll \dim(\Gamma)$. For example, we could try to find a mapping f that maps the actual configurations γ to the reduced configurations φ:

$$f : \Gamma \mapsto \Phi. \tag{7.2}$$

The map f implements an abstraction which is already a part of our modeling approach.

We consider, for now, a discrete-time model. Then we can define two update rules as maps $g : \Gamma \mapsto \Gamma$ for the real system and $h : \Phi \mapsto \Phi$ for the model. They determine how the system develops over time. For a given configuration γ_t at time step t, we determine the configuration of the next time step $t + 1$ by $g(\gamma_t) = \gamma_{t+1}$ and similarly for h. Now we can put everything together as shown in Fig. 7.1. The map g directly determines the evolution in time of the real system. If we are provided with a configuration γ_t and we want to know the configuration that our model predicts, then we need first to map γ_t to the model space Φ using f to obtain φ_t, and then we can apply the update rule h to obtain φ_{t+1}. We can now compare the configurations γ_{t+1} and φ_{t+1} by applying f to γ_{t+1} again. We would like to have

$$h(f(\gamma_t)) \stackrel{!}{=} f(g(\gamma_t)), \tag{7.3}$$

that is, the modeling abstractions implemented by f should be chosen carefully such that the model update rule h can predict the correct model configuration for the next time step. Also note that we could try to define an inverse map $f^{-1} : \Phi \mapsto \Gamma$ that reverses the model abstractions and rebuilds the configuration γ of the real system from the model configuration φ. However, typically f^{-1} cannot usefully be defined because f is surjective, that is, we can have $\gamma_1 \in \Gamma$ and $\gamma_2 \in \Gamma$ with $\gamma_1 \neq \gamma_2$ and $f(\gamma_1) = f(\gamma_2)$. This is also easily accepted by thinking of a possibly extreme abstraction f that reduces a high-dimensional system to a low-dimensional model. Obviously, we would need to throw away a lot of details to achieve that. Once we try to reverse that abstraction, we would need to add all kinds of details without any information about which details were omitted before. For the inverse map f^{-1}, that means that it can, in general, only define an arbitrary representative γ for a model configuration φ.

7.2.2 Why Do We Need Models in Swarm Robotics?

The main objective of modeling in swarm robotics is dimension reduction. We need to abstract away microscopic details to detect key features and general principles of operation in the system. This is also nicely stated by Schweitzer [723]:

> To gain insight into the interplay between microscopic interactions and macroscopic features, it is important to find a level of description that, on the one hand, considers specific features of the system and is suitable for reflecting the origin of the new qualities, but, on the other hand, is not flooded with microscopic details. [...] A commonly accepted theory of agent systems that also allows analytical investigations, is, however, still pending because of the diversity of the various models invented for particular applications.

He is also pointing to the problem that we do not yet have a common model approach that can be applied to any scenario. Instead many different models have been developed and applied to different setups of swarm robotics systems.

The abstraction of microscopic details that fights the "microscopic flood" could in principle also be done automatically in a brute-force approach with methods from machine learning and statistics. Options would be the principal component analysis or feature extraction. However, experience shows that these approaches are not constructive. They reduce the dimensionality, but the obtained reduced configuration space Φ would have, in general, no explanatory power and no meaning.

What we need is a reduced configuration space Φ that still contains and can represent key features of the modeled system. What a key feature is depends on the respective application. In a collective decision-making system, for example, the current status of who is in favor of which option is undoubtedly a key feature, but the agent positions might be of less relevance. In an aggregation scenario, the distance between agents is a key feature, but maybe their internal state is of less importance. In general, we choose whether we want to model time and/or space. If we decide to represent the temporal course of the system, that is, we have a time-dependent model, then we can choose between a discrete- or continuous-time model. Suppose we decide to represent the position of agents or agent densities, that is, we have a spatial model. In that case, we can also choose between a discrete or continuous representation of space. It turns out that many of our models cannot be solved analytically, which means we have to solve them numerically using computers. If you are a computer scientist, then you might want to take a practical position; in this case, we should always think of discrete models. However, the notation of a continuous model is often concise and easy to handle. It also allows for a simple comparison between models, which is essential to quickly determine whether two proposed models are different and how they differ.

Let us consider a simple example using the above developed notation. We consider a binary collective decision-making scenario. Binary means that there are only two options, say, A and B. Furthermore, we assume it is a two-dimensional spatial system and each agent $i \in \{1, 2, \ldots, N\}$ has an internal state representing its current opinion $s_i \in \{A, B\}$. The dimension of the configuration space Γ is $\dim(\Gamma) = 2N + 2N + N = 5N$. We define an extreme model abstraction via map

$$f(\gamma) = \frac{|\{s_i | s_i = A\}|}{N} = \frac{[\text{number of agents in favor of option } A]}{N} = \varphi. \quad (7.4)$$

Here φ represents the key feature, which is the decision-making process. The dimension of the model is $\dim(\Phi) = 1$. We have reduced the $5N$-dimensional system to just one dimension. The important question to ask is now: how does the update rule $(h(\varphi_t) = \varphi_{t+1})$ of the model look like? This is not a trivial question and will be discussed later.

7.3 Local Sampling

When designing and programming controllers for swarm robotics, we face a common challenge because of the lack of information. In many situations, it would be helpful to know about the global picture, for example, the size of the swarm or where the currently most significant cluster of robots is positioned. However, an individual robot does only have local perception and can only collect a bit more information by communicating with its neighbors. It turns out that agents of much higher capabilities face similar challenges in certain situations. For example, everyone who has ever lost in a foreign city knows the problem of only knowing the names of maybe two adjacent streets. Still, without the possession of a city map, that information is of limited use. In fact, on a bigger scale, the human society had a similar problem in the Sixteenth century when trying to locate the Earth in relation to surrounding stars and planets. An example of a regularly recurring problem is predicting the result of votes. Surveying all potential voters would be too expensive; therefore, only a small subset is checked. How to choose this subset usefully is a scientific question. A similar challenge occurs in market research where, for example, the number of customers for a particular product needs to be predicted, of course, also without asking every potential customer.

Inferring the "global picture" in the context of localization is a standard problem in robotics. A common method is SLAM (simultaneous localization and mapping), which generates a global map of the environment while the robot is trying to localize itself within it. Another example is the critical task in RoboCup soccer to localize the ball. The robot team cooperates by sharing each robot's local perception. These local sections of the global picture are then merged to produce a reasonable estimation of the ball's actual position (multi-robot sensor fusion). In RoboCup, broadcasts typically employ this approach, namely global communication, which doesn't scale.

In swarm robotics, the only option is to do local sampling, and information can only be shared with neighbors, but not with everyone in the swarm. In fact, often a robot relies on the information it can gather on itself without acquiring more information through communication. An example is stigmergy, for example, cue-based navigation by pheromones. An agent only knows the local pheromone concentration, which defines a direction and carries the information of where to go. Another example is the alignment rule in flocking, where a robot checks the direction of its neighbors and aligns with them. The underlying assumption of this process is that the local sample of neighbors gives a reasonable estimate of the swarm's dominant direction of motion. Why it is helpful to assume this and why it does not always end in a disaster of a dissolving swarm is nontrivial.

One key ingredient that supports the effectiveness of these local sampling processes is self-organization. An example is the emergence of order in pedestrian flows. Although most of the pedestrians are most likely occupied with other thoughts than how to organize the flow efficiently, often two streams of pedestrians with the same walking direction form. The local sampling here is the detection of the walking direction of the person in front of you. If the person is walking in your direction, you follow, and if the person is approaching you, then you step to the side. The formation of ordered flows based on these simple, local rules is an emergent effect.

7.3.1 Sampling in Statistics

Sampling is, of course, a process that is well-defined in statistics. First, we define the statistical population, which is the set of potential measurements. A sampling is the selection of a subset of elements from such a statistical population. Different types of sampling are known. In a nonprobability sampling, some elements of the population have no chance of getting selected. In cluster sampling, elements are selected in groups. For example, all households in a particular street are chosen or a specific class of students from a school. In accidental sampling, elements are selected that are "close at hand." For example, when a survey is done in a street, then the people that the surveyor accidentally runs into are not particularly selected, but they are just close at hand. Obviously, it is helpful to define a metric that defines the quality of a sampling. The most essential quality of a sampling is representativeness. If the selected subpopulation has "similar properties" as the full population, then the sample is representative. The definition of what "similar" should mean here is crucial, nontrivial, and depends on what is of relevance in a particular situation.

As an example of how tricky it is to pick a good subpopulation and how representative it is, we take the Literary Digest Disaster. The idea was to do a straw poll for the US presidential election in 1936 of Alf Landon against Franklin D. Roosevelt. The vast number of ten million people was polled by mail and phone, and 2.4 million answered. 60% of them were in favor of Alf Landon, and the prediction was that he would gather 370 of 531 electors (remember the uncommon voting system of the US, which is based on electors). The actual result, as the educated reader knows, is, however, very different. It turns out that Landon won only two states (Vermont and Maine, that is, only eight electors). The representativeness of the sampling was of minor quality due to several facts. The sampling was non-probability sampling because the sample was generated from lists of registered car and phone owners, as well as subscribers of the magazine (Literary Digest). These were mainly well-off households, which were typical of the Republican Party (Landon). In addition, most voters answered the poll that showed an above-average interest in the vote. Unfortunately, anti-Roosevelt voters felt more strongly about the election than Roosevelt voters, who subsequently tended not to answer the poll.

7.3.2 Sampling in Swarms

Because the sample in swarm robotics has to be local, it is a non-probability and accidental sampling. Swarm members that are too far away cannot be selected, and the neighbors are chosen because they are "close at hand." Hence, the sample's representativeness is at least questionable. There are several difficulties. The robot might be part of a cluster of robots that is not representable for the swarm. The robot's state and the states of its neighbors might not be independent; that is, they might be correlated. This might also be supported by the robot's behavior, which might actually implement this correlation directly. Finally, the swarm density might be too low and therefore the sample might be tiny, which introduces a strong bias.

We start with a first simple example of local sampling in swarm robotics, similar to the scenario we discussed in Chap. 4. Say, in a robot swarm, robots can have one of two states, black and white, and robots have to estimate the global fraction of black and white robots currently in the swarm. A robot has only a limited sensor range, which defines its neighborhood. See Fig. 7.2 for an example. In total, there are 13 black robots (52%) and 12 white robots (48%). The considered robot (marked with a dot) on the top left can only locally sample. It detects four black robots (66.6%) and two white robots (33.3%, including itself). Obviously, the robot's measured estimate differs from the true situation. However, here the robot is relatively lucky and receives a good estimate. In other areas of the arena, robots may end up with worse estimates.

The second example is more complex. The task is to estimate the size of an area without explicitly measuring it because we assume this is beyond the capabilities of a swarm robot (i.e., no laser scanner, no camera, etc.). How can that be done? An option would be to measure one side length or two, for example, by moving in a straight line with constant speed and taking the time. Then, a square or a rectangular shape of the area is assumed and accordingly calculated. A related approach would be to measure the mean free-path-length, that is, the average time between two collision avoidance events.

It turns out that there is a fascinating example in a natural system of how to solve this task. The observation of the ant species *Leptothorax albipennis* and their search for nest sites has revealed this fact. These ants may follow a procedure that is similar to what is known as Buffon's needle. How this is related to estimating the size of an area will be clear a few lines below. Buffon's needle is a statistical experiment. A needle of length b is dropped randomly onto a plane which is inscribed with parallel straight lines that are s units apart (see Fig. 7.3). We assume $b < s$. The needle has a

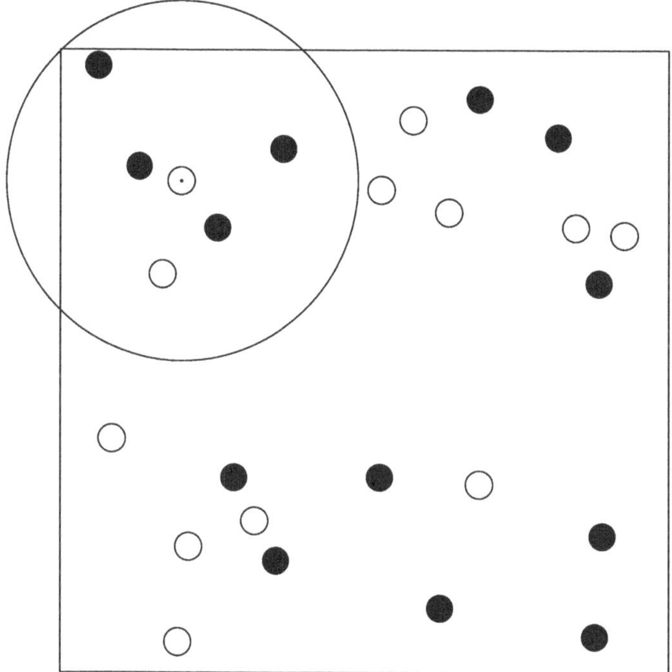

Fig. 7.2 Simple example of local sampling in a robot swarm

Fig. 7.3 Example configuration of Buffon's needle. Length $b = 7$, line spacing $s = 10$, sample: 14 intersections and 11 non-intersections giving a probability $\hat{P} = 14/25 = 0.56$ while the theory gives $P = (2 \cdot 7)/(10\pi) \approx 0.446$

probability $P = 2b/(s\pi)$ of intersecting a line. We do not present a proof here, but you can start additional reading from Ramaley's work [659].

We are interested in the sampling error, which can be interpreted as the representativeness of the sample. Only here representativeness depends exclusively on the size of the sample set because there are no individual differences between samples (in contrast to the above example of voter surveys). First, we note that the needle experiment is a binomial experiment: the needle either intersects a line or does not. Second, we assume that there is an estimated probability $\hat{P}$ that is based on a limited number n of trials (throws of needles). Now we use the binomial proportion confidence interval to give an estimation of the sampling error. The binomial distribution is approximated with a normal distribution. If we choose a 95% confidence level, then we get

$$\hat{P} \pm 1.96\sqrt{\frac{1}{n}\hat{P}(1-\hat{P})}, \tag{7.5}$$

whereas 1.96 is the appropriate percentile of the standard normal distribution (based on a table lookup). The upper bound of a confidence interval following Eq. (7.5) for an assumed correct $\hat{P} = 0.446$ over the number of trials n is given in Fig. 7.4a. It is good news that the interval quickly shrinks with the increasing number of trials. Hence, good estimates can probably be achieved with a few samples only.

The behavior of the ants is described by a variant of Buffon's needle, which is based on randomly scattered lines, see Fig. 7.4b. Say we have two sets of lines. The first set of lines has a total length of L_1 and the second set of lines a total length of L_2. The number of intersections of lines from different sets is n. The area in which these two sets of lines are placed is then estimated by

$$A = 2L_1L_2/(n\pi). \tag{7.6}$$

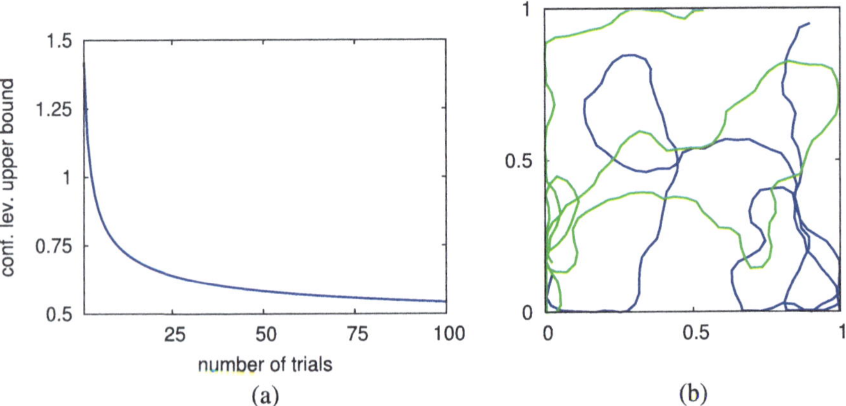

Fig. 7.4 Upper bound of a confidence interval, Eq. (7.5), and two sets of lines. (**a**) Upper bound of a confidence interval following Eq. (7.5) for an assumed correct $\hat{P} = 0.446$ over the number of trials n. (**b**) Two sets of lines and their intersections

The behavior of ants is related in the following way. When their old nest is destroyed, single ants, called scouts, are sent out to explore a region to find potential nest sites of appropriate size. They prefer flat crevices in rocks. A potential nest site is visited at least twice. On the first visit, the scout ant deploys a special pheromone, but not on the second visit. The pheromone trail of the first visit defines the line set L_1 and the scout's trajectory on the second visit defines set L_2.

Mallon and Franks [510] report several experiments. In one of the experiments, an ant colony was given the choice between two nest sites: one of "standard size" and a smaller one of 5/8th (62.5%) of standard size. They chose the standard size nest site 15 times out 15.

Although this method is based on the behavior of a single agent and not on cooperating groups of agents, it still gives a good example of how powerful local sampling can be. Instead of applying high-tech and expensive sensors, the Buffon's-needle approach is practical based on simple methods. In addition, this approach is also flexible because no special hardware is used to do the measurements. Finally, this approach gives a good example of nature's non-intuitive problem-solving in contrast to standard engineering. However, the question remains of how this approach would have been found and designed without knowing the example of the ant behavior.

7.4 Modeling Approaches

In swarm robotics, we are still on the search for an appropriate general modeling technique. That is why many different modeling approaches for swarm robotics have been published. None of them is capable of modeling all typical swarm robotics use cases. Hence, when you want to model a particular swarm system, you have to make a selection of the proper modeling technique. Which one to use is determined by the essential system features relative to the task of the swarm. Sometimes space can be abstracted, but in pattern formation, for example, it is necessary. The following list of models is incomplete but hopefully representative.

7.4.1 *Rate Equation*

A modeling approach, which was popular at least in the beginnings of swarm robotics, is the rate equations. The most important early works on rate equations were published by Martinoli [518, 519] and Lerman [478–482]. Rate equations are usually applied to model chemical systems, in particular chemical reactions and their reaction rate. A simple rate equation is $r = kAB$. This equation models the chemical reaction between two reactants. The concentrations of the chemical species are given by A

and B. k is the rate coefficient. Say we have a chemical reaction $A + B \rightarrow C$, then an ordinary differential equation (ODE) is defined by the reaction rate

$$\frac{dC}{dt} = kAB, \tag{7.7}$$

for concentrations A, B, and C. Writing Eq. (7.7) down like that is supported by the so-called law of mass action, a fundamental law of chemistry. However, it seems also quite intuitive, for example, once we interpret the concentrations A and B as probabilities of encountering these reactants. Then the product of both corresponds to finding them in one spot, which leads to the reaction that, in turn, happens at rate k.

We want to model collective robot systems instead of doing chemistry, hence we have to interpret the rate equations differently. Instead of concentrations of chemical species, we have swarm fractions of robots that are in certain states. Imagine the robots being programmed by finite state machines, and the above A and B would be robots in states A and B of their state machines. Instead of chemical reactions, we have state transitions as effects of robot–robot interactions. A state transition in the robot's state machine may be triggered by encountering another robot that is currently in a particular state. This modeling approach is macroscopic, probabilistic, and non-spatial; that is, we live on a well-mixed assumption.

As the basic concept of rate equations is quite simple, we immediately start with a little example taken from Lerman and Galstyan [479]. They want to model a foraging behavior in robots (see Sect. 5.3.8). In this foraging behavior, robots go for pucks that need to be collected and transported back home. The robots have to avoid obstacles including other robots, and initially they wander around and search for these pucks. For an easy star, we ignore all the other behaviors necessary for foraging, such as puck detection, homing, entering the home, and reverse homing. We focus on two simple states: searching for pucks and avoiding obstacles/robots (see Fig. 7.5).

To model this simplified system, we define $N_s(t)$ as the number of robots in search state at time t. $N_a(t)$ is the number of robots in avoiding the state at time t. We can safely assume a simple conservation law here. The total number of robots stays constant, hence we get

$$N_s(t) + N_a(t) = N. \tag{7.8}$$

Furthermore, we define $M(t)$ as the number of uncollected pucks at time t. We say α_r is the rate coefficient of detecting another robot and α_p is the rate coefficient of detecting a puck. What exactly do these rates define, and where do they come

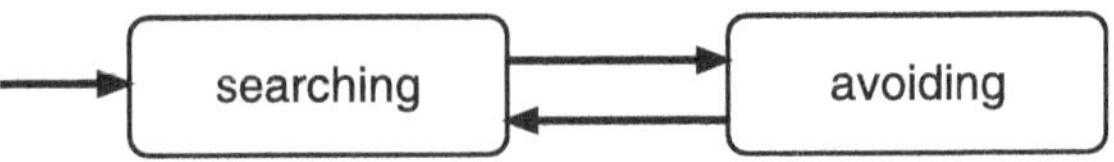

Fig. 7.5 Simple finite state machine for our example on rate equations

from? All specific properties of the robot and how it detects other robots are encoded into α_r and similarly for α_p. Obviously, that can be rather tricky, but it can be, for example, simple geometric considerations, such as opening angles of sensors and their range. If we change the sensors of the robot or we exchange one robot platform for another, then we would like to be able to reflect that in our model. This can be achieved through these rate coefficients.

We are not done yet, as we still need to model time flow. If a searching robot detects an obstacle, such as another robot or a wall, it executes the avoiding behavior for a time period τ, after which it resumes the search. In a usual implementation of a collision avoidance behavior, this time period would actually not be constant; However, for simplicity, we assume this to be the case. Now we have to work our way towards actual rate equations. Let's first consider what is expected to happen once two robots in certain states meet. The number of searching robots N_s decreases when two searching robots detect each other and hence commence avoiding maneuvers. We should also not forget the case when a searching robot detects another robot that is already in avoiding state, then N_s also decreases. We know that N_s increases when robots return from avoiding state and resume searching. That happens at time t if they started the avoiding behavior at time $t - \tau$, that is, τ time ago. Now we could go through similar considerations for N_a. However, we don't need an equation describing the dynamics of the avoiding robots N_a, because we can compute this quantity using the conservation of the total number of robots N. To write down the rate equations, we define fractions of the swarm

$$n_s = N_s/N \tag{7.9}$$

for a fraction of robots in state searching and

$$n_a = N_a/N = 1 - N_s/N = 1 - n_s \tag{7.10}$$

for a fraction of robots in state avoiding. These fractions can also be interpreted as concentrations of robots or as probabilities of finding them.

We start with a term for decreasing n_s. It depends on the rate coefficient α_r, if two searching robots approach each other both make a transition, that is, two at a time, and the other The possibility was that a searching and an avoiding robot would approach each other other.. Hence, we get

$$-2\alpha_r n_s^2 - \alpha_r n_s n_a. \tag{7.11}$$

Instead of n_a we write $1 - n_s$, expand, sum, and get

$$-\alpha_r n_s^2 - \alpha_r n_s. \tag{7.12}$$

that we write as

$$-\alpha_r n_s(n_s + 1). \tag{7.13}$$

What is missing is the term for increasing n_s. Do we really have to think long about what it should be? Actually, not because the number of robots that return from avoiding state at time t corresponds directly to those that left at time $t - \tau$. The latter case is what we have already modeled; we can reuse the above term (7.13) with inverted signs and the different time $t - \tau$. Once we put everything together, including the times, we get

$$\frac{dn_s(t)}{dt} = -\alpha_r n_s(t)(n_s(t) + 1) + \alpha_r n_s(t - \tau)(n_s(t - \tau) + 1). \tag{7.14}$$

This is actually a delay differential equation, that is generally rather difficult to solve. However, that should not bother us here because we can easily use simple numerical solvers. Similarly to n_s and n_a we can also define a fraction of yet uncollected pucks $m(t)$ and define a rate equation for the pucks as well

$$\frac{dm(t)}{dt} = -\alpha_p n_s(t)m(t). \tag{7.15}$$

We can use Eqs. (7.14) and (7.15) to define an arbitrary initial value problem as an example. Say we define $\alpha_r = 0.6$, $\alpha_p = 0.2$, and $\tau = 2.0$. As initial values we set $n_s(0) = 1$ (all robots in search) and $m(0) = 1$ (no puck collected yet). We can solve the equations simply numerically [649]. As a result, we get a model prediction for the swarm and puck fractions as shown in Fig. 7.6.

The initial oscillations for $t < 15$ may look funny, but they should actually be expected for delay differential equations, and they can likely be observed on average in swarm robot experiments. Also note the complete symmetry between n_s and n_a because of our conservation law $n_s = 1 - n_a$. We end our little study here, but for the foraging scenario of Lerman and Galstyan [479] we would need to model many more states.

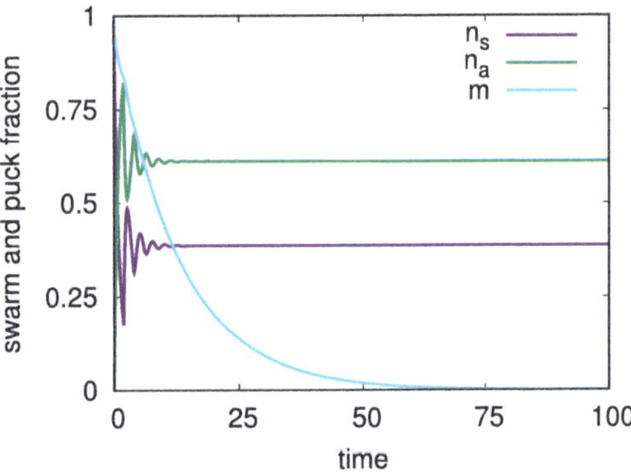

Fig. 7.6 Example for the rate equations approach, numerical solution of an initial value problem ($\alpha_r = 0.6$, $\alpha_p = 0.2$, $\tau = 2.0$, $n_s(0) = 1$, $m(0) = 1$)

7.4.2 *Differential Equations for a Spatial Approach*

The above rate equations approach uses differential equations to model the dynamics of swarm fractions. These are time derivatives. Now, we investigate differential equations to create spatial models, that is, swarm models representing space explicitly. Differential equations allow us to represent space continuously, that is, we achieve much higher precision with them than with the popular but questionable grid-world models. We use the mathematical tool of differential equations here, in particular, stochastic differential equations (SDE) and partial differential equations (PDE).

Before we start, we first have to look into a peculiarity of modeling the motion of swarm robots. Following Rosenblueth et al. [681], the behavior of biological systems, but also of robots, can be classified in several ways. We can distinguish active and passive behavior: a robot actively locomoting or a robot floating in a river. The behavior can be purposeful or random: a robot driving into a room to explore it or a robot turning away from a wall with a random turn. The purposeful–vs–random distinction can be done even more fine-grained [318, 324] as shown in Fig. 7.7. Only a swarm robot provided with proper information has the option of behaving purposefully; otherwise, it has to make a random move. If provided with information, it can still choose to ignore the information and do a random move. Finally, even with information and the intention of using it, physical constraints may require a random move, for example, to avoid collisions. Hence, modeling the purposeful and random behavior of our swarm robots explicitly may turn out to be useful.

We start our journey with a model that, for now, only represents a single robot. To motivate this approach, we interpret a robot's trajectory as a sequence of variables X_t, that is, we assume a discrete time model for the moment. In addition, we say that the swarm robot moves rather erratically, such that it may be fair to say it moves rather randomly. Then we can model the time series X_t as a sequence of random variables. So we have a collection of random variables indexed by a set T representing time:

$$\{X_t : t \in T\}. \tag{7.16}$$

If we assume for now that the robot's motion is completely random, then we can use a fully stochastic model to represent its random motion. Say the robot position at time t is $R(t)$. We model the robot's trajectory with a stochastic differential equation

$$\frac{dR(t)}{dt} = \dot{R}(t) = X_t, \tag{7.17}$$

where X_t is a stochastic process, and it is a random displacement of the robot (a little push). However, this is not a good model because usually our robots do not move completely randomly. We refine our approach by incorporating a stochastic model with drift. We add a deterministic term, for example, a constant c

$$\dot{R}(t) = X_t + c. \tag{7.18}$$

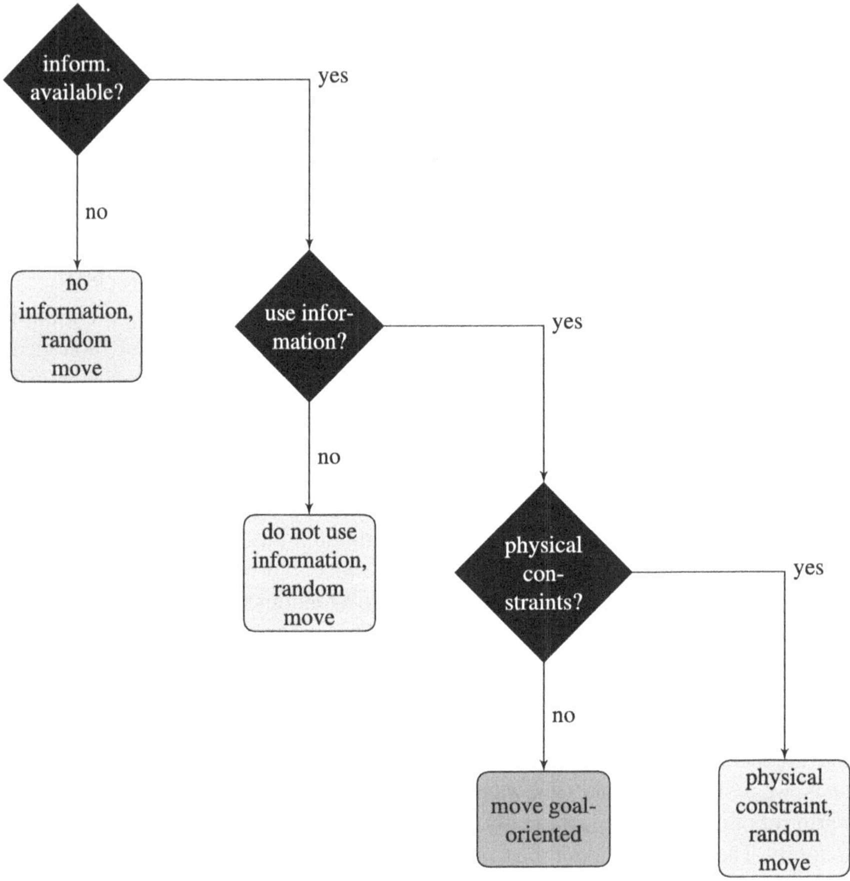

Fig. 7.7 Classes of random robot motion [318, 324]

Our robot is still subject to random displacements in each time step, but also constantly makes progress in a given direction defined by c. For example, we can define a 2-d stochastic model with a drift of $c = 0.1$ in the x-direction, where the robot is started at (10, 10). In Fig. 7.8, we show two realizations (also called Monte Carlo simulations) of the stochastic process. That is implemented by extending the above equation to the 2-d case with robot coordinates R_x and R_y

$$\dot{R}_x(t) = X_t + c_x, \tag{7.19}$$

$$\dot{R}_y(t) = Y_t + c_y. \tag{7.20}$$

This can be rewritten in vector notation as

$$\dot{\mathbf{R}}(t) = \mathbf{F}_t + \mathbf{C}, \tag{7.21}$$

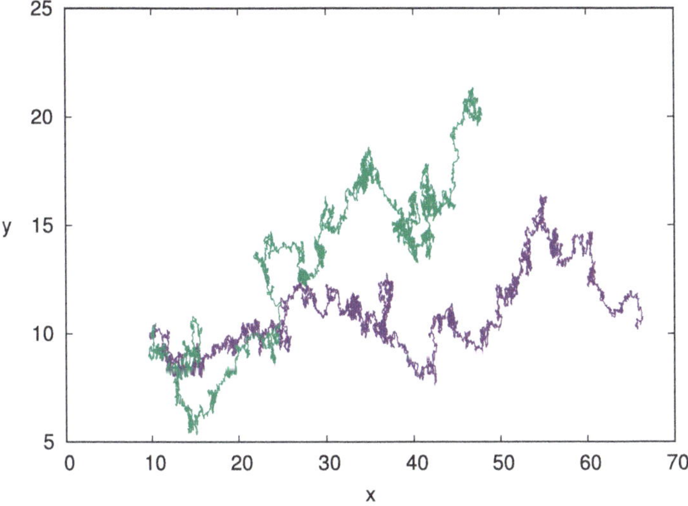

Fig. 7.8 Two realizations of a stochastic model $\dot{R}(t) = X_t + c$ with a drift in x-direction of $c = 0.1$, trajectories are started in (10,10)

where we have substituted $\mathbf{F}_t$ for $\mathbf{X}_t$ for later use and apply definitions

$$\mathbf{R} = \begin{pmatrix} r_x \\ r_y \end{pmatrix}, \quad \mathbf{F}_t = \begin{pmatrix} X_t \\ Y_t \end{pmatrix}, \quad \mathbf{C} = \begin{pmatrix} c_x \\ c_y \end{pmatrix}. \tag{7.22}$$

By further extending Eq. (7.21) we can now introduce the Langevin equation. According to the Langevin equation, the trajectory of a robot is given by

$$\dot{\mathbf{R}}(t) = \mathbf{A}(\mathbf{R}(t), t) + B(\mathbf{R}(t), t)\mathbf{F}(t), \tag{7.23}$$

for a stochastic process **F**(t) (e.g., white noise), **A** takes the robot's position $\mathbf{R}(t)$ and time t as input; it describes and scales the robot's deterministic behavior. B describes and scales the non-deterministic behavior. Note that **A** is a vector and hence also gives directional information, while B is scalar and hence only scales the non-deterministic process defined by $\mathbf{F}(t)$. A possible choice of **A** is, for example,

$$\mathbf{A}(\mathbf{R}(t), t) = \nabla P(\mathbf{R}(t), t), \tag{7.24}$$

which defines a gradient ascent in a potential field P. Originally, the Langevin equation was used to model Brownian motion. We use it as a general and generic model to describe the trajectory of a swarm robot.

Next, we have a look at the so-called Fokker–Planck equation. The Fokker–Planck equation is the macroscopically corresponding piece to the microscopic approach described by the Langevin equation. The most fascinating feature of this pair of equations is that they are one of the very few examples of a mathematical connection between micro- and macro-behaviors. The Fokker–Planck equation is mathematically deducible from the Langevin equation, although this deduction is rather involved. The combination of the Fokker–Planck equation with the Langevin equation establishes a direct mathematical micro-macro link.

The equation goes back to the two physicists Fokker [249] and Planck [641]. For physics, the equation is of interest in the connection with Brownian motion, in particular, Brownian motion with drift. Hence, it describes a diffusion process, and if there are no additional assumptions, then we should expect a so-called heat death in the long run for these systems. For example, if we initially position several particles close to each other and let them diffuse, then we expect them to approach a homogeneous distribution over time.

Probably the easiest to follow deduction is given by Haken [314] while Risken [677] has dedicated a whole book to the equation. Getting the Fokker–Planck equation out of the Langevin equation requires certain assumptions, such as the stochastic process $\mathbf{F}(t)$ being white noise (i.e., Gaussian distributed and zero mean), a modeled particle needs to have received enough collisions within short time intervals, etc. From the point of view of our application here, this is rather a rare special case or even only an abstract idealistic case. The mathematical difficulty of even this special case gives us a hint of maybe not only the mathematical complexity but also the general complexity of creating micro-macro links.

Now we want to have a closer look. The Fokker–Planck equation is a partial differential equation describing the temporal dynamics of a probability density. This density, in turn, describes in the original physical context the probability of finding the particle within a certain area (e.g., in the 1-d case by integrating it over a considered interval). We will interpret it in a slightly different way in the following. The equation is, in direct correspondence to the Langevin equation, a sum of a drift and diffusion term

$$\frac{\partial \rho(\mathbf{r}, t)}{\partial t} = -\nabla(\mathbf{A}(\mathbf{r}, t)\rho(\mathbf{r}, t)) + \frac{1}{2}Q\nabla^2(B^2(\mathbf{r}, t)\rho(\mathbf{r}, t)). \tag{7.25}$$

ρ is a probability density for a single particle at position $\mathbf{r}$ and time t. In our application of swarm robotics, we interpret it instead as the robot density of all coexisting robots of the swarm. That is, when we integrate over an area W

$$s(t) = \int_{\mathbf{r} \in W} \rho(\mathbf{r}, t) \tag{7.26}$$

then we get the expected fraction of the swarm $0 \leq s(t) \leq 1$ within that area at time t. The so-called nabla operator $\nabla = \left(\frac{\partial}{\partial r_1}, \frac{\partial}{\partial r_2}, \ldots\right)$ is based on the spatial derivatives and gives the gradient. It models the "flow" of probability density according to the drift as expected. The drift term pushes density in the direction of the drift $\nabla(\mathbf{A}(\mathbf{r}, t)$. ∇^2 is based on the 2nd spatial derivatives and is known as the Laplacian operator

$$\Delta = \nabla \cdot \nabla = \nabla^2 = \frac{\partial^2}{\partial r_1^2} + \frac{\partial^2}{\partial r_2^2} \tag{7.27}$$

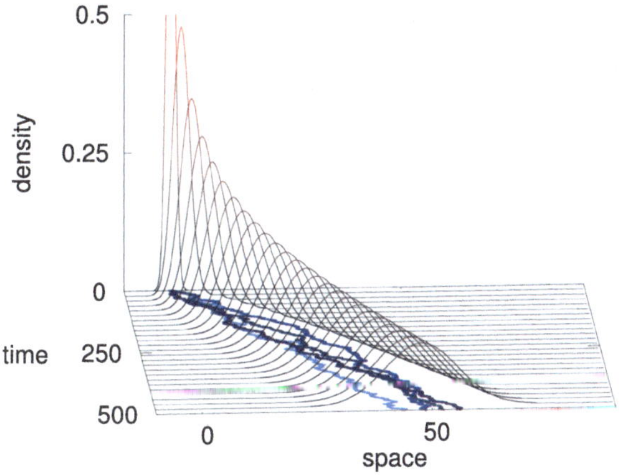

Fig. 7.9 Evolution of a probability density modeled by the Fokker–Planck equation, Eq. (7.25). Example in 1-d space with a drift of $A = 0.1$ and diffusion $B = 0.3$. Shown is the probability density and four example realizations of the corresponding Langevin equation

and models diffusion. This is also with the expected effect, that is, peaks in the probability density are lowered, the slopes around these peaks get flatter, and the valleys in the density are increased. The constant Q in Eq. (7.25) results from special properties of the stochastic process $\mathbf{F}$; one can say it models the strength of the hits on the particle. The first summand is the drift term, and here it models the deterministic part of the robots' behavior. The second summand is the diffusion term,m, and it models the non-deterministic part of the robot's behavior.

In Fig. 7.9 we show examples of how realizations of both the Langevin and the Fokker–Planck equation can look like. This figure also clearly indicates the microscopic property of the Langevin equation on the one hand (particular trajectories of robots) and the macroscopic property of the Fokker–Planck equation, on the other hand (spatial probability densities of the whole swarm).

The micro-macro duality of the Langevin equation and the The Fokker-Planck equation persists in this application as well. Say, we would have found appropriate definitions of $\mathbf{A}$ and B, then we could choose whether we want to use a microscopic model (Langevin) or a macroscopic model (Fokker–Planck). For a microscopic approach, that would mean introducing a Langevin equation for each robot to model the robot's motion. The actual simulation is a realization of these multiple, interacting stochastic processes. Robot interactions need to be handled explicitly. For example, a robot could move differently once it has more than five neighbors. With the Langevin approach, that can only be done by measuring actual distances between robots and calculating neighborhood sizes.

For a macroscopic approach, we need only one Fokker–Planck equation that models the complete swarm at once. Robot interactions can be modeled using the density ρ itself. For example, the above-described behavior that is triggered by more than five neighbors can be translated into a minimal density that triggers that behavior. Also note that such a technique inherently introduces a continuum approach. The density increases continuously from values corresponding to 5 robots, to 5.1 robots,

5.2 robots, etc. This continuum should be interpreted as probabilities, such as interpreting 5.1 as "in nine out of ten cases a robot has five neighbors and in one out of ten cases it has six neighbors."

The Fokker–Planck/Langevin agent modeling approach is inspired by "Brownian agents" of Schweitzer [723]. Hamann [318] applied the approach to robot swarms and Prorok et al. [651] reported a validation by robot experiments. For more details on the Brownian agents, see [234, 352, 722, 724, 725]. Another approach that applies PDE to control robot groups is reported by Milutinovic and Lima [558, 559]. It is an approach strongly influenced by control theory, it makes use of hybrid systems theory, and has a strong theoretical basis.

7.4.3 Network Models

Swarm robotics systems can be interpreted as networks with robots as nodes and edges indicating mutual neighborhood relations. Graph theory provides the more traditional modeling techniques. In addition, there are more recent techniques that can be summarized as "network science." The modeling techniques themselves are similar, but the focus has shifted to more sophisticated topological features and dynamic networks, where edges are allowed to emerge and vanish over time. In swarm robotics, we observe dynamic neighborhoods. A robot may have particular neighbors now, but within a second, they can disappear, and new neighbors may appear.

7.4.3.1 Random Graphs

A graph G is defined as a tuple $G = (V, E)$, for a set of nodes V and a set of edges E. In swarm robotics, nodes represent robots and edges represent a communication connection or that robots are mutually within sensor range. Being mobile robots, our robots move, and hence the nodes should move as well. However, in a graph model, that is usually not supported. Instead, graphs represent a snapshot of the system configuration at a given time. Furthermore, we are usually not interested in one particular graph configuration or topology but rather general properties for a given feature, such as the swarm density. A given swarm density would induce a certain probability of having an edge between two robots/nodes. To investigate such general properties, we generate graphs randomly but with given parameters.

The first model of random graphs was published by Erdős and Rényi [233]. An Erdős–Rényi random graph is an undirected graph $G(X; p)$ with a node set $X = \{0, 1, 2, \dots, N\}$. Each possible edge is included independently with probability p. With probability p, we can emulate the density of the swarm. If p is close to one, many edges are added to the graph and the swarm is dense. If p is close to zero, few edges are added to the graph, and the swarm is sparse. The influence of edge-adding probability is non-linear, for example, if we observe the size of the largest connected components in the graph. For small p, the largest connected component is small. With increasing probability p, the size of the largest connected component increases rapidly until the whole graph is part of the largest connected component. For $N \to \infty$, this rapid increase becomes a phase transition, that is, a discrete jump from almost no nodes being part of the largest connected component to almost all nodes being part of the largest component.

A drawback of random graphs is that they do not represent any spatial structure of the swarm. There is no spatial correlation between nodes. For example, if node A is connected to node B and node B is connected to node C, then it seems likely that node A could also be connected to node C.

Geometric graphs fill this gap and explicitly represent space. A geometric graph is defined as an undirected graph $G(X; r)$ and with node set $X \subseteq \mathbb{R}^d$. In the case of 2-d, we have $X \subset \mathbb{R}^2$ and nodes $x \in X$ and $y \in X$ are connected by an undirected edge if $||x - y|| < r$ for a norm $|| \cdot ||$ on $\mathbb{R}^2$. The norm basically measures the distance between nodes (e.g., Euclidean distance). By the inequality $||x - y|| < r$, we implement the so-called unit disc model, that is, we assume that a robot's sensor range is symmetrical (a disc). For example, in the case of radio communication, it is known to be very different from reality [7, 898]. For our purpose here, it is, however, a sufficient model.

Again, we turn to random graphs, only this time random geometric graphs. A random geometric graph is generated by sampling X from a probability distribution, for example, a uniform distribution. As an example, in a 2-d scenario, we sample randomly uniformly two coordinates $x_1 \in [0, 1]$ and $x_2 \in [0, 1]$ forming $x = (x_1, x_2)$. Edges are determined, as described above, by checking distances between nodes x and y via $||x - y|| < r$. An example is given in Fig. 7.10. Then the edges are not independent anymore as they are in Erdős–Rényi random graphs. Instead, we have

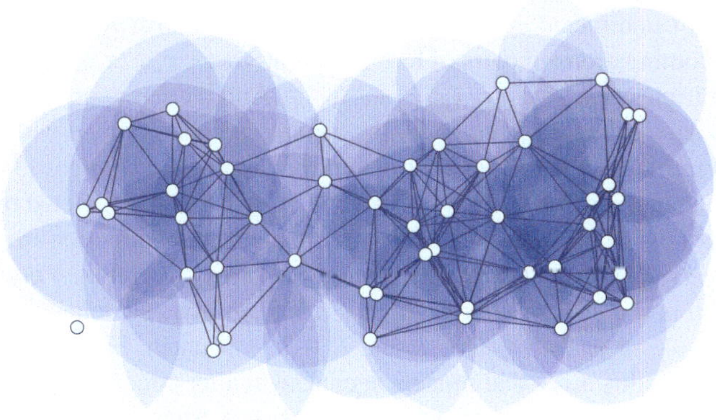

Fig. 7.10 A random geometric graph, blue circles (or disks) indicate sensor range, an edge exists if a neighboring node is located on that disk (created using a script of Friedrich Große)

spatial correlations in the form "edges $e_1 = (A, B)$ and $e_2 = (B, C)$ make edge $e_3 = (A, C)$ more likely" as discussed above. That makes random geometric graphs a more suitable network model for applications of swarm robotics. With random geometric graphs, we can investigate features relevant to swarm robotics, such as average sizes of connected components (groups of robots that cooperate), average number of isolated nodes (robots without any chance to cooperate), and average node degree (possibly a measure for robustness). Random geometric graphs are static models, hence can only represent snapshots of swarm configurations and, in particular, initial conditions.

7.4.4 Network Science and Adaptive Networks

We summarize models that go beyond random graphs and random geometric graphs in network science here. Network models focus more on specific features of the topology, such as complex networks and small world networks [754, 850]. To stay concise here, we skip most of the general network models that have applications across all fields of research. Instead, we focus on a particular subclass called adaptive networks as defined by Gross and Sayama [302]. Adaptive networks differ from most other network models because they allow dynamic updates of their topology. An example application of adaptive networks is the work by Sood and Redner [738] on collective decision-making using the voter model (see Sect. 8.5.2) on heterogeneous graphs.

Here we closely follow the study of Huepe et al. [367]. They present a simple adaptive network model for the locust experiments [115, 886]. Huepe et al. [367] model network motifs, that are sub-networks, in this case, with up to four nodes. We introduce variable x_L for left-goer density (i.e., fraction of the swarm in state left-goer) and variable x_R for right-goer density. Similarly, we introduce variables for densities of undirected edges (i.e., we do not distinguish $R - L$ from $L - R$) between all combinations of pairs of left- and right-goers: x_{RR}, x_{RL}, and x_{LL}. We do the same for triples: x_{RRR}, x_{RRL}, x_{RLL}, and x_{LLL}. And the same for quadruplets that we notate as $x_{R \cdot L \cdot RR}$, $x_{R \cdot R \cdot LL}$, etc. with the considered node framed with dots $(\cdot)$.

Next, we introduce parameters that model the topological changes [367]. $R - L$ links are added randomly at a rate a_{RL} per node. $R - L$ links are deleted randomly at a rate d_{RL} per link. Links between nodes moving in the same direction are added randomly at rate a_{RR} per node and deleted randomly at rate d_{RR} per link. A node switches direction with probability s_{RL} for each $R - L$ link. A node switches direction with probability s_{RRL} for each $L - R - L$ and $R - L - R$ chain. Finally, we introduce noise by a probability s_{noise} that models agents switching directions spontaneously.

The model is written down as a system of ordinary differential equations similar to the rate equations approach, see Huepe et al. [367] for details. For x_R we get

$$\frac{d}{dt} x_R = s_{\text{noise}}(x_L - x_R) + s_{RRL}(x_{RLR} - x_{LRL}). \tag{7.28}$$

For x_{RR} we get

$$\frac{d}{dt}x_{RR} = s_{\text{noise}}(x_{RL} - 2x_{RR}) + s_{RL}(x_{RL} + 2x_{RLR} - x_{RRL}) + s_{RRL}(2x_{RLR} + 3x_{R\cdot L\cdot RR} - x_{R\cdot R\cdot LL}) + a_{RR}x_R^2 - d_{RR}x_{RR}, \quad (7.29)$$

and similarly for x_{LL}. Instead of writing an explicit equation for x_{RL}, we write a sum of all two-node motifs

$$\frac{d}{dt}(x_{RL} + x_{RR} + x_{LL}) = a_{RL}x_Lx_R + d_{RL}x_{RL} + a_{RR}x_R^2 + a_{LL}x_L^2 - d_{RR}x_{RR} - d_{LL}x_{LL}. \quad (7.30)$$

The above ODE system can be closed and simplified by using the following approximations for triplet and quadruplet motifs (see Huepe et al. [367] for details)

$$x_{RLR} = \frac{x_{RL}^2}{2x_L}, \quad (7.31)$$

$$x_{RRL} = \frac{2x_{RL}x_{RR}}{x_R}, \quad (7.32)$$

$$x_{R\cdot L\cdot RR} = \frac{x_{RL}^3}{6x_L^2}, \quad (7.33)$$

$$x_{R\cdot R\cdot LL} = \frac{x_{RL}^2x_{RR}}{x_R^2}, \quad (7.34)$$

and similarly for the symmetric variables (L and R exchanged). Hence, we end up with a system that models only single nodes and pairs of nodes.

See Fig. 7.11a for the results of a numerical integration of the above equations for initial condition $x_R = 0.55$ and $x_L = 0.45$. Shown is the evolution over time

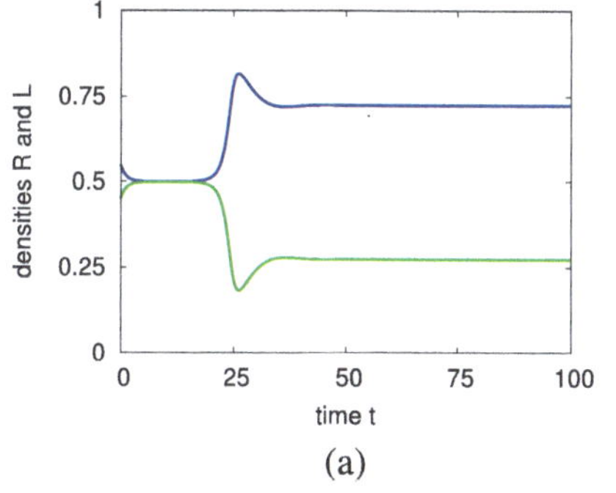

(a)

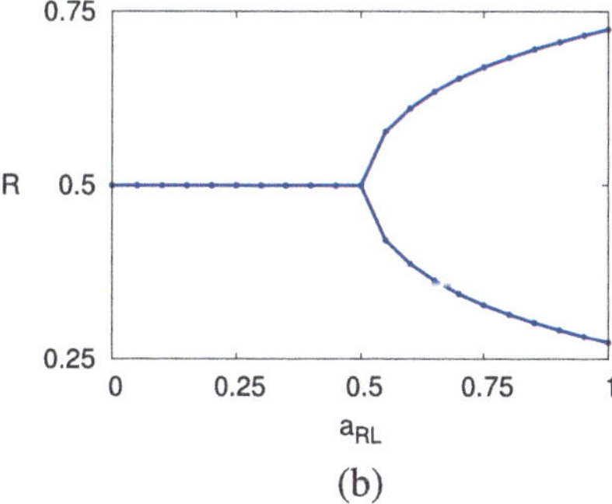

(b)

Fig. 7.11 Adaptive networks population model, numerical results for the evolution over time of densities and a bifurcation diagram; parameters: $d_{RL} = 0.25$, $d_{RR} = 0.1$, $s_{RRL} = s_{RL} = 0.2$, and $s_{\text{noise}} = 0.1$. **a** Evolution over time of densities x_R and x_L ($a_{RL} = 1$) **b** Bifurcation diagram for x_R over parameter a_{RL}

until convergence. See Fig. 7.11b for a bifurcation diagram of x_R over parameter a_{RL} (rate for adding $R - L$ links randomly). The system stays undecided for $a_{RL} < 0.5$. With this simple population-based network model and non-spatial approach, we can analyze the parameters and detect when the system tends to make a collective decision. Huepe et al. [367] have also analyzed how long the system is expected to stay decided.

7.4.5 Swarm Robots as Biological Models

Instead of only stealing ideas from biology and developing bio-inspired methods in robotics, we should also think about giving something back. There is some research on using robots or even swarm robots as biological models. This rather excellent idea is motivated by using robots as real-world models. Biologists, especially in behavioral biology, formulate hypotheses that need to be tested. Testing is not always possible with animal experiments, in particular when one needs to falsify a behavioral model. Abstract mathematical or computational models, as well as simulation, may be efficient but have limited credibility and may miss important features of reality. Advantages of robots as "real-world models" hence include the following. A robot operates in the real world, and the roboticist Rodney Brooks once famously said that "the real world is its own best model." Once you have proved your point in the real world, it is hard to argue that you missed something (unlike in a simulation). The choice of using robots positively constrains the choice of available model, excluding any invalid options [852]. It helps to force you to be concrete in specifying the complete biological system and helps you to produce actually testable hypotheses [851]. Using robots allows us to study complex agent–environment interactions.

An example study is that of Webb and Scutt [853]. They investigate the behavior of crickets. The male crickets sing their cricket song, and female crickets recognize these songs and approach male crickets. This is called "phonotaxis" (i.e., following a sound). "A cricket song typically consists of short bursts [...] of fairly pure tone [...] grouped in various distinctive patterns" [853]. Crickets have one ear in each foreleg and can determine direction-dependent differences between them. The female crickets show a more complex behavior instead of just turning towards the side with the strongest response. Proposed behavioral models are sophisticated and based on known facts from cricket neurophysiology. Webb and Scutt [853] compare a former, complex model that relies on a network that compares the signal amplitude at the two ears. The signal is also analyzed using a low-pass and a high-pass filter. This was compared to a simpler model that is effective due to relative latencies in firing onset and "leaky integration" acts as a low-pass filter, making fast temporal patterns appear continuous. This was tested on a robot, and it may seem reasonable that a physical model of sound propagation, perception, and processing is more reliable than any simulation. The simpler model proved to be effective, hence questioning the more complex model.

7.5 Formal Design Methods

The design of swarm robotics systems is challenging [318], although the individual behaviors are simple. The designer has to program the globally defined task virtually "between the lines" into the individual robots with their local perception. Designing adaptive group behaviors is known to be difficult [524] and similarly the design of emergent behaviors [523, 748]. To avoid designing swarm systems with a naive trial-and-error approach, we have to support the designer in any way during the design process. Options are models that predict the to-be-expected behavior for a given control algorithm or, if possible, even an automatic design process.

In general, there are two big classes of design approaches for any engineered system; one can design a system either in a top-down or bottom-up approach. Crespi et al. [162] specify:

> In the top-down approach, the design process starts with specifying the global system state and assuming that each component has global knowledge of the system, as in a centralized approach. The solution is then decentralized by replacing global knowledge with communication. In the bottom-up approach, on the other hand, the design starts with specifying requirements and capabilities of individual components, and the global behavior is said to emerge out of interactions among constituent components and between components and the environment.

During the design of a swarm system, one typically alternates between these two perspectives. The task is defined globally on the macroscopic level, but the robot behavior and its requirements are defined locally on the microscopic level.

In the following, we discuss multi-scale modeling approaches that address the micro-macro problem. We will shortly go through the options of generating swarm behaviors automatically. These methods are, however, diverse and complex, and subsequently, a detailed discussion is out of this book's scope. Then we discuss applications of classical software engineering in the swarm domain. Finally, we present some of the few formal approaches to programming robot swarms and simulation tools.

7.5.1 *Multi-Scale Modeling for Algorithm Design*

Even in an unsophisticated trial-and-error approach, one would usually not test the control algorithm directly on the robots. Instead, we would study the generated swarm behavior in a simulation. Hence, we already have a multiscale approach as we test our software in an abstract model first and then in the real system (i.e., a system of maximal detail). In addition to the scale "level of detail" we also have the scale micro and macro. As mentioned above, the designer necessarily has to alternate between the micro and the macro level. The levels of detail are not restricted to two (simulation and reality), but can be a whole cascade of increasing complexity as shown in Fig. 7.12. An immediate sophistication of the mere trial-and-error approach can therefore be

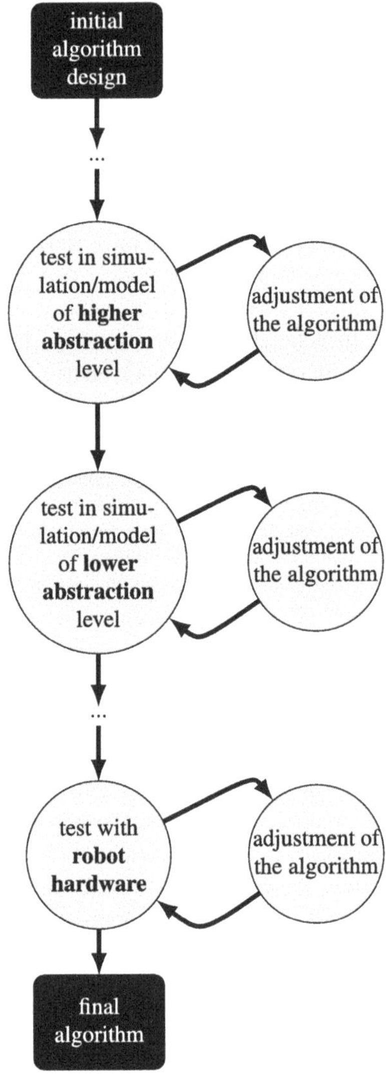

Fig. 7.12 An approach to iterative algorithm design based on a hierarchy of simulations, models, and robotic hardware; schema adapted from [318]

an iterative multi-level or multi-scale approach. The initial algorithm design is tested only in a very abstract model until the designer is satisfied. Only then do we move to the next, more detailed level, adapt the algorithm again until it works properly, and move on. The hope in this approach would be that only minor incremental adaptations are required from step to step, hence simplifying the overall approach. Going directly from an abstract model to reality may introduce fundamental issues that may require heavy changes to the algorithm.

Multi-scale approaches are frequently applied in swarm robotics. Usually, they are not as fine-grained as displayed in Fig. 7.12 but rather restricted to only a few or even only two levels. Typically, two levels often come as a probabilistic microscopic model and a deterministic macroscopic model. For example, Vigelius et al. [825] use a Markov process on the micro level and a Fokker–Planck equation combined with a master equation on the macro level (in addition to simulations and robot experiments). Valentini et al. [805] use Markov chains for the micro level and ODE systems on the macroscopic level. Also, the combination of a Langevin equation on the micro level and a Fokker–Planck equation on the macro level follows this probabilistic–deterministic dichotomy [318], as well as the rate equations approach (probabilistic state machines and ODE system).

7.5.2 *Software Engineering and Verification*

There are studies of software engineering for agent-based systems called agent-oriented software engineering (e.g., see Lind [493]). They are, however, rather focused on software agents. There are only a few software engineering approaches with a specific focus on swarm robotics. A rather rough engineering methodology for swarms is given by Parunak and Brueckner [619]. In the same year Winfield et al. [867] push towards "a new discipline of swarm engineering."

Within software engineering, work is focusing on verification. Verification is the process of checking whether a piece of software meets its specification, that is, whether it is doing what it is supposed to do. Formal verification is the process of proving the correctness of a software (or algorithm). There are only a few studies of formal verification in the context of fully distributed robot systems, as we have them in swarm robotics.

Winfield et al. [868] propose an approach of formal specification for robot swarms for the example of emergent taxis [81], also see Dixon et al. [195]. They use a linear time temporal logic that extends regular logic by a time parameter. The defined logic models the microscopic movements of robots (however, discretized) and their sensor input. The approach has the potential to help in formulating a sophisticated design methodology for swarm robotics.

Winfield et al. [284] propose a method for formal verification of robot swarms based on a formal language called "KLAIM" [176]. KLAIM is a coordination language that allows for defining specifications as accurate models of distributed systems based on a set of primitives. One defines behavioral modules similar to behavior-based robotics (see Sect. 3.5.2). Each module gets input from sensors or other modules and gives outputs to other modules or actuators. The analysis of the specification can be done quantitatively by using the stochastic variant of KLAIM called STOKLAIM [177]. Then one can assign rates to each action that model the duration times of actions. Gjondrekaj et al. [284] use formal tools (stochastic modal logics and model checking) to analyze different performance features of the swarm system. They propose a five-step process: (1) formal model with KLAIM as specification, (2) add stochastic aspects, (3) convert specification into code for the robot, (4) test in simulations, and (5) test on real robots.

Brambilla [100] proposes a top-down approach using property-driven design. The idea is to design robot swarms systematically, such that the swarm will behave correctly by design. He applies methods of model checking and analysis tools, such as Bio-PEPA [146] and KLAIM again. Bio-PEPA is a process algebra designed initially to analyze biochemical networks. It includes tools, such as stochastic simulation, analysis based on ordinary differential equations (ODE), and model checking using PRISM. PRISM is a probabilistic model checker [458]. While the property-driven design approach provides a guideline of how to develop the swarm system, the human designer still needs to be creative to make the transition from macro to micro and to implement the actual robot behavior.

The idea of design patterns is followed by Reina et al. [668], for example, in scenarios of collective decision-making [669]. The underlying idea is that we currently, we cannot find generic design methods for swarm robotics, but we can define solutions for broad classes of problems, that is, design patterns. Reina et al. [669] define a macroscopic model based on rate equations (see Sect. 7.4.1) forming an ODE system and a microscopic model based on probabilistic finite state machines. A micro-macro link (see Sect. 4.4) is established by connecting micro- and macro-parameters. Guided by these models and the known relationship between them, the idea is that a human designer is fully aware of microscopic design choices and to be expected macroscopic effects.

7.5.3 Formal Global-to-Local Programming

With the concept of the macroscopic and microscopic levels in swarm robot systems in mind, the idea of global-to-local programming is easy to grasp. The user should only be required to input a global description of the desired swarm behavior, and some automatic process would then output executable program code for an individual swarm robot. Following the pattern formation example of Yamins and Nagpal [883], we only specify a desired global pattern, which is then processed by a compiler that outputs local rules for individual cells in a grid (see Fig. 7.13 and Sect. 5.3.3). Writing that concept down is so simple compared to how difficult it is to solve this grand challenge of swarm robotics. However, the described but hard problems are the best!

Given that the problem is hard, it seems fine to start simple and simplify as much as possible. That is basically what Yamins and Nagpal [883] are doing, also see [881, 882]. They focus on pattern formation in a one-dimensional grid world, which is more accessible than dynamic robot behaviors. The idea is to define a set of local rules that implement a "virtual self-organizing Turing machine" [883]. The set of rules is defined beforehand, making heavy use of human creativity. Simple rules can be defined for "single-choice patterns" that are constrained and, hence, simplify the system's processing of the initial configuration at runtime. If there is only one option for how to form the pattern (e.g., the very left grid cell has to be blue), then this one option is just implemented by the local rules. More interesting cases are, of course, situations when symmetries need to be broken and the grid cells have

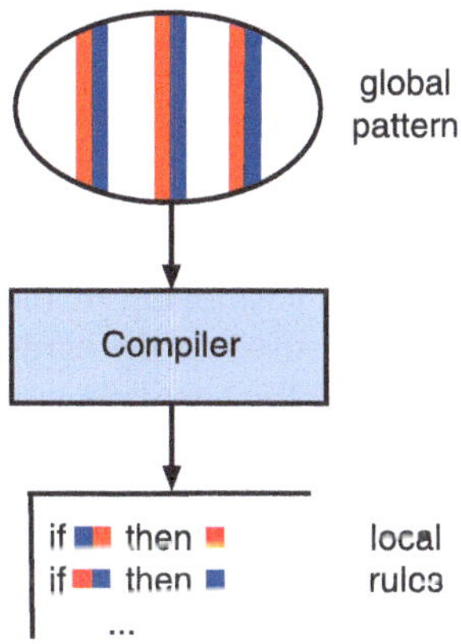

Fig. 7.13 Schema of global-to-local programming following the pattern formation example of Yamins and Nagpal [883]

to collectively agree on which option of the pattern they want to implement today. The symmetry-breaking process is typically sensitive to the initial configuration of the grid. However, still, information about a relative reference frame needs to be propagated through the grid. Yamins and Nagpal [883] use local rules to propagate signals as gradient waves through the grid. With a set of generic local rules at hand, most of the work is already done, and the compiler only needs to automatically translate the global pattern into appropriate instantiations of those generic rules.

A similar but more complex approach is followed by Werfel et al. [857] to implement collective construction (see Sect. 5.3.4). Both the robots and the building material (bricks) are designed such that the whole scenario is simplified to a grid world again; however, three-dimensional now. Again, a user specifies a global pattern; this time, it is the desired construction. A compiler needs to translate this blueprint to local rules for the robots. Werfel et al. [857] represent the local rules by labeling grid cells of a 2-d projection with predetermined paths, along which the robots are allowed to travel. This representation is called "structpath." Each grid cell is labeled by at least one arrow pointing to a neighboring cell. This arrow indicates the allowed motion patterns and, hence, allowed trajectories of the robots. Also, several neighbors can be labeled as allowed destinations. A further simplification is that there is one seed block that defines the entrance to the construction side as well as its position and orientation. Based on this structure, the compiler must adhere to several constraints. For example, robots cannot climb steps that are higher than one brick, and robots cannot put bricks into only one-brick-wide spaces. Deadlocks need to be avoided at all costs. A robot must not get trapped between other robots. For all possible temporal configurations of the construction during the incremental construction process, one needs to avoid robots blocking certain areas by placing a brick at a particular spot too early. In summary, the compiler mostly handles these rather basic constraints while the simplifications and restrictions resolve the actual micro-macro problem.

7.5.4 Simulation Tools

Conducting a swarm robotics experiment is typically expensive because many robots need to be taken care of. Especially in experiments with 50, 100 [805], or even 1000 robots [686] even otherwise simple procedures, such as charging, flashing, or initial positioning of the robots, become tedious tasks. Also, wear, reliability, and robustness of the hardware are issues, of course. Hence, it is useful to do (at least initially) simulations. There are many robot simulators, but not all of them are scalable themselves. Experiments with up to 1,000 robots are challenging, even from a software perspective, in simulation.

A simulation framework that is dedicated to swarm robotic experiments is *ARGoS*[1] [634]. *ARGoS* has a flexible, modular design that is open for different robot platforms, parallelizable, and adjustable to different levels of abstraction. It was under development for more than 13 years, is still maintained, and has been used for many different projects and is currently in use at many institutions in research and teaching.

Other simulators, that are also dedicated to swarm robotics, are the tool *Kilombo* [382] and the formerly commercial but now open-source *Webots* robot simulator.[2] *Kilombo* is specialized to simulate Kilobots, a robot design that is open-hardware but is also commercially distributed [685]. *Webots* is a general-purpose tool to simulate robots and is open for different robot platforms. Another example is *Gazebo* [439] that is a powerful and popular robot simulator. A clear advantage of *Gazebo* is also its full integration into the ROS framework (see Sect. 6.4.1). However, it has only limited use in swarm robotics because it does not scale well with swarm size N (see Sect. 2.7.3). Dimmig et al. [189] provide a comprehensive survey of 44 simulators specialized for Uncrewed Aerial Vehicles (UAVs). The authors categorize simulators by their focus and provide decision-making guidelines to help researchers select the appropriate tool for their application domains, starting with general-purpose simulators like Gazebo before exploring domain-specific options.

With the growing importance of learning-based approaches in robotics, the need for robust simulation tools has increased substantially. Many existing simulators emphasize photorealistic rendering and detailed physics. Yet, they often fall short when it comes to scalability, that is, supporting large numbers of robots operating simultaneously in the same environment. An interesting research track is the application of machine learning to world models. See Sect. 7.6.5 for details.

Tao et al. [771] introduce *ManiSkill3*, a new GPU-parallelized simulator for robotics that emphasizes scalability and efficiency. It supports parallel simulation and rendering, heterogeneous environments, and diverse input modalities, such as point clouds and voxels. With speeds exceeding 30,000 frames per second, *ManiSkill3* enables reinforcement learning and imitation learning (see Sect. 7.6.2) to train from visual input in minutes rather than hours. The framework appears to be somewhat

---

[1] http://www.argos-sim.info/.

[2] https://www.cyberbotics.com/.

more focused on manipulators and humanoids, but mobile robotics is also considered. ManiSkill3 supports sim-to-real and real-to-sim transfer.

NVIDIA Isaac Sim is a GPU-accelerated, open-source robotics simulator built on the NVIDIA Omniverse platform, designed for physically based virtual environments featuring photorealistic rendering and realistic sensor and dynamics simulation using PhysX and RTX technologies.[3] It supports workflows such as synthetic data generation, software-in-the-loop testing, and robot learning through its extension Isaac Lab, and offers seamless integration with ROS 2 (see Sect. 6.4.1).

Beyond these robot simulators, there are several open-source projects for robot simulations and multi-agent simulations. Bettini et al. [77] have developed an open-source, vectorized simulator that is focused on multi-robot learning. VMAS is built in PyTorch and designed to accelerate multi-agent reinforcement learning (MARL) by enabling tens of thousands of parallel 2D simulations on GPUs. Unlike high-fidelity robotic simulators, which become slow at scale, VMAS focuses on fast, high-level coordination tasks while delegating low-level control to other methods. Their framework also includes twelve multi-robot scenarios, such as transport and flocking. Many other multi-agent simulators have been around for many years, but many of them are neither perfectly maintained nor well-documented. Examples are breve [437], player/stage [277], and MASON [503].

Another option is to use game engines as a starting point for custom development of a simple robot simulator. For example, one can program a simple swarm robotics simulator in Python, using *PyGame*, which is a beginner-friendly library for 2D graphics.[4] One can use *Box2D: A 2D Physics Engine for Games*[5] that provides collision detection techniques and allows for rapid prototyping of sensors and robots platforms.

7.6 Quick Overview of Data and Search-Driven Design Methods

7.6.1 Automatic Design in Swarm Robotics

The automatic design of robot controllers for swarms is primarily out of this book's scope, but we quickly go through the most critical approaches. Given that designing swarm robot controllers is challenging, it is reasonable to check automatic approaches that may even treat the whole problem as a black box. The toolbox of artificial intelligence (specifically machine learning) provides many options, and also optimization techniques can be used. We distinguish two classes of approaches here:

[3] https://developer.nvidia.com/isaac/sim.

[4] https://www.pygame.org/.

[5] http://box2d.org/.

(1) data-driven design methods and (2) search-driven design methods. Data-driven approaches learn from logged experience and incorporate methods of machine learning, including reinforcement learning, imitation learning, behavioral cloning, and inverse reinforcement learning. Search-driven approaches rely on iterative exploration of a solution space, usually guided by some objective function, such as evolutionary algorithms, black-box optimization, and Bayesian optimization. Evolutionary and other search algorithms consume evaluation results (comparable to data-driven approaches). Still, they treat each result as an atomic fitness number and immediately discard the trajectory-level information because they learn no model directly from the data. Hence, they are distinguished as optimization-based or search-based synthesis.

Let us now examine learning (see Sect. 7.6.2). There are several learning techniques, one of them is reinforcement learning. Reinforcement learning allows for unsupervised learning, that is, the robot learns by itself and does not require a carefully prepared labeled training data set. Instead, one defines only a reward function that assigns utility values to certain system states, hence, telling the robot which states are desirable and which need to be avoided.

While learning for single robots is already relatively challenging [690, 763], multi-robot learning is even more so [525, 858]. The problem is that the robot needs to remember pairs of experienced states and chosen actions, to exploit previously tested successful actions and to efficiently explore the effects of new actions. While the robot is learning, it is therefore building a data structure of these pairs. This does not necessarily have to happen explicitly (i.e., really storing each state–action pair), but can also be done with approximation techniques [859]. Still, in a multi-robot setting, the state–action pairs either combinatorially explode with increasing swarm size (exponentially more combinations of actions with each additional robot) or the other swarm members are treated as environment, which makes the learning task itself more difficult.

A learning technique that is popular in machine learning is, for example, training artificial neural networks (ANN) by the backpropagation algorithm [687]. This is, however, a supervised learning technique that requires labeled training data. In robotics, that means collecting sensor data and labeling each case with the appropriate action. In certain use cases, the correct action may also depend on the robot's internal state, further complicating the approach. Hence, acquiring labeled training data for robots is a challenging task. Similarly, the techniques of deep learning are often difficult to apply. For example, they are ready for use in applications such as image processing or natural language processing. However, to apply them to robot controller synthesis, they need to be combined with other methods, such as reinforcement learning.

Besides standard learning techniques, there are so-called evolutionary algorithms [223]. They are originally optimization techniques roughly inspired by Darwinian evolution and natural selection (see Sect. 7.6.3). Their application to robotics is called evolutionary robotics [93]. In the standard approach, the idea is to use an ANN, but these are not optimized by learning. Instead, one usually defines a fitness function, similarly to the reward function in reinforcement learning, to evaluate an observed robot behavior. Initially, one creates a population of randomly generated ANNs that are then tested in a robot simulation or even on a real robot. The fitness function should detect early successes in the population; these ANNs are then selected, recombined, and mutated. A new population is created, and the algorithm is iterated over many generations. Although it may seem counterintuitive how random changes to an ANN could result in useful robot behaviors within finite time, the approach is successful [598].

The application of evolutionary robotics to swarm robotics is called evolutionary swarm robotics and can be implemented in a relatively straightforward way [787]. Instead of uploading one ANN to one robot and testing it in an evaluation, the same ANN is uploaded to all swarm robots, and the whole swarm is evaluated. So the standard approach is a homogeneous approach with all robots having the same ANN (but different instances of it). The heterogeneous approach is possible, that is, each robot evolves its own ANN, but it is much more challenging.

7.6.2 *Data-Driven Design Methods*

In data-driven design methods for mobile robotics, controllers are learned directly from data, rather than being programmed or solely optimized through black-box search. These methods exploit experience gathered in simulation or on physical robots to fit models that map sensory inputs s to actions a ($s \rightarrow a$). Supervised learning assumes the availability of a dataset containing input-output pairs (s, a), where each input s is labeled with the correct output a. The learning process aims to minimize a predefined loss function by comparing the model's predictions to these known correct answers. In robotics, this is typically used for perception tasks, such as object recognition and localization. The mapping $s \rightarrow a$ is inferred from a dataset of example pairs (s, a_{best}), such as demonstrations, where the correct action a_{best} is provided by a teacher. The learning process adjusts a policy $\pi(s) = a$ to minimize the discrepancy between the predicted and target actions over the dataset.

Recently, large language models (LLMs) have emerged as robust data-driven systems that can process and generate not only text but also structured control logic, plans, and even executable robot code. In mobile and swarm robotics, large language models are beginning to serve as high-level planners or policy generators. They can translate natural-language task descriptions into structured action sequences or controller parameters. This extends the scope of data-driven design beyond low-level perception and control to full end-to-end autonomy. Here, end-to-end autonomy means that a single learned system can handle the entire pipeline from a high-level human instruction all the way to low-level executable robot actions in an automatic design approach, that is, without the designer manually programming the intermediate steps.

7.6.2.1 Reinforcement Learning

While supervised learning has only limited use for autonomous mobile robots, reinforcement learning (RL) has emerged as a compelling paradigm. Reinforcement learning does not require explicit, correct labels for every situation, which are, anyway, hard to obtain. Instead, the robot interacts with its environment in a trial-and-error manner and receives reward signals indicating how effective its actions are in terms of long-term task performance. The learning process seeks to maximize a cumulative reward over time by discovering an effective policy. This allows RL to handle tasks where the correct action is not known beforehand and where success may only be measurable after a sequence of decisions has been made.

In RL, a robot learns a policy $\pi : s \rightarrow a$ that maps its current state s (sensor data) to an action a (motor commands or behaviors). Learning occurs through repeated interaction with the environment. At each time step t, the robot observes the current state s_t, selects an action a_t according to its policy π, and receives a scalar reward r_t. Over time, the policy is adapted to choose actions that maximize the expected cumulative reward. The cumulative reward can be a finite-horizon undiscounted return $R_\tau = \sum_{t=0}^{T} r_t$ or a infinite-horizon discounted return $R_\tau = \sum_{t=0}^{\infty} \gamma^t r_t$ for discount factor $0 \leq \gamma \leq 1$.

One critical approach is Q-learning [750, 848], which learns the action-value function $Q(s, a)$ that describes the expected cumulative reward if we take action a in state s and follow the best policy thereafter. The Q-learning update is based on the Bellman optimality equation:

$$Q(s_t, a_t) \leftarrow Q(s_t, a_t) + \alpha\big[r_t + \gamma \max_{a'} Q(s_{t+1}, a') - Q(s_t, a_t)\big], \tag{7.35}$$

where α is the learning rate. This equation adjusts the estimate of $Q(s_t, a_t)$ towards the observed reward plus the best predicted value of the next state. In standard (tabular) Q-learning, $Q(s, a)$ would be stored in a table. This works for toy examples that are small, discrete state-action spaces, but becomes infeasible when s is high-dimensional (e.g., actual robot sensor data). Instead of storing $Q(s, a)$ in a lookup table, we approximate it with a parameterized function $\hat{Q}(s, a; \theta)$ (approximation function). Early choices in literature were: linear regression models, tile coding, radial basis function networks, and small multilayer perceptrons. The idea is to govern huge state-action spaces by generalizing Q-values to unseen states (learning a smooth mapping $(s, a) \longmapsto \hat{Q}(s, a; \theta)$).

RL is particularly suited to mobile robotics, where the robot must operate in dynamic, partially observable environments. The optimal action may depend on long-term consequences rather than immediate correctness. In multi-robot learning, however, one needs to design how the learning is organized: is the learning fully distributed and implemented as decentralized learning (heterogeneous swarm, where robots learn individually and share only minimal data), or is it centralized (homogeneous swarm, where all robots share the same controller)? A common strategy is centralized training with decentralized execution [500]. During training, all robots' experiences can be pooled together and shared across robots, which allows the learning algorithm to exploit complete state information and coordination signals that may not be available to an individual robot. Once training is complete, each robot executes its own policy in a fully decentralized way, relying only on its local observations. The difficulty lies in the gap between training and execution. While training can utilize global state information, each robot must act at execution time based only on its limited local observations. This leads to challenges, such as credit assignment, where it is unclear how much each robot contributed to the collective reward [248].

The work by Yasuda and Ohkura [885] is an example of applying reinforcement learning in a multi-robot setup with actual robot experiments. They learn robot controllers for a collective transport task (cf. Sect. 5.3.5).

Kakish et al. [401] present an approach for controlling a leader robot to herd a swarm of follower robots to a desired spatial distribution. The followers move on a graph and respond only to the leader's presence with repulsive behavior. Instead of tracking individual agents, the method employs a mean-field model that describes the swarm as a probability distribution over graph vertices, making the control scalable to large swarms. Two temporal-difference methods (SARSA and Q-learning) are trained to generate leader control policies that minimize the time to reach the target distribution. The policies are learned in simulation (OpenAI Gym) and validated with physical robots on the *Robotarium* testbed, showing successful redistribution of up to 1000 simulated agents and $N = 10$ physical robots. The results demonstrate scalability with swarm size.

7.6.2.2 Deep Learning

Deep reinforcement learning (deep RL) integrates RL with deep neural networks [565], enabling robots to process high-dimensional sensory inputs, such as images, depth maps, or complex proprioceptive data. Instead of relying on manually designed state features, deep RL can learn end-to-end policies from raw sensor streams to control outputs.

In deep RL, a deep ANN is used as a function approximator to represent key components of the RL framework, most commonly the policy, the value function, or both. For policy representation, the deep ANN maps states s (raw or processed sensory inputs) directly to actions a via a policy network $\pi_\theta(s)$, where θ are the ANN weights. In robotics, this enables end-to-end control, for example, from camera pixels (s) to motor control values (a). For value function approximation, the network estimates the value of being in a given state $V(s)$ or taking a given action $Q(s, a)$. The ANN takes a raw state s_t and outputs Q-values $Q(s_t, a; \theta)$ for each possible action a. There is no direct correct label for each action, that is, there is no explicit a_{best} for each s. Instead, the target Q-value for the chosen action is computed from the reward r_t and the maximum predicted Q-value of the next state:

$$y_t = r_t + \gamma \max_{a'} Q(s_{t+1}, a'; \theta) , \tag{7.36}$$

where θ are parameters of an ANN. In Deep Q-Network (DQN) [566] there is a second network with parameters θ^- that are copied from θ only occasionally (e.g., every few thousand steps). By keeping θ^- fixed for multiple updates, the target values y_t change more slowly, reducing the instability that arises when the network is simultaneously learning to predict and to evaluate its own evolving predictions. The loss function is

$$L(\theta) = \mathbb{E}[(y_t - Q(s_t, a_t; \theta))^2] . \tag{7.37}$$

Generally, deep RL has the potential to scale to complex sensory modalities (e.g., onboard cameras for visual navigation). It can learn directly from unstructured data without the need for handcrafted perception pipelines. While RL provides the theoretical framework for learning from interaction, deep RL offers the representational power necessary to handle the complexity of real-world robotics.

Examples of studies in RL for swarm robotics. Hüttenrauch et al. [371] introduce a state-representation method for applying deep reinforcement learning to swarm robotics. Traditional multi-agent RL approaches often concatenate observations from neighbors, which fails to exploit swarm properties (e.g., agent interchangeability and variable swarm size), leading to poor scalability. The authors propose mean feature embeddings, which treat each neighbor's observation as a sample from a distribution and encode it via the empirical mean in a learned feature space. They test different implementations using histograms, radial basis functions, and neural networks in rendezvous and pursuit-evasion tasks under global and local observability, with and without communication. The approach is simple to integrate with existing deep RL algorithms, supports decentralized execution after centralized training, and scales naturally to large swarms.

Sartoretti et al. [697] present a distributed reinforcement learning framework for multi-robot collective construction, extending the single-agent Asynchronous Advantage Actor-Critic (A3C) algorithm to enable homogeneous, collaborative policies. Training is centralized (aggregating experience from multiple agents in a shared, fully observable environment), while execution is decentralized, with each robot independently applying the learned policy without explicit communication. The method is applied to a block-based construction task inspired by Harvard's TERMES robots [627, 857], where agents must gather, place, and remove blocks to assemble user-specified 3D structures, including scaffolding. Simulation results show that the learned policy generalizes to different swarm sizes. The work demonstrates how shared policy learning can generate scalable, communication-free coordination in swarm robotics construction tasks.

Na et al. [587] propose a federated deep reinforcement learning strategy for swarm robotic systems that enables robust collective navigation under limited communication bandwidth. Instead of transmitting raw sensorimotor data to a central server, each robot trains locally and periodically exchanges only neural network weights with a server that aggregates and redistributes them. This soft-weight update reduces communication costs, preserves data privacy, and accommodates robot heterogeneity. In simulation ($N = 4$) and real-robot experiments with only $N = 1$ TurtleBot 2 on navigation and collision avoidance, their approach outperformed several baseline strategies. The approach is particularly suited for swarm deployments in bandwidth-constrained environments.

Chiun et al. [141] present a multi-agent reinforcement learning framework for multi-robot exploration in large-scale indoor environments when robots are equipped with constrained field-of-view sensors, such as cameras. The approach is trained in simulation under the centralized training/decentralized execution paradigm and adapts to different team sizes and sensor ranges without retraining. In extensive tests, their approach outperforms state-of-the-art frontier-based, sampling-based, and learning-based exploration planners. The method is further validated on real drones (Crazyflie 2.1) but with simulated sensor coverage, demonstrating its feasibility for real-world deployment.

7.6.2.3 Imitation Learning

Imitation learning (IL) is a method that directly learns a policy from demonstrations of expert behavior, thereby eliminating the need to define or optimize a reward function explicitly. In contrast to inverse reinforcement learning, which seeks to uncover the hidden objectives that explain expert behavior, imitation learning focuses on reproducing the observed mapping from states to actions. We assume that an expert (e.g., a human operator or a well-designed controller) provides demonstrations in a given environment, and the goal is to learn a policy that mimics this behavior as closely as possible. IL enables robots to learn "how to act" directly from examples, without requiring hand-crafted rewards.

Formally, in IL we have states s and actions a. An expert provides demonstrations in the form of trajectories $\tau = (s_0, a_0, \ldots, s_T, a_T)$. The goal is to learn a policy $\pi_\theta(a|s)$, parameterized by θ, that approximates the expert's policy π^*. A common approach is behavioral cloning, where the problem is cast as supervised learning: given a dataset of state-action pairs $\mathcal{D} = (s_t, a_t)$ from expert trajectories, we minimize the discrepancy between the policy's action distribution $\pi_\theta(a|s)$ and the demonstrated actions.

Schilling et al. [705] demonstrate imitation learning for vision-based swarm flight with quadrotors. They use a flocking algorithm as the expert policy and train a convolutional neural network that maps raw omnidirectional images to velocity commands. Using the DAGGER algorithm [683], the drones iteratively refine their policy to mimic the expert, first in simulation and later in real-world experiments. The resulting vision-based controller achieves collision avoidance and swarm cohesion without requiring position sharing or visual markers, representing a first step towards fully decentralized, vision-based drone swarms.

Zhou et al. [899] propose a centralized approach using graph neural networks to learn swarm behaviors directly from demonstrations of robot trajectories. Trained on models of collective motion, they can both predict the future dynamics of an observed swarm and generate a clone swarm that imitates the demonstrated behavior.

7.6.2.4 Inverse Reinforcement Learning

Inverse reinforcement learning (IRL) is a method to learn the underlying reward function that explains an observed behavior. This is in contrast to directly learning a policy to maximize rewards, as done in standard reinforcement learning. In IRL, we assume that an expert (e.g., a human operator or a well-designed controller) is acting near-optimally with respect to some unknown reward function in a given environment. By observing the expert's state-action trajectories, the goal is to infer this reward function. The reward function can then be used to reproduce similar behavior. This approach could be beneficial in swarm robotics, where designing explicit rewards can be challenging, but demonstrations of desirable collective behaviors (e.g., efficient flocking, cooperative transport) are available. IRL enables robots to learn "what to optimize for" from examples, instead of relying on hand-crafted reward functions.

Formally, in IRL we have states s, actions a, and an unknown reward function $r(s, a)$. An expert provides demonstrations in the form of trajectories $\tau = (s_0, a_0, \ldots, s_T, a_T)$. We assume the expert follows a near-optimal policy π^* that maximizes the expected return $R_\tau = \sum_{t=0}^{T} r(s_t, a_t)$. The IRL problem is to recover a reward function $r_\theta(s, a)$, often parameterized as a linear combination of features $r_\theta(s, a) = \theta^\top \phi(s, a)$, where $\phi(s, a)$ is a vector of hand-designed features that capture relevant aspects of the state-action pair (e.g., distances, alignments, or energy costs), and θ are the weights that determine how strongly each feature contributes to the reward. The goal is to find a reward function such that the expert's behavior has a high likelihood under the induced optimal policy.

In the following, we review several studies that demonstrate how IRL can be applied to multi-robot and biological swarms. These works range from algorithmic advances that make IRL tractable in large multi-agent settings to practical applications where reward functions are inferred from observed collective behaviors in simulations or nature. Together, these studies show both the methodological challenges and the potential of IRL to uncover behavioral policies in natural swarms and transfer them to swarm robotics.

Šošić et al. [837] extend IRL to swarms to learn local reward functions that reproduce observed global behavior. The authors introduce a model that is a subclass of decentralized partially observable MDPs. By exploiting symmetry, they reduce the multi-agent IRL problem to a single-agent case, allowing the use of existing IRL methods. They propose a heterogeneous Q-learning scheme in which some agents act greedily and others explore, improving learning of local policies that approximate the expert's global dynamics. The method is evaluated on the Vicsek model (see Sect. 5.2.5 and Vicsek et al. [824]) of collective motion. They demonstrate that learned local reward functions can generate swarm behaviors that are remarkably similar to those of the expert systems.

Szpirer et al. [767] present two methods for the automatic design of robot swarms that perform composite missions using IRL from demonstrations (also see Gharbi et al. [281]). In their approach, each demonstration consists of example swarm configurations (i.e., robot positions in the arena) that specify the desired spatial organization for the mission at its conclusion. The positions are mapped into a feature vector that encodes the distances between each robot and three landmarks in the arena, as well as the distances to each robot's closest peer. The system learns objective functions from these demonstrations and uses an optimization process to generate control software. The authors compare a fully automatic multi-objective approach with a semi-automatic two-step approach. Experiments in simulation with 20 e-puck robots performing twelve different composite missions show that both methods can outperform a hand-designed baseline.

Wälchli et al. [842] introduce IRL for uncovering the objectives underlying collective swimming behavior. Schooling in fish provides a biological example of collective efficiency in fluid environments, and this study shows how IRL can uncover such objectives. This approach not only deepens our understanding of animal behavior but also has potential future applications in swarm robotics and the design of underwater vehicles. The authors simulate $N = 4$ artificial swimmers in a two-dimensional viscous flow governed by the Navier–Stokes equations. Synthetic data were generated from policies trained to maximize swimming efficiency, and IRL successfully rediscovered reward functions that captured efficient group behavior. Sensitivity analysis revealed that the inferred rewards were strongly linked to shear stress measurements near the swimmers' heads, consistent with how real fish perceive their environment. Using symbolic regression, the authors also derived simple, interpretable reward functions that still produced efficient swimming.

In a similar style of work Pinsler et al. [638] applied IRL to study collective motion in bird flocks. Using GPS trajectory data from homing pigeons, they modeled each bird as an agent in a Markov decision process. They inferred individual reward functions that best explain the observed flocking behavior. So they did not assume that all birds follow the same rules. Their results showed that these learned reward functions can reproduce realistic flocking trajectories and even suggest leader-follower relationships within the group. Such learned rules could be transferred to robot groups for coordinated flight.

7.6.2.5 Large Language Models

We provide a brief overview of early explorations into how large language models (LLMs) can be applied to swarm robotics and multi-robot systems. We introduce approaches that generate executable controller code directly from natural-language instructions. Another approach guides cooperative exploration and navigation using vision-language reasoning. Some enable multi-robot coordination through conversational negotiation with human oversight, while others enhance shared perception by learning decentralized visual-spatial representations.

Ji et al. [388] introduced an end-to-end system that utilizes LLMs to automatically generate, deploy, and refine executable control code for robot swarms from simple natural language instructions. This approach follows a code-as-policy paradigm, meaning it produces white-box, interpretable, and reproducible code to be run on the robots. Their approach has three stages: (1) task analysis, where constraints and a skill library are extracted from user instructions; (2) code generation, guided by a hierarchical skill graph; and (3) code deployment and improvement, which automates rollout to simulated and real-world swarms and integrates feedback from both humans and vision-language models. Demonstrations cover ten swarm tasks (e.g., flocking, coverage, encircling, exploration), achieving high success rates in simulation and real robots, and outperforming existing LLM-based baselines, such as MetaGPT.

Yu et al. [887] introduce a framework for multi-robot cooperative visual target navigation in an exploration task. Their approach is hybrid in the sense that robots process some information locally. Still, they also use a vision-language model (which processes and relates visual information to natural language) as a global planner. Each robot builds a local RGB-D-based point-cloud map. These maps are merged into a global semantic map from which frontiers (unexplored regions) are extracted. Based on visual and textual prompts that encode robot states, map context, and the target object, the approach assigns distinct frontier goals to each robot for coordinated exploration. A local planner generates collision-free paths. Tested in both simulations and with real quadruped robots, the approach outperforms classical multi-robot frontier allocation methods and adapted learning-based baselines.

Hunt et al. [370] present a proof-of-concept system for human-in-the-loop multi-robot coordination using LLMs as conversational agents. Instead of a central planner, multiple robots engage in peer-to-peer, argument-style dialogue to negotiate roles, plan paths, and adapt strategies, with a human supervisor able to intervene at any stage. Their approach utilizes text at every step, maintaining transparency and enabling humans to follow or even modify the plan. In their demonstration, two TurtleBot3 robots, equipped with LIDAR and cameras, act as mobile waste collection units in a mapped environment. Given a high-level task in natural language, the LLM-driven agents discuss and agree on a division of labor, check with the human, and then execute the plan in the real world.

Liu et al. [494] propose Language-Guided Pattern Formation, a framework that combines LLMs with multi-agent reinforcement learning (MARL) to control swarm robots in forming complex shapes. LLMs translate high-level natural language descriptions (e.g., "form a heart shape") into 2D coordinates for the desired pattern. A centralized-training, decentralized-execution MARL method (a permutation-invariant critic variant) then trains decentralized policies that guide robots to the target positions. Robots avoid collisions and do not use inter-robot communication. The approach is validated in simulation and on real robots (they are using the robots of Ichihashi et al. [372] which are similar to Zooids by Le Goc et al. [465]).

Blumenkamp et al. [83] introduce a decentralized visual–spatial foundation model designed for multi-robot applications that operates on pairwise or small-team robots to share visual-spatial information (i.e., semantic scene understanding anchored in spatial coordinates). Their approach enables each robot to estimate the relative pose and bird's-eye-view representation of its surroundings using only monocular camera images. A key element of their system design is the integration of advanced perception modules that jointly address feature extraction, localization, uncertainty estimation, and spatial mapping. DINOv2 is a self-supervised vision transformer that learns general-purpose image representations, enabling robust perception across diverse visual conditions without task-specific retraining. A pose estimator processes these visual features. This computational method infers a robot's position and orientation within a reference frame, specifically from a monocular camera image. In addition to estimating the pose, the system quantifies aleatoric uncertainty, which is the irreducible variability in sensor measurements caused by inherent noise in the environment or the sensing process. A bird's-eye-view predictor then generates a top-down representation of the environment by fusing spatial information from one or multiple viewpoints. Putting everything together, their approach uses a DINOv2 vision transformer to extract visual features, a transformer-based pose estimator that also predicts measurement noise, and a bird's-eye-view predictor that merges top-down scene information from neighboring robots. Trained in simulation on large-scale indoor scene datasets, the approach was validated on heterogeneous robot teams in both indoor and outdoor real-world environments. Experiments show that it achieves sub-meter median pose errors. Demonstrations show, for example, decentralized formation control. A swarm-level coordination algorithm could utilize the output of this approach; however, such coordination logic is outside the scope of that paper.

7.6.3 Search-Driven Design Methods

Search-driven design methods include a range of meta-heuristic optimization techniques that explore significant and complex search spaces by iteratively generating, evaluating, and refining candidate solutions. Instead of relying on analytical gradients or explicit models of the problem, these approaches treat the controller design task as a black box and rely on feedback from swarm-level performance. The general principle is to search broadly for effective robot controllers or parameter settings using population-based or sampling-based strategies, and to refine them gradually across iterations. In the following, we first introduce Evolutionary Design, including specific approaches to evolutionary computation, examples of studies in evolutionary swarm robotics, and the evolutionary answer to imitation learning; we then present Black-box Design, covering Bayesian Optimization with Gaussian Processes, Ant Colony Optimization (ACO), and Particle Swarm Optimization (PSO).

7.6.3.1 Evolutionary Design

Evolutionary algorithms are a family of optimization techniques inspired by the principles of natural evolution, albeit loosely. First efforts include Genetic Algorithms by Holland [361], Evolutionary Strategies by Rechenberg [667], and Genetic Programming by Koza [443]. In swarm robotics, these algorithms are often used to automatically tune parameters of robot controllers, such as the weights of artificial neural networks (ANNs). The approach is population-based, which means that there is always a set of potential solutions. In this context, the population can be, for example, a set of ANN weights. In terms of data structure, the weights of each ANN can be encoded as a floating-point array. Hence, the population is a set of such arrays. It is essential to understand that we have two population concepts here: a population of potential solutions (weights of ANNs) and a population of swarm robots in our simulations. These are different from each other. In the standard approach, each robot receives the same ANN weights, that is, clones of the same ANN to form a homogeneous swarm. The heterogeneous approach, where each robot would try to evolve its own ANN using an evolutionary algorithm, is conceptually and practically much more challenging.

The evolutionary algorithm begins with a randomly initialized population of candidate solutions (see Fig. 7.14). These are evaluated through simulation or real-world experiments by measuring the performance (in evolutionary terminology called fitness) of the resulting swarm behavior. It is important to note that fitness is a global feature in evolutionary swarm robotics, which can, in most cases, not be measured by an individual robot. This is of less importance in simulations where a global overview is easily obtained; however, it is an issue in a decentralized, embodied approach. Based on fitness, better-performing candidates are selected, modified through random mutations (e.g., small changes to weights), and possibly recombined (e.g., one half of the weights array from one 'parent' and one half of the weights from another 'parent') to form the next generation. This iterative cycle of evaluation, selection, and

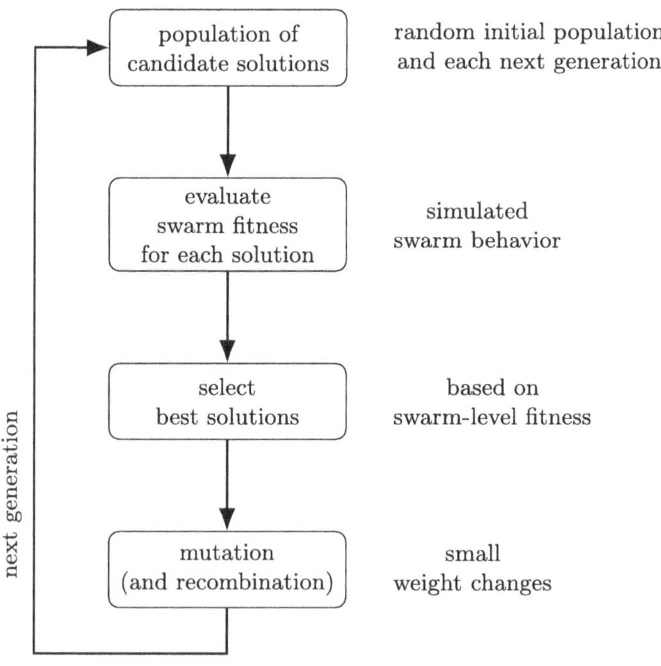

Fig. 7.14 General iterative approach of an evolutionary algorithm

variation continues over many generations in a potentially computationally intensive approach, gradually improving the swarm's performance. Evolutionary algorithms are beneficial when the desired behavior is challenging to specify manually. The problem is treated as a black box and solved in this meta-heuristic approach. In cases when one has domain knowledge about the problem (i.e., an expert may know a potential solving mechanism), evolutionary algorithms are not the best choice.

Specific approaches to evolutionary computation. There are numerous approaches to evolutionary computation, as well as various families of algorithms. Here, we focus on two that were used to implement significant steps forward in evolutionary swarm robotics: NEAT and novelty search.

Stanley and Miikkulainen [744] developed the NEAT algorithm: NeuroEvolution of Augmenting Topologies. This is a bottom-up approach to train ANNs that not only optimizes their weights but also their architecture (e.g., by adding artificial neurons and links). Starting from minimal structures, NEAT incrementally adds complexity only when beneficial, which allows it to search efficiently in low-dimensional spaces. Key innovations include protection of structural innovations via 'speciation' and explicit fitness sharing within such 'species.' This enables NEAT to discover compact and practical neural controllers.

Lehman and Stanley [469, 470] introduce novelty search, which is an alternative to traditional objective-based optimization in evolutionary algorithms. The key idea is to ignore the objective function entirely and instead reward behavioral novelty, that is, how different a candidate solution behaves compared to past observed behaviors. This approach addresses two common challenges of evolutionary computation: (1) deception in fitness landscapes, where local optima mislead objective-based search; (2) open-ended evolution, where the goal is to continually generate new and increasingly complex solutions without a predefined endpoint. The authors demonstrate that novelty search, combined with NEAT, can outperform traditional fitness-based search. The method works by encouraging exploration in behavior space rather than convergence toward a fixed goal. While objective-based evolutionary algorithms are already more explorative than learning techniques that follow error gradients, novelty search is even more explorative. This idea led to MAP-Elites [579], where one tries to collect a diverse set of behaviors rather than solving a specific problem at hand. The high degree of exploration is, hence, resolved by an automatic behavior selection mechanism at runtime.

Mouret and Clune [579] introduce an evolutionary search algorithm called MAP-Elites, designed to illuminate rather than optimize search spaces. Instead of finding just one optimal solution, MAP-Elites explores how high-performing solutions are distributed across a user-defined feature space. It does so by finding and recording the best solution for each cell in this discretized space, where each cell corresponds to a combination of features (e.g., ground contact times for each leg of a legged robot). MAP-Elites reveals how performance varies across different types of solutions, aiding in understanding tradeoffs and structure in complex spaces. It often outperforms traditional optimization algorithms in both global performance and the discovery of diverse, reliable solutions. The authors demonstrate the effectiveness of MAP-Elites in domains such as neural network evolution, simulated soft robotics, and a real-world soft robotic arm.

Examples of studies in evolutionary swarm robotics. There are many successful applications of methods of evolutionary computation to swarm robotics in studies of evolutionary swarm robotics. Lacking a categorization, we proceed through the most important approaches chronologically. Examples are the evolution of aggregation behaviors [788] and the evolution of direct robot–robot communication [789], which usually turns out to be very challenging. Instead of ANN, it is also possible to use other representations of robot controllers. For example, finite state machines were used by König et al. [441] in a navigation task. Ferrante et al. [241, 242] evolve task specialization and task partitioning with standard evolutionary algorithms, but a system of behavioral rules instead of an ANN. Francesca et al. [251] present a dedicated approach to the automatic design of robot swarms based on predefined behavioral modules and probabilistic finite state machines. In their paper, Francesca et al. [251] compare the performance of their automatic design approach to manual design of robot behaviors in five tasks and find advantages for the automatic approach.

Gomes et al. [286] explore how novelty search can be used to evolve neural network controllers for robot swarms. Instead of rewarding solutions based on how well they perform a predefined task, novelty search rewards behavioral diversity. The study uses two tasks: aggregation and resource sharing. They find that this approach leads to diverse and often simpler solutions compared to fitness-based methods. The authors also evaluate hybrid methods that combine novelty and fitness, showing that they can further enhance performance.

Gomes et al. [288] continued to address a key limitation in evolutionary swarm robotics: the tendency to converge prematurely to stable but suboptimal states. The authors continued to integrate novelty search to encourage behavioral diversity, rather than optimizing toward fixed fitness goals. They evaluate three approaches: novelty at the team level, at the individual robot level, and a combined method. Across several multi-robot tasks, including a predator-prey pursuit scenario, the team-level novelty approach consistently outperformed the others. It improved the solution quality and scaled well with the number of robots.

Nitschke and Didi [596] explore how to improve the quality of evolved behaviors in swarm robotics using evolutionary methods and policy transfer. They study RoboCup keep-away soccer as a benchmark, where robots must coordinate to keep a ball away from opponents. The authors compare different evolutionary search strategies, including objective-based, novelty-based, and hybrid methods that combine these approaches. They demonstrate that combining novelty search with traditional objective-based evolution and utilizing policy transfer (i.e., transferring solutions from simpler to more complex tasks) significantly enhances the effectiveness of evolved behaviors. The best results are achieved when hybrid search methods are combined with policy transfer, resulting in the development of compact yet high-performing neural controllers for increasingly complex swarm tasks.

Hasselmann et al. [342] present a systematic empirical comparison of prominent neuro-evolutionary methods used for the automatic offline design of swarm robotic controllers. They evolve neural networks in simulation for five representative swarm missions (xor-aggregation, homing, foraging, shelter, directional-gate) and then deploy them on real robots to evaluate their robustness to the reality gap. While all methods performed well in simulation, their performance dropped significantly on real robots, often failing to outperform a simple random walk. The more straightforward method 'EvoStick' performed comparably to more sophisticated algorithms like CMA-ES, xNES, and NEAT. A modular design approach called 'AutoMoDe-Chocolate' by the authors with a deliberately restricted design space, showed much greater robustness across the reality gap.

Mkhatshwa and Nitschke [564] explore how co-evolution of robot behavior and morphology affects swarm performance over a set of environments that require varying levels of cooperation. By comparing five different evolutionary methods, the authors assess the value of maintaining population diversity in both behavior and morphology (they call it "body-brain quality-diversity"). They find that a method that explicitly promotes diversity, such as EDQD-M based on EDQD [340], a method that hybridizes MAP-Elites [579] and mEDEA [104], outperforms others in complex tasks. Maintaining both behavioral and morphological diversity in swarm robotics allows robots to evolve diverse and suitably complex combinations of body and controller. When many individually adapted robots interact in a swarm, they give rise to effective collective behaviors, even in increasingly challenging environments.

The evolutionary answer to imitation learning. Similar to imitation learning and IRL, there are efforts using methods of evolutionary computation to obtain robot controllers from observation [232]. Alharthi et al. [11] propose a method to infer swarm controllers directly from video demonstrations of swarm behavior. The authors use an evolutionary algorithm to evolve behavior tree controllers, guided by a fitness function that compares eight high-level swarm motion metrics between the observed video and simulated behaviors. Their approach requires no prior knowledge of the swarm type or a labeled dataset, and emphasizes human-readable rules rather than opaque models. Experiments with simulated swarms show that the method can accurately reconstruct the original controllers in most cases (75% exact matches, with an average Jaccard similarity of 0.868). This work represents an early step toward the automated and interpretable extraction of local swarm rules from observed collective dynamics, with applications in both swarm robotics design and the analysis of biological swarms.

An unexpected similar approach is 'ghost swarms' by Alharthi et al. [12]. Building on the above idea, they investigate whether one can recover swarm controllers not from the motion of the robots themselves but from the environmental imprints left behind by their collective activity. Their method extracts human-readable rules from demonstrations that only show these imprints, such as traces or patterns in the environment (e.g., movement of objects), without tracking individual robots. Using evolutionary search, the system learns controllers from a single demonstration, without making assumptions about the controller's structure.

7.6.3.2 Black-Box Design

Black-box optimization refers to a class of optimization methods that operate without requiring knowledge of the objective function's internal structure (e.g., gradients). Instead, they treat the function as a black box: they input candidate solutions, observe the outputs (fitness or performance), and use that feedback to generate new candidates. In that sense, black-box optimization is a broader category and includes evolutionary computation as a subset.

Bayesian optimization with Gaussian processes. Bayesian Optimization is a sample-efficient, model-based black-box optimization technique that is particularly suitable when evaluations are costly, such as in real-world experiments. Instead of directly optimizing the actual objective function (e.g., swarm performance), Bayesian Optimization builds a probabilistic surrogate model, typically a Gaussian Process, based on previously evaluated candidates. This surrogate not only predicts the expected performance of new candidates but also quantifies uncertainty. At each iteration, the method selects new candidate solutions by optimizing an acquisition function (such as expected improvement or upper confidence bound), which balances exploitation of areas expected to perform well and exploration of uncertain regions. This strategic sampling makes Bayesian Optimization particularly effective when only a limited number of expensive evaluations can be performed. In robotics and swarm robotics, it is used to automatically tune controller parameters or design components with minimal trial-and-error, often in conjunction with simulation-to-reality pipelines
[95, 169].

Ant Colony Optimization (ACO). ACO is a classical bio-inspired metaheuristic introduced by Dorigo and Stützle [199], Dorigo et al. [198], taking remote inspiration from the foraging behavior of ants. In ACO, multiple artificial agents explore a graph representing the solution space of a combinatorial problem, such as the well-known Traveling Salesman Problem. Each edge in the graph is associated with two values: a pheromone level, representing accumulated experience, and a visibility, typically the inverse of the edge length, serving as a heuristic. Ants probabilistically choose their next step based on a combination of pheromone concentration and visibility. Over time, positive feedback emerges: frequently used paths accumulate more pheromone, making them more attractive to other ants. To prevent convergence to suboptimal solutions, pheromone evaporation reduces the influence of unused paths, introducing negative feedback. After each iteration, agents deposit pheromone in proportion to the quality of their tour (i.e., shorter paths result in more pheromone being deposited). This iterative process of exploration, exploitation, reinforcement, and evaporation allows the swarm to discover high-quality solutions without central coordination collectively. As partially discussed in Sect. 5.3.3, stigmergy and, in particular, pheromones are challenging to implement physically in swarm robotics. Hence, ACO can barely serve as an inspiration for directly implementing a controller for swarm robots. Instead, it is often used as a centralized path planner for swarms [103, 504].

Particle Swarm Optimization (PSO). PSO is a roughly biologically inspired black-box optimization method, initially proposed by Kennedy and Eberhart [419], and inspired by the flocking behavior of birds. In PSO, a set of candidate solutions, called particles, collectively move through the search space in pursuit of the optimal solution. Each particle evaluates the quality of its current position and adjusts its movement based on three components: its own previous best position, the best position found by its neighbors, and its current direction of travel. Neighborhoods can be defined either geographically (as in classic flocking models) or socially (fixed interaction partners). This decentralized coordination makes PSO especially relevant for distributed optimization tasks, including those in swarm robotics. When it is applied to swarm robotics, there is always the problem of how to deal with the global best position that all robots cannot know in the fully decentralized approach of a swarm robotics system. Hereford [355] derived a distributed variant of PSO that eliminates "the needs for a central agent to coordinate the movements of the robots and has limited inter-robot communications requirements." Simulation results indicate that the approach applies to robot swarms and scales well. Similarly, Pugh and Martinoli [655] explore a modified PSO algorithm for unsupervised learning in multi-robot systems, where each robot manages a single particle and communicates only locally. Using realistic simulations with Khepera robots, the authors demonstrate that PSO can efficiently evolve neural network controllers for obstacle avoidance and aggregation behaviors, even in the presence of restricted communication and without a central supervisor. The approach scales well.

7.6.4 *Challenge of Reality Gap*

Most of the methods discussed above require repeated testing of candidate policies or controllers. Ideally, these evaluations would take place directly on the robots, but this is usually impractical due to limited availability, high costs, and wear on the hardware. Instead, one relies on simulators. However, this introduces a significant challenge known as the reality gap: the inevitable discrepancy between performance in simulation and in the real world. A natural reaction to this problem is to improve the simulator to reduce the gap; yet, in practice, it cannot be eliminated. As a result, robot controllers developed through intense simulations may perform well in the simulator but not as well, or not at all, in the real world. This is usually due to some hard-to-simulate features, such as friction, actuator dynamics, or temporally correlated sensor noise. Even minor inaccuracies or hidden bugs can be discovered and used as 'loopholes,' producing controllers that succeed in simulation only by relying on these artifacts, with little chance of working outside the simulated environment. The reality gap problem has been recognized in evolutionary robotics for a long time, and methods have been developed to address it [380].

Koos et al. [442] introduced a transferability approach for evolutionary robotics, designed to mitigate the reality gap. The method explicitly acknowledges limitations of simulators and combines two complementary processes: (1) optimization in simulation to explore a wide range of candidate controllers, and (2) a learned transferability function that predicts how well a solution found in simulation will perform on a real robot. This function is inferred from a limited set of real-world tests. Experiments with a four-legged robot show that this approach successfully evolves controllers while requiring only a small number of physical trials.

Salvato et al. [694] review methods for transferring controllers trained in simulation to real robots. The survey identifies three main strategies to mitigate this problem: domain randomization, which exposes controllers to varied simulated conditions to increase robustness; adversarial reinforcement learning, which employs adversarial agents to generate challenging perturbations; and transfer learning, which reuses models or policies from simulation to accelerate adaptation in real robots. While these approaches show promise, most results remain task-specific and difficult to compare systematically. The authors emphasize that a general, task-independent method for reliable sim-to-real transfer is still missing and suggest future directions such as meta-learning and probabilistic model-based techniques.

7.6.5 Modern Approaches for Robot World Models

The World Models approach by Ha and Schmidhuber [311] gives a robot a simpler world model of its environment. They demonstrate how agents can learn compact internal models of their environments and then train policies within these models, effectively "dreaming" to improve performance. The approach combines a variational autoencoder (VAE) to compress high-dimensional sensory input, a mixture density RNN (MDN-RNN) to predict future latent states, and a lightweight controller that selects actions based on these representations. A VAE is a neural network that learns to take high-dimensional data, like camera images, and compress it into a much smaller and more useful representation. This compressed version is called a latent vector, which stores the essential information about the image in just a few numbers. An MDN-RNN then uses these latent vectors, together with the robot's actions, to predict how the environment will change over time. A lightweight controller then chooses actions based only on this compact representation, allowing the agent to train efficiently in its own imagined world before transferring the learned policy back to reality. Using this architecture, agents learned to drive cars in a racing benchmark in some cases, training entirely within their own generated "dream environments" and successfully transferring the learned policies back to the real environment. This work demonstrated that world models facilitate data-efficient reinforcement learning and laid the groundwork for subsequent approaches.

Another impressive effort is that of DeepMind in collaboration with others. Bruce et al. [111] introduce a new class of foundation world models that can generate fully interactive environments from nothing more than large-scale Internet video data.

Genie is a spatiotemporal transformer-based generative world model. Its architecture combines a video tokenizer (VQ-VAE [809]) and a spatiotemporal Transformer, an autoregressive dynamics model (decoder-only Transformer), and a latent action model (spatiotemporal Transformer), scaling up to 11 billion parameters. A spatiotemporal transformer (ST-Transformer) is a neural network that processes videos by attending to both the spatial structure within each frame and the temporal changes across frames [879]. Trained without action labels, *Genie* learns a latent action space that allows users or agents to control video trajectories on a frame-by-frame basis. Demonstrations show that *Genie* can transform text prompts, images, or even hand-drawn sketches into playable environments and can also learn control dynamics without requiring labeled actions.

Genie was applied to robotic manipulation scenarios using data from Google's RT-1 project [108], which features a robotic arm performing everyday tasks. From more than 200k episodes of simulated and real demonstrations, *Genie* learned to control the arm consistently. It discovered latent actions, such as up, down, or left, without any explicit action labels. The model could also reproduce physical effects, such as the deformation of objects. For mobile robotics, such an approach could one day allow robots to acquire control strategies directly from large video corpora. In swarm robotics, this raises the possibility of training collective behaviors from vast amounts of unlabeled multi-agent video data. Also see the newer version online.[6]

7.7 Further Reading

There are not too many options for books on modeling swarm robot systems. One option is the book by Valentini [800], which is, however, focused on collective decision-making (see Chap. 8). Another option is the book by Hamann [318], which covers the concept of swarm models based on Fokker–Planck equations. Other, more specific options are the book by Galam [265] on sociophysics (opinion dynamics) and the book by Gross and Sayama [302] on adaptive networks. For the important concept of Multi-Agent Reinforcement Learning (MARL) see the book by Albrecht et al. [8].

7.8 Exercises

Exercise 7.1 (*Beyond Binary Decision-Making*) In Eq. (7.4) we gave a modeling abstraction for a binary decision-making system:

$$f(\gamma) = \frac{|\{s_i | s_i = A\}|}{N} = \frac{[\text{number of agents in favor of option } A]}{N} = \varphi. \tag{7.38}$$

[6] https://deepmind.google/discover/blog/genie-3-a-new-frontier-for-world-models/.

What is the analogon of that for a collective decision-making system with three options (A, B, C)?

Exercise 7.2 (*Buffon's Needle*) Implement Buffon's needle for needle length $b = 0.7$ and line spacing $s = 1$. Note that you do not have to simulate the whole setting in a big 2-d plane. Instead, you can reduce the situation to a single line segment of width s because the middle of the needle will always end up between two lines. What remains is simple geometry that determines whether the needle intersects with one of the two lines based on the needle's angle.

(a) Test your program by checking whether it estimates good approximations to π according to $\pi = 2b/(sP)$.
(b) An experiment consists of n needles being thrown. Run experiments with $n \in \{10, 20, \dots, 990, 1000\}$ trials. For each setting of n, run the experiment 10,000 times (i.e., $10{,}000 \times n$) and calculate the standard deviation of the resulting set of intersection probabilities. Plot the standard deviation over the number of trials n.
(c) For a maximum of $n = 100$ trials, plot the currently measured probability of many experiments over the trial number n along with the Binomial proportion 95%-confidence interval.[7]
(d) Measure the ratio of experiments for which the true probability[8] is outside of the 95%-confidence interval based on the measured probability. Plot this ratio over the number of trials n. Interpret your results.

Exercise 7.3 (*Local Sampling in a Swarm*) We have a robot swarm of size N with the robots uniformly distributed over the unit square (robot positions (x, y) with $x \in [0, 1]$, $y \in [0, 1]$). Each robot is either black or white with equal probability. The local sampling is done in the following way. A particular robot perceives the color of all robots (including itself) in the neighborhood. The neighborhood is defined by the sensor range r. A robot belongs to the neighborhood of another robot if their distance is less than the sensor range r. From the local ratio of black robots, the robot estimates the overall number of black robots in the swarm.

Implement this scenario. Do experiments for swarm sizes $N \in [2, 200]$ each with sensor ranges $r \in [0.02, 0.5]$. For each setting of N and r do 1000 independent experiments and calculate the mean and standard deviation of the robot's estimate of black robots in the swarm. Plot mean and standard deviation for some interesting settings. Try to think of the consequences that these measurements would have for an implementation of a robot swarm whose effectiveness is directly dependent on these local samplings.

[7] From Sect. 7.3.2 we know that the Binomial proportion 95%-confidence interval is defined by $\hat{P} \pm 1.96\sqrt{\frac{1}{n}\hat{P}(1-\hat{P})}$.

[8] From Sect. 7.3.2 we know that the true probability is $P = 2b/(s\pi)$.

Exercise 7.4 (*Dimension Reduction and Modeling*) We implement the locust simulation (continuous space, discrete time). The locusts live on a ring of circumference $C = 1$ (positions $x_0 = 0$ and $x_1 = 1$ are identical). The locusts move with a speed of 0.001 either to the left ($v = -0.001$) or to the right ($v = +0.001$). They have a perception range of $r = 0.045$. A locust switches its direction in one of two situations:

(1) The majority of locusts within its perception range have an opposite direction to that of the considered locust.
(2) A locust spontaneously switches its direction with a probability $P = 0.015$ per time step.

Initially, the locusts are uniformly randomly distributed and are moving left or right with equal probability. We simulate a swarm of size $N = 20$ for 500 time steps.

(a) Implement and test your simulation. Plot the number of left-going locusts over time for one run.
(b) We want to take the number of left-going locusts L as our modeling approach. Hence, we implement the dimension reduction of our model by averaging over several system configurations and summarizing them in groups of equal left-goer number. Create a histogram of the observed transitions $L_t \to L_{t+1}$ (change in the number of left-goers within one time step) with your simulation by doing 1000 sample runs of 500 time steps each. For example, you can use a 2-d array $A[\cdot][\cdot]$ of integers and an entry of this array $A[L_t][L_{t+1}]$ is increased by one whenever a transition $L_t \to L_{t+1}$ is observed. Plot the histogram.
(c) In addition, count also the occurrences $M[L]$ of each model state L and use these to normalize the histogram entries: $A[i][j]/M[i]$. That way, we get approximations of the transition probabilities. Use these approximations $P_{i,j} = A[i][j]/M[i]$ to sample evolutions of L_t over time t. Plot one such trajectory of L. How does this compare to the plot done in (a)?

Exercise 7.5 (*Rate Equations*) We use the rate equations model of searching and avoiding:

$$\frac{dn_s(t)}{dt} = -\alpha_r n_s(t)(n_s(t) + 1) + \alpha_r n_s(t - \tau_a)(n_s(t - \tau_a) + 1)$$
$$\frac{dm(t)}{dt} = -\alpha_p n_s(t) m(t)$$

(a) Use a tool of your choice to calculate the temporal course of this system of ordinary differential equations (a simple forward integration in time can also be implemented from scratch for this system). Notice that we have delay equations. How should the delays be treated, especially early in the simulation ($t < \tau_a$)? Use the following setting for the parameters: $\alpha_r = 0.6$, $\alpha_p = 0.2$, $\tau_a = 2$, $n_s(0) = 1$, $m(0) = 1$. Calculate the values of n_s and m for $t \in (0, 50]$ and plot them. Interpret your result.

(b) Now we want to extend the model. In addition to *searching* and *avoiding* we introduce a third state: *homing* (n_h). Robots that have found a puck do a transition to the state *homing* in which they stay for a time $\tau_h = 15$. We assume that for unspecified reasons, robots in state *homing* do not interfere with each other or with robots of any other state (assumption: no avoidance behavior for robots in state *homing* necessary). After the time of τ_h, they have reached the home base and do a transition back to *searching*. Add an equation for n_h and edit the equation of n_s accordingly. Calculate the values of n_h, n_s, and m for $t \in (0, 160]$ and plot them. In a second calculation, reset the ratio of pucks at time $t = 80$ to $m(80) = 0.5$ and plot the results. Interpret your result.

Exercise 7.6 (*Adaptive Networks*) We numerically solve the ODE system given for adaptive networks in Sect. 7.4.4. Go carefully through Eqs. (7.28)–(7.34), write down the ODE system, and formulate an initial value problem (e.g., $x_R(t_0 = 0) = 0.6$ and $x_L(t_0 = 0) = 0.4$). Implement your own simple numerical integration method (e.g., forward integration in time) or use a standard mathematical software tool to integrate the equations numerically.

(a) Plot the results and recreate Fig. 7.11a. Interpret the results: what can be said about the occurring edges in the network?
(b) Extend your program and do the necessary scans to recreate the bifurcation diagram shown in Fig. 7.11b. You need to integrate your system for different initial conditions to observe all fixed points. Basically, you have to run your integration for an appropriate number of time steps and plot the final observed system state. Repeat this process for a range of different parameter values. What does the result mean for our collectively deciding swarm robotics system?

Exercise 7.7 (*General Network Models*) Construct two different network models for a swarm of robots and compare their properties.

(a) Construct a set of Erdős–Rényi random graphs $G(20, p)$ with $N = 20$ nodes, where each possible edge is added independently with probability $p = 0.2$.
(b) Construct a set of random geometric graphs with $N = 20$ nodes placed uniformly at random inside a unit square $[0, 1] \times [0, 1]$. Connect two nodes by an edge if their Euclidean distance is less than the communication radius $r = 0.3$.
(c) For both sets of graphs, calculate the average degree (the mean number of neighbors per node) and the average number of connected components across all graphs for each set. Optionally, also determine the size of the largest connected component.
(d) Compare the two models. Which model tends to have more isolated nodes? Which shows stronger local clustering or spatial correlations? To calculate a metric of spatial correlations, you can choose to use the clustering coefficient (measure of transitivity): For a node i, we define its clustering coefficient as $C_i = \frac{2E_i}{k_i(k_i-1)}$, where k_i is the degree of node i, and E_i is the number of edges among its neighbors. To calculate E_i, let $\mathcal{N}(i)$ be the neighbor set of i and

$k_i = |\mathcal{N}(i)|$. Count how many unordered pairs $\{j, k\} \subset N(i)$ are connected: $E_i = \left|\{ \{j, k\} \subset N(i) : (j, k) \in E \}\right|$. The average clustering coefficient $\langle C \rangle$ across all nodes measures how strongly neighbors of a node are also connected.

(e) Discuss why the Erdős–Rényi model does not capture spatial correlations between robots, while the random geometric graph does. Reflect on when an Erdős–Rényi model might still be useful in swarm robotics and when the geometric model is essential.

Exercise 7.8 (*Hill Climber as Simplified Evolutionary Algorithm*) Artificial evolution is already a caricature of natural evolution, and the following algorithm is yet another caricature of evolutionary computation. Nevertheless, it is an interesting starting point because it implements a straightforward optimization strategy. The goal is to 'evolve' the target string "`charles darwin was always seasick`" which consists of 33 characters (lowercase letters and spaces). We work with only one individual (no population). The algorithm proceeds as follows. We start with a random string of the correct length composed of lowercase letters and spaces (uniform distribution over all letters and spaces). In each generation, one character is chosen uniformly at random and replaced by a random character (again, lowercase letters or spaces). This mutation is accepted only if the new string's fitness is not lower than that of the previous string; otherwise, the algorithm backtracks to the previous string. Hence, this is a simple hill-climbing strategy. The fitness function is defined as the number of correct characters (correct letters or spaces in the correct position).

(a) Implement this program. Print the current string in each generation to visualize the optimization process.
(b) Explain why this problem can be considered a "good-natured" optimization problem, especially in terms of the implied fitness landscape.
(c) Provide a mathematical estimation of how many generations are needed on average to find the correct string. Then test your prediction empirically by running many independent trials and averaging the number of generations required.

Exercise 7.9 (*Classical Optimization with Evolutionary Algorithms*) In evolutionary robotics, the fitness of a robot controller is usually defined indirectly via the generated behavior and the corresponding fitness function. In standard optimization problems, however, the fitness is (often) directly defined via the objective function. In this task, we implement an evolutionary algorithm to solve the three-dimensional Ackley problem. The objective function to minimize is

$$f(x, y, z) = -20\exp\left(-0.2\sqrt{\tfrac{1}{3}(x^2+y^2+z^2)}\right) - \exp\left(\tfrac{1}{3}\left(\cos(2\pi x)+\cos(2\pi y)+\cos(2\pi z)\right)\right) + 20 + \exp(1)\,, \tag{7.39}$$

with the search domain defined as $x, y, z \in [-32.768, 32.768]$. To translate this minimization problem into a maximization problem, we define fitness as

$$F(x, y, z) = \frac{1}{f(x, y, z) + 1} .$$

(a) Implement a complete evolutionary algorithm of your choice to solve this problem. Choose a genetic representation, a population size (> 1, no hill climbers), a selection method, a replacement strategy (e.g., generational replacement, elitism), a mutation rate, and decide whether you want to include recombination. Initialize the population uniformly over the full search interval $[-32.768, 32.768]^3$.
(b) Debug your implementation. Verify that the selection is working correctly and ensure that you are optimizing the correct objective function.
(c) Record the best and average fitness values in each generation and plot their progression over time. Interpret the behavior you observe.
(d) Experiment with different algorithmic parameters. Which settings yield good performance? Which settings lead to poor results?
(e) Compare the performance of your evolutionary algorithm to a simple hill climber. Additionally, test the algorithm on higher-dimensional versions of the Ackley problem.

Exercise 7.10 (*Classical Optimization with a Reinforcement Learning Agent*) Reinforcement learning can also be applied to standard optimization problems. In this task, we consider the one-dimensional variant of the Ackley problem. The domain is $x \in [-5, 5]$ and the objective function is

$$f(x) = -20 \exp\Big(-0.2|x|\Big) - \exp\Big(\cos(2\pi x)\Big) + 20 + \exp(1) .$$

We reformulate the problem as an episodic RL task: a single agent starts at a random x-position in $[-5, 5]$ and can choose one of three discrete actions in each step: move left, move right, or stay. The environment returns a reward equal to the negative of the objective function value, we have $r_t = -f(x_t)$. Episodes end after a fixed number of steps.

(a) Implement the environment as described above.
(b) Implement tabular Q-learning with discretized states (e.g., divide $[-5, 5]$ into 100 bins).
(c) Train the agent until convergence. Plot the best trajectory the agent follows.
(d) Compare performance with different discretizations of the state space (e.g., 50, 100, 200 bins). How does resolution affect convergence?

Exercise 7.11 (*Classification with Evolutionary Algorithms*) Statistical classification is a classical problem in machine learning. It is an example of supervised learning (a set of already classified examples is provided). It hence differs from typical situations in evolutionary robotics (where classifying sensor input with appropriate actuator outputs is precisely what we do not want to specify by hand). In the following, the task is to evolve a classifier for a simple problem with two-dimensional features. For this purpose, we evolve a small ANN without a hidden layer (single-layer perceptron).

(a) Generate a training set of two-dimensional feature vectors. Use the following procedure: half of the data points belong to class 0 and are distributed in a cluster around (0.25, 0.25), the other half belong to class 1 and are distributed around (0.75, 0.75). For each point (x, y), use random numbers $r_1, r_2 \in [0, 1]$ and compute

$$x = \mu_x + 0.25\sqrt{\frac{-2\ln(r_1)}{\ln 2}}\cos(2\pi r_2), \qquad y = \mu_y + 0.25\sqrt{\frac{-2\ln(r_1)}{\ln 2}}\sin(2\pi r_2),$$

with $(\mu_x, \mu_y) = (0.25, 0.25)$ for class 0 and $(\mu_x, \mu_y) = (0.75, 0.75)$ for class 1. Store the data in a file with four columns: index, class, x, and y.

(b) Implement a simple ANN that has two input neurons, one output neuron, and a bias neuron, but no hidden layer. Three weights fully describe this ANN. As an activation function, we use

$$\varphi(x) = \frac{2}{1 + \exp(-2x)} - 1\,.$$

An input is classified as 'class 0' if $\varphi < 0$ and as 'class 1' if $\varphi > 0$. To compute the fitness of an ANN, evaluate its output for all entries of the data set and count how many are classified correctly.

(c) Evolve an ANN that solves the problem satisfactorily. Output the three weights of the best evolved ANN and plot the dividing line implemented by the ANN,

$$y = \frac{w_0}{w_2} - \frac{w_1}{w_2}x\,,$$

where w_0 is the bias weight, w_1 is the input weight of x, and w_2 the input weight of y.

(d) Generate a more complex training set with overlapping classes and attempt to evolve an ANN that classifies it. For this task, you will need to add a hidden layer, and bias neurons are also recommended.

Exercise 7.12 (*Classification with Reinforcement Learning*) We revisit the two-class classification problem, but we now use reinforcement learning. A dataset of two-dimensional feature vectors is generated with two clusters (class 0 around (0.25, 0.25) and class 1 around (0.75, 0.75), as in the previous task). Each episode presents the agent with one data point (x, y), normalized to $[0, 1]^2$. The agent must output one of two discrete actions: predict class 0 or predict class 1. The reward is $+1$ for a correct prediction and -1 for an incorrect one.

(a) Generate the dataset of 2D feature vectors.

(b) Implement an RL agent using a linear function approximator (perceptron) that maps the two inputs (x, y) to one output action. The state is a two-dimensional input $s = (x, y)$, and a linear model with a softmax function represents the policy. The parameters are the bias weight w_0 together with the feature weights w_1 for x and w_2 for y. The policy is therefore given by

$$\pi_\theta(a|s) = \text{softmax}(z), \quad z = w_0 + w_1 x + w_2 y \,, \tag{7.40}$$

where $\theta = (w_0, w_1, w_2)$.

(c) Train the agent using a policy gradient algorithm. Implement the REINFORCE algorithm [860] with the perceptron weights as the policy parameters. For each training episode, present the agent with a randomly chosen data point (x, y) from the dataset. The perceptron outputs a probability distribution over the two possible actions (class 0 or class 1), obtained by passing the linear combination of inputs through a softmax function. For the vector of scores $z = (z_1, z_2)$, the softmax of component i is defined as

$$\text{softmax}(z_i) = \frac{e^{z_i}}{\sum_{j=1}^{K} e^{z_j}} \,. \tag{7.41}$$

Sample an action from this distribution, compare it to the actual class label, and assign a reward $r = +1$ for a correct classification or $r = -1$ for an incorrect classification. Update the policy parameters after each episode using the REINFORCE update rule

$$\Delta\theta = \alpha\, r\, \nabla_\theta \log \pi_\theta(a|s) \,, \tag{7.42}$$

where α is the learning rate, $s = (x, y)$ is the input, a is the chosen class, and $\pi_\theta(a|s)$ is the probability of action a given input s. Probability of taking action $a = 1$ is defined by

$$p \equiv \pi_\theta(a{=}1 \mid s) = \frac{1}{1 + e^{-z}} \,. \tag{7.43}$$

The resulting updates of $\theta = (w_0, w_1, w_2)$ if the taken action is $a = 1$ are:

$$\Delta w_0 = \alpha\, r\, (1 - p) \,, \tag{7.44}$$

$$\Delta w_1 = \alpha\, r\, (1 - p)\, x \,, \tag{7.45}$$

$$\Delta w_2 = \alpha\, r\, (1 - p)\, y \,. \tag{7.46}$$

If the taken action is $a = 0$ (so $\pi_\theta(a{=}0 \mid s) = 1 - p$), we get

$$\Delta w_0 = -\alpha\, r\, p\,, \tag{7.47}$$

$$\Delta w_1 = -\alpha\, r\, p x\,, \tag{7.48}$$

$$\Delta w_2 = -\alpha\, r\, p y\,. \tag{7.49}$$

Train for a sufficient number of episodes until the classification accuracy reaches a stable level.

(d) Evaluate the trained policy on the full dataset and compute the classification accuracy. Compare it to the result of a supervised logistic regression trained directly on the dataset.

Exercise 7.13 (*Behavioral Attractors–Ring Problem*) We consider the 'ring problem' [597]. A simulated agent lives on a circular stripe divided into 40 cells: 20 on the left half and 20 on the right half. The cells in each half are numbered from 0 to 19 in a different random order. The agent moves from cell to cell, perceives the current cell's number, and decides to move either clockwise (CW) or counterclockwise (CCW).

The agent's neural network is simple. It has 20 input units encoding the number of the current cell and one output neuron representing the action (CW or CCW). If the robot is at a cell with number i, then all input units are 0 except for input unit i, which is 1. This single active input is multiplied by the corresponding weight w_i, transformed by an activation function of your choice, and the output is compared to a threshold to generate motor control: CW or CCW.

Initially, the robot is placed at a random position on the ring. The simulation should run for a sufficient number of steps, allowing the robot to reach the left half of the ring and remain there. Define a fitness function that rewards the robot for spending more time on the left half of the ring. The fitness should be averaged over multiple repetitions with different initial positions.

(a) Implement the ring world with 40 cells, each labeled randomly with numbers 0–19 on the left and right halves.
(b) Implement the neural network controller with 20 input units and one output neuron as described above. Define the activation function and output threshold of your choice.
(c) Define a fitness function that rewards the agent for spending time on the left half of the ring. Run simulations with multiple initial positions and average the results.
(d) Use evolutionary computation to evolve the weights $w_0, \ldots, w_{19}$ of the controller.
(e) Output the evolved solution: for each of the 40 ring cells, list whether the agent moves CW or CCW. Verify that the evolved controller solves the problem perfectly.

Exercise 7.14 (*Behavioral Attractors–Ring Problem with RL*) We reconsider the 'ring problem' with reinforcement learning. The agent lives on a circular stripe of 40 cells, with 20 on the left half and 20 on the right half, labeled 0–19 in random order. At each step, the agent perceives the current cell index i and must choose one of two actions: move clockwise (CW) or counterclockwise (CCW).

The reward is defined as $+1$ if the agent is located on the left half of the ring and -1 if it is on the right half.

(a) Implement the ring environment with 40 cells and random labeling.
(b) Implement Q-learning with a table of size 40 states $\times$ 2 actions.
(c) Train the agent for sufficiently many episodes with random initial states $s_0 \sim \text{Unif}(S)$ and fixed horizon H (no terminal states). Use ε-greedy exploration:

$$\pi(a \mid s) = \begin{cases} 1 - \varepsilon_t + \frac{\varepsilon_t}{2}, & a = \arg\max_{a'} Q(s, a'), \\ \frac{\varepsilon_t}{2}, & \text{otherwise}, \end{cases} \tag{7.50}$$

with

$$\varepsilon_t = \max\left(\varepsilon_{\min}, \varepsilon_0 e^{-t/\tau}\right) . \tag{7.51}$$

After executing $(s_t, a_t, r_{t+1}, s_{t+1})$, update Q with learning rate $\alpha \in (0, 1]$ and discount $\gamma \in [0, 1)$:

$$\delta_t = r_{t+1} + \gamma \max_{a'} Q(s_{t+1}, a') - Q(s_t, a_t), \tag{7.52}$$

and

$$Q(s_t, a_t) \leftarrow Q(s_t, a_t) + \alpha\, \delta_t . \tag{7.53}$$

Track the average return per episode $\frac{1}{H} \sum_{t=0}^{H-1} r_{t+1}$ and the fraction of time on the left half $\frac{1}{H} \sum_{t=0}^{H-1} L(s_{t+1})$ for

$$L(s) = \begin{cases} 1 & , \text{if } s \in S_L \\ 0 & , \text{if } s \in S_R \end{cases} . \tag{7.54}$$

(d) Test the trained policy by starting the agent at every possible position and verifying that it converges to staying on the left half.

Exercise 7.15 (*Survival by Autonomous Recharging*) We implement a 'navigate and recharge' task [597]. A robot is placed in an empty arena (no walls). In one corner, there is a lamp, and the floor in that corner is painted black within a quadrant. This black area represents the charging area. The robot has a simulated battery level $b \in [0, b_{\max},]$ that is immediately recharged $b = b_{\max})$ once the robot enters the black area. The battery discharges at a fast linear rate.

The robot is equipped with the following sensors: three proximity sensors (front, 90° left, 90° right), two ambient light sensors (front and back, sensing a light intensity field as in the Braitenberg vehicles), one ground sensor (discrete, 0 for black and 1 for non-black), and one battery-level sensor. The controller is implemented as an ANN with one hidden layer. You may choose any number of hidden neurons (preferably between 2 and 5). Optionally, implement a recurrent ANN by providing each hidden neuron with its activation from the previous time step (as an additional input multiplied by a dedicated weight).

The fitness function is defined as

$$F = V(1 - i) , \tag{7.55}$$

with $0 \leq V, i \leq 1$, where V is the average rotation speed of both wheels and i is the average activation level of the proximity sensors ($i \approx 0$ means no close-by objects, $i \approx 1$ means close-by objects). Whenever the robot is positioned in the charging area, it receives no fitness for that period.

(a) Implement the robot simulation with the given sensor setup, battery model, and charging mechanism.
(b) Design and implement the ANN controller with a hidden layer. Optionally, extend it to a recurrent ANN.
(c) Define suitable parameters for robot speed, discharge rate, size of the charging area, maximum battery level b_{max}, arena dimensions, proximity sensor range, and evaluation time. Ensure these parameters relate in a meaningful way to provoke non-trivial behaviors.
(d) Base the fitness evaluation on multiple independent trials with random initial robot positions. Consider adding small amounts of noise to the motor control values to avoid overly simple solutions.
(e) Run evolutionary optimization of the ANN controller. Due to the difficulty of the problem, prepare for long computations. Experiment with different parameter settings until a satisfactory behavior is evolved.
(f) Test your best evolved controller from various starting positions. Does it reliably reach the charging area from various initial positions?

Chapter 8
Collective Decision-Making

They exist in loose swarms [...]. However, they will unite in moments of danger, or to be more precise, in the event of any sudden change that constitutes a threat to their survival

—Stanisław Lem, The Invincible,

But the amoebas are certainly creative on an individual basis. [...] A thought is probably only taken into consideration if the impulse behind it is strong enough, that's to say if enough yrr are trying to introduce it into the collective at the same time.

—Frank Schätzing, The Swarm

the Emperor ...was interested in having you advance fictionalized predictions that might stabilize his dynasty ...I ask only that you perfect your psychohistorical technique so that mathematically valid predictions, even if only statistical in nature, can be made.

—Isaac Asimov, Foundations

Abstract We study methods of collective decision-making—an important capability for a swarm to become autonomous. This chapter introduces collective decision-making as a key capability for enabling autonomous swarms to act intelligently at the macroscopic level. It reviews foundational ideas from individual and group decision-making, including challenges such as consensus formation and the Byzantine Generals Problem, and surveys a broad set of modeling approaches (e.g., urn models, voter and majority rule dynamics, Hegselmann–Krause, Kuramoto, and Ising models) to guide model selection. Practical implementations are illustrated through experiments with Kilobots and e-pucks, highlighting the processes of consensus formation and the tradeoff between speed and accuracy. The chapter also addresses swarm-specific challenges, including local communication limits, decision-making under uncertainty, optimal stopping strategies (e.g., the $1/e$ rule, Chao1), and the influence of contrarian agents. Together, these perspectives provide both theoretical and practical foundations for designing and analyzing self-organizing swarm systems.

H. Hamann, *Swarm Robotics*,
https://doi.org/10.1007/978-3-032-10584-4_8

8.1 Introduction

8.1.1 Learning Objectives

After reading this chapter, you should be able to:

- Understand why collective decision-making is essential for autonomous swarms and how it enables macroscopic intelligence.
- Grasp key concepts of individual decision-making, including rationality, utility, and decision-making under uncertainty.
- Explore group decision-making dynamics, such as consensus, voting methods, and challenges like groupthink or voting paradoxes.
- Analyze collective motion (e.g., flocking, locust alignment) as a decision process at the swarm level.
- Compare major collective decision-making models, including:
  - Traditional urn models (Ehrenfest, Eigen, Swarm Urn)
  - Opinion dynamics models (Voter, Majority Rule, Hegselmann–Krause, Axelrod, Sznajd, Sociophysics)
  - Synchronization models (Kuramoto, Swarmalator)
  - Statistical physics models (Ising, Fiber Bundle)
- Recognize challenges specific to robot swarms, such as local communication, heterogeneous information, deadlocks, and the consensus and Byzantine Generals problems.
- Learn strategies for optimal stopping, including heuristic thresholds, the Chao1 estimator, and optimal stopping theory.
- Apply density estimation methods, such as Kernel Density Estimation, to assess option quality and inform distributed stopping decisions.
- Examine hardware implementations and understand the speed–accuracy tradeoff.
- Interpret implicit decision-making in behaviors like BEECLUST aggregation.
- Analyze the role of contrarian agents in shaping opinion dynamics, preventing consensus, or sustaining exploration in dynamic environments.

8.1.2 Guiding Questions

- What is collective decision-making, and why is it essential for autonomous swarms?
- How does decision-making in swarms differ from humans, especially under uncertainty, and what strategies exist?
- What challenges arise in achieving consensus, including the Byzantine Generals Problem and the consensus problem in distributed control?

- How do known versus unknown options affect swarm decisions, and what stopping strategies can be applied?
- How do swarm size and option number influence difficulty, and what early phases of decision-making are important?
- What are the main approaches to group decision-making, and how do consensus methods compare with voting-based approaches and their paradoxes?
- How do animals achieve consensus efficiently, and what lessons can be drawn for robot swarms?
- What are the key phases of a decision-making process, and how does option quality matter?
- How do classic models, such as urn models, voter and majority rule, Hegselmann–Krause, Kuramoto, Ising, etc., capture collective decision-making dynamics?
- How have collective decision-making processes been implemented on robot swarms, such as Kilobot site selection, e-puck collective perception, or implicit decisions in BEECLUST aggregation?

8.1.3 From Individual to Collective Autonomy

One of the major objectives in swarm robotics is to create autonomous swarms. Autonomy here means to make independent decisions and hence having the possibility of behaving intelligently. The individual robot certainly should be autonomous; however, the swarm as a whole should be autonomous. Not only can an individual take independent decisions, but also the whole swarm. Therefore, collective decision-making is an essential, if not the most essential, capability of a swarm.

Each of us knows that decision-making can be a challenging process. Even if you have to decide, say, about a particular career choice, it can be agonizing. Decision-making in a small group of people is even more challenging ("Should we go to the lake or to the movies?"), and decision-making on the scale of millions of people can be horrifying, as we know from our democratic elections. For a swarm of robots, decision-making is, in some ways, easier and, in other ways, harder. On the one hand, we will program the robots to follow a common goal. Hence, they have a significant advantage over human society, where we celebrate diversity even though it complicates our decision-making. In terms of a common objective, we synchronize the swar, but, on the other hand, we still face difficulties. In human decision-making, such as elections, we assume that several global channels of communication exist. Of course, not all information is public, not everyone has the same amount of information, and we don't have complete transparency, but a lot of information is shared globally. In a robot swarm, we don't have a global channel; instead, robots can only communicate with their neighbors. This creates several problems. Robots need to learn that there is a collective decision-making process currently running. Robots may have different information in sub-populations and hence disagree on what the best option is, even though they share the same objectives. The swarm may run into deadlock situations, possibly even without noticing it. Finally, the robots may not know whether the decision has been taken already.

8.1.4 Next Steps

In this chapter, we first look into decision-making in general and how it is done in groups. Then we try to understand collective motion as an example of collective decision-making and go through several models. Collective decision-making, as it is implemented in swarm robotics, takes influence from many different fields of research. Usually, we explicitly include randomness into the system to ensure a minimum amount of exploration. This makes the swarm system inherently stochastic. Models from statistical physics can represent collective decision-making. Models from the field of opinion dynamics are, of course, relevant. Furthermore, models from material science and models of standardization can also be applied to collective decision-making. While it would be convenient to pick the best of these options and stick to it, we want to gather broader knowledge and learn from all of them. The literature on modeling and analyzing collective decision-making has grown rapidly in the last ten years. This research could benefit from better and more diverse benchmarking, as well as inspiration from real-world challenges. It is still a young, developing field, and it will pay off to stay explorative in terms of models.

8.2 Decision-Making

Any autonomous robot has to decide at some time. A decision is the selection between possible actions. Decision-making is the cognitive process of selecting an action from several alternatives. The perfect decision is an optimal decision, which means that no alternative decision results in a better outcome.

Decision-making can be based on certainty, that is, all current circumstances that influence the decision-making are known. The environments in which our robots typically operate are, however, not that simple. Instead, we assume that the robot has to decide under uncertainty, which is the actual art of decision-making. Decision-making under uncertainty means that the current state and/or the result of the robot's actions are not fully known. To still have the chance to act intelligently, the robot needs to know preferences between different possible outcomes of its actions. We assume that the robot can estimate the outcome of its actions based on the incomplete information about the current state of the environment. Each outcome has a certain utility, which is a measure of the quality of how useful a certain outcome is. A rational decision is an optimal decision under a set of constraints such as uncertainty. An agent

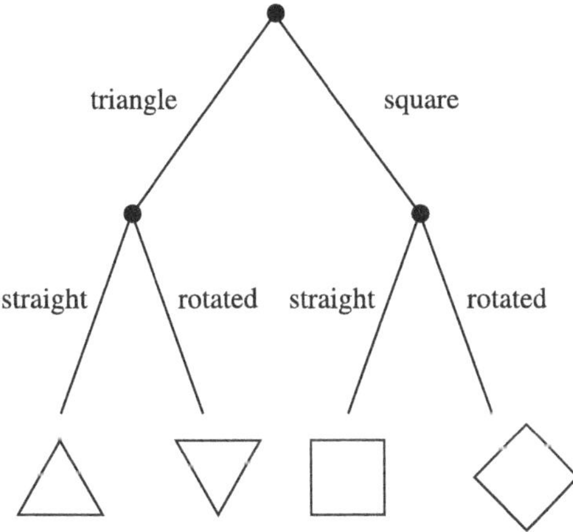

Fig. 8.1 An example of a decision tree (CC0 1.0)

is rational if and only if it chooses the action that yields the highest expected utility, averaged over all the possible outcomes of that action. This is called the principle of maximum expected utility.

A common way of organizing and representing decision-making is through decision trees. An example is shown in Fig. 8.1.

In the following, we formulate a general model of decision-making. The robot has m possible alternative actions $A_1, \ldots, A_m$ available and there are n possible initial states $s_1, \ldots, s_n$. The robot knows a utility $N_{ij} \in \mathbb{R}$ for each action A_i and state s_j. These utilities define the utility matrix $N = (N_{ij})_{1 \leq i \leq m, 1 \leq j \leq n}$. As discussed above, decision-making under certainty is easy, and hence the strategy under certainty about the current state (i.e., j is given) is clear: select $\hat{i} \in \{1, \ldots, m\}$ so that $N_{\hat{i}j} = \max_i N_{ij}$. However, we assume uncertainty that the robot is not sure about the current state of the environment, but might know probabilities of being in a certain state. We define three strategies of decision-making under uncertainty.

Using the cautious strategy (max-min-utility) the robot selects $\hat{i} \in \{1, \ldots, m\}$ so that $N_{\hat{i}j} = \max_i \min_j N_{ij}$. The minimally possible utility (i.e., the assumed worst possible state j) is maximized over all possible strategies i.

Using the high-risk strategy (max-max-utility), the robot selects $\hat{i} \in \{1, \ldots, m\}$ so that $N_{\hat{i}j} = \max_i \max_j N_{ij}$. The maximally possible utility (i.e., assumed best possible state j) is maximized over all possible strategies i.

Using the alternative cautious strategy (min-max-utility), the robot selects $\hat{i} \in \{1, \ldots, m\}$ so that $R_{\hat{i}j_i} = \min_i \max_j R_{ij}$ whereas $N_j = \max_i N_{ij}$ (maximal utility in state j) and risk matrix $R_{ij} = N_j - N_{ij}$ (risk of action i in state j). The maximally possible risk (i.e., assumed most risky state j) is minimized over all possible strategies i.

8.3 Group Decision-Making

We have learned that decision-making under uncertainty is already a difficult task. In swarm robotics, we have, however, multiple robots that cooperate and consequently also have to make decisions collectively. Group decision-making or collaborative decision-making is needed when a group of individuals chooses from alternatives. In contrast to a single agent making a decision, group decision-making is more complex and might show effects of social influence and, for example, group polarization.

In consensus decision-making, the idea is to avoid "winners" and "losers" of the decision. While the majority approves the decision, the minority agrees to go along with that decision. This is implemented by giving the minority the right to veto any decision. Obviously, this procedure is complicated, as everyone can tell who has taken part in a consensus decision-making process. process within a diverse group.

More common, of course, are voting-based methods. A strict method based on voting is to require a majority (>50%) for a decision, even if there are more than two options. Plurality, in contrast, allows decisions based on less than 50%, that is, just the largest block decides.

There are several challenges in group decision-making. An example is "Groupthink" (1972, Irving Janis). Groupthink is a psychological phenomenon within a group of people who minimize conflict and reach a consensus decision without critical evaluation of alternative ideas.

The difference between individual decision-making and group decision-making is that not just one agent decides, but that a set of individual decisions exists. How these individual decisions (micro-level) can be generally transformed into a decision of the group turns out to be a general problem.

Each voter is represented by a relation R_i of its own. $N(x, y)$ is the number of voters with xRy. We define the majority relation M by $xMy \Leftrightarrow N(x, y) \geq N(y, x)$. That way M is always reflexive and connex (counting votes always works) but not always transitive.

There are several so-called voting paradoxes. A minimal example of such a voting paradox is the following. We have three candidates $X = \{a, b, c\}$ and three voters $I = \{1, 2, 3\}$. Each voter i specifies their preferences by a relation R_i:

$$R_1 = \{(a, b), (b, c), (a, c), (a, a), (b, b), (c, c)\} , \tag{8.1}$$

$$R_2 = \{(c, a), (a, b), (c, b), (a, a)(b, b), (c, c)\} , \tag{8.2}$$

$$R_3 = \{(b, c), (c, a), (b, a), (a, a), (b, b), (c, c)\} . \tag{8.3}$$

These preferences determine a majority relation that has a cycle:

$$M = \{(a, b), (b, c), (c, a), (a, a), (b, b), (c, c)\} . \tag{8.4}$$

The majority is said to prefer a over b, b over c, and c over a, which is obviously a contradiction.

The scientific field that analyzes the above-stated problem of how to aggregate individual preferences to achieve a group decision is called social choice theory (also known as the theory of collective choice). The list of results from this research includes the above-mentioned voting paradox. Additional results are Arrow's impossibility theorem and the Ostrogorski paradox. Arrow's impossibility theorem states that with more than two candidates, no rank order voting system can convert the ranked preferences of individuals into a community-wide ranking while also meeting a specific set of (democratic) criteria. The Ostrogorski paradox refers to the problem of combining preferences in sets. There is a distortion of voters' attitudes when they are voting for collections of factual issues, as it is common in the form of party platforms.

8.3.1 *Group Decision-Making in Control Theory*

As mentioned above, distributed control theory has solving methods for the consensus problem. These are algorithms or protocols designed to ensure that a group of agents reaches a consensus on a particular state. This is usually done by defining a consensus dynamics, that is, mathematical models describing how the states of robots evolve under a specific protocol designed to solve the consensus problem. Each robot i updates its state based on the states of its neighbors $\mathcal{N}_i$, and we model this with a system of differential equations. For simplicity, let's consider a first-order consensus protocol (i.e., a robot's state depends only on its current state and the states of its neighbors, without involving higher-order derivatives, such as velocity or acceleration). We define

$$\dot{x}_i = -\sum_{j \in \mathcal{N}_i} a_{ij}(x_i - x_j) \, , \tag{8.5}$$

with $\dot{x}_i$ is the time derivative of the state of robot i, $\mathcal{N}_i$ is the set of neighbors of robot i, a_{ij} is a weight representing the influence that robot j has on robot i, typically defined as $a_{ij} = a_{ji} \geq 0$, and $a_{ii} = 0$ for all i. The weight matrix $A = [a_{ij}]$ is the adjacency matrix of a graph that describes the communication network between robots. For each robot j that is not a neighbor of robot i, we need to set $a_{ij} = a_{ji} = 0$ because they cannot communicate. The non-zero weights are chosen based on the respective control problem and can be dynamic. For example, we can use a calculated error as a weight. Say the error between the states of two agents i and j at time t is

$$e_{ij}(t) = |x_j(t) - x_i(t)| \, . \tag{8.6}$$

We can define the weights a_{ij} as a function of the error

$$a_{ij}(t) = g(e_{ij}(t)) \, , \tag{8.7}$$

where $g(\cdot)$ is a function that determines how the weight varies with the error. This is actually a proportional control approach (P control). We can choose $g(e_{ij}(t))$ to be a linear function of the error, such as

$$g(e_{ij}(t)) = K_p e_{ij}(t) , \tag{8.8}$$

where $K_p > 0$ is the proportional gain. This ensures that the influence of neighbor j on robot i increases proportionally to the magnitude of the error. The state update for each agent then becomes proportional to the weighted difference between its current state and the states of its neighbors, driving the system toward consensus.

The above-mentioned strong assumption of having one connected component is part of a tradeoff. The payoff is that control theory can give guarantees and prove properties. Consensus protocols are analyzed in terms of their convergence and stability. Behavior-based approaches, as usually applied in swarm robotics, do not allow for that.

8.3.2 Group Decision-Making in Animals

From the knowledge that group decisions are complex and often paradoxical, we infer that decision-making in groups of animals is a challenge. In addition, these animal groups often rely on consensus decisions because an opposing minority might provoke the separation of the swarm, which jeopardizes the survival of the swarm. For example, animal groups have to decide which new nest site or resting place to choose, or in which direction they want to move in a flock. Hence, the question arises of how animals can achieve consensus decisions efficiently. There are a few simplifying factors. Typically, the members of the swarm have no conflicting interests. The well-being of the swarm is of interest to its members. There is also only a soft requirement for optimality. Approximations are often sufficient. The time requirements are equally soft, and hence, a slow iterative process might be acceptable.

An example is nest choice in honeybees. Only about 5% of the swarm disperses as so-called scout bees, which explore potential nest sites in the vicinity. Once they have found such a potential nest site, they return and advertise their findings by dancing to other scouts. The length of these dances is correlated with the quality of a site, which generates a positive feedback (good sites are advertised by long dances, which are observed by more bees). Eventually, the scouts reach a consensus and lead the remaining 95% of the swarm to the respective nest site. These 95% do not contribute to the decision process itself.

A second example is the determination of travel routes in groups of navigating birds. Although the scientific knowledge is limited, it seems that in relatively small groups, more experienced birds contribute more during the decision-making process. However, a majority decision might occur. In relatively large groups, self-organizing processes seem likely.

Although many different species rely on consensus decision-making processes that have been sufficient to guarantee their survival, these systems are not completely robust and might result in fatal disasters. An example of a bad collective decision is circular milling in ants. In particular species, the pheromone trails might form a loop that traps a group of ants. Unfortunately, they are not able to detect this situation and keep circling until eventually they die.

Collective decision-making in natural swarms is based on typical swarm principles, such as local communication and agents interacting only with direct neighbors or via the environment. Consequently, no direct voting process can be implemented that would be similar to voting in a group of human beings. Instead, the process is asynchronous because several spatially separated agents have to start to find a decision at the same time. These local decisions based on local perception might cause deadlocks or oscillations in the process. An example of a deadlock is shown in Fig. 8.2.

Because agents have to operate without global knowledge, it is possible that in one region of the swarm a subgroup of agents starts to converge on an option that is different from the chosen option of another subgroup at a different region. Based on a statically defined process, this might result in a deadlock. This process is similar to crystallization, where impurities (particularly line defects) emerge because crystallization starts at several close-by regions. The adaptive behavior of swarms is, however, different from the static process of crystallization. It is a dynamic and stochastic system, and therefore, these impurities do not stay forever. Even a perfect tie with 50% for option A and 50% for option B will eventually be overcome. This is due to fluctuations that disturb the system and drive it out of local optima (undesired states of limited use).

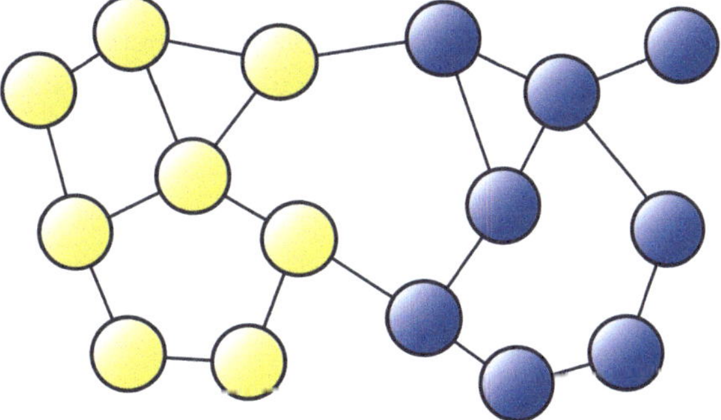

Fig. 8.2 Deadlock situation in a collective decision process of a swarm. Each agent has the correct state following a local majority rule, but the resulting global state is an undecided 50–50 state

8.4 Challenges and Context of Collective Decision-Making

Collective decision-making in robot swarms is shaped by both technical challenges and their embodiment. Challenges include the consensus problem, unreliable communication, and unknown option spaces. Spatial arrangement and interaction geometry strongly influence outcomes, and collective motion can be viewed as a form of decision-making process. In the following, we address challenges, spatial embodiment, and motion as decision-making.

8.4.1 *Speed-Versus-Accuracy Tradeoff*

The ideal of collective decision-making is to reach a decision instantaneously and without error, thereby maximizing both speed and accuracy. In practice, however, these goals conflict: decisions made quickly are often less reliable, while highly accurate decisions usually take more time. Importantly, the reverse is not guaranteed, as spending more time does not always yield better outcomes. Decision-making appears to have natural limits on speed and accuracy, a phenomenon known as the speed–accuracy tradeoff, which can be observed across individuals, groups, and swarms.

The speed–accuracy tradeoff in collective decision-making describes the tension between making a decision quickly versus making it correctly. In the context of collective decision-making, we define speed as the rate at which decisions are made.

Speed in Collective Decision-making

Speed in collective decision-making refers to the time it takes for a swarm to reach consensus, that is, the time until all, or a threshold fraction, of robots commit to one option.

The threshold fraction can be a low number of robots in the swarm (e.g., θN with $\theta = 0.1$) that are tolerated from joining the consensus. This is often desirable because a low fraction of robots enables the swarm to stay adaptive to dynamic environments. In the context of collective decision-making, we define accuracy as follows.

Accuracy in Collective Decision-making

Accuracy in collective decision-making refers to the frequency with which the swarm's decision aligns with the objectively better option in independent runs.

Accuracy is often measured as the probability of a correct consensus over repeated trials. We define the speed-versus-accuracy tradeoff.

Speed-Versus-Accuracy Tradeoff

The speed-versus-accuracy tradeoff in collective decision-making is the principle that speed and accuracy cannot be maximized simultaneously. Optimizing for one necessarily comes at the cost of the other: faster decisions tend to be less accurate, while more accurate decisions require more time.

The tradeoff arises because mechanisms that prompt the swarm to make decisions faster (e.g., strong positive feedback, low decision thresholds) also increase the risk of converging on the wrong option. Conversely, mechanisms that increase accuracy (e.g., more sampling, higher thresholds, weaker feedback) typically slow down convergence. Many swarm models (e.g., voter model, majority rule) reveal that optimizing one dimension usually comes at the cost of the other. This is in good correspondence with the respective findings in ants [255] and for the decision-making of humans [87]. When designing a robot controller for collective decision-making in a robot swarm, one has to decide whether accuracy or speed is more relevant for the respective application. The speed-versus-accuracy tradeoff has been treated in literature, for example, by Valentini et al. [805].

8.4.2 *Technical and Algorithmic Challenges*

8.4.2.1 Consensus Problem Byzantine Generals Problem

In related fields, the problem of collective decision-making is often known as the consensus problem. In distributed computing, one of the most prominent papers to mention is that of Lamport et al. [461], who introduced the Byzantine Generals Problem. This is about the challenge of achieving reliable communication at the system level, even with unreliable components. The Byzantine Generals Problem is a thought experiment to illustrate the challenges of reaching agreement (consensus) in a system where some components may fail or even act maliciously. Besides distributed systems, it is also known in control theory and generally in computer science. The challenge is that group members need to communicate indirectly, and some may even act maliciously. The scenario is hypothetical and invented. The name Byzantine was probably chosen metaphorically (Byzantine Empire). Some generals must decide to attack or retreat, but they rely on messages to coordinate, and some may send false information. The goal is for all loyal generals to agree on a single plan and act together, despite the presence of traitors. These thoughts suggest that achieving consensus in a group becomes more challenging when some participants cannot be trusted, for example, in the case of technically unreliable robots. Solutions to this problem form the foundation of many systems that require reliability despite failures, such as fault-tolerant robots and secure communication protocols. A simple algorithm solving this issue is the Oral Messages Algorithm that ensures consensus among N generals despite up to F traitors. The commander sends their proposed value to all generals. Each general forwards this message to others, ensuring that everyone hears from all participants. Decisions are then made based on the majority of messages received. If there's a tie or conflicting information, a default value is used. For more than one traitor ($F > 1$), the process is repeated recursively to filter out inconsistencies, ensuring agreement as long as $N > 3F$.

8.4.2.2 Consensus Problem in Distributed Control

In distributed control, the consensus problem addresses how a group of robots, each with its own state, can agree on a standard value through local interactions [604]. We assume a graph that defines neighborhoods and each robot i has a state x_i. Considered robot states x_i are typically values related to motion control, such as position, velocity, or heading. The purpose of the consensus problem is often formation control. The network is represented as a graph $G = (V, E)$, where

$$V = \{v_1, v_2, \dots, v_n\} \quad \text{(set of robots)},$$
$$E = \{(v_i, v_j) \mid v_i, v_j \in V\} \quad \text{(set of communication links)}.$$

Usually, it is assumed that the graph is one connected component (i.e., there is a path from any node to any other node). This is a fundamentally strong assumption that, most of the time, does not hold in swarm robotics. The goal of the consensus problem in control theory is to design a protocol such that all robots in the network converge to a common value, called the consensus value x^*. We assume swarm size N and $x_i(t) \in \mathbb{R}$ represent the state of robot i at time t. The objective is to ensure $\lim_{t\to\infty} x_i(t) = x^*, \quad \forall i \in \{1, 2, \dots, N\}$. Modern studies address time-varying graphs, where communication links between robots change dynamically. See Sect. 8.3.1 for more details.

8.4.2.3 Known or Unknown Options

A substantial body of literature exists in swarm robotics on methods for solving best-of-n problems. In most of these benchmarks, the number of options n is known by the robots from the start. That knowledge simplifies the problem. The swarm may still need to explore the quality of each option, but it does not need to examine whether there are more and potentially better options elsewhere. Once the number of options n is initially unknown, we need to implement a general exploration strategy that not only compares known options but also discovers new ones. While the goodness of estimation for each option quality can be assessed, the swarm cannot know how many more options there are to explore.

When should the swarm stop exploring? Our goal is to algorithmically determine a stopping point at which most of the relevant options have likely already been discovered. We model the number of explored options n as a function of time t. The number of new options found in a considered time interval Δt is $\Delta n(t) = n(t) - n(t - \Delta t)$. A simple strategy could be: If $\Delta n(t) \approx 0$ over multiple intervals, this suggests that few new options are being found. The exploration saturates, and we can stop. There should be a diminishing return effect (i.e., less and less exploration success over time, also see Sect. 2.4) similar, for example, to species accumulation curves in ecology [150, 292] or Heaps' Law in linguistics [501]. Basically, we would expect a sublinear function similar to $n(t) = -\exp(-t)$ that would, in turn, imply a function identical to $\Delta n(t) = \exp(-t)$. An observed exponential decline in the discovery rate can be exploited to implement a stopping condition. If $\Delta n(t)$ drops significantly compared to earlier discovery rates, stop exploration.

A more sophisticated approach is using the so-called Chao1 method by Chao [136]. It is inspired by the idea that some options may be harder to find than others. If the swarm is still struggling to locate items (i.e., they have been seen only once so far), there are likely more that the swarm has not encountered yet. We introduce two additional variables: the number of options seen exactly once $f_1(t)$ and the number of options seen exactly twice $f_2(t)$. For $f_2(t) > 0$, the estimate of the number of unseen options by Chao1 is

$$\hat{n}_{\text{Chao1}}(t) = n(t) + \frac{f_1^2(t)}{2 f_2(t)} . \tag{8.9}$$

If the swarm has seen many things only once, it seems likely the swarm is still in its discovery phase. Note that there are many other ways to model that situation [229, 777]. Also consider that the motion of the robots can be optimized to find options in space (e.g., patchy vs. uniform).

8.4.2.4 Number of Options and Optimal Stopping

The number n of available options to choose from influences the difficulty of the decision-making problem. The difficulty of the decision-making problem can be partially assessed by comparing the number of options n to the swarm size N. If there are few options $n \ll N$ compared to the swarm size, then the decision-making problem is considered easy with respect to the number of options. If there are many options $n \approx N$ compared to the swarm size, then the decision-making problem is already more complex, as it might also require exploration. Even if the number of options n is initially known, the problem of visiting all or many of them to explore their quality remains. To be efficient, this may require solving a task allocation problem.

If there are very many options $n \gg N$, the problem is expected to be significantly more difficult than in situations of $n \approx N$ or $n < N$. With so many options $O_i, i \in \{0, 1, \dots, n-1\}$ and assuming that options have qualities $q(O_i)$, exploration becomes the main challenge. It may be infeasible to explore all n options, so the swarm must explore probabilistically. The key question is then when to stop exploring.

A related class of problems is the so-called optimal stopping problems treated by optimal stopping theory [732]. These problems are focused on choosing the right time to trigger a certain action to maximize the outcome of the particular application. A prominent example is the secretary problem. The secretary problem involves an observer who must select the best candidate from a total of n alternatives, presented sequentially. For each candidate, the observer must decide immediately whether to accept or reject them, with no option to return to previously dismissed candidates. The objective is to maximize the probability of selecting the overall best candidate. Freeman [256] introduced a well-known strategy that estimates an effective stopping rule for this problem. The method advises rejecting the first n/e candidates (where e is Euler's number, the base of the natural logarithm), and then selecting the next candidate who is better than all previous ones. This approach is known as the $1/e$ strategy, as it yields a probability of $1/e \approx 37\%$ for successfully selecting the best candidate.

Given the numerous options, we can employ a statistical approach to analyze the data. We assume these qualities are drawn from an unknown distribution $q \sim p(q)$. We need to estimate the expected value of exploring an unseen option, identify likely high-quality options, and decide when to stop. One approach to solving this is Kernel Density Estimation (KDE). KDE is a non-parametric method to estimate the probability density function of a random variable based on a finite set of data points. We assume having obtained a set of observed qualities $\{q(O_{i_1}), q(O_{i_2}), \dots, q(O_{i_m})\}$ for $m \ll n$. We use KDE to estimate a density $\hat{p}(q)$ of option qualities:

$$\hat{p}_h(q) = \frac{1}{mh} \sum_{j=1}^{m} K\left(\frac{q - q_j}{h}\right) , \tag{8.10}$$

for a kernel function K and a bandwidth h, which controls smoothness. Based on estimated density $\hat{p}$ and a best-so-far option quality q^*, we can estimate the probability that a new option will outperform the best so far:

$$P(q_{\text{new}} > q^*) = \int_{q^*}^{\infty} \hat{p}_h(q)\, dq . \tag{8.11}$$

One can define a threshold probability P_{tresh}, once we have done quite some exploration ($m \gg 0$) and we find $P(q_{\text{new}} > q^*) < P_{\text{tresh}}$ then we stop. What remains is that we want to implement KDE in a distributed manner. We could follow Hu et al. [365] to implement that.

The influence of the number of options must be distinguished from the magnitude effect in decision-making. Magnitude-sensitivity describes the phenomenon that the performance of decision-making varies with the magnitude (reward value or stimulus) of the available options [640]. Specifically, in high-magnitude conditions, individuals tend to respond more quickly but also more randomly, increasing the likelihood of errors in situations when one alternative is objectively correct.

8.4.2.5 Continuous Option Space

Most of the swarm robotics literature on collective decision-making focuses on discrete options rather than estimating a continuous value. This relatively strong bias towards discrete problems is unfortunate. However, many problems in swarm robotics can be modeled as discrete decision-making problems (e.g., selecting a nest site or following a trail). Continuous option space is considered as the estimation problem in control theory [604]. The interesting variant here is decentralized environmental estimation [476]. Each robot gathers only partial and possibly noisy information about a continuous environmental variable (such as temperature, concentration, or elevation) and refines its local estimate by integrating information received from neighboring robots. This process enables the swarm to form a spatially distributed estimate. The key challenge lies in designing local rules for communication and estimation that ensure convergence to an accurate global estimate, despite limited communication ranges, asynchronous updates, and sensor noise. A standard approach from control theory is distributed Kalman filtering [122].

Raoufi et al. [662] solve collective estimation for a continuous environmental feature in unbounded environments. The robot swarm estimates the mean value of a spatially distributed environmental variable (e.g., light intensity). Once that is achieved, the robots aggregate at the corresponding contour in space. The system combines a dispersal algorithm that preserves network connectivity, a consensus algorithm based on local averaging, and a bio-inspired gradient-following behavior (consensus-based taxis).

8.4.2.6 Neglected Context of Collective Decision-Making

Most literature in swarm robotics on collective decision-making has focused on the core aspects, that is, effective algorithms to reach fast decisions. However, the context and logistics of collective decision-making have been somewhat overlooked. Khaluf et al. [430] argue that two critical phases need more attention: sensing the actual need to start a decision and exploring available options (see above). Drawing on biological inspiration, the authors emphasize the importance of how stimuli trigger decision-making and how individuals actively explore options. The authors propose a more comprehensive framework that includes, for example, stimulus detection and option discovery. They argue that neglecting the early phases limits a system's autonomy, coherence, and responsiveness in real-world applications.

8.4.3 Embodied Geometry of Decisions

In most collective decision-making scenarios in the literature on swarm robotics, the spatiality of the swarm is often not considered a crucial aspect of the scenario. It is known that spatial correlations need to be avoided (e.g., echo chambers in the form of robot clusters), but this is often countered by provoking a well-mixed system (e.g., corralled random walks of the robots). Sridhar et al. [743] investigate how different animals, from insects to fish, make spatial decisions while moving, that is, they embed them into (virtual) space and focus on the impact of embodiment and motion. They show that common geometrical principles underlie both individual and collective decision-making. For fruit flies, locusts, and zebrafish, the authors demonstrate that organisms faced with multiple options spontaneously reduce the task to a sequence of binary decisions. These decisions arise from neural dynamics characterized by local excitation and global inhibition, resulting in bifurcations in movement trajectories. A key finding is that the decision process is not only considered in terms of time, but also as a phenomenon that occurs in synchronization with movement. Both spatial and temporal features of this embodied situation directly impact the decision-making process, whether it is made by an individual or a collective. The lesson learned here is that swarms may rely more on cognitive processes than previously assumed, and that collective decision-making should be considered more often as a process inherently embedded in space.

8.4.4 Motion as a Decision Process

Also, some collective phenomena that intrinsically rely on continuous features can be reduced to a collective decision. The consensus on a common direction α in a flock can be viewed as a decision out of an infinity of alternatives $\alpha \in [0^\circ, 360^\circ)$. The swarm's motion can also artificially be restricted to a pseudo-1D setting on a ring (see Fig. 8.3) which leaves only two discrete options: clockwise motion and counterclockwise motion.

As an example, we take the behavior of the desert locust, *Schistocerca gregaria*, which shows collective motion, often called "marching bands," in the growth stage of a wingless nymph. The collective motion is density-dependent, and individuals seem to change their direction as a response to neighbors [115]. An additional interesting feature is a spontaneous switching behavior that is observed in these swarms. In comparatively small swarms, the locusts tend to switch the direction of their collective motion even if they were mostly aligned before.

In the experiments, it is observed that groups of locust nymphs (immature form of some invertebrates, particularly insects, which undergoes gradual metamorphosis (hemimetabolism) before reaching its adult stage) with relatively low densities are highly aligned and March in one direction around the ring-shaped arena for up to 2 or 3 h. Then they spontaneously switch their preferred direction within only a

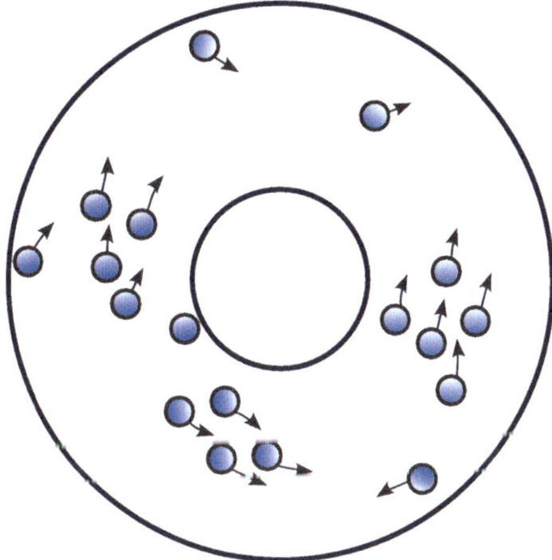

Fig. 8.3 Schematic representation of locusts moving in a ring-shaped arena

few minutes and March in the opposite direction, again for a number of hours. In experiments with higher densities, marching groups travel in the same direction for the full 8-hour duration of the experiment.

The essentials of this behavior are grasped by measuring only the percentage of locusts that are in one of the two states (clockwise or counterclockwise) counter-clockwise motion). These swarm fractions can be measured over a specific period, which results in a diagram as shown in Fig. 8.4.

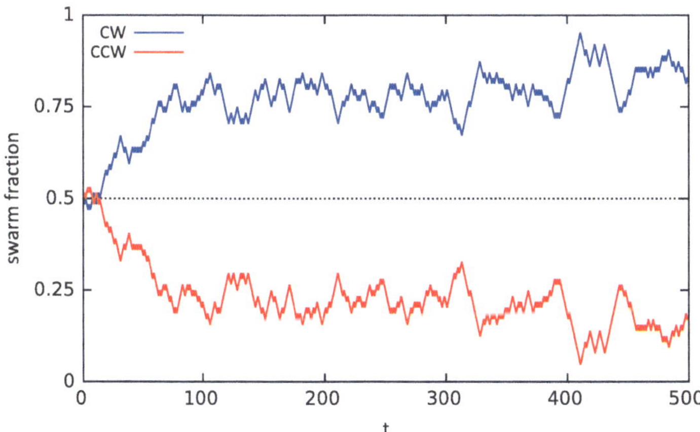

Fig. 8.4 Percentage of clockwise (CW) and counterclockwise (CCW) moving agents over time (discrete time steps of the model)

Yates et al. [886] give the distribution of the average velocities over many independent experiments, which is a symmetric bimodal distribution (two peaks, one for a positive and one for a negative speed). It is essential to understand that this does not correspond to a symmetric distribution of positive and negative velocities (i.e., clockwise and counterclockwise motion) within a single experiment, but gives the average over several experiments. Hence, it cannot show the broken symmetry of the direction of motion seen in a particular experiment, but instead shows a bimodal distribution.

8.5 Models for Collective Decision-Making Processes

As mentioned in the introduction to this chapter, for modeling collective decision-making in swarm robotics, we can take inspiration from many different fields. The one, perfectly generic model seems not yet to exist. Hence, we try to gather a relatively broad knowledge about model candidates to enable us to select the most appropriate model for a given scenario. These models differ a lot in their complexity, which means that in the ideal case, we can use straightforward models when appropriate.

In the following, we will try to reuse a common notation as much as possible. We assume that the swarm has to decide over a set of options $O = \{O_1, O_2, \dots, O_m\}$ with $m > 1$ options. A robot i has a defined opinion o_i at any time and a neighborhood $\mathcal{N}_i$ of robots. The set $\mathcal{N}_i$ contains all indices of robot's i neighbors and omits the considered robot's index i itself. A robot i is member of a neighborhood group $\mathcal{G}_i$ of robots. The set $\mathcal{G}_i$ contains all indices of robot's i neighbors and also includes the considered robot's index i itself. In a typical case, the task for the swarm is to find a consensus on one of the options O (i.e., 100% agreement in the swarm of one option).

A collective decision-making process consists of an iteration over up to three phases: exploration, dissemination, and opinion switch (see Fig. 8.5). The options O may have different qualities $q(O_i)$. In a special case, all may have the same quality $\forall i, j : q(O_i) = q(O_j)$, hence, the task is a symmetry-breaking problem. Then all options are equal in their utility, but the swarm still has to find a consensus on any of

Fig. 8.5 Collective decision-making as a process of iterations over three phases: exploration, dissemination, and opinion switch

them. During the exploration phase, a robot i may explore an area associated with its current opinion o_i (e.g., a potential construction site) and gather information about its quality $q(o_i)$.

During the dissemination phase, the robot signals its opinion to its neighbors, either by explicit messaging (e.g., via radio or infrared communication) or by giving cues (e.g., lighting an RGB LED). The duration for which the robot disseminates may vary according to the quality of its opinion. For example, it could be directly proportional to the quality (i.e., doubled dissemination time for doubled quality). This correlation of opinion quality and dissemination duration can then trigger a positive feedback loop. More robots perceive opinions of high quality, they switch to views of high quality, and then disseminate high-quality opinions themselves.

During the opinion switch phase, the robots follow a decision-making rule, such as the voter model or the majority rule, to switch their opinion. The robots do not have to follow these three phases of exploration, dissemination, and opinion switch in a synchronized way. Instead, each robot can follow its own clock except for keeping the correct duration of the dissemination phase in the case that it is correlated with opinion quality.

Many of the following models are microscopic models, that is, they define a behavior for individual robots. These models cannot be used directly to make predictions about the expected behavior of the swarm. A lot of research actually revolves around that very question of how to determine the global behavior based on a given local decision-making rule. More details about achieving consensus in robot swarms are given by Valentini [800].

8.5.1 Urn Models

To get a better understanding of collective decision-making processes like the locust system, we develop a model of it. Our guiding principle will be simplicity. Therefore, we focus on a simple binary decision process between options A and B. We say there is no bias between these two options, that is, A and B are of equal utility. Consequently, it does not matter on which of the two our system converges as long as it does converge on one. Due to our simplicity principle, we restrict the model to a single system variable $s(t)$, which gives, without loss of generality, the swarm fraction that is in favor of option A. Say we start with a perfect tie: $s(0) = 0.5$. What should be our next step? Well, we are interested in how $s(t)$ changes over time. Furthermore, it is our educated guess that this change will depend on s itself, which can be described by a function $\Delta s(s(t))$. This function Δs could, for example, be measured in a simulation. However, our objective is to get a better understanding of the underlying principles in collective decision-making systems. Hence, we want to model a process that defines Δs by itself.

At first, we consider the influence of space in decision-making systems like the locust system. It seems likely that spatial features influence the system behavior. For example, it might be the case that neighboring agents are more likely to share the same opinion than randomly chosen agents. However, we have decided to go for a model with just one variable, and consequently, we have to develop a non-spatial model. Our model ignores agent positions, which means that our model is based on the well-mixed assumption. While one might think that a non-spatial model makes as many assumptions about space as an atheist about God, it turns out that a non-spatial model assumes space to be irrelevant, which is equivalent to the assumption of a well-mixed state. There are no correlations between agents based on spatial features. As discussed above, this assumption is most likely wrong, but we stick to our simplicity motto and willingly accept the resulting shortcomings of the model.

8.5.1.1 Ehrenfest Urn Model

Next, we have an intriguing idea. Given we have a well-mixed state and we are working on fractions of a swarm, what about modeling the system as a lottery drawing? In the lottery community, well-mixed devices are eminently respected. Hence, we decided to represent robots by marbles in an urn. The distribution of these marbles follows the robot states, which is $s(t)$. Urn models are well-known in statistics, and they have also been used as models for a variety of systems. For example, there are the Pólya urn models [508, 644] and there is also the famous urn model by Ehrenfest and Ehrenfest [222], which is interestingly also known as the dog-flea model. They followed a serious inquiry as they defined the model to give at least an explanation of the second law of thermodynamics. They attempted to support the controversial ideas of Ludwig Boltzmann, who sadly committed suicide the year before. Here is a definition of the drawing process according to the Ehrenfest urn model. We have an urn filled with N marbles of two colors: blue and yellow. Initially, all are blue. Now we start drawing. Whenever we draw a blue marble, we replace it with a yellow marble. If we draw a yellow marble, we replace it with a blue one (see top of Fig. 8.6).

As an example, we take $N = 64$ marbles and initially all are blue, that is, $s(0) = 1$. Then, we repeat the above drawing process multiple times. If we keep track of the fraction of blue marbles in the urn over all rounds, we can plot a diagram as shown in Fig. 8.6. The diagram shows several independent runs along with an empirically obtained average. The empirical average indicates an exponential decrease. In the following we analyze the system in more detail.

To formalize this process, we keep track of how the number of blue marbles (without loss of generality) changes depending on how many blue marbles were in the urn at that time. We can do this empirically, or we can actually compute the average expected "gain" in terms of blue marbles. For example, say at time t we have $B(t) = 16$ blue marbles in the urn and a total of $N = 64$ marbles. The probability of drawing a blue marble is therefore $P_B = \frac{16}{64} = 0.25$. The case of drawing a blue marble has to be weighted by -1 because this is the change in terms of blue marbles in that case. The probability of drawing a red marble is $P_R = \frac{48}{64} = 0.75$ which is weighted by $+1$.

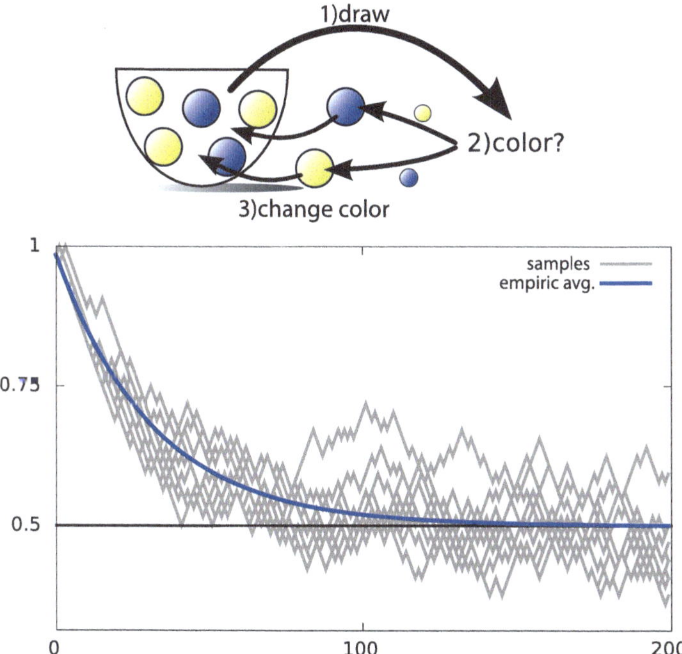

Fig. 8.6 Ehrenfest urn model, top: rules of drawing, bottom: ratio of blue marbles in the urn (vertically) over rounds of iterated drawing (horizontally), initially all marbles are blue

Hence, the expected average change ΔB of blue marbles per round depending on the current number of blue marbles $B = 16$ is $\Delta B(B = 16) = 0.25(-1) + 0.75(+1) = 0.5$. This can be done for all possible states yielding

$$\Delta B(B) = -2 \cdot \frac{B}{64} + 1. \tag{8.12}$$

Hence, the average dynamics of this game is given by $B(t+1) = B(t) + \Delta B(B(t))$.

The recurrence $B_t = B_{t-1} - 2\frac{B_{t-1}}{64} + 1$ can be solved by generating functions [293]. For $B_0 = 0$ we obtain the generating function

$$G(z) = \sum_t \left(\sum_{k \le t} \left(\frac{62}{64} \right)^k \right) z^t. \tag{8.13}$$

The tth coefficient $[z^t]$ of this power series is the closed form for B_t. We get

$$[z^t] = B_t = \sum_{k \le t} \left(\frac{62}{64} \right)^k = \frac{1 - \left(\frac{62}{64}\right)^t}{1 - \frac{62}{64}}. \tag{8.14}$$

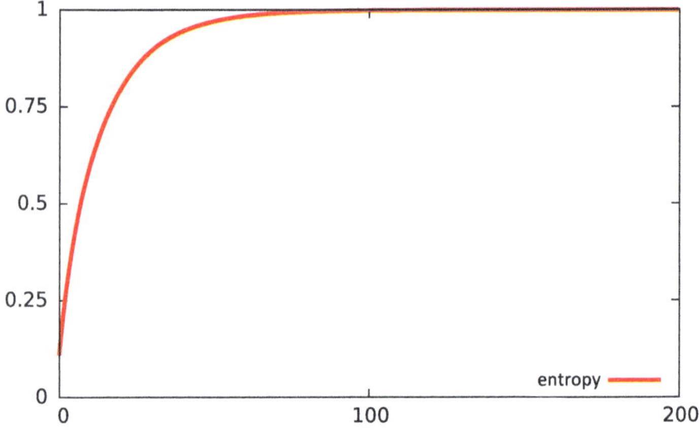

Fig. 8.7 Monotonic increase in Shannon entropy (vertical axis, Eq. (8.15)) over rounds (horizontal axis) for the Ehrenfest urn model

Hence, for initializations $B_0 = 0$ and the symmetrical case $B_0 = 64$ the system converges on average rather fast to the equilibrium $B = 32$. The actual dynamics of this game are, of course, a stochastic process which can, for example, be modeled by $B(t+1) = B(t) + \Delta B(B(t)) + \xi(t)$, for a noise term ξ.

As mentioned above, the Ehrenfest urn model was designed as a toy model for the application in statistical physics, in particular, to analyze the entropy production of diffusion processes. To complete this short discussion of the Ehrenfest model, we check how the entropy of this system evolves over several rounds. At first, we need a definition of entropy for this system. There are several options for defining entropy. Here, we choose the Shannon entropy, which we define based on $s(t)$

$$H(t) = -s(t)\log_2(s(t)) - (1 - s(t))\log_2(1 - s(t)). \tag{8.15}$$

By substituting $s(t) = \frac{1-(\frac{62}{64})^t}{1-\frac{62}{64}}$ we get the monotonically increasing entropy as shown in Fig. 8.7. This monotonic increase was proven by Kac [395].

8.5.1.2 Eigen Urn Model

The Ehrenfest model does not serve as a good model for collective decision-making systems because the model converges on average to the undecided state. This is just consequential, as it is a model of diffusion, while we are interested in self-organizing systems. Fluctuations away from the equilibrium should be enforced. Hence, we basically need to invert the model. It turns out that this was done by Eigen and Winkler [226]. Eigen's model was defined to show the effect of positive feedback. It works as follows. We have an urn filled with N marbles of two colors: blue and yellow.

Initially, there are 50 contrast to the Ehrenfest model, we draw with replacement now. We draw a marble, notice its color, and put it back into the urn. Whenever we draw a blue marble, we replace a yellow one from the urn with a blue marble. If we draw a yellow marble, we replace a blue one from the urn with a yellow marble.

The expected average change of blue marbles per round changes according to

$$\Delta B(B) = \begin{cases} 2\frac{B}{64} - 1, & \text{for } B \in [1, 63] \\ 0, & \text{else} \end{cases} . \tag{8.16}$$

In contrast to the Ehrenfest model, fluctuations are enforced, and the distribution of marbles is driven to extremes, as shown in Fig. 8.8. The Eigen model has two special configurations. For $B = 0$ and $B = 64$ we can only draw blue ($B = 0$) or yellow ($B = 64$) marbles, while we would actually be asked to replace one of the other colors from the urn, which is impossible in these cases. Therefore, once one of these two states, $B = 0$ or $B = 64$, is reached, we stay there forever. Compared to typical collective decision-making systems, that seems not appropriate because these extreme situations of perfect consensus are typically not achieved.

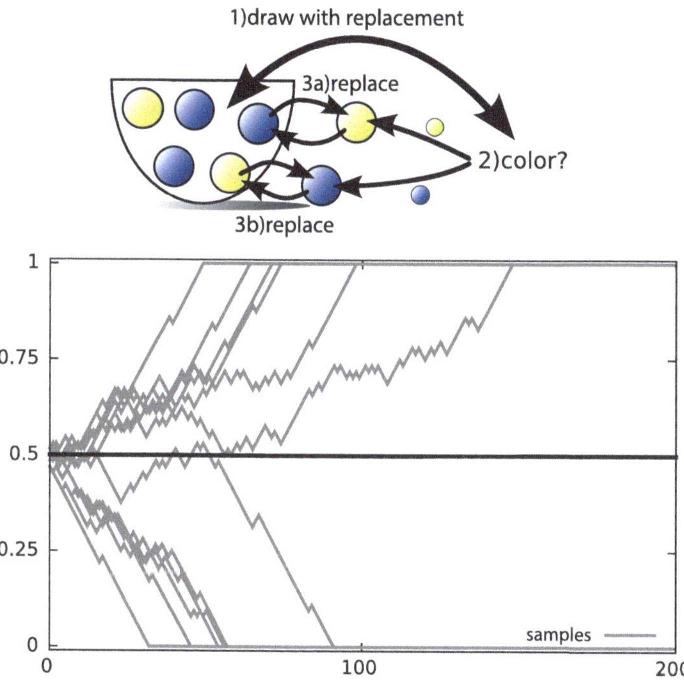

Fig. 8.8 Eigen urn model, top: rules of drawing, bottom: ratio of blue marbles in the urn (vertically) over rounds of iterated drawing (horizontally), initially 50% of the marbles are blue

8.5.1.3 Swarm Urn Model

To fix the disadvantage of the Eigen model, we develop a variant of an urn model that avoids perfect consensus and is more appropriate for collective decision-making in robot swarms [319]. We start from the Eigen model, but include an explicit choice between positive and negative feedback. This choice is based on a probability for positive feedback P_{FB}. The drawing process has several steps: draw with replacement, noticing the color, determine whether in this round we have positive feedback (probability P_{FB}) or negative feedback (probability $1 - P_{\text{FB}}$), and a final step that exchanges marbles of the respective colors (see Fig. 8.10, top).

Before we have a look at typical trajectories, we introduce the underlying concept of a positive feedback probability P_{FB}. Considering the Ehrenfest model again, we notice that we have a constant positive feedback probability of $P_{\text{FB}} = 0$ (see Fig. 8.9a). Considering the Eigen model again, we notice that we have a constant positive feedback probability of $P_{\text{FB}} = 1$ (see Fig. 8.9b). To fix the problem of converging on a consensus in the Eigen model, we need a positive feedback probability that varies depending on the current system state. We need a high rate of positive feedback close to $s \approx 0.5$ to get away from the undecided state based on fluctuations. We need a lower rate of positive feedback, that is, primarily negative feedback close to $s \approx 0$ and $s \approx 1$ to stay away from the consensus states. Out of many possible choices of functions we arbitrarily choose $P_{\text{FB}}(s) = 0.75 \sin(\pi s)$ (see Fig. 8.9c). Examples of resulting trajectories are given in Fig. 8.10 (bottom). As desired the trajectories stay away from the extreme states $s \in \{0, 1\}$ because once they get close negative feedback dominates ($P_{\text{FB}} < 0.5$).

Finally, we have a look at the resulting expected changes of the state variable s in one round: $\Delta s(s(t))$ (see Fig. 8.11). For the swarm model, we get

$$\Delta s(s) = 4\left(P_{\text{FB}}(s) - \frac{1}{2}\right)\left(s - \frac{1}{2}\right).$$

For the Ehrenfest model, we have one fixed point $s^* = 0.5$. For the Eigen model we have two fixed points: $s_1^* = 0$ and $s_2^* = 1$. For the swarm model, we have two

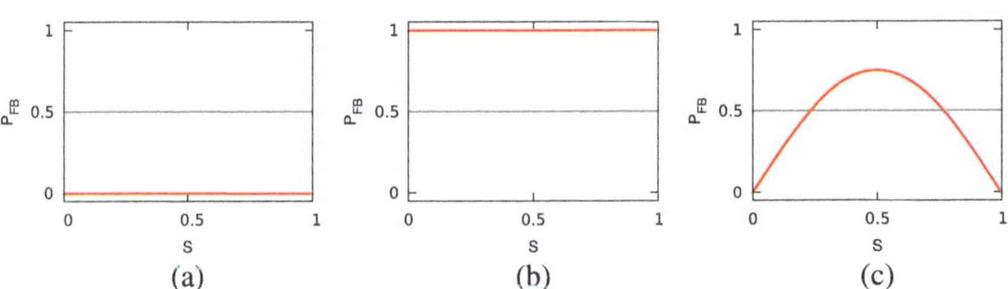

Fig. 8.9 Functions of positive feedback probability $P_{\text{FB}}(s)$ for the **a** Ehrenfest ($P_{\text{FB}}(s) = 0$), **b** Eigen ($P_{\text{FB}}(s) = 1$), and **c** swarm urn model ($P_{\text{FB}}(s) = 0.75 \sin(\pi s)$)

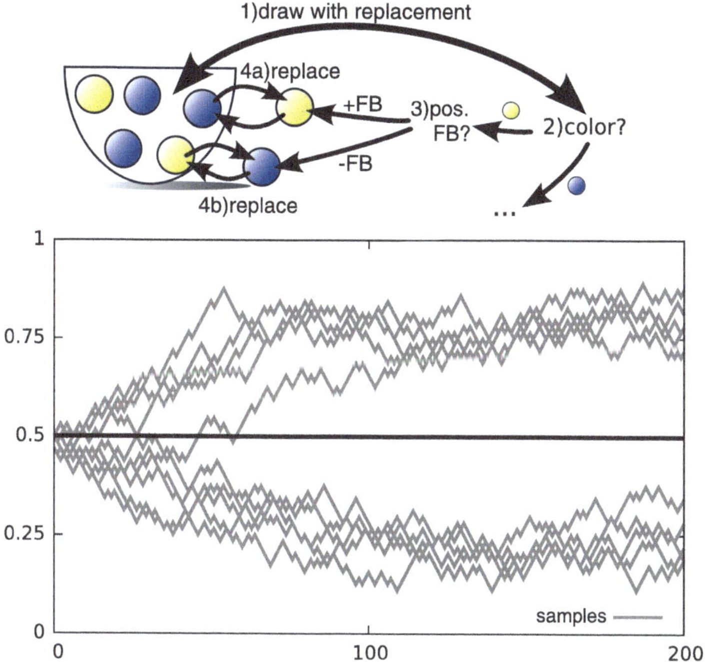

Fig. 8.10 Swarm urn model, top: rules of drawing, bottom: ratio of blue marbles in the urn (vertically) over rounds of iterated drawing (horizontally), initially 50% of marbles are blue

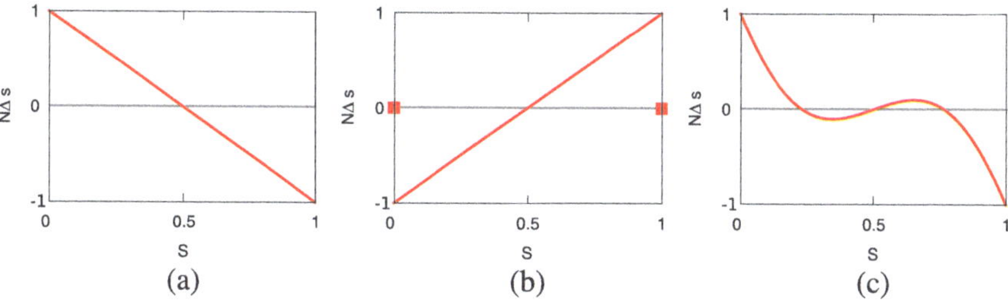

Fig. 8.11 Expected changes of the state variable s in one round: $\Delta s(s(t))$, for the **a** Ehrenfest, **b** Eigen, and **c** swarm model

fixed points: $s_1^* \approx 0.23$ and $s_2^* \approx 0.77$. Between $s \approx 0.23$ and $s \approx 0.77$, the swarm is driven away from $s = 0.5$ by positive feedback. Between $s \approx 0.23$ and $s = 0.5$, the swarm is driven towards $s = 0$. Between $s = 0.5$ and $s \approx 0.77$, the swarm is driven towards $s = 1$. Between $s = 0$ and $s \approx 0.23$, the swarm is driven towards $s = 0.5$ as well as between $s \approx 0.77$ and $s = 1$, the swarm is driven towards $s = 0.5$.

With the swarm urn model, we now have a simple macroscopic model of collective decision-making that has some potential to represent even microscopic aspects, such as one robot persuading another robot. This urn model explicitly represents feedbacks and also helps to get a better understanding of positive and negative feedback as introduced in Sect. 1.3.1. Despite its simplicity, the swarm urn model has the potential to apply to scenarios of swarm robotics.

8.5.2 *Voter Model*

The voter model [147] is a simple model of collective decision-making that has fantastic properties. A robot i considers its neighbors' current opinions o_j with $j \in \mathcal{N}_i$. It picks one neighbor j at random and switches its own opinion to the opinion of that neighbor (if the robot's previous opinion is identical with the opinion of the picked neighbor, then it just keeps its opinion). Now you may ask yourself what kind of pointless behavior that is, picking the opinions of random neighbors. Indeed, you don't want to adopt that as your approach to the next elections. However, the purpose of this model is not just to give a baseline of the simplest behavior; instead, it turns out that the voter model is competitive. The voter model implements a slow decision-making process, but it is very accurate, that is, it helps the swarm to converge on the correct decision [805]. This high degree of accuracy based on randomly picking neighbors may be counterintuitive, but it can be understood in the way that it is slow, while not paying much attention to the current state of the system. If there is a majority for a particular opinion, then it is more likely to be picked. However, there is no determined process that would enforce the win of a local majority. Hence, the decision process is relatively robust.

8.5.3 *Majority Rule*

The majority rule is another simple model of collective decision-making. In contrast to the voter model, it seems more intuitive. A robot i checks its neighborhood group $\mathcal{G}_i$ and counts the occurrence w_j of each option O_j. The robot then switches its opinion to the most frequent option O_k with $k = \text{argmax } w_j$, that is, the majority within its neighborhood group. Compared to the voter model, the majority rule is less accurate but faster [805]. Intuitively, this can be understood in the following way. Say, we have an extreme situation where each robot has an atypical sensor range that covers the complete swarm. Say, there are only two options O_1 and O_2, and initially each robot has a 50–50 chance of having one of them as its opinion. Then the overall distribution of the two options in the swarm is also close to a 50–50 situation. Hence, the initial majority is determined randomly, but all robots would update their opinion and follow that randomly determined initial opinion. In 50% of the cases, this initial majority will be the worst option of the two. With the voter model, even a robot with this extremely long sensor range would only randomly pick one of the robots out of the swarm and switch to its opinion. If dissemination is correlated to the quality of an option, the swarm would still have a chance to make a correct collective decision.

8.5.4 Cross-Inhibition Model

Reina et al. [671] developed a mathematical model of how honeybee swarms choose among n potential nest sites. They extended binary-choice models to the best-of-n problem ($n > 2$), demonstrating how swarms can reliably reach consensus even when many alternatives of varying quality exist. Their key innovation is the use of inhibitory interactions (stop signals) between agents committed to different sites. This mechanism is not only important in bees [726] but also widespread in natural systems, such as neural circuits [88] and cellular decision-making.

Inhibition works differently from classic consensus models, such as the majority rule and the voter model. In the majority rule, each agent aligns with the opinion held by the majority of its neighbors, which ensures rapid convergence but may easily lock the swarm into a wrong decision if early fluctuations favor an inferior option. In the voter model, an agent copies the state of one randomly chosen neighbor. This produces slow diffusion of opinions. The voter model is also susceptible to stubborn or contrarian agents, whose fixed or opposing opinions can disproportionately slow down or bias the consensus process. By contrast, the inhibition model adds negative feedback: an agent committed to one option can actively reduce support for competing options by sending stop signals to other agents. This prevents overcrowding on inferior options.

In the inhibition model of collective decision-making, each agent (bee or robot) can be either uncommitted or committed to one of n options. Agents switch states through two types of transitions: (1) spontaneous transitions or (2) interaction transitions. There are two spontaneous transitions:

(1a) discovery: an uncommitted agent finds an option and commits to it, with a probability proportional to the option's quality.
(2a) abandonment: a committed agent may give up and return to the uncommitted state, which is more likely if the option is of low quality.

There are two interaction transitions:

(2a) recruitment (positive feedback): a committed agent recruits an uncommitted one to the same option (probability is proportional to the quality of the recruiter's option).
(2a) cross-inhibition (negative feedback): a committed agent can send an inhibitory signal to an agent committed to a different option (probability is proportional to the quality of the sender's option), causing it to revert to the uncommitted state.

This mix of positive feedback (recruitment) and negative feedback (inhibition) enables the swarm to make decisions quickly and to prevent lock-in to poor or equal-quality options. Note that in this model, the swarm does not know beforehand how many options exist or what their qualities are. Instead, options are progressively discovered by uncommitted agents, with discovery simplified as a transition to commitment at a rate proportional to the option's quality. The model demonstrates that swarms can solve not only simple binary choices but also the best-of-n decision problem, where multiple options of varying quality exist. The key finding is that a successful swarm depends on the balance between spontaneous behavior (discovery and abandonment) and social interaction (recruitment and cross-inhibition). The best strategy is a time-dependent increase in signaling: start with low interaction (to let quality drive commitment), then gradually raise interaction (to force consensus when needed). Unlike, for example, the voter model, the inhibition model actively suppresses competing options, allowing the group to avoid indecision and reach consensus more robustly.

8.5.5 Bayes Bots

Most approaches to collective decision-making in swarm robotics do not use statistical modeling. However, that would allow for a principled way to handle uncertainty from noisy sensors and limited data. Bayesian inference enables agents to combine evidence, resulting in more reliable decisions. The Bayes Bots approach, as presented by Ebert et al. [218], applies the rigor of Bayesian decision theory. Their approach enables simple, sparsely connected robots to make accurate collective decisions through local Bayesian updates, with decision speed adapting to the task's difficulty. They introduce a decentralized Bayesian algorithm for collective decision-making in swarm robotics. Each robot behaves as a Bayesian estimator, updating a posterior distribution of the environment based on local observations and information exchanged with neighbors. A decision is made when the posterior probability mass passes a predefined threshold, after which robots may reinforce their choice through positive feedback. They also find that properly tuned priors and thresholds can balance the speed-versus-accuracy tradeoff.

The standard approach of Bayes Bots struggles with lock-in states, similar to the voter model. Pfister and Hamann [628] have proposed a solution for dynamically changing option qualities.

8.5.6 Hegselmann and Krause

A popular model across many different fields of research is the Hegselmann–Krause model [347]. Considering its simplicity, it is rather surprising that it was published as late as 2002. The robots choose from a continuum of possible options x defined on an interval $x \in [0, L]$ with $L \geq 1$. So the robots literally position themselves, although the interval $[0, L]$ needs not to be directly associated with positions in real-world space. The idea is that each robot moves to the mass center of its neighbor group. The update rule in each time step for a robot i is given by

$$x_i = \frac{1}{|\mathcal{G}_i|} \sum_{j \in \mathcal{G}_i} x_j, \tag{8.17}$$

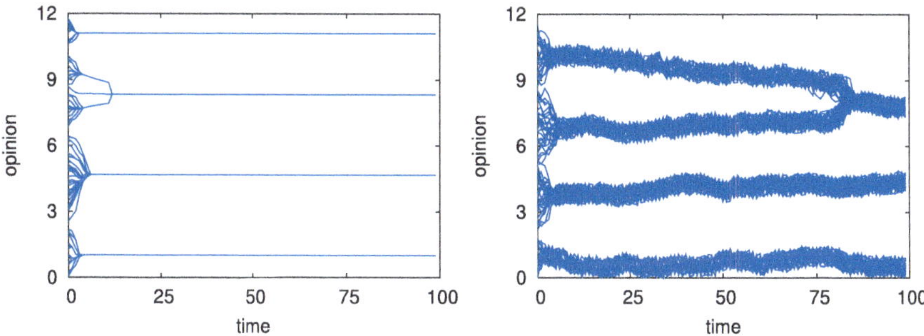

Fig. 8.12 Examples of trajectories in "opinion space" for (left) the original Hegselmann–Krause model, Eq. (8.17) and (right) the variant with exploration, Eq. (8.18); used parameter setting: $L = 12$, swarm size $N = 100$, initial robot positions sampled from a random uniform distribution on $[0, 12]$

where $\mathcal{G}_i = \{1 \leq j \leq N : ||x_i - x_j|| \leq 1\}$. Furthermore, one assumes that all agents simultaneously update their position, which does not mean that it is illegitimate to define and investigate an asynchronous Hegselmann–Krause variant. The robots are greedily following this model. They quickly gather with their neighbors, but don't explore whether there are more peers just behind their horizon. As a result, one gets many small clusters of aggregated robots for reasonable values of L, swarm size N, and initial distributions of opinions (see the left part of Fig. 8.12). To get a more useful model for swarm robotics, one can add noise term ε_i [208], that can, for example, be sampled from a random uniform distribution on the interval $\varepsilon_i \in [-0.5, 0.5]$ yielding

$$x_i = \frac{1}{|\mathcal{G}_i|} \sum_{j \in \mathcal{G}_i} x_j + \varepsilon_i. \tag{8.18}$$

The noise term ε_i then implements explorative behavior that enables robots to (temporarily) move out of these small clusters and eventually join other clusters (see the right part of Fig. 8.12).

8.5.7 *Kuramoto Model*

The Kuramoto model is usually used to describe synchronization processes by coupled oscillators [455]. A robot is represented by an oscillator (e.g., a harmonic oscillator from classical mechanics or even simpler) and robots "communicate" by being coupled to other robots. Each robot $i \in \{1, 2, \dots, N\}$ has as an internal state its current phase $\theta_i \in [0, 2\pi]$. In the standard Kuramoto model, the phase is updated by

$$\frac{d\theta_i}{dt} = \omega_i + \frac{K}{N} \sum_{j=1}^{N} \sin(\theta_j - \theta_i), \tag{8.19}$$

for the robot's preferred frequency ω_i and coupling strength K. The standard model, however, has global coupling, that is, each robot talks to all other robots. Instead, we prefer local coupling

$$\frac{d\theta_i}{dt} = \omega_i + \frac{K}{|\mathcal{G}_i|} \sum_{j \in \mathcal{G}_i} \sin(\theta_j - \theta_i), \tag{8.20}$$

with $\mathcal{G}_i = \{1 \leq j \leq N : \text{dist}(\theta_i, \theta_j) \leq r\}$, for some (sensor) range r, and dist$(\cdot)$ respects the torus-like concept of radians

$$\text{dist}(x, y) = \min(\|x - y\|, 2\pi - \|x - y\|). \tag{8.21}$$

Initially, all robots start with a random phase. The coupling introduces a feedback pushing towards synchronization. So this is similar to a popular experiment with metronomes (usually an annoying tool to make music students keep the pace). For example, one can put five asynchronous metronomes on a board and the board on two beverage cans. The board then starts to move slightly back and forth, imposing a force on all metronomes, basically a global coupling, that synchronizes them. With better synchronization, the board moves even more and imposes a bigger force on the metronomes (positive feedback). In the end, all metronomes are synchronized.

Many variants of the Kuramoto model are possible and have been proposed. One can weight the coupling by the distance, one can add noise, one can impose a structure in the form of a 1D or 2D array of oscillators, etc. An example of an application of the Kuramoto model in robotics is given by Moioli et al. [570]. They have used the model as a benchmark.

8.5.8 *Swarmalator*

Given we have models of synchronizing (e.g., Kuramoto model, Sect. 8.5.7) and desynchronizing oscillators (see Sect. 5.2.2) but also models of aggregating and segregating mobile agents (see Sect. 5.3.1), an option is to put them together in a model where the phase of oscillators and the location of agents who "carry" the oscillator depend on each other. This was done by O'Keeffe et al. [612] with their "swarmalators" that generalize swarms and oscillators. They define mobile oscillators whose locations affect their phase dynamics and vice versa. The key idea is to join an aggregation model with a Kuramoto model. We have N swarmalators (or robots). We define the position of robot i as $\mathbf{x}_i$ and its phase as θ_i. Coupling constants K and J govern the strength of spatial and phase interactions, respectively. ω_i is the natural frequency of robot i. A robot's velocity $\dot{\mathbf{x}} = \frac{d\mathbf{x}_i}{dt}$ and phase change $\dot{\theta} = \frac{d\theta_i}{dt}$ are both influenced by all other robots. The update rules of a swarmalator are

$$\frac{d\mathbf{x}_i}{dt} = \frac{1}{N}\sum_{j\neq 1}^{N}\left[\frac{\mathbf{x}_j - \mathbf{x}_i}{|\mathbf{x}_j - \mathbf{x}_i|}\left(1 + J\cos(\theta_j - \theta_i)\right) - \frac{\mathbf{x}_j - \mathbf{x}_i}{|\mathbf{x}_j - \mathbf{x}_i|^2}\right], \tag{8.22}$$

$$\frac{d\theta_i}{dt} = \omega_i + \frac{K}{N}\sum_{j\neq 1}^{N}\frac{\sin(\theta_j - \theta_i)}{|\mathbf{x}_j - \mathbf{x}_i|}. \tag{8.23}$$

The factor $\left(1 + J\cos(\theta_j - \theta_i)\right)$ modulates the spatial interaction based on the phase difference between robots, which introduces the coupling between spatial and phase dynamics. The sum in the update of phase θ governs how the phase is affected by the phases of others, with the influence modulated by their spatial separation.

If for the spatial strength of interactions we have $K > 0$, the phase interactions among robots work to reduce their phase differences. If we have $K < 0$, these interactions aim to increase the phase disparities between them. For strength of phase interactions $J > 0$ swarmalators favor proximity to others with similar phases. When $J < 0$, the opposite occurs: swarmalators are more likely to be drawn toward others with differing phases in space. If we set $J = 0$, swaramalators are "phase-agnostic" [612] and they are spatially attracted independent of their phase.

We have defined in Eq. 8.23 an individual natural frequency ω_i, but here we choose a homogeneous swarm with natural frequency $\omega = 1$. For random initial conditions ($\mathbf{x}$ uniformly randomly from a box with side length two and θ uniformly randomly from between $-\pi$ and π), we can solve Eqs. 8.22 and 8.23 numerically. Following O'Keeffe et al. [612], there are five relevant stationary states (see Fig. 8.13). In the static synchronous state, all swarmalators share identical phases and converge into a single, tightly bound spatial cluster, exhibiting complete synchronization. In the static asynchronous state, swarmalators possess randomly distributed phases θ and remain uniformly dispersed in space without forming any coherent clusters. In the static phase wave, the phases θ are organized in a smooth gradient that correlates with the swarmalators' spatial positions, creating a stable wave-like spatial pattern. In the splintered phase wave, swarmalators segregate into multiple distinct spatial clusters, each maintaining internally synchronized phases θ that differ from those of other clusters. The active phase wave is a dynamic state in which swarmalators continuously move through space while their internal phases propagate in a synchronized, wave-like pattern across the entire system.

Ceron et al. [133] introduce a generalized swarmalator model where agents can have clockwise or counterclockwise circular orbits, interact locally rather than globally, and possess varying natural frequencies. These modifications lead to complex collective behaviors such as vortex lattices, beating clusters, interacting phase waves, and splintered phase waves, many of which resemble natural phenomena. The study suggests potential uses in swarm robotics.

By integrating swarmalator dynamics, robots can achieve synchronized behaviors, maintaining cohesive and adaptable spatial arrangements. For example, swarmalators can enable robots to spread out efficiently across an area while synchronizing their sensors to optimize information sharing and coverage.

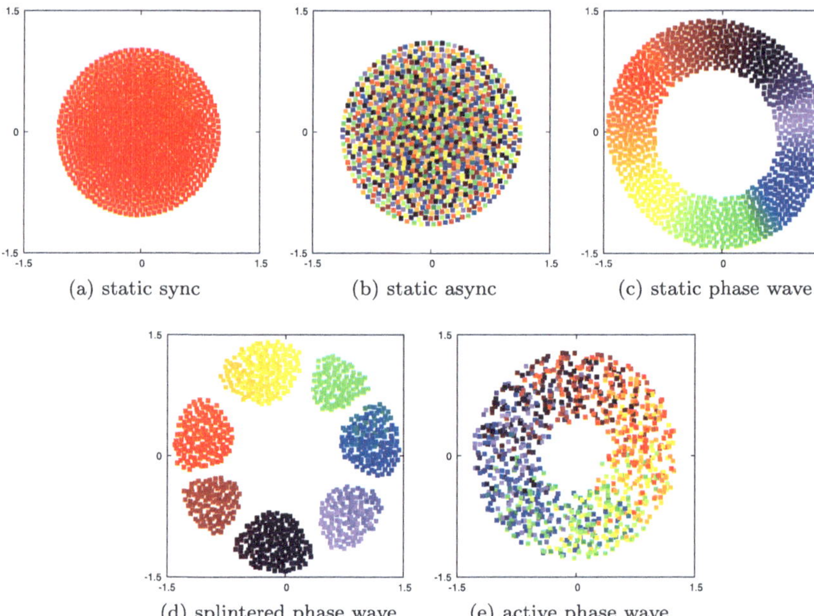

Fig. 8.13 Scatterplots of common stationary system states in the (x,y) plane for the Swarmalator model. $N = 1000$ swarmalators, $T = 2000$ or $T = 10,000$ (for splintered phase wave) steps, step size $dt = 0.05$, natural frequency homogeneous ($\omega_i = 0, \forall i$). **a** Static sync state for $J = 0.1$ and $K = 1.0$, **b** Static async state for $J = 0.1$ and $K = -1.0$, **c** static phase wave for $J = 1.0$ and $K = 0.0$, **d** splintered phase wave for $J = 1.0$ and $K = -0.1$, **e** active phase wave for $J = 1.0$ and $K = -0.75$

Barciś and Bettstetter [55] present a system of swarm robots inspired by swarmalators, integrating temporal synchronization and spatial swarming into a unified, discrete-time coordination model suitable for real-world robotic applications. They address practical constraints, such as communication delays and physical limitations of robots, ensuring robust behavior under real-world conditions. Their model introduces controlled versions of swarmalator patterns, including static sync, static async, splintered phase waves, and static phase waves. Simulations and experiments with ground robots and drones validate the model, demonstrating stable space-time patterns and resilience to delays and message drops.

8.5.9 Axelrod Model

Axelrod's model of the dissemination of culture [41] is an abstraction of cultural diffusion. We have an $L \times L$ square lattice of cells. In each cell, there lives an agent with a culture that is characterized by a list of n features (integers on a defined interval). All agents are initialized to a random culture. The update in each time step is done in the following way. First, an agent i is randomly selected. Second, a neighbor j of agent i is chosen randomly. Agent i and j interact with each other with probability $P = s/n$, where s is the number of cultural features that are the same for agent i and j. Agent i selects a cultural feature of agent j that differs from its own and overwrites its own value with that of agent j. This process is continued until either all neighboring agents share the same culture or have entirely different cultures (i.e., $P = 0/n = 0$). The main message of Axelrod's model is that "local convergence can lead to global polarization"and that "simple mechanisms of change can give counterintuitive results" (large territories with little polarization) [41].

8.5.10 Ising Model

Our starting point is the well-known Ising model that was originally developed as a toy model to investigate ferromagnetism and phase transitions [884]. The Ising model is popular across fields, and physicists make fun of those applying it to alternative systems, quoting from Sznajd-Weron and Sznajd [765]:

> The Ising spin system is undoubtedly one of the most frequently used models of statistical mechanics. Recently, this model has also become the most popular physics export article to "other branches of science" such as biology, economy or sociology. There are two main reasons for that: first verbalized by Nobel prize winner Peter. B. Medawar—"physics envy"—a syndrome appearing in some researchers who would like to have such beautiful and relatively simple models as physicists have (for example the Ising model).

The original Ising model is defined on a lattice, that is, we have points positioned regularly and, for example, connected to four neighbors each (square lattice, see Fig. 8.14). Usually, one assumes that this lattice is static, that is, each point keeps its neighbors for ever. In addition to the lattice, each of these points has a so-called spin s which is binary and can only take one of two values: $+1$ or -1. The standard

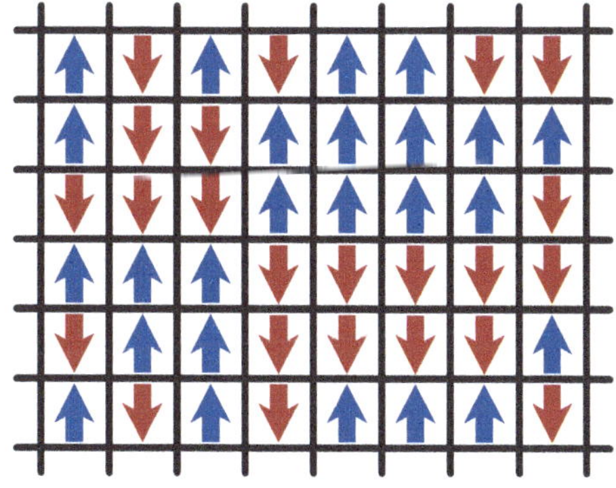

Fig. 8.14 Example configuration of a 2D Ising model. Red arrows represent spin $s = +1$ and blue arrows represent spin $s = -1$

definition of the Ising model is given via its Hamiltonian H, which gives the overall energy of the system. We have

$$H = -J \sum_{\langle ij \rangle} s_i s_j - B \sum_i s_i, \tag{8.24}$$

for a coupling constant J, an external magnetic field with constant B, the term $\langle ij \rangle$ gives all pairs of neighboring spins as defined by the lattice, and we have N spins, hence $i \in \{1, 2, \dots, N\}$. Here, we do not need an external field and set $B = 0$ which yields

$$H = -J \sum_{\langle ij \rangle} s_i s_j. \tag{8.25}$$

Interestingly, the Hamiltonian does not describe the local behavior of individual spins; instead, the physicist is initially satisfied by knowing the global system state only. The cultural clash here is that swarm roboticists have to design robot behaviors, whereas physicists find readily engineered systems that need to be modeled. The actual physical system described by the Ising model (i.e., ferromagnetic systems) changes its spins spontaneously based on fluctuations and local interactions. Hence, the behavior of spins is not far from the behavior of our robots that also explore their environment by random actions and cooperate with neighboring robots.

Next, we define the behavior of the spins via a Monte Carlo simulation (Markov chain Monte Carlo simulations, [75]). While physicists use that simulation only to calculate certain properties of the spin system and do not want to describe the actual behavior of spins by it, we take the local rules of the simulation as a definition of the agent's behavior. We use the numerical method of the Metropolis algorithm as our definition of the agent behavior.

The Metropolis algorithm defines a so-called single-spin-flip dynamics and tries to flip a single spin per time step that implements the transition from a configuration v to the next configuration w. For that we choose a random site i with homogeneous selection probability $g(i) = 1/N$ and calculate the change in energy $\Delta E = E_w - E_v$ that a considered flip would cause. Then we accept the flip: (i) if $\Delta E \leq 0$, with a probability of one; and (ii) if $\Delta E > 0$, only with a defined acceptance probability A depending on ΔE defined by

$$A(v, w) = \exp(-\beta(E_w - E_v)) = \exp(-\beta \Delta E), \tag{8.26}$$

with $\beta = (kT)^{-1}$, for Boltzmann constant k and temperature T. For simplicity and because we do not want to simulate actual physical systems, we usually set $k = 1$. A negative energy difference ($\Delta E < 0$) corresponds to a majority decision of agents and a positive energy difference ($\Delta E > 0$) corresponds to a minority decision. For $T > 0$, we have a non-zero probability of doing a minority decision and for $T = 0$ minority decisions have zero probability. The temperature defines whether there is free energy available to do explorative actions, that is, actions that do not decrease the overall energy.

An interesting setup could be the following. We initialize with $T = \infty$, which gives us a random, uncorrelated initialization of spins. Then we set $T = 0$, that is, we cool the system, because we say that we want a homogeneous system state. As a result, and according to Eq. (8.26) we do not accept spin-flips that would increase the energy (i.e., a minority decision). Hence, we have a system that optimizes locally greedily. The interested reader can continue from here with a paper by Galam [263] that connects an Ising model directly with the group decision-making.

8.5.11 Fiber Bundle Model

The fiber bundle model goes back to Peires [625]. A modern overview is, for example, given by Raischel et al. [658]. We assume a material that has N parallel fibers in a regular lattice. We investigate the reaction of the bundle set to a force that pulls these fibers. Each fiber i is elastic and stretches until it breaks at an individual breaking threshold θ_i. Each fiber's individual threshold θ_i is an independent identically distributed random variable. Choosing a distribution is a modeling decision. A typical choice is a Weibull distribution

$$P(\theta) = 1 - \exp\left(-\left(\frac{\theta}{\lambda}\right)^m\right), \tag{8.27}$$

where m is the Weibull index and λ is the scale parameter. Once a particular fiber has failed, its load has to be shared by the remaining intact fibers. We distinguish two ways of sharing the load. In global load sharing, the load is equally redistributed over all intact fibers. In local load sharing, the entire load of the failed fiber is redistributed equally over its local neighborhood. For example, we can choose a von Neumann neighborhood and distribute the load among the four (not yet broken) neighbors (north, south, east, west). An interesting effect observed in this model is the occurrence of avalanches of breakings. We can count how many fibers break in an event of such an avalanche, which defines the avalanche size Δ. The distribution D of avalanches of size Δ follows a power law distribution with an exponent $5/2$ [658]

$$D(\Delta) \propto \Delta^{-5/2}. \tag{8.28}$$

In the context of swarm robotics, we can interpret the fiber bundle model as a time model for collective decision-making. Say the task is not to choose from several options but to decide when to take a single option (e.g., escape or when to start construction, etc.). Then "breaking" would be re-interpreted as a positive event of a robot's individual decision of wanting to take or taking action now. That decision then possibly triggers decisions in its neighborhood. The individual breaking threshold θ_i would correspond to the robot's individual sensitivity. An avalanche would correspond to the rapid propagation of a decision through the swarm. The power law distribution of avalanche sizes Δ would give us a hint of how many robots we should expect to make decisions per time.

A drawback of the fiber bundle model is that fibers are stationary, and their neighborhoods do not change. The breaking threshold θ_i introduces an interesting heterogeneity that may have positive effects on the robot swarm, but this relation is not clear.

8.5.12 Sznajd Model

The Sznajd model is also known as USDF model ("united we stand, divided we fall"). An individual i has an opinion $S_i = -1$ for no or $S_i = 1$ for yes and an individual has two neighbors. In each time step, two individuals S_i and S_{i+1} are randomly chosen. The opinion of their neighbors S_{i-1} and S_{i+2} is then updated following these two rules

- if $S_i S_{i+1} = 1$ then $S_{i-1} = S_i$ and $S_{i+2} = S_i$,
- if $S_i S_{i+1} = -1$ then $S_{i-1} = S_{i+1}$ and $S_{i+2} = S_i$.

This system always ends up either in a consensus state or a stalemate state (disagreement). A power law for the decision time distribution with an exponent of -1.5 was found.

8.5.13 Bass Diffusion Model

The Bass diffusion model [59] is an abstraction of how consumers adopt an innovative product over time after its first release. The underlying assumption is that news about the product spreads in a word-of-mouth fashion from past adopters to not-yet-adopters. An adoption of a product is the first-time purchase of that product. This is related to Arthur's concept of "lock-ins" [31] that occur, for example, when there is a "format war" between mutually incompatible proprietary formats (e.g., the videotape format competition between Betamax and VHS). However, with the Bass diffusion model, we only focus on one product and how it spreads throughout the market. The word-of-mouth process with adopters (consciously or unconsciously) persuading not-yet-adopters can be interpreted as a collective decision-making process with two options: to buy or not to buy. With the Bass diffusion model, we can study the process over time.

The fraction of the market that adopts the new product at time t is $f(t)$ and the fraction of the market that has adopted the new product already at time t is $F(t)$.

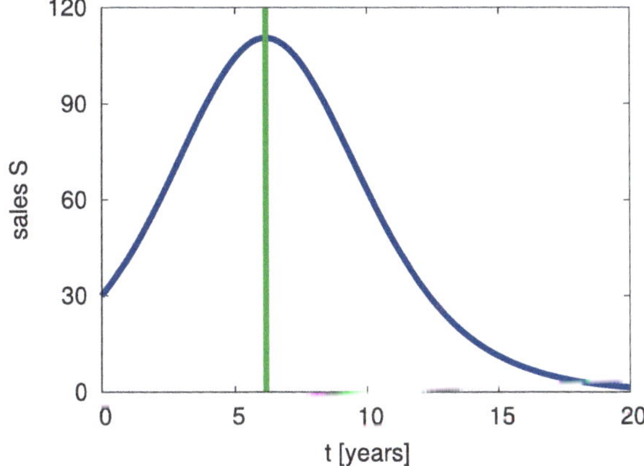

Fig. 8.15 Sales in the bass diffusion model, market size $N = 1000$, coefficient of innovation $p = 0.03$, coefficient of imitation $q = 0.38$, time of peak sales $t^* \approx 6.19$

For those that have not yet adopted, the fraction of the potential market adopting at time t is linearly proportional to the number of previous adopters. We have

$$\frac{f(t)}{1 - F(t)} = p + qF(t), \tag{8.29}$$

where p is the coefficient of innovation, q is the coefficient of imitation. For a (potential) market of size N, we get sales $S(t) = Nf(t)$, which can be expressed depending on p, q, and N only [59].

$$S(t) = N\frac{(p+q)^2}{p}\frac{e^{-(p+q)t}}{\left(1 + \frac{q}{p}e^{-(p+q)t}\right)^2}. \tag{8.30}$$

The time of peak sales t^* is given by $t^* = \frac{\ln q - \ln p}{p+q}$. An example plot of sales over the years is shown in Fig. 8.15. The market size N can be interpreted as the swarm size, and $S(t)$ specifies how many robots have been convinced about the option "buy" at time t.

8.5.14 *Sociophysics and Contrarians*

Galam [265] calls the set of models that he developed over several decades "sociophysics." He focuses on collective decision-making, democratic voting, and opinion dynamics. Out of the many works of his [264–268], we select a model of collective decision-making that incorporates so-called contrarians [264, 265]. "A contrarian is someone who deliberately decides to oppose the prevailing choice of others, whatever this choice is" [264]. In a robot swarm, a contrarian could be a faulty robot or a sabotaging intruder. It could also be part of a second robot group deliberately added to the swarm (hence making it a heterogeneous swarm) to keep the swarm from converging on a consensus.

The idea of the contrarian model is to assume a certain fraction $0 \leq a \leq 1$ of contrarians in a population. The system state is given by $0 \leq s \leq 1$, that is the fraction of agents in favor of option O_1 for a binary decision problem $O = \{O_1, O_2\}$. Depending on the current system state s_t, the contrarians are always against the current majority. We want to investigate the evolution of this system, that is, a time series $s_0, s_1, s_2, \ldots$. The process of how an agent switches opinion is based on a simple assumption. We say that in each time step, arbitrary groups $\mathcal{G}$ of agents form. These groups have a defined constant size of $|\mathcal{G}| = r$. Any possible spatial correlations between agents and their opinions are ignored. We also assume that within this group, a majority rule determines the opinion switch. Now we are interested in the resulting update rule for the system state s. We start without contrarians, that is, $a = 0$. We have to do the combinatorics and look through all possible combinations of groups of size r with two opinions o_1 and o_2. We sum over all combinations that result in a majority decision for option o_1 and get the update rule

$$s_{t+1} = \sum_{m=\frac{r+1}{2}}^{r} \binom{r}{m} s_t^m (1 - s_t)^{r-m}. \tag{8.31}$$

To add the influence of contrarians, we have to include a basically inverse term. It accounts for those groups where a majority would have emerged for opinions o_2 but due to contrarians, was overruled. The fraction of contrarians a then weights the two terms. We get

$$\begin{aligned} s_{t+1} =& (1 - a) \sum_{m=\frac{r+1}{2}}^{r} \binom{r}{m} s_t^m (1 - s_t)^{r-m} \\ &+ a \sum_{m=\frac{r+1}{2}}^{r} \binom{r}{m} (1 - s_t)^m s_t^{r-m}. \end{aligned} \tag{8.32}$$

For example, with a group size $r = 3$, we get

$$\begin{aligned} s_{t+1} =& (1 - a)(s_t^3 + 3s_t^2(1 - s_t)) \\ &+ a((1 - s_t)^3 + 3(1 - s_t)^2 s_t). \end{aligned} \tag{8.33}$$

The resulting functions $s_{t+1}(s_t)$ for varied $a \in \{0, 0.05, 0.1, 0.25\}$ are given in Fig. 8.16. The crossings of function $s_{t+1}(s_t)$ with the diagonal $s_{t+1} = s_t$ are the fixed points. Depending on the contrarian fraction a there is a qualitative change. For $a \in \{0, 0.05, 0.1\}$ the values $s_1^* \approx 0$ and $s_2^* \approx 1$ are stable fixed points while $s_3^* = 0.5$ is an unstable fixed point. However, for $a = 0.25$ there is only one stable fixed point $s_1^* = 0.5$, which means the swarm converges on the undecided state of 50%

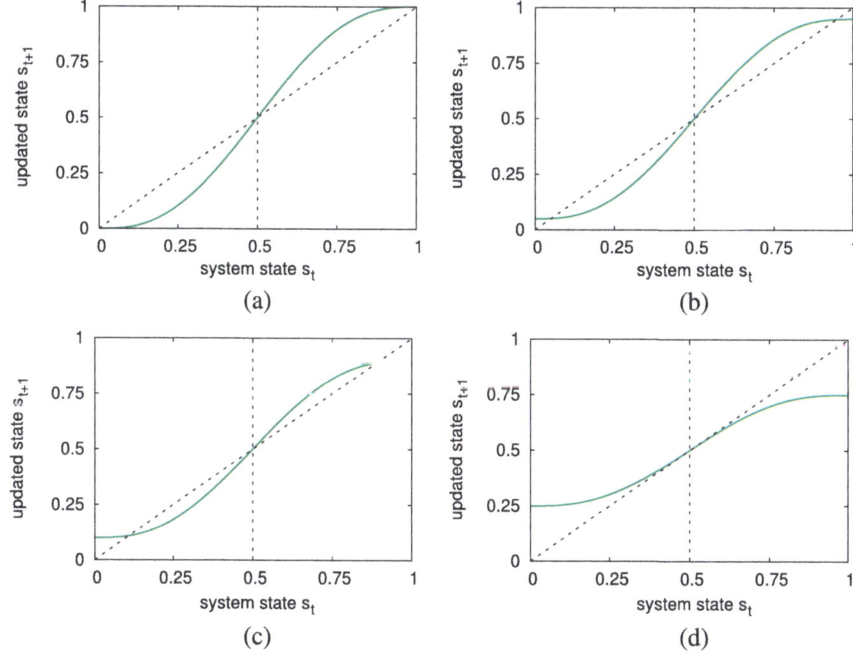

Fig. 8.16 Sociophysics and contrarians [264], change of system states s (fraction of agents in favor of opinion o_1) depending on the fraction of contrarians a in the swarm, group size $r = 5$, Eq. (8.32). **a** Contrarians $a = 0$ **b** Contrarians $a = 0.05$ **c** Contrarians $a = 0.1$ **d** Contrarians $a = 0.25$

in favor of opinion o_1 and 50% in favor of opinion o_2. A relatively small fraction of contrarians is enough to obstruct the decision-making process. In addition, the steepness of the function $s_{t+1}(s_t)$ determines how much time the swarm requires to converge (steeper is faster). A small fraction of contrarians $a = 0.05$ already slows down the process.

As mentioned above, the use of contrarians in robot swarms can be to avoid consensus. In a dynamic environment, it is desirable always to have a small percentage of robots that have a different opinion from the majority. For example, these robots keep visiting and exploring areas corresponding to their opinion and would hence detect a dynamic increase in quality. An adaptive swarm has to avoid consensus or implement exploration in an alternative way. The concept of contrarians is similar to what Merkle et al. [548] call "anti-agents." In an aggregation scenario, for example, anti-agents try to prevent clustering. Merkle et al. [548] suggest to use anti-agents to prevent undesired emergent effects in a swarm (see Sect. 5.2.7).

8.6 Implementations

8.6.1 Decision-Making with 100 Robots

Only recently have robot swarms gotten considerably bigger. Since 2014, reported hardware experiments in swarm robotics went up to 1000 robots [686]. Here we have a look at a swarm of 100 Kilobots [685] that are supposed to explore two potential target areas and collectively decide which one the swarm should choose. In a series of studies, Valentini et al. have investigated the impact of the voter model and the majority rule on collective decision-making in robot swarms [801–803, 805]. The task for the robot swarm is to choose between the blue site and the red site (see Fig. 8.17). Initially, the robots are positioned within the middle of the arena, they are initialized to a blue or red opinion with equal probabilities, and they know nothing about how desirable these areas are. The robots know, however, how to navigate to them. Behind the red site, there is a light beacon (if you look closely, you can see the respective shadows of the robots in Fig. 8.17c). The robots know that when they approach the light beacon (phototaxis) then they navigate towards and eventually onto the red site. If they approach areas of lower light intensity (anti-phototaxis), then they navigate towards the blue site. Once the robots are on a site, they can explore it. In this experiment, which is emulated by IR messages that can only be received by robots on that site. The message contains the actual site quality values. Here, the red site has a higher quality than the blue site; therefore, we expect a consensus on the red site.

The robots follow the procedure of collective decision-making in three phases, as discussed above (Fig. 8.5). A probabilistic finite state machine controls the robots (Fig. 8.17a) that has an exploration, dissemination, and switching state for both opinions (red and blue). In the exploration state, the robots navigate to the respective site of their current opinion, explore its quality, and navigate back into the white area in the middle. During the following dissemination stage, the robots move randomly within the white area and send messages containing their current opinion as local broadcasts into their neighborhood, that is, the area around them that is reached by their infrared signal (a few centimeters, whereas the overall arena is of size 100×190 cm). Finally, in the switch state, the robots move randomly within the white area, listen for messages from their neighbors for a while, count the number of messages of either of the two opinions, and then switch. Two switching rules were investigated. Using the voter model, the robot picks a random message and switches to the opinion disseminated in that message. Using the majority rule, the robot compares the count of "blue messages" to "red messages" and switches to the majority.

Independent of whether the voter model or majority rule is used, based on the description given so far, there would be no bias towards any of the two sites because the measured site quality has no influence yet. This bias is implemented by varying the time spent in the dissemination state depending on the measured quality of the site. A higher quality results in more dissemination time. The mean time is directly

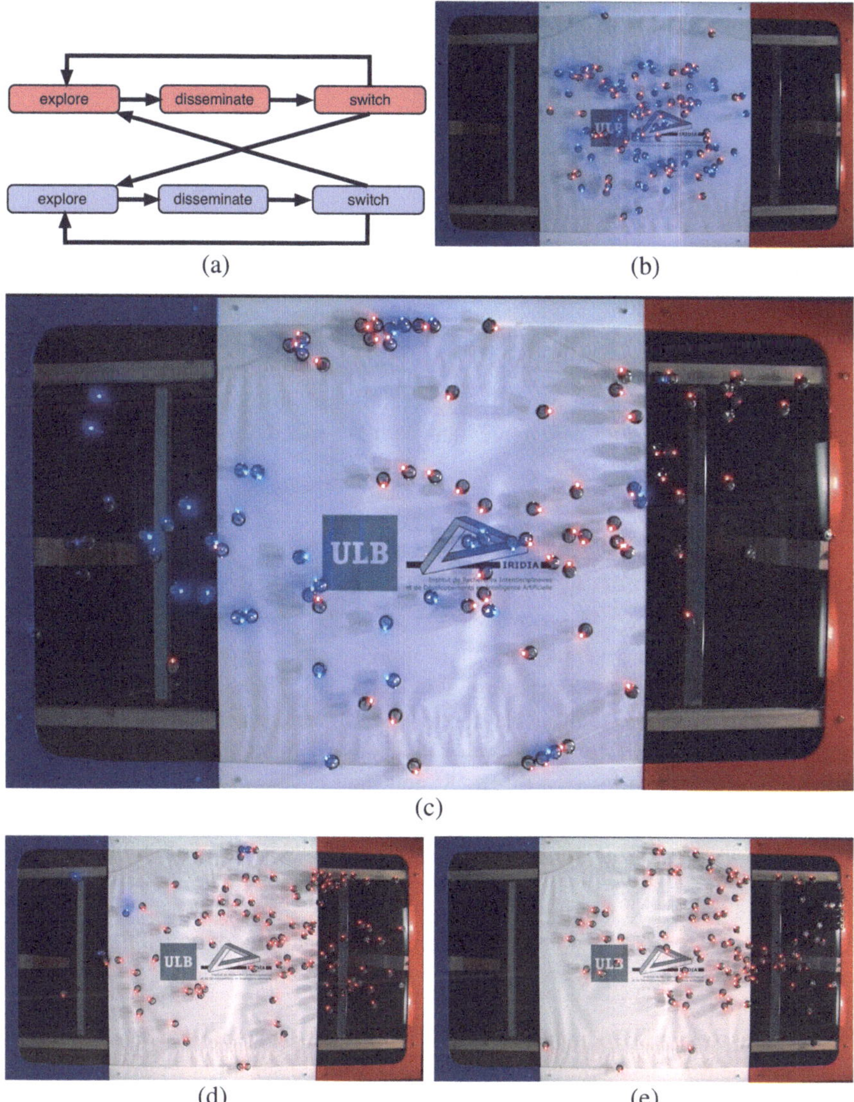

Fig. 8.17 Collective decision-making with 100 Kilobots [805]. **a** Finite state machine of robot behavior **b** Initial configuration of robot experiment **c** Early configuration of robot experiment **d** Intermediate configuration **e** Late configuration

proportional to the quality. This is called modulated positive feedback because the robots themselves continuously influence the degree of positive feedback.

An experiment with these 100 robots takes 90 min, which may seem quite long. The Kilobots cannot move fast with their little vibrating legs. Also, the exploration phase was emulated to have a certain minimal duration, although the robots could perform it much faster technically. As seen in Fig. 8.17 and as reported by Valentini et al. [805], the approach is effective. In the end, the swarm reaches consensus on red. This scenario could be an initial phase, for example, of swarm construction. Initially, the swarm has to agree on where to start to build, to focus all the power of the swarm on one building site.

The difference between the voter model and the majority rule is that the voter model turns out to be more accurate (i.e., the correct consensus is found more frequently), while the use of a majority rule results in a faster decision-making process [805]. This is in complete agreement with the speed-versus-accuracy tradeoff (see Sect. 8.4.1).

8.6.2 Collective Perception as Decision-Making

Although Valentini et al. [804] call their approach "collective perception," it can also be seen as a collective decision-making scenario. The exploration phase is of even more interest here than in the above example with the Kilobots. The robots have to decide about a global feature of their environment. They move on a floor with white and black tiles (see Fig. 8.18) and they have a ground sensor to detect the brightness. The task for the swarm is to determine whether black or white tiles are in the majority. In contrast to the above example, the robots do not navigate to a certain area depending on their opinion; instead, they can permanently explore. The frequency of black and white tiles is a global feature of the environment. The task difficulty can be regulated by choosing either a clear majority of one of the tile types (e.g., 66% black as in Fig. 8.18a) or a relation that is closer to the 50–50-state (e.g., 52% black as in Fig. 8.18a).

In this experiment, 20 e-puck robots [576] were used. The robots do random walks and measure the time for how long their ground sensor detected the tiles of their current opinion (e.g., in the case of opinion "black," the robot measures the time for how long its ground sensor detected black tiles). Again, the dissemination time is then modulated according to that measurement, which serves as an option quality here. Valentini et al. [804] tested again the voter model and the majority rule. In addition, a method called "direct comparison" was used, which would directly compare the quality measures to determine whether the robot should switch its opinion. All three approaches are effective. The overall results are mixed. The majority rule results in fast decisions as expected. The direct-comparison method works well in rather small swarms but does not scale well with increasing swarm size and also

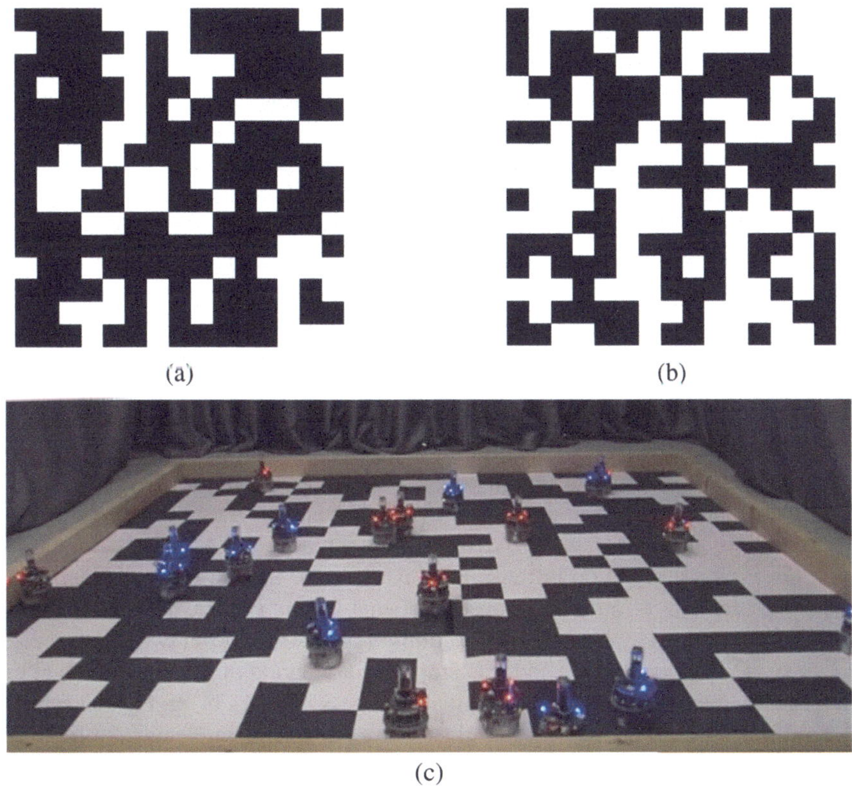

Fig. 8.18 Collective perception as collective decision-making: the robots must determine whether black or white tiles are in the majority. Two environments are considered: **a** easy (66% black) **b** hard (52% black) **c** robot experiment with 20 e-puck robots [804]

has difficulties with harder tasks. The voter model resides arguably between the two other approaches, but the results are not totally clear [804].

The unique aspect of this scenario is that collective decision-making is combined with collective perception. Similarly to what possibly happens in our brain when we look at Fig. 8.18, the robots have to estimate the majority collectively. The reported approach [804] obviously tries to keep the communication between robots simple, but more information could be shared. For example, robots could calculate averages over the measured qualities. The random walk could be replaced by a dispersion approach while still allowing the robots to mix. That way, the robots may cover more area and may gather more representative samples.

8.6.3 Aggregation as Implicit Decision-Making

We can interpret aggregation behaviors in robot swarms as collective decision-making. At least implicitly, the swarm agrees on an area to meet. The number of options is potentially infinite because space is continuous. The decision-making aspect becomes clearer once the robots have to select from a given set of possible aggregation areas. If all of these areas have the same utility or if at least the best ones are of equal utility, then the swarm faces a symmetry-breaking problem. The robot swarm has to find a consensus on which of the equivalent areas it wants to choose. Even if all areas are of different utility, the decision-making can still be implemented by a collective approach.

A popular example of an aggregation algorithm for swarms is BEECLUST [32, 34, 422, 423, 710, 712]. It implements an adaptive aggregation behavior, that is, not all positions in space are of equal utility for aggregation. Instead, a value can be measured at any point in space to determine this spot's utility. For example, this feature can be light, temperature, or the concentration of a chemical. Initially, the algorithm was inspired by the aggregation behavior of young honeybees. A key concept of the algorithm is that it relies on infrequent measurements. Instead of implementing a gradient ascent, that is, a robot follows the spatial feature's slope, the spots of high utility are determined by the social interactions of robots. Whenever two robots meet, they measure the local feature and stay stopped for a period of time proportional to the measured feature. In a positive feedback process, clusters of aggregated robots form mostly close to areas of high utility. For examples of implementations with different robot platforms, see Fig. 8.19. The details of this approach are discussed later in Chap. 9.

By interpreting BEECLUST as a collective decision-making approach, we find several unique properties compared to a standard decision-making approach. process as described above. Instead of having a few clearly defined options O, the robots can, in principle, choose from infinitely many positions in continuous space. However, by forming clusters, they create virtual, discrete options that an external observer can at least distinguish. If you like, aggregating robots mark positions in the "decision space" as options and try to attract other robots to their opinion.

8.7 Further Reading

A long text that is fully dedicated to collective decision-making in swarm robotics is that of Valentini [800]. Especially, the introduction of the concepts is a great read, and the overall approach is rather theoretically oriented. A shorter and related text is that of Valentini et al. [806]. Out of the domain of swarm robotics but still focused on collective decision-making and opinion dynamics are the works of Serge Galam on, his "sociophysics" [264–268]. A paper of similar style is that of Castellano et al. [127], which covers a wide range of models related to collective decision-making from the statistical physics perspective.

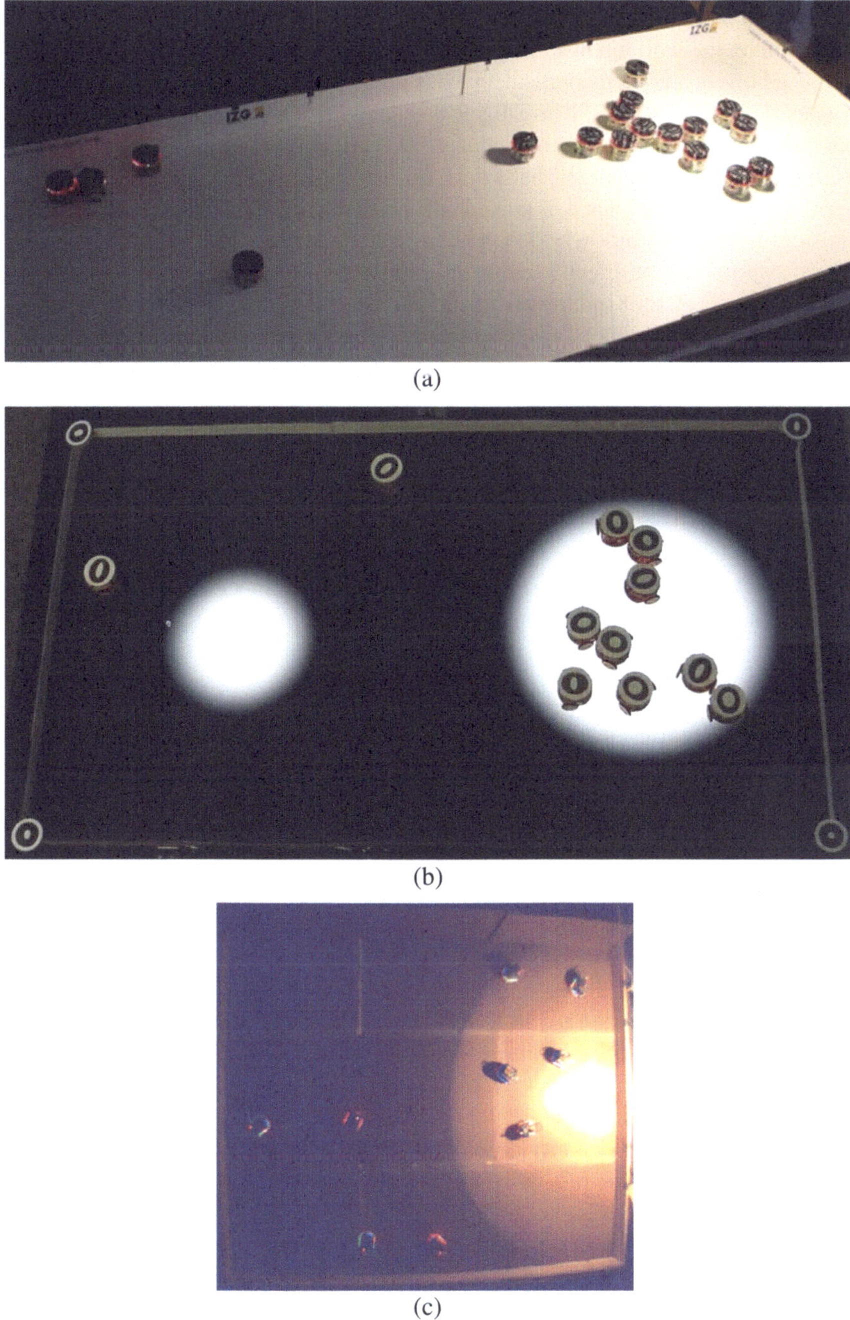

Fig. 8.19 Implementations of the BEECLUST algorithm on different robot platforms. **a** e-pucks (Ralf Mayet and Thomas Schmickl) **b** Colias robots on a 42 inch (107 cm) LCD screen by Arvin et al. [35] **c** Thymio II by Wahby et al. [838]

Additionally, there are several specialized but relevant highlights in the literature. For example, the work on discrimination of resources by Campo et al. [119]. Scheidler [703] offers a great theoretical analysis for the influence of latencies. Similarly, Montes de Oca et al. [578] investigate latencies in a swarm. With focus on animal behavior, the work of Couzin et al. [159, 160] is highly relevant. Also Szopek et al. [766] report interesting behaviors in honeybees and Dussutour et al. [216] report the essential influence of noise in decision-making. An example of collective decision-making in a bio-hybrid system is given by Halloy et al. [316]. Influencing natural swarms with artificial swarms is an intriguing future option for pest control. Finally, Reina et al. [668] address software-engineering approaches for programming collectively deciding robot swarms. In another paper, Hasegawa et al. [341] study the connection between individual threshold-based yes/no—decisions of individual swarm members and the resulting macroscopic decision of the swarm.

8.8 Exercises

Exercise 8.1 (*Estimating When to Stop Exploring*) The number of available options n is unknown at the start. The swarm must explore the environment to discover options. However, continued exploration comes at a cost, so it is important to determine when to stop exploring.

(a) Simulate discovery over time: Implement a simple simulation in which a swarm of robots explores an environment with an unknown number of total options n. Option non-spatial: Simulate the law of diminishing returns with an exponential decay: probability of discovering an option $p_{\text{discover}}(t) = p_0 \cdot e^{-kt}$, for initial discovery probability p_0 and decay rate k. In each time step, draw for each of your N robots a Bernoulli trial with $p_{\text{discover}}(t)$. You need to identify options. You could say that all options have an equal probability of being found. However, a more interesting approach could be to assign each option a different probability, with some being easier to find and others being harder to find. Option spatial: Simulate a grid world. Uniformly randomly place n_{true} distinct options at grid cells. N robots start at random locations and perform random walks (see Sect. 5.2.4). An option is discovered when a robot visits a cell that contains one. Alternatively, you could let all robots start from the same grid cell to make the discovery of options far away more difficult.

(b) Implement a stopping rule based on declining discovery rate. Track the cumulative number of discovered options $n(t)$. Plot $\Delta n(t)$ over time. Define a heuristic threshold (e.g., $\Delta n(t) < \varepsilon$ over several intervals) and stop when that condition is met. Discuss how this threshold affects the completeness of exploration.

(c) Implement the Chao1 estimator. Track the cumulative number of discovered options $n(t)$, and the number of options seen exactly once $f_1(t)$ and twice $f_2(t)$. At each time step, compute the Chao1 estimate $\hat{n}_{\text{Chao1}}(t) = n(t) + \frac{f_1^2(t)}{2 f_2(t)}$. Stop exploration when the estimated total number of options $\hat{n}_{\text{Chao1}}(t)$ is close (e.g., 95%) to the currently discovered number $n(t)$.

(d) Compare the Chao1 estimator swarm to the one based on a heuristic threshold. Who is performing better? Measure the percentage of total options that have been discovered. Consider the number of exploration time steps done until stopping. Define a cost function that considers the tradeoff between the cost of time and the cost of non-explored options.

Exercise 8.2 (*Optimal Stopping with Many Options Based on Quality*) You are given a swarm of size N tasked with selecting the best option among $n \gg N$ alternatives O_i, where each option has an unknown quality $q(O_i) \sim p(q)$, drawn from an unknown underlying distribution. The swarm can sequentially explore options and observe their quality values, but cannot explore all n options due to time constraints. Instead, it must estimate the distribution of option qualities and decide when to stop studying and select the best found.

(a) Conceptual questions: Explain why the decision problem becomes harder as n increases compared to N. Can this problem be solved with methods developed for the classic secretary problem? Outline how the 1/e strategy could be used here. What success probability does it offer?

(b) Density estimation and stopping: Suppose the swarm has already explored $m \ll n$ options and obtained observed quality values: $\{q_1, q_2, \ldots, q_m\}$. Use the Kernel Density Estimation (KDE) method to find the estimated density $\hat{p}_h(q)$ of option qualities. You can use Gaussian kernels. To choose the bandwidth h, consider using standard selection methods, such as Silverman's rule of thumb, likelihood cross-validation, or plug-in methods like the Sheather–Jones estimator. Given a current best observed quality $q^* = \max\{q_1, \ldots, q_m\}$, define how to compute the probability that a new unseen option will be better than q^*. Define a stopping rule based on a threshold probability P_{thresh}. Explain how this rule helps the swarm decide when to stop exploring.

(c) Propose how KDE could be implemented in a distributed fashion in a swarm, referencing ideas from Hu et al. [365]. What challenges do you foresee?

Exercise 8.3 (*Swarmalators*) Implement the swarmalator system as described in Sect. 8.5.8. Implement a simple solver for the ODE system, such as the Euler method. Try to reproduce all relevant stationary states. For the splintered phase wave, the system may show a longer transient; that is, you may require more iterations to find it.

(a) Try different values of parameters J and K. For the parameter space defined by the (J, K) plane, what are the boundaries between the different regions that can be associated with each of the stationary states?

(b) Test different initializations of individual natural frequencies ω_i, that is, inhomogeneous swarmalators. Start simple with two sub-populations of natural frequencies $\omega_i \in \{0, 1\}$. Next, try something like $\omega_i = \frac{i}{N}$ or similar. What differences do you observe?

Exercise 8.4 (*Urn Model for Locust Scenario*) We use the locust simulation again. We work with the following parameters: circumference $C = 0.5$, speed of 0.01, perception range of $r = 0.045$, swarm size of $N = 50$, spontaneous direction switch with a probability $P = 0.15$ per time step.

In your simulation, you still have to keep track of the left-goers L_t at a time step t and count them. We are interested in the average change of L_t in dependence on itself. That is, we are searching for a function $\Delta L(L)$. To measure this function, we simulate the locusts for 100 time steps. The locusts need these 100 time steps to organize themselves. We simulate an additional 20 time steps while we keep track of L_t. At each time step $t \in [101, 120]$ we "store" the change of L within one time step $\Delta L = L_t - L_{t-1}$, for example, by adding it to an array. It's an array because we want to measure ΔL in dependence on L itself. Hence, you can have a variable *deltaSum*[] in your program and for each time step $t \in [101, 120]$ you do: $deltaSum[L_{t-1}] += deltaSum[L_{t-1}] + (L_t - L_{t-1})$. Also, keep track of how often you have added something to an entry (e.g., a variable *count*[]) in order to normalize your measurements. Repeat these runs many times—up to 50,000 samples could actually be useful. Finally, write your data of the measured function $\Delta L(L)$ to a file for later use.

(a) Plot the data for $\Delta L(L)$. What is the mathematical and domain-specific meaning of the positions L^* with $\Delta L(L^*) = 0$?

(b) Next, we fit the swarm urn model to this data. Remember the equations

$$P_{\text{FB}}(s, \varphi) = \varphi \sin(\pi s), \tag{8.34}$$

$$\Delta s(s) = 4\left(P_{\text{FB}}(s, \varphi) - \frac{1}{2}\right)\left(s - \frac{1}{2}\right), \tag{8.35}$$

whereas P_{FB} is the probability of positive feedback and $\Delta s(s)$ is the expected average change of the considered state variable. Here, these are the left-goers, and hence s is the ratio of left-goers $s = L/N$. In addition, we have to introduce a scaling constant c to the idealized equation, which yields

$$\Delta s(s) = 4c\left(P_{\text{FB}}(s, \varphi) - \frac{1}{2}\right)\left(s - \frac{1}{2}\right). \tag{8.36}$$

Fit this function to your data, that is, try to find appropriate values for the scaling constant c and the feedback intensity φ. Be sure to keep φ within a reasonable interval: $\varphi \in [0, 1]$.

c. Plot your data together with the fitted function. Also plot the resulting probability of positive feedback P_{FB}. What can be said about these results?

Exercise 8.5 (*Voter Model for Binary Consensus*) We use the classical voter model for a binary decision-making problem. We work with the following parameters: swarm size $N = 50$, initial opinions split evenly between two options A and B, and asynchronous updates. At each discrete time step t, the system state is given by the opinions $o_i(t) \in \{A, B\}$ for each agent $i \in \{1, \ldots, N\}$.

The update dynamics are defined as follows. First, one agent i is selected uniformly at random. Then a neighbor j of i is selected uniformly at random from the neighborhood $\mathcal{N}_i$. Finally, i copies the opinion of j, while all other agents remain unchanged:

$$o_i(t+1) = o_j(t), \qquad o_k(t+1) = o_k(t) \text{ for all } k \neq i \; . \tag{8.37}$$

We consider two different interaction graphs: (1) The ring topology R_N, where the neighborhood of agent i is given by

$$\mathcal{N}_i = \{(i-1) \bmod (N+1),\ (i+1) \bmod (N+1)\} \; , \tag{8.38}$$

so each agent interacts only with its immediate predecessor and successor. (2) The complete graph K_N, where each agent interacts with all other agents, that is, $\mathcal{N}_i = \{1, \ldots, N\} \setminus i$. In your simulation, you have to keep track of the "magnetization"

$$m(t) = \frac{1}{N} \sum_{i=1}^{N} x_i(t) \; , \tag{8.39}$$

which is the fraction of A-agents at time t for

$$x_i(t) = \begin{cases} 1 & \text{if } o_i(t) = A \\ 0 & \text{if } o_i(t) = B \end{cases} \; . \tag{8.40}$$

We are interested in the consensus time

$$T_{\text{cons}} = \min t \geq 0 : o_1(t) = o_2(t) = \cdots = o_N(t) \; , \tag{8.41}$$

and in the probability of ending in all-A consensus as a function of the initial magnetization $m(0)$.

(a) Implement the voter model for R_N and K_N with $N = 50$. Plot the magnetization $m(t)$ over time for both graphs and compare the speed of convergence.
(b) Run many repetitions to estimate the expected consensus time $\mathbb{E}[T_{\text{cons}}]$ for both graphs. Plot $\mathbb{E}[T_{\text{cons}}]$ as a function of N for $N \in \{10, 20, 50, 100\}$. What is the scaling law you observe?
(c) Estimate the probability of ending in all-A consensus as a function of initial magnetization $m(0)$. Does this probability equal $m(0)$?

(d) Introduce a small fraction $p = 0.05$ of zealots for opinion A, that is, agents that never change their opinion. Simulate again and measure how this affects both consensus probability and consensus time.
(e) Now repeat all over with the majority rule instead.

Exercise 8.6 (*Kuramoto Model for Synchronization*) We use the Kuramoto model of coupled oscillators to study synchronization in swarms. Each of the N oscillators (or robots) has a phase $\theta_i(t) \in [0, 2\pi)$ at time t. The dynamics are defined as

$$\dot{\theta}_i(t) = \omega_i + \frac{K}{N} \sum_{j=1}^{N} \sin\left(\theta_j(t) - \theta_i(t)\right) , \tag{8.42}$$

where ω_i is the natural frequency of oscillator i and $K \geq 0$ is the global coupling strength. We are interested in the degree of synchronization measured by the Kuramoto order parameter

$$r(t)e^{i\psi(t)} , \tag{8.43}$$

where $r(t) \in [0, 1]$ quantifies the coherence of the oscillators:

$$r(t) = \left| \frac{1}{N} \sum_{j=1}^{N} e^{i\theta_j(t)} \right| , \tag{8.44}$$

and $\psi(t)$ is the average phase:

$$\psi(t) = \arg\left(\frac{1}{N} \sum_{j=1}^{N} e^{i\theta_j(t)} \right) . \tag{8.45}$$

In your simulation, you have to integrate the differential equations for $\theta_i(t)$ over time using, for example, Euler's method with a sufficiently small time step Δt.

We use the following parameters: swarm size $N = 100$, natural frequencies ω_i drawn from a normal distribution $\mathcal{N}(0, 1)$, and different values of the coupling strength K. Initialize all phases $\theta_i(0)$ uniformly at random in $[0, 2\pi)$.

(a) Simulate the Kuramoto model for $K \in \{0, 0.5, 1.5, 3.0\}$. Plot the order parameter $r(t)$ as a function of time. What happens to synchronization as K increases?
(b) Compute the long-term average synchronization level $\bar{r} = \frac{1}{T} \int_{t_0}^{t_0+T} r(t)dt$ after a transient t_0. Plot $\bar{r}$ as a function of K for $K \in [0, 5]$. Do you observe a transition from incoherence to synchrony?
(c) How does the distribution of natural frequencies ω_i affect synchronization? Compare a narrow distribution (e.g., $\omega_i \sim \mathcal{N}(0, 0.1^2)$) to a broad one (e.g., $\omega_i \sim \mathcal{N}(0, 2^2)$).
(d) Discuss briefly how this model relates to synchronization in robot swarms (e.g., firefly-inspired flashing, coordinated movement, consensus in phase variables).

Exercise 8.7 (*Balancing Speed and Accuracy in Collective Choice*) We simulate a collective decision-making system and test mixtures of robots that follow the majority rule and those that follow the voter model. The system consists of a swarm of N randomly moving robots that tries to make a collective decision. Each robot has a given number of neighbors. Due to the movement, the neighbors change at each time step. To achieve consensus, K out of N robots follow the majority rule (MR), while $N - K = L$ agents follow the Voter Model (VM). We do not want to analyze the spatial aspects of the swarm, so you may assume that each robot has five random neighbors. The swarm consists of $N = 101$ agents and does 2000 rounds of decision-making in each independent simulation run.

The robots have to decide between two options: A and B. Option A is better than option B (without loss of generality) and thus the overall correct choice. The decision-making process of each robot follows these rules:

- If the robot voted A, it does nothing with a probability of $1 - P_{\text{qual}}$.
- If the robot voted A, do the following with probability P_{qual}. Always do it if the robot voted for B:
  - with probability $P_{\text{dec}} = 0.7$:
    check neighbors and change the decision if appropriate
  - otherwise: switch opinion spontaneously with a probability of $P_{\text{spont}} = 0.1$ or do nothing

(a) Implement the decision-making process for a swarm of $N = 101$ robots. Test all possible different ratios of MR- to VM-using robots, from $K = N$ and $L = 0$ to $K = 0$ and $L = N$. Also test all possible initialization ratios for the opinion of the swarm, from "0 robots vote for A" to "N robots vote for A." Use $P_{\text{qual}} = 0.6$. Do 100 independent runs for each ratio. To speed up the simulation, you may stop each run after a consensus has been reached (i.e., at least $N - 10$ or at most 10 robots voted for A).

    (1) Measure how many times the swarm has decided correctly (at least $N - 10$ robots voted for A). Plot the average accuracy of all runs over the opinion initialization ratio. Plot each MR/VM-ratio as one line.
    (2) Measure when (i.e., in which round of the process) the swarm has decided on an opinion. Plot the speed over the opinion initialization ratio. Plot each MR/VM-ratio as one line. We define the speed as the inverse of the average decision time over all runs.
    (3) Construct a speed-versus-accuracy diagram by plotting speed of convergence over accuracy of the collective decision. Determine the average speed and accuracy for four different probabilities P_{qual} (roughly between 0.5 and 0.9) over many different initializations and independent runs for ten different MR-VM-ratios each. Plot your results and interpret them.

(b) Now, we implement a centralized "observer" (for now, feel free to consider a fully distributed approach later). This observer has global information and monitors the swarm. If the process is too slow, the observer replaces VM-robots with MR-robots. If the decisions are too inaccurate, the observer should replace MR-robots with VM-robots.
Define minimum values for both accuracy and speed. Do 100 independent runs for an initial MR-VM-mix and let the observer measure speed and accuracy. If at least one of the parameters does not meet the defined minimum values, the observer should intervene. Then the process should start over again (100 runs, observe, possibly intervene, etc.). After a few runs, adjust the system's minimum requirements and observe the results.

Chapter 9
Case Study: Adaptive Aggregation

> *"A smart machine will first consider which is more worth its while: to perform the given task or, instead, to figure some way out of it."*
>
> *—Stanisław Lem, The Futurological Congress*

Abstract We try the impossible and summarize almost all we have learned in this book. In this little case study, we try to integrate all the different techniques that one requires to design a system of swarm robots. The task is rather simple, and we focus on adaptive aggregation, that is, the swarm has to aggregate at a particular spot determined by environmental features. We follow a biological role model and apply standard modeling techniques of swarm robotics. Finally, the control algorithm and the micro–macro model are verified against robot experiments.

9.1 Introduction

9.1.1 Learning Objectives

After reading this chapter, you should be able to:

- Understand the design process of swarm systems for adaptive aggregation at environmentally determined locations.
- Identify biological role models, such as honeybee thermotaxis, for developing robust swarm behaviors.
- Explain the principles of the BEECLUST algorithm, which uses positive feedback (e.g., waiting times) to promote clustering without goal-oriented behavior or gradient detection.
- Comprehend how micro–macro models, based on Langevin and Fokker–Planck equations, simplify individual dynamics into collective spatial models.

H. Hamann, *Swarm Robotics*,
https://doi.org/10.1007/978-3-032-10584-4_9

- Recognize the importance of validating algorithms and models with real-world experiments, as shown with robots aggregating under light.
- Apply validated models to optimize swarm design and refine control parameters for adaptation to dynamic environments.

9.1.2 Guiding Questions

- What core problem does this case study address in swarm robotics, and how does it relate to real-world applications?
- How does honeybee thermotaxis serve as a biological role model for adaptive aggregation in robot swarms?
- What are the key mechanisms of the BEECLUST algorithm, and how does positive feedback (e.g., waiting times) enable aggregation without gradient sensing or goal-oriented behavior?
- What alternative aggregation methods were considered, and why are they arguably less suitable regarding sensor requirements, environmental conditions, or collective dynamics?
- How are individual robot dynamics modeled microscopically with Langevin equations, and what simplifications (e.g., random walk, stop/move transitions) represent BEECLUST behavior?
- How is collective behavior represented macroscopically with Fokker–Planck equations, and how are discrete state transitions translated into continuous stopping rates and delays?
- How were the algorithm and model validated experimentally using robots under dynamic light fields, and what metrics were applied to measure aggregation?
- What were the main findings on model accuracy, and how can the validated model optimize swarm design and refine control parameters such as waiting times?

9.2 Use Case

Let's imagine we are several years into the future, and meanwhile, you have founded a successful software company that produces robot control software. Of course, you specialize in swarm robotics. By emulating some of the software engineering speak, we say we go through a swarm robotics use case. Say, a customer orders software from you and specifies it in the following way. The software controls a robot swarm, which aggregates at a particular spot determined by sensor input and stays flexible in a dynamic environment. That sensory input could be anything, but say it is temperature. Then we want aggregation at the warmest spot. Possible scenarios depend on the distribution of temperature, such as shown in Fig. 9.1. However, much

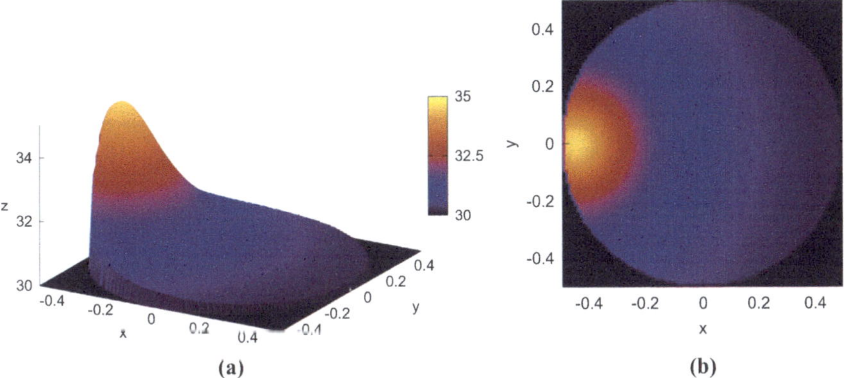

Fig. 9.1 Spatial temperature distribution for aggregation at the warmest spot. **a** Side view, **b** Top view

more complex scenarios are possible. The temperature distribution could have local optima, many scattered spots instead of just one well-defined, smooth peak. There could also be systematic plateaus that give no information via the gradient (i.e., the spatial derivative is zero, and we cannot guess in what direction we may find a peak).

9.3 Alternative Solutions

You may want to guess an efficient approach immediately, but we first go through a few alternative solutions.

9.3.1 Ad Hoc Approach

A simple ad hoc approach may be to let robots drive randomly around while measuring. A robot could keep track of the most significant measured values. After some time, a robot stops when approaching a considerable value again. In this approach, we would ignore the multi-robot setting. As an add-on, we could share the most significant measured values with neighboring robots whenever they pass by. This approach has the disadvantage that we make heavy use of the sensor (which may cost additional energy). We also require a relatively robust and reliable sensor, because a faulty sensor reading may give us a far-off maximum. Once the robot tries to find that maximum again, we run into a deadlock, as the robot will not find it. Also note that the simplified version of this approach—just stopping once we find a relatively high value after some time—is directly related to the so-called secretary problem [256].

9.3.2 Gradient Ascent

A slightly more developed approach could be to implement a gradient ascent, that is, robots try to detect the spatial derivative of the temperature either at a given position or approximated between two nearby points. This can arguably be called the standard engineering approach. We can include the multi-robot setup by allowing robots to communicate promising directions that they have determined to their neighbors. It comes with several requirements. First, the approach only works if the gradient is well defined and the temperature distribution is relatively smooth, that is, there are no or only a few sharp peaks. For a temperature distribution, that is, maybe a fair assumption. However, second, we require that there are no local optima. Otherwise, robots get trapped at mediocre peaks. Third, we also require reliable sensors to detect the gradient. We either need two sensors to do two measurements at different positions at the same time, or we need memory to measure over time by moving the robot in between. Especially the approach with two sensors is questionable. We need double the hardware if we choose to put two sensors on the robot. Also, depending on the size of the swarm robot, we may not be able to place the sensors at a sufficiently long distance from each other (e.g., consider the Kilobot's 33 mm diameter, see Sect. 3.6.1).

9.3.3 Positive Feedback

Maybe it is helpful to design our approach as an inherently collective approach from the very start. In addition, we can also follow the definition of self-organization (see Sect. 1.3) and add an essential ingredient: positive feedback. Putting both together, we can get something like the following. Each robot searches for a good spot (e.g., a high temperature that it hasn't seen for a while) and once it finds the spot, the robot hangs around there for a while. If nobody joins the robot, it gives up and tries to find another spot. This is collective by design, and we have positive feedback because clusters of stopped robots may form and grow. A drawback, however, may be that robots spend too much time in globally uninteresting regions. Also, maybe too many robots may stay stopped lonely with low chances of ever seeing another robot, for example, if the swarm density is relatively low (cf. stick pulling, see Sect. 5.3.6).

9.4 Biological Inspiration: Honeybees

To elaborate on the positive feedback idea, we check a biological system for inspiration. As it turns out, there is an animal behavior that resembles our desired robot behavior nicely. Young honeybees happen to aggregate in clusters in the bee hive, influenced by the temperature. Here, "young" means that their age is below 24 h, that is, these honeybee ladies just got out of their cocoon and left their honeycomb. They can neither fly nor sting—a fact that is very much welcomed by honeybee researchers. So they only walk within the beehive. Young honeybees prefer a specific temperature, which helps them to dry their wings. They stay in the hive, which is usually totally dark and has regions with different temperatures. It was observed that young honeybees aggregate at spots of 36 °C [766]. These young honeybees are known to show reactions to social stimuli, that is, they change their behavior if they meet other bees.

Biologists have designed an experiment to learn more about the social aspects of the young bees' behavior. The young honeybees are put into an area of controlled conditions. It is a round area covered with honeycombs, it is all dark, and has a carefully adjusted temperature distribution. There is an optimal temperature (36 °C) spot on the left-hand side. The lowest temperature on the right-hand side is only slightly colder and has 30 °C. The overall temperature distribution is relatively homogeneous, and the induced gradient information is rather "flat," that is, it is supposed to be difficult for a bee to determine a direction of increasing temperature [417, 766].

The experiment is monitored and data is recorded with an infrared camera and an array of temperature sensors in the floor below the honeycombs. The young honeybees are initially placed in the center of the area and then left for up to 2 h. The experiment is repeated several times with either a single bee, a few bees, or many bees. The fascinating result is that a single bee cannot find the optimum spot. Instead, she keeps walking around as shown in Fig. 9.2 (computer simulation). A few bees are inefficient but able to find the optimum. A big group of bees finds the optimum efficiently. We have an actual swarm effect because we observe a significantly better

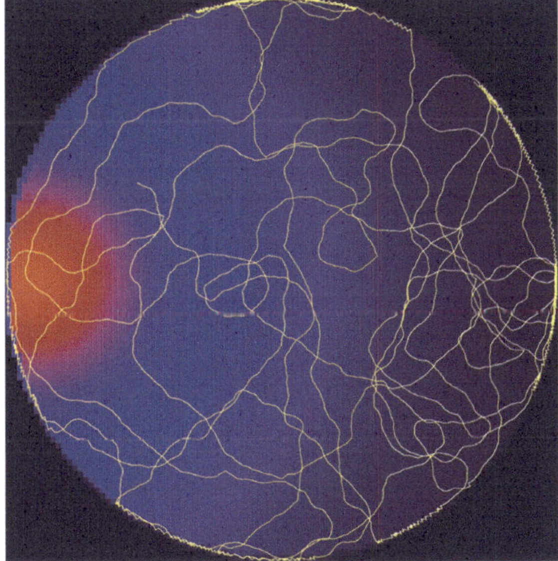

Fig. 9.2 Example trajectory of a young honeybee navigating in a circular area with a "flat" temperature distribution between 36 °C (left) and 30 °C (right). The shown trajectory is a computer simulation

individual performance with increasing group size (cf. Chap. 2). As a result, we can catch the hypothesis that some social interaction is a crucial ingredient of this behavior.

If we want to find out about this missing link—what social interaction between bees helps them in finding the optimum?—then we have to look deeper into the biological approach. A challenge is that there is no known reductionist approach for determining an animal's behavior. There is no way of learning much from cutting the bee's brain into pieces. There is, of course, neuroscience, but it cannot give us the underlying algorithm of honeybee behavior. Instead, biologists define a behavioral model as a hypothesis. For our little experiment here, the behavioral model can look like this [712]:

step 1 Move straightforward
step 2 Ostacles or robots around?

- (**a**) in case of an obstacle: turn away, return to step 1;
- (**b**) in case of a robot: stop, measure temperature, wait for some time (the longer the higher the temperature), U-turn, return to step 1.

This is actually the BEECLUST algorithm [422, 710, 712] as discussed in Sect. 5.3.1. With this approach, robots stay stopped longer in warmer areas, hence increasing the chance of creating a growing cluster of robots once other robots join. We get a positive feedback effect (see Sect. 1.3) because bigger clusters are more likely to be found by robots. In addition, once a cluster has reached a certain size, robots may get trapped within a cluster. These robots are physically blocked; even if they want to leave, they cannot. We notice that the positive feedback depends on two features, one feature that we can control ourselves (waiting time) and another feature we cannot control because it is a natural effect (physically blocked robots). The above behavioral model ignores that there could be areas that are too hot for the bees (or our robots). In a variation, we could wait for shorter times if the measured temperature is very high.

9.5 Model

Next, we want to model this scenario. We need to determine whether the honeybee behavior is actually a good inspiration for our scenario and check whether the approach is efficient. We have at least one crucial parameter, the waiting time depending on temperature, that we need to configure; a model helps to do that in a structured approach. Also, for the sake of exercise, of course, it is worth looking into a modeling approach again. But first, we discuss a few alternative sources of inspiration for a modeling approach.

9.5.1 Modeling Aggregation: Interdisciplinary Options

We have not yet developed a generic modeling approach in swarm robotics. Modeling a swarm robotics system is still a kind of art. Scientifically speaking, that is problematic but also an interesting challenge. We should accept the challenge as an opportunity to work in an interdisciplinary way. At the same time, adapting models from other fields of research also helps to estimate the difficulty of the modeling problem by detecting analogies to other modeling challenges. Aggregation is a perfect example.

Besides investigations of honeybees, there are many more models for aggregation processes in several species: bark beetles [182], ants [183, 783], and cockroaches [22, 183, 316, 385, 477]. Here, we do not go into further details concerning biological models.

What is even more interesting about modeling aggregation is that a diverse selection of research fields contributes. Especially within physics, there are several relevant concepts. Nucleation [706] investigates how a material transforms from one thermodynamic phase to another. Thermodynamic phases are gas, liquid, solid, and plasma. An example of nucleation is water droplets that form in supersaturated vapor (e.g., water droplets on your beer bottle in summer). Although this may seem unrelated, nucleation is used as a model to describe aggregation scenarios, such as traffic jams [47].

Another example from thermodynamics is Ostwald ripening [609, 836] that describes the process following the nucleation. The system does not go immediately into an equilibrium state, but instead several clusters (e.g., of water) form. Ostwald ripening basically describes the competition between the different clusters that grow and dissolve. Vinković and Kirman [828] report an interesting study that combines Ostwald ripening with the Schelling model (a model of segregation in human populations). Similar methods can also be used to model city formation and growth [517].

Obviously, physics intensively investigates systems subject to gravitation, and also clustering can be an effect of gravitation [344, 561]. Examples are star formation [444] and galaxy formation [742]. Particles or—more general—matter subject to gravitation from deterministic systems that are different from swarms in an important feature. Gravitational systems are fully determined by the initial state and cannot break symmetries, in contrast to groups of autonomous agents. Imagine a homogeneous distribution of stationary small particles in 2-d space. Now we add two big stationary masses separated by a certain distance. The two masses attract the small particles. However, we can immediately draw a virtual line in the middle between the two big masses. Particles in the left half approach the left mass, and particles in the right half approach the right mass. Suppose we ignore any complex dynamics as observed in three-body problems [57]. In that case, we can conclude that the small particles will equally separate and aggregate at either one of the two masses. Swarms of robots are different from that; we can program them to deliberately break such symmetries and aggregate at just one spot [326]. In pattern formation processes, in turn, gravity can also serve itself as a means to break symmetries, for example, in the case of pattern formation in cells [769].

In summary, one can say that there are plenty of options to get inspiration for swarm robot aggregation models. From a practical engineering point of view, we can be happy with whatever works. Any model that successfully supports us in designing the swarm robot system has its place. Whether there is one generic model approach that always works is an unanswered question of swarm robotics.

9.5.2 *Spatial Model*

In the following, we want to reuse the micro–macro approach of Sect. 7.4.2 and create a microscopic model for an individual robot together with a macroscopic model for the whole swarm based on the Langevin equation and the Fokker–Planck equation. We define a few modeling objectives. The model should

- represent the difference between the ineffective single behavior with "flat gradient information" and the effective behavior of a swarm;
- represent parameters of the agent algorithm directly in the model;
- represent spatial features;
- show quantitative agreement with robot swarm experiments.

Only if the model can distinguish between the ineffective single behavior and the effective swarm behavior can it display a significant swarm effect resulting from social interaction. A modeling approach in swarm robotics is only helpful if the model can, despite its abstraction level, be related to implementation aspects of the control algorithms. Here, we investigate an adaptive aggregation scenario, and therefore spatial features are of importance. Often, models are sufficient if they only show a qualitative agreement (e.g., the model reflects only general features of reality) with robot experiments. However, if we can achieve a quantitative agreement (e.g., model predictions match directly with experimental measurements) with a reasonable overhead, then we should do so.

Microscopic Model

We construct our microscopic model, the Langevin equation, step by step, to model the behavior of an individual robot. We start from the general Langevin equation, Eq. (7.23), and replace $\mathbf{A}(\mathbf{R}(t))$ already with a gradient $\nabla P(\mathbf{R}(t))$, yielding

$$\dot{\mathbf{R}}(t) = \alpha \nabla P(\mathbf{R}(t)) + B\mathbf{F}(t). \tag{9.1}$$

We have to define the parameters. Remember that $\mathbf{R}$ is the robot's position, here in 2-d. We use the potential field P to model the temperature field. We define a gradient ascent depending on parameter α. $\nabla P(\mathbf{R}(t))$ gives the gradient of the temperature field at the robot's current position. With $\alpha \in [0, 1]$ we can regulate the intensity of the robot's goal-oriented behavior.

As before, $\mathbf{F}$ represents the stochastic process that models the robot's random behavior. We assume that the robot goes intentionally or unintentionally in random directions. That may seem strange as it does not directly correspond to step 1 of the behavioral model: "move straight forward." However, it turns out to be a viable modeling assumption and simplification. An alternative modeling approach could be the Lévy flight [511], which allows bigger jumps that could represent straight motion. Here, we stick to the Brownian motion approach, which is also called Wiener process, which in turn is a special case of the Lévy flight. As before, B scales the intensity of the random behavior.

In Fig. 9.3, you see example trajectories using Eq. (9.1) and the effect of varying parameter α. The temperature distribution is still the same as in Fig. 9.1, that is, the desired warm spot is at the left border. For $\alpha = 0.01$, we see basically no bias toward the optimum. Then, with increasing α, the robot approaches the optimum more and more directly and stays closer to it. Note that we have not yet defined a condition for stopping. Hence, the robot continues to move at all times.

Next, we want to model the "flat gradient information" situation. The question is what the correct setting for parameter α is. In the case of the young honeybee experiments, the situation is somewhat unclear [417, 766]. Some honeybees approach the warm spot more or less directly (so-called "goal finders"), but they vary in their intensity of goal-directedness. At the same time, there exist other young bees that walk around randomly or walk along the boundaries [417, 766]. To simplify our approach, we are not going to model these diverse results of the biological system and focus instead on possible robot implementations. We go for the rather extreme assumption of ignoring the temperature gradient and the goal-oriented behavior completely. We set $\alpha = 0$. This is a strong assumption because to achieve the desired aggregation at the warm spot, we then fully rely on an emergent, macroscopic effect. Suppose we can still manage to aggregate the swarm without individual goal-oriented behavior. In that case, we have also simplified the (hardware) requirements for the robot swarm because robots do not need to detect differences in the temperatures anymore. We set the individual goal-oriented behavior off with $\alpha = 0$ and we get

$$\dot{\mathbf{R}}(t) = \alpha \nabla P(\mathbf{R}(t)) + B\mathbf{F}(t) = B\mathbf{F}(t), \tag{9.2}$$

which means we have a pure random walk. From a modeling perspective, we should appreciate that it simplified our model considerably.

Now we have to step out of the modeling framework that a mere Langevin equation offers us, because we want to model the stopping behavior of robots. Robots should be allowed to make a state transition from moving to staying stopped. Hence, we identify two states: a robot is either *moving* or *stopped*. For the state *stopped*, we need to define the waiting times finally. We define a function, which gives the corresponding

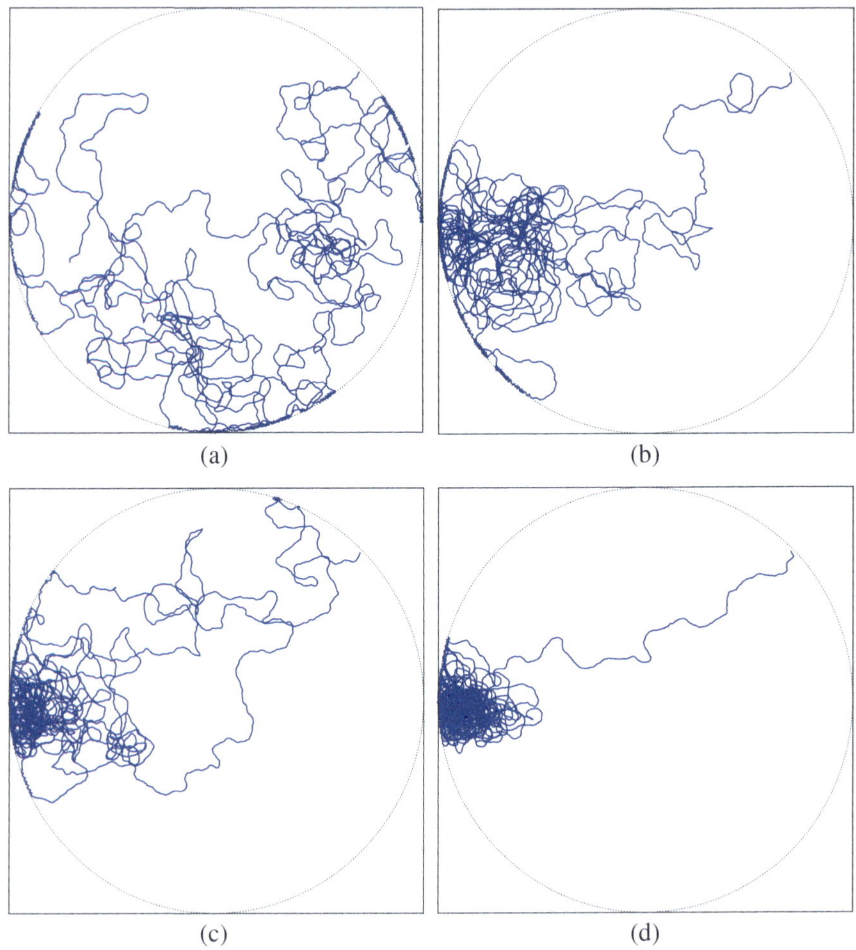

Fig. 9.3 Example trajectories sampled from the Langevin equation, Eq. (9.1), with four settings of parameter $\alpha \in \{0.01, 0.025, 0.05, 0.1\}$ regulating the intensity of the robot's goal-oriented behavior. The desired warm spot is at the left border. **a** $\alpha = 0.01$, **b** $\alpha = 0.025$, **c** $\alpha = 0.05$, **d** $\alpha = 0.1$

waiting time for a given position $\mathbf{R}$ and its associated temperature $P(\mathbf{R})$. Following the standard approach of BEECLUST [710], we define a waiting time function

$$w(\mathbf{R}) = \frac{w_{\max} P^2(\mathbf{R})}{P^2(\mathbf{R}) + c}, \tag{9.3}$$

for maximal waiting time $w_{\max}$, temperature $P(\mathbf{R})$ at position $\mathbf{R}$, and a scaling constant c. See Fig. 9.4 for an example of a waiting time function inspired by the setup based on a light distribution instead of temperature as described by Schmickl and Hamann [710].

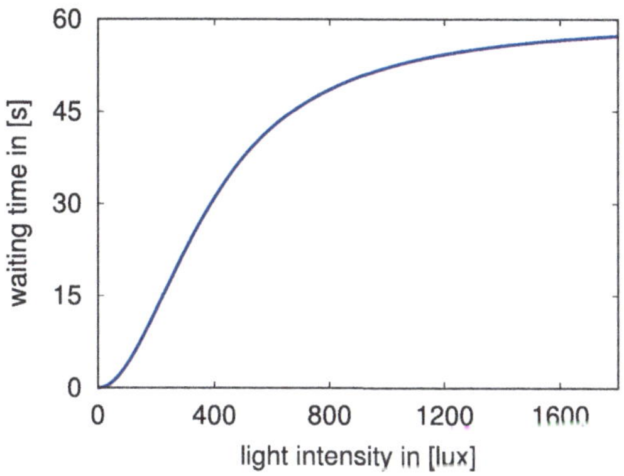

Fig. 9.4 Example waiting time function, Eq. (9.3), for a setting working with light measurements instead of temperature [710]. Parameter setting is $w_{\text{max}} = 60$ and $c = 1.5 \times 10^5$

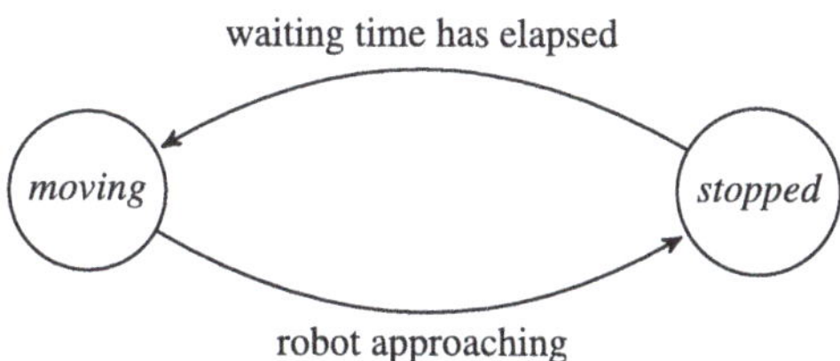

Fig. 9.5 Finite state machine modeling our microscopic modeling approach

We use the waiting time function $w(\mathbf{R})$ to determine for how long a robot has to stay in *stopped* state before it switches back to *moving*. We can model our approach with a finite state machine, see Fig. 9.5. The transition *moving*→*stopped* is triggered by another robot in proximity. We model other robots with additional instances of the microscopic Langevin model. In the moment of that transition, the robot measures the temperature (or light) and determines its waiting time. The waiting time of our model operates via the robot position $\mathbf{R}$, while the robot itself does not know its position, it can certainly measure the local temperature. The backward transition *stopped*→*moving* is triggered by an elapsed waiting time.

Each of the states is associated with an appropriate Langevin equation that models the robot's dynamics in that state. As described above, for state *moving* we have

$$\dot{\mathbf{R}}_m(t) = B\mathbf{F}(t). \tag{9.4}$$

For state *stopped* it is even simpler, as we have

$$\dot{\mathbf{R}}_s(t) = 0. \tag{9.5}$$

This trivial equation does what it is supposed to: the robot should not move while *stopped*.

Our microscopic model is complete now. The dynamics of the robots are modeled by simple Langevin equations, which we have actually simplified to mere stochastic processes (random walk). The model is not purely mathematical because we have to administrate the state transitions. Also, we have a model instance for each robot, so we have to administrate position, state, and remaining waiting time for each robot. In addition, we have a computational overhead because we need to check distances between robots, which generally scales $\mathcal{O}(N^2)$ for swarm size N (unless a domain-specific, special data structure is created). All of that is, however, typical for a microscopic model that is by nature always close to a robot simulation.

Macroscopic Model

We construct the macroscopic counterparts to the microscopic model. Thanks to the micro–macro link between the Langevin equation and the Fokker–Planck equation, we can do that in a schematic approach. For each microscopic state, we define a corresponding probability density. We define $\rho_m(\mathbf{r}, t)$ as the density of moving robots within an infinitesimally small area located at $\mathbf{r}$ at time t. Equivalently, we define $\rho_s(\mathbf{r}, t)$ as density of stopped robots at $\mathbf{r}$ at time t. For both densities, we need a model of motion that is directly determined by the corresponding Langevin equation. For state *moving* we get

$$\dot{\mathbf{R}}_m(t) = B\mathbf{F}(t) \;\Rightarrow\; \frac{\partial \rho_m(\mathbf{r}, t)}{\partial t} = B^2 \nabla^2 \rho_m(\mathbf{r}, t). \tag{9.6}$$

That is a mere diffusion equation because we deleted any goal-oriented behavior already in the Langevin equation. For state *stopped* we get the expected simple result of

$$\dot{\mathbf{R}}_s(t) = 0 \;\Rightarrow\; \frac{\partial \rho_s(\mathbf{r}, t)}{\partial t} = 0. \tag{9.7}$$

In similarity to the finite state machine shown in Fig. 9.5, we need a counterpart for the macroscopic model. The Fokker–Planck approach is continuous and forces us to redefine the transitions. We can do that by extending our Fokker–Planck equation to a so-called master equation (see, for example, the book of Schweitzer [723]). A master equation models the flow of probability density between different states almost as if it would model flows of liquids. So we have to think about how much density flows from the *moving* density to the *stopped* density and back. As said, the microscopic and inherently discrete transitions “waiting time elapses” and “robot approaching” need to be translated into continuous counterparts.

In the case of the transition *moving*→*stopped*, we cannot determine individual robots approaching each other because we deal only with probability densities. We have to determine a probabilistic approach, basically asking: given density $\rho_m(\mathbf{r}, t)$, how likely is it that (at least) two robots approach each other. To keep this modeling approach here simple, we make a relatively strong assumption saying that the probability of two robots approaching each other is directly proportional to the density of moving robots $\rho_m(\mathbf{r}, t)$. Hence, we can define a stopping rate φ. We say that the product $\rho_m(\mathbf{r}, t)\varphi$ gives us the correct density flow of stopping robots—a rather crude approach hiding all details of how an effective model can be created this way [318, 713].

It turns out, we can model the flow for *stopped→moving* with a so-called delay equation (similar to what we did with rate equations in Sect. 7.4.1). So with the waiting time function, we can determine a time delay $w(\mathbf{r})$. That means that robots $\rho_m(\mathbf{r}, t)\varphi$ leaving state *moving* spend $w(\mathbf{r})$ time in a waiting loop until they arrive back in state *moving* again. The resulting finite state machine for the macroscopic model is shown in Fig. 9.6.

Actually, the density of stopped robots ρ_s is mathematically unnecessary. Robots leaving state *moving* spend some time in "nirvana" and come back again. One can also say, they stay in a waiting loop for time $w(\mathbf{r})$ (almost like an airplane waiting for clearance to land). So we delete ρ_s and use time delays instead as shown in Fig. 9.7. The swarm fraction that is subtracted from $\rho_m(\mathbf{r}, t_0)$ at time t_0 is added to $\rho_m(\mathbf{r}, t_1)$ again at time $t_1 = t_0 + w(\mathbf{r})$.

The resulting macroscopic model is completely described by the time evolution of the density of robots in state *moving*, given by

$$\frac{\partial \rho_m(\mathbf{r}, t)}{\partial t} = B^2 \nabla^2 \rho_m(\mathbf{r}, t) - \rho_m(\mathbf{r}, t)\varphi + \rho_m(\mathbf{r}, t - w(\mathbf{r}))\varphi, \tag{9.8}$$

with stopping rate φ as described above, and we would need to determine B experimentally. Note that we are not considering explicitly moving robots approaching stopped robots (e.g., something of the form $\rho_m \rho_s$). This can be interpreted as a linearization approach for simplification, but it may alternatively be incorporated into the stopping rate φ.

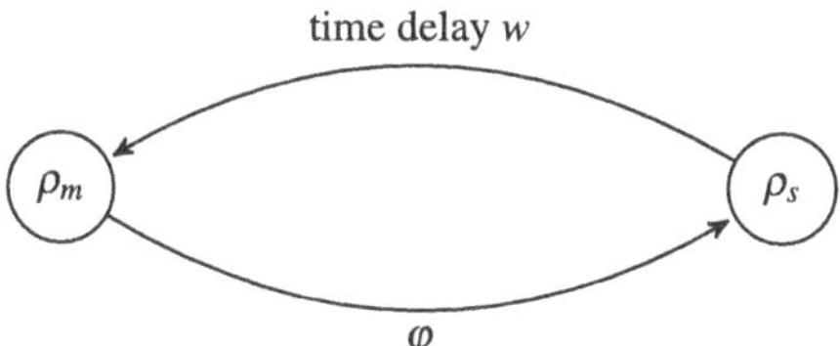

Fig. 9.6 Finite state machine modeling our macroscopic modeling approach

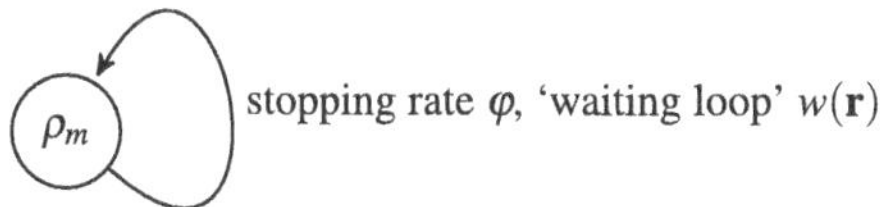

Fig. 9.7 Simplified finite state machine modeling our macroscopic approach with a single robot density $\rho_m(\mathbf{r}, t)$ and time delays

As said, it is not necessary, but for completeness and a better understanding, we also give the density of robots in state *stopped*

$$\frac{\partial \rho_s(\mathbf{r}, t)}{\partial t} = \rho_m(\mathbf{r}, t)\varphi - \rho_m(\mathbf{r}, t - w(\mathbf{r}))\varphi. \tag{9.9}$$

With Eq. (9.8) we now have a model. In a usual use case, we apply it in the following way. We input a relevant initial state, which is here a spatial distribution of robots in state *moving*. Due to the time delay $t - w(\mathbf{r})$ we also have to define values for the recent past before "time zero" of our model to reach a complete description of an initial state. Typically, one would set that history to $\rho_m(\mathbf{r}, t) = 0$ for $t < 0$ to keep things simple. That is, we assume nobody ever stopped in this system before. Mathematically, we define an initial value problem that we need to solve now. Many PDEs, and especially PDE with time delays, are not easily solved analytically. We restrict ourselves to a numerical solution that we do not want to describe in detail here. A simple approach is a forward integration in time with the Euler method. For more details, see, for example, Press et al. [649] and Hamann [318].

9.6 Verification

Finally, we want to verify our model, Eq. (9.8), and our algorithm (see Sect. 9.4) with robot experiments. We follow closely the study by Schmickl et al. [713] and the similar study by Kernbach et al. [422].

In the robot experiments, the Jasmine [383] robot was used. Jasmine is a simple and small swarm robot ($30 \times 30 \times 25\,\text{mm}^3$) with a differential drive. It has six infrared sensors pointing all around the robot. On top, they have two sensors measuring the ambient light. The swarm has a size of 15 Jasmine robots that operate on a rectangular area of $150 \times 100\,\text{cm}^2$. A light distribution replaces the spatial temperature distribution of the young honeybees. There is one lamp at each of the two short sides of the arena. The lamps are operated in three modes each: off, dimmed, and bright. The robot experiment is structured in four phases to test the robot swarm's reaction to a dynamic environment. This approach is also helpful to verify the models. In each of the four phases, the two lights—left, right—are set to

- off, dimmed;
- bright, dimmed;
- dimmed, bright;
- dimmed, off.

Each phase lasts 3 min (180 s). The robot controller is an implementation of the behavioral model for the young honeybees described in Sect. 9.4. A notable detail is that the robots do not reliably detect each other. Both the detection of obstacles (walls) and the detection of robots are based on infrared. So robots have to discriminate between the two. This is done by sophisticated timing. If the robot currently measures distances with its infrared sensors, then incoming infrared signals are interpreted as reflections. If the robot is currently not measuring distances but receiving infrared signals, then it assumes a nearby robot.

Six independent experiments with 15 Jasmine robots over the whole duration of 720 s were done. Initially, the robots were distributed uniformly randomly all over the rectangular area and were set to state *moving*. Similarly, the macroscopic Fokker–Planck model, Eq. (9.8), was initialized to a uniform distribution with all densities in state *moving*. The efficiency of the aggregation behavior in the robot experiments was measured by counting stopped robots beneath the two lights, for details see Schmickl et al. [713]. In the case of the Fokker–Planck model, the initial value problem was solved by forward integration in time. To measure the equivalent of aggregated robots at any time, the density of the respective area beneath the lamps was integrated; for details see Schmickl and Hamann [710]. The model was fitted using the single free parameter B in Eq. (9.8), that is, the intensity of random behavior. Note that the environment influences this intensity of random behavior, the robot's behavior, and the robot–robot encounters. The more often robots approach each other, the more often they turn, which influences their distribution in space. That distribution is modeled here exclusively by a diffusion process. Hence, we need to find the parameter B experimentally, which is done here post hoc by merely fitting the model to the robot experiments. Although fitting a free parameter of course helps to get a better agreement between model and experiment, the results given in Fig. 9.8 show still an unexpectedly good match between the robot experiments and the Fokker–Planck approach.

Initially, robots aggregate at the dimmed right light. After 180 s, the left light is turned on to bright mode, gathers quickly aggregated robots, and the cluster at the right light is reduced. Then the right light is increased to bright mode, and the left light is diminished to dimmed mode. As expected, the right light gathers additional robots again, and the left light loses robots. Finally, starting from $t = 540$, more robots gather at the left light, and the cluster at the right light dissolves.

The model represents all dynamics qualitatively correctly, and the quantitative agreement is also good. Note that the free parameter B basically influences the speed of the diffusion process but has a limited influence on these qualitative aspects. For more details about the spatial predictions of the model, see, for example, Correll and Hamann [153] and Schmickl et al. [713].

In an actual application, we can use the model, for example, to refine the waiting time function $w(\mathbf{R})$. We may have specific requirements, such as a quick cluster formation or a clear distinction between cluster sizes at dimmed lights versus bright lights. We can use the model to test quickly many different setups, quantify their performance automatically, and pick appropriate parameters. Also, different temperatures or light fields can be tested. The results gained by the model, however, need to be verified in simulations and eventually in hardware experiments. The model may be of high quality, but it is still an abstraction that can, in principle, not represent all real situations.

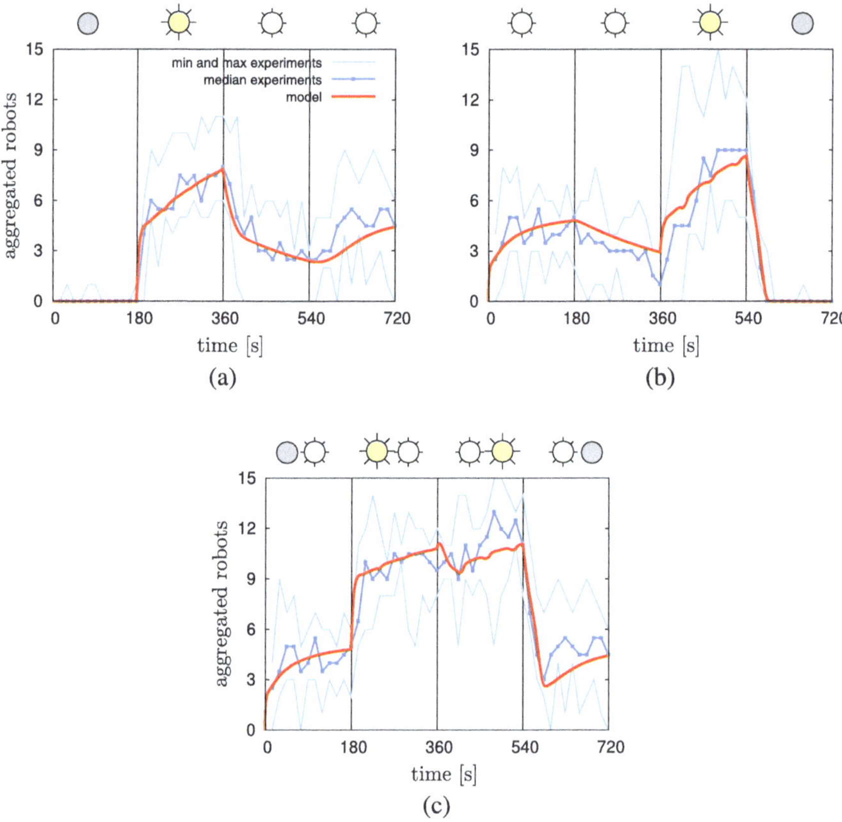

Fig. 9.8 Number of aggregated robots beneath the left, the right, and any lamp. The blue lines give results (min., median, and max.) of six independent robot experiments with 15 Jasmine robots. The red line gives the result of a Fokker–Planck model given in Eq. (9.8) [318, 713]. **a** Aggregated robots at the left light, **b** Aggregated robots at the right light, **c** Sum of aggregated robots at both lights

9.7 Short Summary

In this chapter, we tried to go through the whole swarm robot design process in one effort. We started from the specification by our customer, and some bio-inspiration luckily helped us in defining a practical algorithm. We have found a control algorithm that makes the swarm adaptive to a dynamic environment. We have developed a spatial model that can model where robots tend to aggregate depending on environmental features and parameters of the algorithm. The model is in good agreement with robot experiments, and we could use it to optimize the waiting time function and the stopping rate. So let's hope our customer is satisfied.

9.8 Further Reading

Where to go from here? Actually, I'm convinced that you can make your way alone from here. Feel free to explore the rich swarm robotics literature for yourself. Still, here are some hints about literature directly related to this chapter's content. The literature on BEECLUST is big and versatile. Starting from Schmickl and Hamann [710], you can read the more recent works by Arvin et al. [34, 35] and Wahby [838] as well as Sadeghi Amjadi et al. [691]. Sion et al. [734] study how an informed minority can influence aggregation in robot swarms. There are also interesting studies of several interacting swarms by Meister [543] and Bodi [84, 85]. Wahby et al. [840] propose a mechanism that adapts the waiting times online to changing swarm densities. There are many other interesting works on aggregation in robots and animals: Trianni et al. [788], Soysal and Şahin [741], Na et al. [586], and Garnier et al. [270].

9.9 Exercises

Exercise 9.1 (*Your own BEECLUST experiments*) Design and implement your own BEECLUST study. Use your favorite robot hardware or work in simulation only. You can try to reproduce the scenario with dynamic light settings or experiment with a higher number of potential aggregation sites. What about testing the robustness of BEECLUST against adversarial robots? You can also be creative and follow your own ideas.

Exercise 9.2 (*Stopping rate φ*) Design and implement an experiment to measure the actual stopping rate φ in BEECLUST. You could start by estimating it based on geometric arguments (e.g., the area covered by proximity sensors), probabilistic arguments (e.g., sensor model and the probability of false negatives), and collision theory. Initiate your empirical test, gather the data, statistically analyze it, and compare it to your theoretical estimations.

Exercise 9.3 (*Parameter sensitivity analysis*) Investigate how sensitive the aggregation behavior is to variations in the waiting time function w. Start from parameterizing the function differently by systematically varying the parameters $w_{\max}$ and c. Then also implement different functional forms (e.g., linear, exponential). Quantify the resulting aggregation performance (e.g., speed of aggregation, adaptability to dynamic conditions). Compare your findings to the original BEECLUST formulation and discuss which parameterizations yield the best and most robust outcomes.

Exercise 9.4 (*Symmetry breaking*) Consider a setup with two equally attractive aggregation sites (i.e., identical light sources). Simulate the swarm using BEECLUST and analyze the conditions under which the swarm converges to a single site versus splitting between two sites. Vary swarm size and noise levels in perception or motion. Discuss the role of stochastic fluctuations and positive feedback in enabling symmetry breaking and reflect on parallels to collective decision-making.

Exercise 9.5 (*Swarm density variation*) Study the effect of swarm density on the success of aggregation. Run simulations with varying numbers of robots in the same arena size, ranging from very sparse to highly crowded conditions. Discuss how density-dependent effects may influence the robustness and scalability of BEECLUST.

Epilogue

Reality is that which, when you stop believing in it, doesn't go away.

—*Philip K. Dick,*

How to Build a Universe That Doesn't Fall Apart Two Days Later

The future is already here—it's just not evenly distributed.

—*William Gibson, Neuromancer*

The absolute joy of discovery isn't in what we've already figured out, but in the questions that are still open. I hope that this book has sparked your curiosity and inspired you to try out your own ideas. Although years have passed between the first and second editions, swarm robotics remains a young and rapidly evolving field with much still to be discovered. In the past eight years, it has advanced quickly. With the rise of robotics, we can expect more implementations and real-world applications of swarm systems. It is an exciting time, full of promise and uncertainty. We have no idea what we are capable of, and therefore, we cannot know what we will become. Swarm robotics has reached a point where theory and practice are converging. The next decade will bring new scientific insights and real deployments in the air, in water, and on the ground. The challenge is no longer if swarm systems can work, but how to design them responsibly, robustly, and at scale. These questions will define the next decade, and I have dedicated my career to exploring them. I hope this book has given you both a foundation and inspiration to think bigger: about what swarms can do, the limits they face, and how we might shape their future.

H. Hamann, *Swarm Robotics*,
https://doi.org/10.1007/978-3-032-10584-4

References

1. Feasibility of a mars multi-rover mission. Technical Report JPL 760-160, Jet Propulsion Laboratory, California Institute of Technology, Pasadena, CA, February 1977. Technical Report, February 28
2. R. Abbott, Emergence, entities, entropy, and binding forces, in *The Agent 2004 Conference on: Social Dynamics: Interaction, Reflexivity, and Emergence*. (Argonne National Labs and University of Chicago, 2004)
3. H. Abelson, D. Allen, D. Coore, C. Hanson, G. Homsy, T. Knight, R. Nagpal, E. Rauch, G. Sussman, R. Weiss, Amorphous computing. Commun. ACM **43**(5), 74–82 (2000)
4. A. Adamatzky, *Physarum Machines: Computers from Slime Mould* (World Scientific, 2010)
5. J. Aguilar, D. Monaenkova, V. Linevich, W. Savoie, B. Dutta, H.-S. Kuan, M.D. Betterton, M.A.D. Goodisman, D.I. Goldman, Collective clog control: optimizing traffic flow in confined biological and robophysical excavation. Science **361**(6403), 672–677 (2018)
6. G. Aguzzi, R. Casadei, M. Viroli, Macroswarm: a field-based compositional framework for swarm programming, in *International Conference on Coordination Languages and Models*. (Springer, Berlin, 2023), pp. 31–51
7. N. Akhtar, O. Ozkasap, S.C. Ergen, Vanet topology characteristics under realistic mobility and channel models, in *Wireless Communications and Networking Conference (WCNC)* (IEEE, 2013), pp. 1774–1779
8. S.V. Albrecht, F. Christianos, L. Schäfer, *Multi-agent Reinforcement Learning: Foundations and Modern Approaches* (MIT Press, 2024)
9. M. Alhafnawi, E.R. Hunt, S. Lemaignan, P. O'Dowd, S. Hauert, MOSAIX: a swarm of robot tiles for social human-swarm interaction, in *International Conference on Robotics and Automation (ICRA)* (IEEE, 2022), pp. 6882–6888
10. M. Alhafnawi, M. Gomez-Gutierrez, E.R. Hunt, S. Lemaignan, P. O'Dowd, S. Hauert, Express yourself: enabling large-scale public events involving multi-human-swarm interaction for social applications with MOSAIX (2024). arXiv:2411.09975
11. K. Alharthi, Z.S. Abdallah, S. Hauert, Automatic extraction of understandable controllers from video observations of swarm behaviors, in *International Conference on Swarm Intelligence (ANTS)* (Springer, 2022), pp. 41–53

H. Hamann, *Swarm Robotics*,
https://doi.org/10.1007/978-3-032-10584-4

12. K. Alharthi, Z.S. Abdallah, S. Hauert, Ghost swarms: Learning swarm rules from environmental changes alone, in *Genetic Programming: 28th European Conference (EuroGP)* (Springer, Berlin, 2025), pp. 1–17. ISBN 978-3-031-89990-4. https://doi.org/10.1007/978-3-031-89991-1_1
13. W.C. Allee, Co-operation among animals. Am. J. Sociol. **37**(3), 386–398 (1931)
14. M. Allwright, N. Bhalla, H. El-faham, A. Antoun, C. Pinciroli, M. Dorigo, SRoCS: leveraging stigmergy on a multi-robot construction platform for unknown environments, in *International Conference on Swarm Intelligence* (Springer, Cham, 2014), pp. 158–169
15. M. Allwright, N. Bhalla, M. Dorigo, Structure and markings as stimuli for autonomous construction, in *2017 18th International Conference on Advanced Robotics (ICAR)* (IEEE, 2017), pp. 296–302
16. M. Allwright, W. Zhu, M. Dorigo, An open-source multi-robot construction system. HardwareX **5**, e00050 (2019). ISSN 2468-0672. https://doi.org/10.1016/j.ohx.2018.e00050, https://www.sciencedirect.com/science/article/pii/S2468067218300786
17. A. Almansoori, M. Alkilabi, E. Tuci, On the evolution of adaptable and scalable mechanisms for collective decision-making in a swarm of robots. Swarm Intell. **18**(1), 79–99 (2024)
18. J. Altmann, D. Suter, Survey of the status of small and very small missiles. Technical Report Report No. 2, Chair Experimental Physics III, TU Dortmund University, February 2022. Preventive Arms Control for Small and Very Small Armed Aircraft and Missiles
19. J. Altmann, D. Suter, Small and very small armed aircraft and missiles: trends in technology and preventive arms control. Technical Report Report No. 3, Chair Experimental Physics III, TU Dortmund University, March 2023. Preventive Arms Control for Small and Very Small Armed Aircraft and Missiles
20. J. Altmann, M. Pilch, D. Suter, Small armed aircraft and missiles: dangers for international security. Die Friedens-Warte **95**(3–4), 316–339 (2022). ISSN 0340-0255. https://doi.org/10.35998/fw-2022-0014, https://biblioscout.net/article/10.35998/fw-2022-0014
21. G.M. Amdahl, Validity of the single processor approach to achieving large scale computing capabilities, in *AFIPS Conference Proceedings* (ACM, 1967), pp. 483–485
22. J.-M. Ame, C. Rivault, J.-L. Deneubourg, Cockroach aggregation based on strain odour recognition. Anim. Behav. **68**, 793–801 (2004)
23. C. Anderson, J.J. Boomsma, J.J. Bartholdi, Task partitioning in insect societies: bucket brigades. Insectes Sociaux **49**, 171–180 (2002a)
24. C. Anderson, G. Theraulaz, J.-L. Deneubourg, Self-assemblages in insect societies. Insectes Soc. **49**(2), 99–110 (2002)
25. P.W. Anderson, More is different. Science **177**(4047), 393–396 (1972)
26. J.D. Arbuckle, *Self-assembly and Self-repair by Robot Swarms* (University of Southern California, 2007)
27. D.J. Arbuckle, A.A.G. Requicha, Self-assembly and self-repair of arbitrary shapes by a swarm of reactive robots: algorithms and simulations. Auton. Robots **28**(2), 197–211 (2010). ISSN 1573-7527. https://doi.org/10.1007/s10514-009-9162-7
28. R.C. Arkin, *Behavior-Based Robotics* (MIT Press, 1998)
29. R.C. Arkin, M. Egerstedt, Temporal heterogeneity and the value of slowness in robotic systems, in *2015 IEEE International Conference on Robotics and Biomimetics (ROBIO)* (IEEE, 2015), pp. 1000–1005
30. S. Arseneault, D. Vielfaure, G. Beltrame, RASS: Risk-aware swarm storage, in *Proceedings of the 21st International Conference on Autonomous Agents and Multiagent Systems (AAMAS)* (2022), pp. 1521–1523
31. W.B. Arthur, Competing technologies, increasing returns, and lock-in by historical events. Econ. J. **99**(394), 116–131 (1989)
32. F. Arvin, A.E. Turgut, S. Yue, Fuzzy-based aggregation with a mobile robot swarm, in *Swarm Intelligence (ANTS'12)*, Lecture Notes in Computer Science, vol. 7461 (Springer, Berlin, 2012), pp. 346–347. ISBN 978-3-642-32649-3. https://doi.org/10.1007/978-3-642-32650-9_39

33. F. Arvin, J. Murray, C. Zhang, S. Yue, Colias: an autonomous micro robot for swarm robotic applications. Int. J. Adv. Robot. Syst. **11**(7), 113 (2014)
34. F. Arvin, A.E. Turgut, F. Bazyari, K.B. Arikan, N. Bellotto, S. Yue, Cue-based aggregation with a mobile robot swarm: a novel fuzzy-based method. Adapt. Behav. **22**(3), 189–206 (2014b)
35. F. Arvin, A.E. Turgut, T. Krajník, S. Yue, Investigation of cue-based aggregation in static and dynamic environments with a mobile robot swarm. Adapt. Behav. **24**(2), 102–118 (2016). https://doi.org/10.1177/1059712316632851
36. F. Arvin, J. Espinosa, B. Bird, A. West, S. Watson, B. Lennox, Mona: an affordable open-source mobile robot for education and research. J. Intell. Robot. Syst. **94**, 761–775 (2019)
37. M. Asadi, A. Nowé, J. Ghofrani, Congestion-aware multi-agent path planning for pick-up and delivery tasks, in *Proceedings of the Genetic and Evolutionary Computation Conference (GECCO)* (2025), pp. 1523–1531
38. S. Atakishiyev, M. Salameh, H. Yao, R. Goebel, Explainable artificial intelligence for autonomous driving: a comprehensive overview and field guide for future research directions. IEEE Access (2024)
39. F. Augugliaro, S. Lupashin, M. Hamer, C. Male, M. Hehn, M.W. Mueller, J.S. Willmann, F. Gramazio, M. Kohler, R. D'Andrea, The flight assembled architecture installation: cooperative construction with flying machines. IEEE Control Syst. **34**(4), 46–64 (2014)
40. T. Aust, M.S. Talamali, M. Dorigo, H. Hamann, A. Reina, The hidden benefits of limited communication and slow sensing in collective monitoring of dynamic environments, in *International Conference on Swarm Intelligence* (Springer, Berlin, 2022), pp. 234–247
41. R. Axelrod, The dissemination of culture: a model with local convergence and global polarization. J. Conflict Resolut. **41**(2), 203–226 (1997)
42. T. Baca, M. Petrlik, M. Vrba, V. Spurny, R. Penicka, D. Hert, M. Saska, The MRS UAV system: pushing the frontiers of reproducible research, real-world deployment, and education with autonomous unmanned aerial vehicles. J. Intell. Robot. Syst. **102**(1), 26 (2021)
43. J. Bachrach, J. Beal, Programming a sensor network as an amorphous medium, in *Distributed Computing in Sensor Systems (DCOSS'06, extended abstract)* (2006)
44. J. Bachrach, J. Beal, J. McLurkin, Composable continuous-space programs for robotic swarms. Neural Comput. Appl. **19**, 825–847 (2010)
45. Y. Bai, P.T.T. Ngoc, H.D. Nguyen, D.L. Le, Q.H. Ha, K. Kai, Y.X.S. To, Y. Deng, J. Song, N. Wakamiya et al., Swarm navigation of cyborg-insects in unknown obstructed soft terrain. Nat. Commun. **16**(1), 221 (2025)
46. T. Bailey, H. Durrant-Whyte, Simultaneous localization and mapping (SLAM): Part II. IEEE Robot. Autom. Mag. **13**(3), 108–117 (2006). https://doi.org/10.1109/MRA.2006.1678144
47. P. Bak, *How Nature Works: The Science of Self-organized Criticality* (Springer Science & Business Media, 2013)
48. P. Bak, C. Tang, K. Wiesenfeld, Self-organized criticality. Phys. Rev. A **38**(1), 364–374 (1988). https://doi.org/10.1103/PhysRevA.38.364
49. B. Balázs, T. Vicsek, G. Somorjai, T. Nepusz, G. Vásárhelyi, Decentralized traffic management of autonomous drones. Swarm Intell. **19**, 29–53 (2024)
50. T. Balch, Hierarchic social entropy: an information theoretic measure of robot group diversity. Auton. Robots **8**(3), 209–238 (2000). ISSN 0929-5593
51. P. Ball, Forging patterns and making waves from biology to geology: a commentary on Turing (1952) 'the chemical basis of morphogenesis'. Philos. Trans. R. Soc. Lond. B Biol. Sci. **370**(1666) (2015). ISSN 0962-8436. https://doi.org/10.1098/rstb.2014.0218
52. M. Ballerini, N. Cabibbo, R. Candelier, A. Cavagna, E. Cisbani, I. Giardina, V. Lecomte, A. Orlandi, G. Parisi, A. Procaccini, M. Viale, V. Zdravkovic, Interaction ruling animal collective behavior depends on topological rather than metric distance: evidence from a field study. Proc. Natl. Acad. Sci. **105**(4), 1232–1237 (2008)
53. F. Baluška, M. Levin, On having no head: cognition throughout biological systems. Front. Psychol. **7**, 902 (2016). ISSN 1664-1078. https://doi.org/10.3389/fpsyg.2016.00902, https://www.frontiersin.org/article/10.3389/fpsyg.2016.00902

54. P. Baran, Reliable digital communications systems using unreliable network repeater nodes. Technical report, The RAND Corporation, Santa Monica, CA, 1960
55. A. Barciś, C. Bettstetter, Sandsbots: robots that sync and swarm. IEEE Access **8**, 218752–218764 (2020)
56. A. Barciś, M. Barciś, C. Bettstetter, Robots that sync and swarm: a proof of concept in ROS 2, in *International Symposium on Multi-Robot and Multi-Agent Systems (MRS)* (IEEE, 2019), pp. 98–104
57. J. Barrow-Green, *Poincaré and the Three Body Problem* (London Mathematical Society, American Mathematical Society, London, 1997)
58. P. Bartashevich, L. Knopf, P. Romanczuk, Transient milling dynamics in collective motion with visual occlusions, in *International Conference on Simulation of Adaptive Behavior* (Springer, Berlin, 2024), pp. 157–168
59. F.M. Bass, A new product growth for model consumer durables. Manag. Sci. **15**(5), 215–27 (1969)
60. R. Bastien, P. Romanczuk, A model of collective behavior based purely on vision. Sci. Adv. **6**(6), eaay0792 (2020)
61. L. Bauersfeld, K. Muller, D. Ziegler, F. Coletti, D. Scaramuzza, Robotics meets fluid dynamics: a characterization of the induced airflow below a quadrotor as a turbulent jet. IEEE Robot. Autom. Lett. (2024)
62. L. Bayindir, A review of swarm robotics tasks. Neurocomputing **172**(C), 292–321 (2015). http://dx.doi.org/10.1016/j.neucom.2015.05.116
63. L. Bayindir, E. Şahin, A review of studies in swarm robotics. Turk. J. Electr. Eng. Comput. Sci. **15**, 115–147 (2007). http://journals.tubitak.gov.tr/elektrik/issues/elk-07-15-2/elk-15-2-2-0705-13.pdf
64. G. Beauchamp, E. Fernández-Juricic, The group-size paradox: effects of learning and patch departure rules. Behav. Ecol. **16**(2), 352–357 (2004). ISSN 1045-2249. https://doi.org/10.1093/beheco/arh169
65. R. Beckers, O.E. Holland, J.-L. Deneubourg, From local actions to global tasks: stigmergy and collective robotics, in *Artificial Life IV* (1994), pp. 189–197
66. S. Beeby, N. White, *Energy Harvesting for Autonomous Systems* (Artech House, 2010)
67. G. Bekey, R. Ambrose, V. Kumar, D. Lavery, A. Sanderson, B. Wilcox, J. Yuh, Y. Zheng, *Robotics: State of the Art and Future Challenges* (World Scientific, 2008)
68. I. Belykh, R. Jeter, V. Belykh, Foot force models of crowd dynamics on a wobbly bridge. Sci. Adv. **3**(11) (2017). https://doi.org/10.1126/sciadv.1701512
69. M. Ben-Ari, F. Mondada, *Elements of Robotics* (Springer Nature, 2017)
70. M.Y.B. Zion, Y. Caba, A. Modin, P.M. Chaikin, Cooperation in a fluid swarm of fuel-free micro-swimmers. Nat. Commun. **13**(1), 184 (2022)
71. M.Y.B. Zion, J. Fersula, N. Bredeche, O. Dauchot, Morphological computation and decentralized learning in a swarm of sterically interacting robots. Sci. Robot. **8**(75), eabo6140 (2023)
72. G. Beni, From swarm intelligence to swarm robotics, in *Swarm Robotics—SAB 2004 International Workshop*, ed. by E. Şahin, W.M. Spears. Lecture Notes in Computer Science, vol. 3342 (Springer, Santa Monica, CA, 2005), pp. 1–9. https://doi.org/10.1007/978-3-540-30552-1_1
73. M.R. Benjamin, H. Schmidt, P.M. Newman, J.J. Leonard, Nested autonomy for unmanned marine vehicles with MOOS-IvP. J. Field Robot. **27**(6), 834–875 (2010)
74. A. Berea, I. Cohen, M.R. D'Orsogna, K. Ghosh, N. Goldenfeld, C.J. Goodnight, G. Hampson, A.M. Hein, L.A. McGraw, A.D. Sarwate, J. Urton, IDR team summary 6, in *Collective Behavior From Cells to Societies*. (The National Academies Press, Washington, D.C., 2014)
75. B.A. Berg, A. Billoire, Markov chain Monte Carlo simulations, in *Wiley Encyclopedia of Computer Science and Engineering*. ed. by B.W. Wah (Wiley, New York, 2007). https://doi.org/10.1002/9780470050118.ecse696
76. S. Berman, Q. Lindsey, M.S. Sakar, V. Kumar, S.C. Pratt, Experimental study and modeling of group retrieval in ants as an approach to collective transport in swarm robotic systems. Proc. IEEE **99**(9), 1470–1481 (2011). ISSN 0018-9219. https://doi.org/10.1109/JPROC.2011.2111450

77. M. Bettini, R. Kortvelesy, J. Blumenkamp, A. Prorok, VMAS: a vectorized multi-agent simulator for collective robot learning, in *International Symposium on Distributed Autonomous Robotic Systems (DARS)* (Springer, 2022), pp. 42–56
78. M. Bhattacharya, M. Penica, E. O'Connell, M. Southern, M. Hayes, Human-in-loop: a review of smart manufacturing deployments. Systems **11**(1) (2023). ISSN 2079-8954. https://doi.org/10.3390/systems11010035, https://www.mdpi.com/2079-8954/11/1/35
79. T. Biancalani, L. Dyson, A.J. McKane, Noise-induced bistable states and their mean switching time in foraging colonies. Phys. Rev. Lett. **112**, 038101 (2014)
80. J.D. Bjerknes, A. Winfield, On fault-tolerance and scalability of swarm robotic systems, in *Distributed Autonomous Robotic Systems (DARS 2010)*, ed. by A. Martinoli, F. Mondada, N. Correll, G. Mermoud, M. Egerstedt, M. Ani Hsieh, L.E. Parker, K. Støy. Springer Tracts in Advanced Robotics, vol. 83 (Springer, Berlin), pp. 431–444. ISBN 978-3-642-32722-3. https://doi.org/10.1007/978-3-642-32723-0_31
81. J.D. Bjerknes, A. Winfield, C. Melhuish, An analysis of emergent taxis in a wireless connected swarm of mobile robots, in *IEEE Swarm Intelligence Symposium*. ed. by Y. Shi, M. Dorigo (IEEE Press, Los Alamitos, CA, 2007), pp.45–52
82. O. Bjurling, R. Granlund, J. Alfredson, M. Arvola, T. Ziemke, Drone swarms in forest firefighting: a local development case study of multi-level human-swarm interaction, in *Proceedings of the 11th Nordic Conference on Human-Computer Interaction: Shaping Experiences, Shaping Society* (2020)
83. J. Blumenkamp, S. Morad, J. Gielis, A. Prorok, CoViS-Net: A cooperative visual spatial foundation model for multi-robot applications (2024). arXiv:2405.01107
84. M. Bodi, R. Thenius, T. Schmickl, K. Crailsheim, How two cooperating robot swarms are affected by two conflictive aggregation spots, in *Advances in Artificial Life: Darwin Meets von Neumann (ECAL'09)*, Lecture Notes in Computer Science, vol. 5778 (Springer, Berlin, 2011a), pp. 367–374
85. M. Bodi, R. Thenius, M. Szopek, T. Schmickl, K. Crailsheim, Interaction of robot swarms using the honeybee-inspired control algorithm BEECLUST. Math. Comput. Model. Dyn. Syst. **18**, 87–101 (2011)
86. F. Boem, L. Sabattini, C. Secchi, Decentralized state estimation for heterogeneous multi-agent systems, in *2015 54th IEEE Conference on Decision and Control (CDC)* (IEEE, 2015), pp. 4121–4126
87. R. Bogacz, E. Brown, J. Moehlis, P. Holmes, J.D. Cohen, The physics of optimal decision making: a formal analysis of models of performance in two-alternative forced-choice tasks. Psychol. Rev. **113**(4), 700 (2006a)
88. R. Bogacz, E. Brown, J. Moehlis, P. Holmes, J.D. Cohen, The physics of optimal decision making: a formal analysis of models of performance in two-alternative forced-choice tasks. Psychol. Rev. **113**(4), 700 (2006b)
89. M. Boldrer, V. Krátký, V. Walter, M. Saska, Swarming in the wild: a distributed communication-less Lloyd-based algorithm dealing with uncertainties (2025). arXiv:2504.18840
90. E. Bonabeau, Predicting the unpredictable. Harvard Bus. Rev. **80**(3) (2002)
91. E. Bonabeau, M. Dorigo, G. Theraulaz, *Swarm Intelligence: From Natural to Artificial Systems* (Oxford University Press, New York, NY, 1999)
92. M. Bonani, V. Longchamp, S. Magnenat, P. Rétornaz, D. Burnier, G. Roulet, F. Vaussard, H. Bleuler, F. Mondada, The marXbot, a miniature mobile robot opening new perspectives for the collective-robotic research, in *IEEE/RSJ International Conference on Intelligent Robots and Systems (IROS)* (IEEE, 2010), pp. 4187–4193
93. J.C. Bongard, Evolutionary robotics. Commun. ACM **56**(8), 74–83 (2013). https://doi.org/10.1145/2493883
94. F. Bonnet, L. Cazenille, A. Gribovskiy, J. Halloy, F. Mondada, Multi-robot control and tracking framework for bio-hybrid systems with closed-loop interaction, in *IEEE International Conference on Robotics and Automation (ICRA)* (2017), pp. 4449–4456. https://doi.org/10.1109/ICRA.2017.7989515

95. D.M. Bossens, D. Tarapore, Rapidly adapting robot swarms with swarm map-based Bayesian optimisation, in *IEEE International Conference on Robotics and Automation (ICRA)* (IEEE, 2021), pp. 9848–9854
96. J.F. Boudet, J. Lintuvuori, C. Lacouture, T. Barois, A. Deblais, K. Xie, S. Cassagnere, B. Tregon, D.B. Brückner, J.C. Baret, H. Kellay, From collections of independent, mindless robots to flexible, mobile, and directional superstructures. Sci. Robot. **6**(56), eabd0272 (2021). https://doi.org/10.1126/scirobotics.abd0272
97. S. Boyd, A. Ghosh, B. Prabhakar, D. Shah, Randomized gossip algorithms. IEEE Trans. Inf. Theory **52**(6), 2508–2530 (2006)
98. D. Bozhinoski, M. Birattari, Towards an integrated automatic design process for robot swarms. Open Res. Eur. **1**, 112 (2022)
99. V. Braitenberg, *Vehicles: Experiments in Synthetic Psychology* (MIT Press, 1984)
100. M. Brambilla, *Formal methods for the design and analysis of robot swarms*. Ph.D. thesis, Université Libre de Bruxelles, 2014
101. M. Brambilla, C. Pinciroli, M. Birattari, M. Dorigo, A reliable distributed algorithm for group size estimation with minimal communication requirements, in *International Conference on Advanced Robotics (ICAR)* (2009)
102. M. Brambilla, E. Ferrante, M. Birattari, M. Dorigo, Swarm robotics: a review from the swarm engineering perspective. Swarm Intell. **7**(1), 1–41 (2013). ISSN 1935-3812. https://doi.org/10.1007/s11721-012-0075-2
103. M. Brand, M. Masuda, N. Wehner, X.-H. Yu, Ant colony optimization algorithm for robot path planning, in *International Conference on Computer Design and Applications* (IEEE, 2010), pp. V3–436
104. N. Bredeche, J.-M. Montanier, Environment-driven embodied evolution in a population of autonomous agents, in *International Conference on Parallel Problem Solving from Nature* (Springer, Berlin, 2010), pp. 290–299
105. N. Bredeche, E. Haasdijk, A. Prieto, Embodied evolution in collective robotics: a review. Front. Robot. AI **5**, 12 (2018)
106. C.M. Breder, Equations descriptive of fish schools and other animal aggregations. Ecology **35**(3), 361–370 (1954)
107. A. Bricard, J.-B. Caussin, N. Desreumaux, O. Dauchot, D. Bartolo, Emergence of macroscopic directed motion in populations of motile colloids. Nature **503**(7474), 95–98 (2013)
108. A. Brohan, N. Brown, J. Carbajal, Y. Chebotar, J. Dabis, C. Finn, K. Gopalakrishnan, K. Hausman, A. Herzog, J. Hsu, et al., RT-1: robotics transformer for real-world control at scale (2022). arXiv:2212.06817
109. R. Brooks, A robust layered control system for a mobile robot. IEEE J. Robot. Autom. **2**(1), 14–23 (1986)
110. R.A. Brooks, Intelligence without representation. Artif. Intell. **47** (1991)
111. J. Bruce, M. Dennis, A. Edwards, J. Parker-Holder, Y. (Jimmy) Shi, E. Hughes, M. Lai, A. Mavalankar, R. Steigerwald, C. Apps, Y. Aytar, S. Bechtle, F. Behbahani, S. Chan, N. Heess, L. Gonzalez, S. Osindero, S. Ozair, S. Reed, J. Zhang, K. Zolna, J. Clune, N. De Freitas, S. Singh, T. Rocktäschel, Genie: generative interactive environments, in *Proceedings of the 41st International Conference on Machine Learning* (ICML'24, 2024)
112. A. Brutschy, G. Pini, C. Pinciroli, M. Birattari, M. Dorigo, Self-organized task allocation to sequentially interdependent tasks in swarm robotics. Auton. Agents Multi-Agent Syst. **28**(1), 101–125 (2014). ISSN 1387-2532. https://doi.org/10.1007/s10458-012-9212-y
113. J. Buck, E. Buck, F. Case, F.E. Hanson, Control of flashing in fireflies. J. Comput. Physiol. A **144**, 287–298 (1981)
114. J. Buck, E. Buck, Biology of synchronous flashing of fireflies. Nature **211**, 562–564 (1966)
115. J. Buhl, D.J.T. Sumpter, I.D. Couzin, J.J. Hale, E. Despland, E.R. Miller, S.J. Simpson, From disorder to order in marching locusts. Science **312**(5778), 1402–1406 (2006). https://doi.org/10.1126/science.1125142
116. S.A. Burden, T. Libby, K. Jayaram, S. Sponberg, J.M. Donelan, Why animals can outrun robots. Sci. Robot. **9**(89), eadi9754 (2024). https://doi.org/10.1126/scirobotics.adi9754, https://www.science.org/doi/abs/10.1126/scirobotics.adi9754

117. E. Buss, T.-L. Rabbel, V. Horvat, M. Krizmancic, S. Bogdan, M. Wahby, H. Hamann, Phytonodes for environmental monitoring: stimulus classification based on natural plant signals in an interactive energy-efficient bio-hybrid system, in *Proceedings of the 2022 ACM Conference on Information Technology for Social Good GoodIT '22* (Association for Computing Machinery, New York, NY, USA, 2022), pp. 258–264. ISBN 9781450392846. https://doi.org/10.1145/3524458.3547266
118. S. Camazine, J.-L. Deneubourg, N.R. Franks, J. Sneyd, G. Theraulaz, E. Bonabeau, *Self-Organizing Biological Systems* (Princeton University Press, Princeton, NJ, 2001)
119. A. Campo, S. Garnier, O. Dédriche, M. Zekkri, M. Dorigo, Self-organized discrimination of resources. PLoS ONE **6**(5), e19888 (2011)
120. G. Caprari, P. Balmer, R. Piguet, R. Siegwart, The autonomous microbot 'alice': a platform for scientific and commercial applications, in *Proceedings of the Ninth International Symposium on Micromechatronics and Human Science* (Nagoya, Japan, 1998), pp. 231–235
121. G. Caprari, A. Colot, R. Siegwart, J. Halloy, J.-L. Deneubourg, Animal and robot mixed societies: building cooperation between microrobots and cockroaches. IEEE Robot. Autom. Mag. (2005). https://doi.org/10.1109/MRA.2005.1458325
122. R. Carli, A. Chiuso, L. Schenato, S. Zampieri, Distributed kalman filtering based on consensus strategies. IEEE J. Sel. Areas Commun. **26**(4), 622–633 (2008)
123. H. Carlsson, E. Van Damme, Global games and equilibrium selection. Econometrica: J. Econ. Soc. 989–1018 (1993)
124. J. Carneiro, K. Leon, Í. Caramalho, C. Van Den Dool, R. Gardner, V. Oliveira, M.-L. Bergman, N. Sepúlveda, T. Paixão, J. Faro et al., When three is not a crowd: a crossregulation model of the dynamics and repertoire selection of regulatory CD4+ T cells. Immunol. Rev. **216**(1), 48–68 (2007)
125. R.A. Carrasco, F. Núñez, A. Cipriano, Fault detection and isolation in cooperative mobile robots using multilayer architecture and dynamic observers. Robotica **29**(4), 555–562 (2011)
126. M. Casiulis, E. Arbel, Y. Lahini, S. Martiniani, N. Oppenheimer, M. Yah Ben Zion, A geometric condition for robot-swarm cohesion and cluster-flock transition (2024). arXiv:2409.04618
127. C. Castellano, S. Fortunato, V. Loreto, Statistical physics of social dynamics. Rev. Mod. Phys. **81**, 591–646 (2009)
128. E.C. Ferrer, The Blockchain: A new framework for robotic swarm systems (2016). arXiv:1608.00695
129. J.-B. Castillo-Sánchez, E. González-Parada, J.-M. Cano-García, Swarm robot communications in ROS 2: an experimental study. IEEE Access **12**, 142930–142943 (2024). https://doi.org/10.1109/ACCESS.2024.3470254
130. A. Cavagna, A. Cimarelli, I. Giardina, G. Parisi, R. Santagati, F. Stefanini, M. Viale, Scale-free correlations in starling flocks. Proc. Natl. Acad. Sci. **107**(26), 11865–11870 (2010)
131. A. Cavalcanti, R.A. Freitas Jr., Nanorobotics control design: a collective behavior approach for medicine. IEEE Trans. NanoBiosci. **4**(2), 133–140 (2005)
132. L. Cazenille, Y. Chemtob, F. Bonnet, A. Gribovskiy, F. Mondada, N. Bredeche, J. Halloy, How to blend a robot within a group of zebrafish: Achieving social acceptance through real-time calibration of a multi-level behavioural model, in *Biomimetic and Biohybrid Systems: 7th International Conference, Living Machines* (Springer, Berlin, 2018), pp. 73–84
133. S. Ceron, K. O'Keeffe, K. Petersen, Diverse behaviors in non-uniform chiral and non chiral swarmalators. Nat. Commun. **14**(1), 940 (2023)
134. A.A. Chafik, J. Gaber, S. Tayane, M. Ennaji, J. Bourgeois, T. El Ghazawi. From conventional to programmable matter systems: a review of design, materials, and technologies. ACM Comput. Surv. **56**(8), 1–26 (2024)
135. L. Chaimowicz, V. Kumar, Aerial shepherds: coordination among UAVs and swarms of robots, in *Distributed Autonomous Robotic Systems*, vol. 6 (Springer, Berlin, 2007), pp. 243–252
136. A. Chao, Nonparametric estimation of the number of classes in a population. Scand. J. Stat. 265–270 (1984)
137. B. Chazelle, An algorithmic approach to collective behavior. J. Stat. Phys. **158**(3), 514–548 (2015)

138. J. Chen, M. Gauci, M.J. Price, R. Groß, Segregation in swarms of e-puck robots based on the Brazil nut effect, in *Proceedings of the 11th International Conference on Autonomous Agents and Multiagent Systems (AAMAS 2012)* (Richland, SC, IFAAMAS, 2012), pp. 163–170
139. L. Chen, A. Avizienis, N-version programming: a fault-tolerance approach to reliability of software operation, in *Proceedings of the 8th IEEE International Symposium on Fault-Tolerant Computing*, vol. 1 (1978), pp. 3–9
140. X. Chen, A. Marjovi, J. Huang, A. Martinoli, Particle source localization with a low-cost robotic sensor system: algorithmic design and performance evaluation. IEEE Sens. J. **20**(21), 13074–13085 (2020)
141. J. Chiun, S. Zhang, Y. Wang, Y. Cao, G. Sartoretti, MARVEL: multi-agent reinforcement learning for constrained field-of-view multi-robot exploration in large-scale environments (2025). arXiv:2502.20217
142. A.L. Christensen, R. O'Grady, M. Dorigo, From fireflies to fault-tolerant swarms of robots. IEEE Trans. Evol. Comput. **13**(4), 754–766 (2009). ISSN 1089-778X. https://doi.org/10.1109/TEVC.2009.2017516
143. D.J. Christensen, U.P. Schultz, K. Stoy, A distributed and morphology-independent strategy for adaptive locomotion in self-reconfigurable modular robots. Robot. Auton. Syst. **61**(9), 1021–1035 (2013)
144. T.H. Chung, R. Daniel, DARPA OFFSET: a vision for advanced swarm systems through agile technology development and experimentation. Field Robot. **3**(1), 97–124 (2023)
145. P. Chvykov, T.A. Berrueta, A. Vardhan, W. Savoie, A. Samland, T.D. Murphey, K. Wiesenfeld, D.I. Goldman, J.L. England, Low rattling: a predictive principle for self-organization in active collectives. Science **371**(6524), 90–95 (2021)
146. F. Ciocchetta, J. Hillston, Bio-PEPA: a framework for the modelling and analysis of biological systems. Theor. Comput. Sci. **410**(33), 3065–3084 (2009)
147. P. Clifford, A. Sudbury, A model for spatial conflict. Biometrika **60**(3), 581–588 (1973)
148. M.E. Clynes, *Cyborgs and Space, Astronautics* (The Cyborg Handbook, 1960)
149. E.A. Codling, M.J. Plank, S. Benhamou, Random walk models in biology. J. R. Soc. Interface **5**(25), 813–834 (2008)
150. R.K. Colwell, J.A. Coddington, Estimating terrestrial biodiversity through extrapolation. Philos. Trans. R. Soc. Lond. Ser. B Biol. Sci. **345**(1311), 101–118 (1994)
151. M. Coppola, K.N. McGuire, C. De Wagter, G.C.H.E. De Croon, A survey on swarming with micro air vehicles: fundamental challenges and constraints. Front. Robot. AI **7**, 18 (2020)
152. C. Coquet, A. Arnold, P.-J. Bouvet, Control of a robotic swarm formation to track a dynamic target with communication constraints: analysis and simulation. Appl. Sci. **11**(7) (2021). ISSN 2076-3417. https://doi.org/10.3390/app11073179, https://www.mdpi.com/2076-3417/11/7/3179
153. N. Correll, H. Hamann, Probabilistic modeling of swarming systems, in *Springer Handbook of Computational Intelligence*. ed. by J. Kacprzyk, W. Pedrycz (Springer, Berlin, 2015), pp.1423–1431
154. N. Correll, M. Schwager, D. Rus, Social control of herd animals by integration of artificially controlled congeners, in *Proceedings of the 10th International Conference on Simulation of Adaptive Behavior: From Animals to Animats*, Lecture Notes in Computer Science, vol. 5040 (Springer, Berlin, 2008), pp. 437–446
155. N. Correll, B. Hayes, C. Heckman, A. Roncone. *Introduction to Autonomous Robots: Mechanisms, Sensors, Actuators, and Algorithms* (MIT Press, 2022)
156. A. Costanzo, C.K. Hemelrijk, Spontaneous emergence of milling (vortex state) in a Vicsek-like model. J. Phys. D Appl. Phys. **51**(13), 134004 (2018)
157. I.D. Couzin, N.R. Franks, Self-organized lane formation and optimized traffic flow in army ants. Proc. R. Soc. Lond. Ser. B Biol. Sci. **270**(1511), 139–146 (2003)
158. I.D. Couzin, J. Krause, R. James, G.D. Ruxton, N.R. Franks, Collective memory and spatial sorting in animal groups. J. Theor. Biol. **218**, 1–11 (2002). https://doi.org/10.1006/jtbi.2002.3065

159. I.D. Couzin, J. Krause, N.R. Franks, S.A. Levin, Effective leadership and decision-making in animal groups on the move. Nature **433**, 513–516 (2005)
160. I.D. Couzin, C.C. Ioannou, G. Demirel, T. Gross, C.J. Torney, A. Hartnett, L. Conradt, S.A. Levin, N.E. Leonard, Uninformed individuals promote democratic consensus in animal groups. Science **334**(6062), 1578–1580 (2011). ISSN 0036-8075. https://doi.org/10.1126/science.1210280, http://science.sciencemag.org/content/334/6062/1578
161. J.W. Crandall, M.A. Goodrich, D.R. Olsen, C.W. Nielsen, Validating human-robot interaction schemes in multitasking environments. IEEE Trans. Syst. Man Cybern. Part A Syst. Humans **35**(4), 438–449 (2005)
162. V. Crespi, A. Galstyan, K. Lerman, Top-down vs bottom-up methodologies in multi-agent system design. Auton. Robot. **24**(3), 303–313 (2008)
163. D. Crisp, A. Pathare, R.C. Ewell, The performance of gallium arsenide/germanium solar cells at the Martian surface. Acta Astronaut. **54**(2), 83–101 (2004)
164. J. Crutchfield, The calculi of emergence: computation, dynamics, and induction. Physica D **75**(1–3), 11–54 (1994)
165. J. Crutchfield, Is anything ever new?, in *Complexity: Metaphors, Models, and Reality*, ed. by G. Cowan, D. Pines, D. Melzner. SFI Series in the Sciences of Complexity proceedings, vol. 19 (Addison-Wesley, Reading, MA, 1994b), pp. 479–497
166. N.A. Cruz, J.C. Alves, Autonomous sailboats: an emerging technology for ocean sampling and surveillance, in *OCEANS 2008* (IEEE, 2008)
167. E. Şahin, Swarm robotics: from sources of inspiration to domains of application, in *Swarm Robotics—SAB 2004 International Workshop*, ed. by E. Şahin, W.M. Spears. Lecture Notes in Computer Science, vol. 3342 (Springer, Berlin, 2005), pp. 10–20
168. F. Cucker, S. Smale, Emergent behavior in flocks. IEEE Trans. Autom. Control **52**(5), 852–862 (2007)
169. A. Cully, J. Clune, D. Tarapore, J.-B. Mouret, Robots that can adapt like animals. Nature **521**, 503–507 (2014). https://doi.org/10.1038/nature14422
170. R. da Silva Guerra, H. Aonuma, K. Hosoda, M. Asada, Behavior change of crickets in a robot-mixed society. J. Robot. Mechatron. **22**, 526–531 (2010)
171. M.J. Daigle, X.D. Koutsoukos, G. Biswas, Distributed diagnosis in formations of mobile robots. IEEE Trans. Robot. **23**(2), 353–369 (2007)
172. V. Darley, Emergent phenomena and complexity, in *Artificial Life IV*. ed. by R. Brooks, P. Maes (1994), pp. 411–416
173. S. Das, M. Ciarchi, Z. Zhou, J. Yan, J. Zhang, R. Alert, Flocking by turning away. Phys. Rev. X **14**(3), 031008 (2024)
174. B. Davis, Reinforced random walk. Prob. Theory Relat. Fields **84**(2), 203–229 (1990)
175. M. De Marchi, N. Bombieri, Orchestration-aware optimization of ROS 2 communication protocols, in *Design, Automation and Test in Europe Conference and Exhibition (DATE)* (IEEE, 2024)
176. R. De Nicola, G.L. Ferrari, R. Pugliese, KLAIM: a kernel language for agents interaction and mobility. IEEE Trans. Softw. Eng. **24**(5), 315–330 (1998)
177. R. De Nicola, J. Katoen, D. Latella, M. Loreti, M. Massink, Model checking mobile stochastic logic. Theor. Comput. Sci. **382**(1), 42–70 (2007)
178. J. Degesys, I. Rose, A. Patel, R. Nagpal, DESYNC: self-organizing desynchronization and TDMA on wireless sensor networks, in *International Conference on Information Processing in Sensor Networks* (2017), pp. 11–20
179. J. Deguet, Y. Demazeau, L. Magnin, Elements about the emergence issue: A survey of emergence definitions. Complexus **3**(1–3), 24–31 (2006)
180. J. Delcourt, N.W.F. Bode, M. Denoël, Collective vortex behaviors: diversity, proximate, and ultimate causes of circular animal group movements. Q. Rev. Biol. **91**(1) (2016). https://doi.org/10.1086/685301
181. A. Demers, D. Greene, C. Hauser, W. Irish, J. Larson, S. Shenker, H. Sturgis, D. Swinehart, D. Terry, Epidemic algorithms for replicated database maintenance, in *Proceedings of the Sixth Annual ACM Symposium on Principles of Distributed Computing, PODC '87* (Association

for Computing Machinery, New York, NY, USA, 1987), pp. 1–12. ISBN 089791239X. https://doi.org/10.1145/41840.41841
182. J.-L. Deneubourg, J.-C. Gregoire, E. Le Fort, Kinetics of larval gregarious behavior in the bark beetle *Dendroctonus micans* (Coleoptera: Scolytidae). J.Insect Behav. **3**(2), 169–182 (1990)
183. J.-L. Deneubourg, A. Lioni, C. Detrain, Dynamics of aggregation and emergence of cooperation. Biol. Bull. **202**, 262–267 (2002)
184. B. Deng, Machine ethics: the robot's dilemma. Nature (2015). http://dx.doi.org/10.1038/523024a
185. J. Deseigne, O. Dauchot, H. Chaté, Collective motion of vibrated polar disks. Phys. Rev. Lett. **105**, 098001 (2010)
186. M.R. Devlin, S. Kim, O. Campàs, E.W. Hawkes, Material-like robotic collectives with spatiotemporal control of strength and shape. Science **387**(6736), 880–885 (2025)
187. B. Di Lorenzo, G.C. Maffettone, M. di Bernardo, A continuification-based control solution for large-scale shepherding (2024). arXiv:2411.04791
188. C. Dimidov, G. Oriolo, V. Trianni, Random walks in swarm robotics: an experiment with Kilobots, in *Swarm Intelligence (ANTS 2016)* (Springer, Berlin, Heidelberg, 2016), pp. 185–196. ISBN 978-3-319-44427-7. https://doi.org/10.1007/978-3-319-44427-7_16
189. C.A. Dimmig, G. Silano, K. McGuire, C. Gabellieri, W. Hšnig, J. Moore, M. Kobilarov, An overview and in-depth systematic comparisons. IEEE Robot. Autom. Mag. Surv. Simul. Aerial Robots (2024)
190. H. Ding, H. Hamann, Sorting in swarm robots using communication-based cluster size estimation, in *Ninth International Conference on Swarm Intelligence (ANTS 2014)*, ed. by M. Dorigo, M. Birattari, S. Garnier, H. Hamann, M.M. de Oca, C. Solnon, T. Stützle. Lecture Notes in Computer Science, vol. 8667 (Springer, Berlin, 2014), pp. 262–269
191. J.P. Distefano, S. Chowdhury, E. Esfahani, Exploring human-swarm interaction dynamics in cyber-physical systems: a physiological approach. J. Integr. Design Process Sci. 10920617241292155 (2024)
192. M.D. Soorati, H. Hamann, Robot self-assembly as adaptive growth process: collective selection of seed position and self-organizing tree-structures, in *IEEE/RSJ International Conference on Intelligent Robots and Systems (IROS)* (IEEE, 2016), pp. 5745–5750. https://doi.org/10.1109/IROS.2016.7759845
193. M.D. Soorati, M.K. Heinrich, J. Ghofrani, P. Zahadat, H. Hamann, Photomorphogenesis for robot self-assembly: adaptivity, collective decision-making, and self-repair. Bioinspiration Biomimetics **14**(5), 056006 (2019). https://doi.org/10.1088/1748-3190/ab2958
194. M.D. Soorati, J. Clark, J. Ghofrani, D. Tarapore, S.D. Ramchurn, Designing a user-centered interaction interface for human–swarm teaming. Drones **5**(4), 131 (2021)
195. C. Dixon, A. Winfield, M. Fisher, Towards temporal verification of emergent behaviours in swarm robotic systems, in *Towards Autonomous Robotic Systems (TAROS)* (2011), pp. 336–347
196. M. Dorigo, D. Floreano, L.M. Gambardella, F. Mondada, S. Nolfi, T. Baaboura, M. Birattari, M. Bonani, M. Brambilla, A. Brutschy, D. Burnier, A. Campo, A.L. Christensen, A. Decugnière, G. Di Caro, F. Ducatelle, E. Ferrante, A. Förster, J. Guzzi, V. Longchamp, S. Magnenat, J. Martinez Gonzales, N. Mathews, M. Montes de Oca, R. O'Grady, C. Pinciroli, G. Pini, P. Rétornaz, J. Roberts, V. Sperati, T. Stirling, A. Stranieri, T. Stützle, V. Trianni, E. Tuci, A.E. Turgut, F. Vaussard, Swarmanoid: a novel concept for the study of heterogeneous robotic swarms. IEEE Robot. Autom. Mag. **20**(4), 60–71 (2013)
197. M. Dorigo, E. Şahin, Guest editorial: swarm robotics. Auton. Robot. **17**(2–3), 111–113 (2004)
198. M. Dorigo, T. Stützle, *Ant Colony Optimization* (The MIT Press, 2004). ISBN 0-262-04219-3
199. M. Dorigo, V. Maniezzo, A. Colorni, Ant system: optimization by a colony of cooperating agents. IEEE Trans. Syst. Man Cybern. Part B (Cybernetics) **26**(1), 29–41 (1996)
200. M. Dorigo, E. Bonabeau, G. Theraulaz, Ant algorithms and stigmergy. Futur. Gener. Comput. Syst. **16**(9), 851–871 (2000)

201. M. Dorigo, V. Trianni, E. Sahin, R. Groß, T.H. Labella, G. Baldassarre, S. Nolfi, F. Mondada, J.-L. Deneubourg, D. Floreano, L.M. Gambardella, Evolving self-organizing behaviors for a swarm-bot. Auton. Robots **17**, 223–245 (2004). https://doi.org/10.1023/B:AURO.0000033972.50769.1c, http://www.swarm-bots.org
202. M. Dorigo, E. Tuci, V. Trianni, R. Groß, S. Nouyan, C. Ampatzis, T.H. Labella, R. O'Grady, M. Bonani, F. Mondada, SWARM-BOT: Design and implementation of colonies of self-assembling robots, in *Computational Intelligence: Principles and Practice*. ed. by G.Y. Yen, D.B. Fogel (IEEE Press, Los Alamitos, CA, 2006), pp. 103–135
203. M. Dorigo, M. Birattari, M. Brambilla, Swarm robotics. Scholarpedia **9**(1), 1463 (2014). http://dx.doi.org/10.4249/scholarpedia.1463
204. M. Dorigo, G. Theraulaz, V. Trianni, Swarm robotics: past, present, and future [point of view]. Proc. IEEE **109**(7), 1152–1165 (2021)
205. M. Dorigo, A. Pacheco, A. Reina, V. Strobel, Blockchain technology for mobile multi-robot systems. Nat. Rev. Electr. Eng. **1**(4), 264–274 (2024)
206. Y.T. dos Passos, X. Duquesne, L.S. Marcolino, Congestion control algorithms for robotic swarms with a common target based on the throughput of the target area. Robot. Auton. Syst. **159**, 104284 (2023)
207. G.Y. Dosieah, A. Özdemir, M. Gauci, R. Groß, Moving mixtures of active and passive elements with robots that do not compute, in *International Conference on Swarm Intelligence (ANTS)* (Springer, Berlin, 2022), pp. 183–195
208. I. Douven, A. Riegler, Extending the Hegselmann-Krause model I. Log. J. IGPL **18**(2), 323–335 (2009)
209. D. Qiang, V. Faber, M. Gunzburger, Centroidal Voronoi tessellations: applications and algorithms. SIAM Rev. **41**(4), 637–676 (1999)
210. M. Duarte, V. Costa, J. Gomes, T. Rodrigues, F. Silva, S.M. Oliveira, A.L. Christensen, Evolution of collective behaviors for a real swarm of aquatic surface robots. PLOS ONE **11**(3), 1–25 (2016a). https://doi.org/10.1371/journal.pone.0151834
211. M. Duarte, V. Costa, J. Gomes, T. Rodrigues, F. Silva, S.M. Oliveira, A.L. Christensen, Unleashing the potential of evolutionary swarm robotics in the real world, in *Proceedings of the 2016 on Genetic and Evolutionary Computation Conference Companion, GECCO '16 Companion* (ACM, New York, NY, USA, 2016b), pp. 159–160. ISBN 978-1-4503-4323-7. https://doi.org/10.1145/2908961.2930951
212. F. Ducatelle, G.A. Di Caro, L.M. Gambardella, Cooperative self-organization in a heterogeneous swarm robotic system, in *Proceedings of the 12th Conference on Genetic and Evolutionary Computation (GECCO)* (ACM, 2010), pp. 87–94
213. G. Dudek, M. Jenkin, E. Milios, D. Wilkes, A taxonomy for swarm robots, in *Proceedings of IEEE/RSJ International Conference on Intelligent Robots and Systems (IROS)*, vol. 1 (IEEE, 1993) pp. 441–447
214. J.W. Durham, R. Carli, F. Bullo, Pairwise optimal discrete coverage control for gossiping robots, in *49th IEEE Conference on Decision and Control (CDC)* (IEEE, 2010), pp. 7286–7291
215. A. Dussutour, V. Fourcassié, D. Helbing, J.-L. Deneubourg, Optimal traffic organization in ants under crowded conditions. Nature **428**, 70–73 (2004)
216. A. Dussutour, M. Beekman, S.C. Nicolis, B. Meyer, Noise improves collective decision-making by ants in dynamic environments. Proc. R. Soc. Lond. B **276**, 4353–4361 (2009)
217. F. Dyson, A meeting with Enrico Fermi. Nature **427**, 297 (2004). https://doi.org/10.1038/427297a
218. J.T. Ebert, M. Gauci, F. Mallmann-Trenn, R. Nagpal, Bayes bots: collective Bayesian decision-making in decentralized robot swarms, in *IEEE International Conference on Robotics and Automation (ICRA)* (IEEE, 2020), pp. 7186–7192
219. A.M. Edwards, R.A. Phillips, N.W. Watkins, M.P. Freeman, E.J. Murphy, V. Afanasyev, S.V. Buldyrev, M.G.E. da Luz, E.P. Raposo, H.E. Stanley, et al., Revisiting Lévy flight search patterns of wandering albatrosses, bumblebees and deer. Nature **449**(7165), 1044–1048 (2007)

220. R. Eftimie, Hyperbolic and kinetic models for self-organized biological aggregations and movement: a brief review. J. Math. Biol. **65**(1), 35–75 (2012)
221. M. Egerstedt, *Robot Ecology: Constraint-Based Design for Long-Duration Autonomy* (Princeton University Press, 2021)
222. P. Ehrenfest, T. Ehrenfest, Über zwei bekannte Einwände gegen das Boltzmannsche H-Theorem. Physikalische Zeitschrift **8**, 311–314 (1907)
223. Á.E. Eiben, J.E. Smith, *Introduction to Evolutionary Computing*. Natural Computing Series (Springer, 2003)
224. M. Eigen, P. Schuster, A principle of natural self-organization. Naturwissenschaften **64**(11), 541–565 (1977). ISSN 0028-1042. https://doi.org/10.1007/BF00450633
225. M. Eigen, P. Schuster, *The Hypercycle: A Principle of Natural Self Organization* (Springer, 1979)
226. M. Eigen, R. Winkler, *Laws of the Game: How the Principles of Nature Govern Chance* (Princeton University Press, Princeton, NJ, 1993). ISBN: 978-0-691-02566-7
227. A. Einstein, Über die von der molekularkinetischen Theorie der Wärme geforderte Bewegung von in ruhenden Flüssigkeiten suspendierten Teilchen. Ann. Phys. **17**, 549–560 (1905)
228. H. El-Fiqi, B. Campbell, S. Elsayed, A. Perry, H.K. Singh, R. Hunjet, H.A. Abbass, The limits of reactive shepherding approaches for swarm guidance. IEEE Access **8**, 214658–214671 (2020). https://doi.org/10.1109/ACCESS.2020.3037325
229. J.E. Schmitz, S. Rahmann, A comprehensive review and evaluation of species richness estimation. Briefings Bioinform. **26**(2), bbaf158 (2025). ISSN 1477-4054. https://doi.org/10.1093/bib/bbaf158
230. W. Elmenreich, B. Heiden, G. Reiner, S. Zhevzhyk, A low-cost robot for multi-robot experiments, in *12th International Workshop on Intelligent Solutions in Embedded Systems (WISES)* (IEEE, 2015), pp. 127–132
231. M.R. Endsley, E.O. Kiris, The out-of-the-loop performance problem and level of control in automation. Hum. Factors **37**(2), 381–394 (1995)
232. M.D. Erbas, L. Bull, A.F.T. Winfield, On the evolution of behaviors through embodied imitation. Artif. Life **21**(2), 141–165 (2015)
233. P. Erdős, A. Rényi, On random graphs. Publ. Math. Debrecen **6**(290–297), 156 (1959)
234. U. Erdmann, W. Ebeling, L. Schimansky-Geier, F. Schweitzer, Brownian particles far from equilibrium. Eur. Phys. J. B Condens. Matter Complex Syst. **15**(1), 105–113 (2000)
235. C. Eschke, M.K. Heinrich, M. Wahby, H. Hamann, Self-organized adaptive paths in multi-robot manufacturing: reconfigurable and pattern-independent fibre deployment, in *IEEE/RSJ International Conference on Intelligent Robots and Systems (IROS)* (2019), pp. 4086–4091. https://doi.org/10.1109/IROS40897.2019.8967951
236. V. Faber, O.M. Lubeck, A.B. White Jr., Superlinear speedup of an efficient sequential algorithm is not possible. Parallel Comput. **3**(3), 259–260 (1986)
237. N.T. Fadai, S.T. Johnston, M.J. Simpson, Unpacking the Allee effect: determining individual-level mechanisms that drive global population dynamics. Proc. R. Soc. A **476**(2241), 20200350 (2020)
238. Z. Fang, J. Qin, J. Qin, Q. Ma, R. Han, Q. Liu, Eye movement-based human—swarm interaction for coverage control of mobile robots with constraints. IEEE Trans. Ind. Electron. (2024)
239. N. Farrow, J. Klingner, D. Reishus, N. Correll, Miniature six-channel range and bearing system: algorithm, analysis and experimental validation, in *IEEE International Conference on Robotics and Automation (ICRA)* (IEEE, 2014), pp. 6180–6185
240. J. Ferber, *Multi-Agent Systems: An Introduction to Distributed Artificial Intelligence* (Addison-Wesley, New York, 1999)
241. E. Ferrante, E. Duéñez-Guzmàn, A.E. Turgut, T. Wenseleers, Evolution of task partitioning in swarm robotics, in *Proceedings of the Workshop on Collective Behaviors and Social Dynamics of the European Conference on Artificial Life (ECAL 2013*. ed. by V. Trianni (MIT Press, 2013)
242. E. Ferrante, A.E. Turgut, E. Duéñez-Guzmàn, M. Dorigo, T. Wenseleers, Evolution of self-organized task specialization in robot swarms. PLoS Comput. Biol. **11**(8), e1004273 (2015). https://doi.org/10.1371/journal.pcbi.1004273

243. M. Ferro, F.N. Piñan Basualdo, P.R. Giordano, S. Misra, C. Pacchierotti, Experimental evaluation of haptic shared control for multiple electromagnetic untethered microrobots. IEEE Trans. Autom. Sci. Eng. (2024)
244. P. Flajolet, A.M. Odlyzko, Random mapping statistics, in *Workshop on the Theory and Application of of Cryptographic Techniques* (Springer, Berlin, 1989), pp. 329–354
245. *Flora robotica*. Project website (2017). http://www.florarobotica.eu
246. D. Floreano, H, Lipson, *From Individual Robots to Robot Societies* (2021)
247. D. Floreano, C. Mattiussi, *Bio-inspired Artificial Intelligence: Theories, Methods, and Technologies* (MIT Press, Cambridge, MA, USA, 2008)
248. J. Foerster, G. Farquhar, T. Afouras, N. Nardelli, S. Whiteson, Counterfactual multi-agent policy gradients, in *Proceedings of the AAAI Conference on Artificial Intelligence*, vol. 32 (2018)
249. A.D. Fokker, Die mittlere Energie rotierender elektrischer Dipole im Strahlungsfeld. Ann. Phys. **348**(5), 810–820 (1914)
250. L.R. Ford Jr., D.R. Fulkerson, *Flows in Networks* (Princeton University Press, 2015)
251. G. Francesca, M. Brambilla, A. Brutschy, L. Garattoni, R. Miletitch, G. Podevijn, A. Reina, T. Soleymani, M. Salvaro, C. Pinciroli, V. Trianni, M. Birattari, An experiment in automatic design of robot swarms: Automode-vanilla, evostick, and human experts, in *Ninth International Conference on Swarm Intelligence (ANTS 2014)*, ed. by M. Dorigo, M. Birattari, S. Garnier, H. Hamann, M.M. de Oca, C. Solnon, T. Stützle. Lecture Notes in Computer Science, vol. 8667 (xxxx, xxx, 2014), pp.25–37
252. A. Francis, S. Li, C. Griffiths, J. Sienz, Gas source localization and mapping with mobile robots: a review. J. Field Robot. **39**(8), 1341–1373 (2022)
253. N.R. Franks, A.B. Sendova-Franks, Brood sorting by ants: distributing the workload over the work-surface. Behav. Ecol. Sociobiol. **30**(2), 109–123 (1992)
254. N.R. Franks, A. Wilby, B.W. Silverman, C. Tofts, Self-organizing nest construction in ants: sophisticated building by blind bulldozing. Anim. Behav. **44**, 357–375 (1992)
255. N.R. Franks, A. Dornhaus, J.P. Fitzsimmons, M. Stevens, Speed versus accuracy in collective decision making. Proc. R. Soc. Lond. B **270**, 2457–2463 (2003)
256. P.R. Freeman, The secretary problem and its extensions: a review. Int. Stat. Rev./Revue Internationale de Statistique **51**(2), 189–206 (1983)
257. R. Frei, G.D.M. Serugendo, The future of complexity engineering. Centreal Eur. J. Eng. **2**(2), 164–188 (2012). https://doi.org/10.2478/s13531-011-0071-0
258. E. Freund, On the design of multi-robot systems, in *Proceedings of the IEEE International Conference on Robotics and Automation*, vol. 1 (IEEE, 1984), pp. 477–490
259. E. Freund, H. Hoyer, Real-time pathfinding in multirobot systems including obstacle avoidance. Int. J. Robot. Res. **7**(1), 42–70 (1988)
260. R. Friedman, D. Gavidia, L. Rodrigues, A.C. Viana, S. Voulgaris, Gossiping on MANETs: the beauty and the beast. SIGOPS Oper. Syst. Rev. **41**(5), 67–74 (2007). ISSN 0163-5980. https://doi.org/10.1145/1317379.1317390
261. R. Fujisawa, S. Dobata, Lévy walk enhances efficiency of group foraging in pheromone-communicating swarm robots, in *Proceedings of the 2013 IEEE/SICE International Symposium on System Integration* (IEEE, 2013), pp. 808–813
262. R.M. Fuoss, L. Onsager, Conductance of strong electrolytes at finite dilutions. Proc. Natl. Acad. Sci. **41**(5), 274–283 (1955)
263. S. Galam, Rational group decision making: a random field Ising model at T=0. XXPhys. A **238**(1–4), 66–80 (1997)
264. S. Galam, Contrarian deterministic effect on opinion dynamics: the "hung elections scenario". Physica A **333**(1), 453–460 (2004). https://doi.org/10.1016/j.physa.2003.10.041
265. S. Galam, Sociophysics: a review of Galam models. Int. J. Modern Phys. C **19**(3) (2008)
266. S. Galam, S. Moscovici, Towards a theory of collective phenomena: Consensus and attitude changes in groups. Eur. J. Soc. Psychol. **21**(1), 49–74 (1991). https://doi.org/10.1002/ejsp.2420210105

267. S. Galam, S. Moscovici, Towards a theory of collective phenomena. II: Conformity and power. Eur. J. Soc. Psychol. **24**(4), 481–495 (1994)
268. S. Galam, S. Moscovici, Towards a theory of collective phenomena. III: Conflicts and forms of power. Eur. J. Soc. Psychol. **25**(2), 217–229 (1995)
269. L. Garattoni, M. Birattari, Autonomous task sequencing in a robot swarm. Sci. Robot. **3**(20), eaat0430 (2018)
270. S. Garnier, J. Gautrais, M. Asadpour, C. Jost, G. Theraulaz, Self-organized aggregation triggers collective decision making in a group of cockroach-like robots. Adapt. Behav. **17**(2), 109–133 (2009)
271. S. Garnier, T. Murphy, M. Lutz, E. Hurme, S. Leblanc, I.D. Couzin, Stability and responsiveness in a self-organized living architecture. PLoS Comput. Biol. **9**(3), e1002984 (2013). https://doi.org/10.1371/journal.pcbi.1002984
272. B. Gates, A robot in every home. Sci. Am. **296**(1), 58–65 (2007)
273. M. Gauci, J. Chen, T.J. Dodd, R. Groß, Evolving aggregation behaviors in multi-robot systems with binary sensors, in *Distributed Autonomous Robotic Systems: The 11th International Symposium* (Springer, Berlin, 2014a), pp. 355–367
274. M. Gauci, J. Chen, W. Li, T.J. Dodd, R. Gross, Clustering objects with robots that do not compute, in*Proceedings of the 2014 International Conference on Autonomous Agents and Multi-agent Systems* (International Foundation for Autonomous Agents and Multiagent Systems, 2014b), pp. 421–428
275. M. Gauci, R. Nagpal, M. Rubenstein, Programmable self-disassembly for shape formation in large-scale robot collectives, in *13th International Symposium on Distributed Autonomous Robotic Systems (DARS 16)* (2016)
276. M. Gauci, M.E. Ortiz, M. Rubenstein, R. Nagpal, Error cascades in collective behavior: a case study of the gradient algorithm on 1000 physical agents, in *Proceedings of the 16th Conference on Autonomous Agents and MultiAgent Systems* (International Foundation for Autonomous Agents and Multiagent Systems, 2017), pp. 1404–1412
277. B. Gerkey, R.T. Vaughan, A. Howard, The player/stage project: Tools for multi-robot and distributed sensor systems, in *Proceedings of the 11th International Conference on Advanced Robotics (ICAR 2003)* (2003), pp. 317–323
278. V. Gerling, S. Von Mammen. Robotics for self-organised construction, in *2016 IEEE 1st International Workshops on Foundations and Applications of Self* Systems (FAS*W)* (2016), pp. 162–167. https://doi.org/10.1109/FAS-W.2016.45
279. C. Gershenson, D. Helbing, When slower is faster. Complexity **21**(2), 9–15 (2015)
280. N. Gershfeld, T.M. Paine, M.R. Benjamin, Adaptive and collaborative bathymetric channel-finding approach for multiple autonomous marine vehicles. IEEE Robot. Autom. Lett. **8**(7), 4028–4035 (2023)
281. I. Gharbi, J. Kuckling, D.G. Ramos, and Mauro Birattari. Show me what you want: Inverse reinforcement learning to automatically design robot swarms by demonstration. In *IEEE International Conference on Robotics and Automation (ICRA)*, pages 5063–5070, 2023. https://doi.org/10.1109/ICRA48891.2023.10160947
282. F.H. Giddings, *Sociology [a Lecture Delivered at Columbia University in the Series on Science, Philosophy and Art, February 26, 1908]* (Columbia University Press, 1908)
283. A. Gierer, H. Meinhardt, A theory of biological pattern formation. Biol. Cybern. **12**(1), 30–39 (1972)
284. E. Gjondrekaj, M. Loreti, R. Pugliese, F. Tiezzi, C. Pinciroli, M. Brambilla, M. Birattari, M. Dorigo, Towards a formal verification methodology for collective robotic systems, in *Formal Methods and Software Engineering* (Springer, 2012), pp. 54–70
285. I. Goldstein, A. Pauzner, Demand–deposit contracts and the probability of bank runs. J. Finance **60**(3), 1293–1327 (2005)
286. J. Gomes, P. Urbano, A.L. Christensen, Evolution of swarm robotics systems with novelty search. Swarm Intell. **7**(2), 115–144 (2013)
287. J. Gomes, P. Mariano, A.L. Christensen, Cooperative coevolution of partially heterogeneous multiagent systems, in *Proceedings of the 2015 International Conference on Autonomous*

Agents and Multiagent Systems (International Foundation for Autonomous Agents and Multiagent Systems, 2015), pp. 297–305
288. J. Gomes, P. Mariano, A.L. Christensen, Novelty-driven cooperative coevolution. Evol. Comput. **25**(2), 275–307 (2017)
289. G. Gompper, R.G. Winkler, T. Speck, A. Solon, C. Nardini, F. Peruani, H. Löwen, R. Golestanian, U.B. Kaupp, L. Alvarez, et al., The 2020 motile active matter roadmap. J. Phys. Condens. Matter **32**(19), 193001 (2020)
290. R.J. González. *War Virtually: The Quest to Automate Conflict, Militarize Data, and Predict the Future* (University of California Press, 2022)
291. D.M. Gordon, The organization of work in social insect colonies. Nature **380**, 121–124 (1996). https://doi.org/10.1038/380121a0
292. N.J. Gotelli, R.K. Colwell, Quantifying biodiversity: procedures and pitfalls in the measurement and comparison of species richness. Ecol. Lett. **4**(4), 379–391 (2001)
293. R. Graham, D. Knuth, O. Patashnik, *Concrete Mathematics: A Foundation for Computer Science* (Addison-Wesley, Reading, MA, 1998). ISBN: 0-201-55802-5
294. A. Grama, A. Gupta, G. Karypis, V. Kumar, *Introduction to Parallel Computing* (Addison-Wesley Professional, 2003)
295. P.-P. Grassé, La reconstruction du nid et les coordinations interindividuelles chez bellicositermes natalensis et cubitermes sp. la théorie de la stigmergie: essai d'interprétation du comportement des termites constructeurs. Insectes Soc. **6**, 41–83 (1959)
296. B.D. Greenshields, J.R. Bibbins, W.S. Channing, H.H. Miller, A study of traffic capacity, in *Highway Research Board Proceedings*, vol. 14 (Washington, DC, 1935), pp. 448–477
297. A. Gribovskiy, J. Halloy, J.-L. Deneubourg, H. Bleuler, F. Mondada, Towards mixed societies of chickens and robots, in *2010 IEEE/RSJ International Conference on Intelligent Robots and Systems (IROS)* (IEEE, 2010), pp. 4722–4728
298. G. Grimmett, *Percolation, Grundlehren der mathematischen Wissenschaften*, vol. 321 (Springer, Berlin, Germany, 1999)
299. R. Groß, M. Dorigo, Evolution of solitary and group transport behaviors for autonomous robots capable of self-assembling. Adapt. Behav. **16**(5), 285–305 (2008)
300. R. Groß, M. Bonani, F. Mondada, M. Dorigo, Autonomous self-assembly in swarm-bots. IEEE Trans. Robot. **22**(6), 1115–1130 (2006)
301. R. Groß, S. Magnenat, F. Mondada, Segregation in swarms of mobile robots based on the Brazil nut effect, in *IEEE/RSJ International Conference on Intelligent Robots and Systems (IROS 2009)* (IEEE, 2009), pp. 4349–4356
302. T. Gross, H. Sayama, *Adaptive Networks: Theory, Models, and Data* (2009)
303. M. Guillermo, Morphogens and synaptogenesis in drosophila. J. Neurobiol. **64**(4), 417–434 (2005) https://doi.org/10.1002/neu.20165
304. J. Guiochet, M. Machin, H. Waeselynck, Safety-critical advanced robots: a survey. Robot. Auton. Syst. **94**, 43–52 (2017)
305. N.J. Gunther, A simple capacity model of massively parallel transaction systems, in *CMG National Conference* (1993), pp. 1035–1044
306. N.J. Gunther, *Guerrilla Capacity Planning: A Tactical Approach to Planning for Highly Scalable Applications and Services* (Springer, 2007)
307. N.J. Gunther, P. Puglia, K. Tomasette, Hadoop super-linear scalability: The perpetual motion of parallel performance. ACM Queue **13**(5), 46–55 (2015)
308. W. Guo, X. Wang, X. Zheng, Lane formation in pedestrian counterflows driven by a potential field considering following and avoidance behaviours. Physica A: Stat. Mech. Appl. **432**, 87–101 (2015). ISSN 0378-4371. https://doi.org/10.1016/j.physa.2015.03.020, https://www.sciencedirect.com/science/article/pii/S0378437115002666
309. J.L. Gustafson, Reevaluating Amdahl's law. Commun. ACM **31**(5), 532–533 (1988). ISSN 0001-0782. https://doi.org/10.1145/42411.42415
310. J.L. Gustafson, Fixed time, tiered memory, and superlinear speedup, in *Proceedings of the Fifth Distributed Memory Computing Conference (DMCC5)* (1990), pp. 1255–1260
311. D. Ha, J. Schmidhuber, World models (2018). arXiv:1803.10122

312. E.C. Haas, M. Fields, C. Stachowiak, S. Hill, K. Pillalamarri, Extreme scalability: designing interfaces and algorithms for soldier-robotic swarm interaction, year 2. Army Res. Lab. (2010)
313. G. Habibi, W. Xie, M. Jellins, J. McLurkin, Distributed path planning for collective transport using homogeneous multi-robot systems, in *Distributed Autonomous Robotic Systems: The 12th International Symposium*. ed. by N.-Y. Chong, Y.-J. Cho (Springer, Japan, Tokyo, 2016), pp. 151–164. ISBN 978-4-431-55879-8. https://doi.org/10.1007/978-4-431-55879-8_11
314. H. Haken, *Synergetics—An Introduction* (Springer, Berlin, Germany, 1977)
315. H. Haken, *Synergetics: Introduction and Advanced Topics* (Springer, Berlin, Germany, 2004)
316. J. Halloy, G. Sempo, G. Caprari, C. Rivault, M. Asadpour, F. Tâche, I. Saïd, V. Durier, S. Canonge, J.M. Amé, C. Detrain, N. Correll, A. Martinoli, F. Mondada, R. Siegwart, J.-L. Deneubourg, Social integration of robots into groups of cockroaches to control self-organized choices. Science, **318**(5853), 1155–1158 (2007). https://doi.org/10.1126/science.1144259
317. H. Hamann, Modeling and investigation of robot swarms. Master's thesis, University of Stuttgart, Germany, 2006
318. H. Hamann, *Space-Time Continuous Models of Swarm Robotics Systems: Supporting Global-to-Local Programming* (Springer, Berlin, Germany, 2010)
319. H. Hamann, Towards swarm calculus: Urn models of collective decisions and universal properties of swarm performance. Swarm Intell. **7**(2–3), 145–172 (2013). https://doi.org/10.1007/s11721-013-0080-0
320. H. Hamann, Superlinear scalability in parallel computing and multi-robot systems: shared resources, collaboration, and network topology, in *Architecture of Computing Systems—ARCS 2018*. ed. by M. Berekovic, R. Buchty, H. Hamann, D. Koch, T. Pionteck (International Publishing, Springer, Cham, 2018a), pp. 31–42. ISBN 978-3-319-77610-1
321. H. Hamann, Opinion dynamics with mobile agents: contrarian effects by spatial correlations. Front. Robot. AI **5**(63) (2018b). https://doi.org/10.3389/frobt.2018.00063
322. H. Hamann, A. Reina, Scalability in computing and robotics. IEEE Trans. Comput. **71**(6), 1453–1465 (2021)
323. H. Hamann, H. Wörn, An analytical and spatial model of foraging in a swarm of robots, in *Swarm Robotics—Second SAB 2006 International Workshop*, ed. by E. Şahin, W. Spears, A.F.T. Winfield. Lecture Notes in Computer Science, vol. 4433 (Springer, Berlin, 2007), pp. 43–55
324. H. Hamann, H. Wörn, A framework of space-time continuous models for algorithm design in swarm robotics. Swarm Intell. **2**(2–4), 209–239 (2008). https://doi.org/10.1007/s11721-008-0015-3
325. H. Hamann, M. Szymanski, H. Wörn, Orientation in a trail network by exploiting its geometry for swarm robotics, in *IEEE Swarm Intelligence Symposium, Honolulu, USA, April 1-5*. ed. by Y. Shi, M. Dorigo (IEEE Press, Los Alamitos, CA, 2007), pp. 310–315
326. H. Hamann, B. Meyer, T. Schmickl, K. Crailsheim, A model of symmetry breaking in collective decision-making, in *From Animals to Animats 11*, ed. by S. Doncieux, B. Girard, A. Guillot, J. Hallam, J.-A. Meyer, J.-B. Mouret. Lecture Notes in Artificial Intelligence, vol. 6226 (Springer, Berlin, Germany, 2010a), pp. 639–648. https://doi.org/10.1007/978-3-642-15193-4_60
327. H. Hamann, J. Stradner, T. Schmickl, K. Crailsheim, A hormone-based controller for evolutionary multi-modular robotics: from single modules to gait learning, in *Proceedings of the IEEE Congress on Evolutionary Computation (CEC'10)* (2010b), pp. 244–251
328. H. Hamann, T. Schmickl, K. Crailsheim, A hormone-based controller for evaluation-minimal evolution in decentrally controlled systems. Artif. Life **18**(2), 165–198 (2012a). https://doi.org/10.1162/artl_a_00058
329. H. Hamann, T. Schmickl, K. Crailsheim, Self-organized pattern formation in a swarm system as a transient phenomenon of non-linear dynamics. Math. Comput. Model. Dyn. Syst. **18**(1), 39–50 (2012)
330. H. Hamann, T. Schmickl, H. Wörn, K. Crailsheim, Analysis of emergent symmetry breaking in collective decision making. Neural Comput. Appl. **21**(2), 207–218 (2012c). https://doi.org/10.1007/s00521-010-0368-6

331. H. Hamann, I. Karsai, T. Schmickl, Time delay implies cost on task switching: a model to investigate the efficiency of task partitioning. Bull. Math. Biol. **75**(7), 1181–1206 (2013). https://doi.org/10.1007/s11538-013-9851-4
332. H. Hamann, G. Valentini, Y. Khaluf, M. Dorigo, Derivation of a micro-macro link for collective decision-making systems: Uncover network features based on drift measurements, in *13th International Conference on Parallel Problem Solving from Nature (PPSN 2014)*, ed. by T. Bartz-Beielstein. Lecture Notes in Computer Science, vol. 8672 (Springer, Berlin, 2014), pp. 181–190. https://doi.org/10.1007/978-3-319-10762-2_18
333. H. Hamann, M. Wahby, T. Schmickl, P. Zahadat, D. Hofstadler, K. Støy, S. Risi, A. Faina, F. Veenstra, S. Kernbach, I. Kuksin, O. Kernbach, P. Ayres, P. Wojtaszek, Flora robotica—mixed societies of symbiotic robot-plant bio-hybrids, in *Proceedings of IEEE Symposium on Computational Intelligence (IEEE SSCI 2015)* (IEEE, 2015), pp. 1102–1109. https://doi.org/10.1109/SSCI.2015.158
334. H. Hamann, M.D. Soorati, M.K. Heinrich, D.N. Hofstadler, I. Kuksin, F. Veenstra, M. Wahby, S.A. Nielsen, S. Risi, T. Skrzypczak, P. Zahadat, P. Wojtaszek, K. Støy, T. Schmickl, S. Kernbach, P. Ayres, Flora robotica—an architectural system combining living natural plants and distributed robots (2017). arXiv:1709.04291
335. H. Hamann, T. Aust, A. Reina, Guerrilla performance analysis for robot swarms: degrees of collaboration and chains of interference events, in *Swarm Intelligence*. ed. by M. Dorigo, T. Stützle, M.J. Blesa, C. Blum, H. Hamann, M.K. Heinrich, V. Strobel (Springer International Publishing, Cham, 2020), pp. 134–147
336. M.H. Hansell, *Animal Architecture and Building Behaviour* (Longman, 1984)
337. K. Harada, P. Corradi, S. Popesku, J. Liedke, Heterogeneous multi-robot systems, in *Symbiotic Multi-Robot Organisms: Reliability, Adaptability, Evolution*, ed. by P. Levi, S. Kernbach. Cognitive Systems Monographs, vol. 7 (Springer, Berlin, 2010), pp. 79–163
338. C.E. Harriott, A.E. Seiffert, S.T. Hayes, J.A. Adams, Biologically-inspired human-swarm interaction metrics. Proc. Hum. Factors Ergon. Soc. Ann. Meet. **58**(1), 1471–1475 (2014)
339. A.G. Hart, C. Anderson, F.L. Ratnieks, Task partitioning in leafcutting ants. Acta Ethologica **5**, 1–11 (2002)
340. E. Hart, A.S.W. Steyven, B. Paechter, Evolution of a functionally diverse swarm via a novel decentralised quality-diversity algorithm, in *Proceedings of the Genetic and Evolutionary Computation Conference* (2018), pp. 101–108
341. E. Hasegawa, N. Mizumoto, K. Kobayashi, S. Dobata, J. Yoshimura, S. Watanabe, Y. Murakami, K. Matsuura, Nature of collective decision-making by simple yes/no decision units. Sci. Rep. **7**(1), 14436 (2017)
342. K. Hasselmann, A. Ligot, J. Ruddick, M. Birattari, Empirical assessment and comparison of neuro-evolutionary methods for the automatic off-line design of robot swarms. Nat. Commun. **12**(1), 4345 (2021)
343. S. Hauert, S. Leven, M. Varga, F. Ruini, A. Cangelosi, J.-C. Zufferey, D. Floreano, Reynolds flocking in reality with fixed-wing robots: Communication range versus maximum turning rate, in *IEEE/RSJ International Conference on Intelligent Robots and Systems* (IEEE, 2011), pp. 5015–5020
344. S. Hawking, W. Israel, *300 Years of Gravitation* (Cambridge University Press, Cambridge, UK, 1987)
345. A.T. Hayes, How many robots? Group size and efficiency in collective search tasks, in *Distributed Autonomous Robotic Systems*, vol. 5, ed. by H. Asama, T. Arai, T. Fukuda, T. Hasegawa (Springer, Tokyo, 2002), pp. 289–298. ISBN 978-4-431-65941-9, https://doi.org/10.1007/978-4-431-65941-9_29
346. A.T. Hayes, A. Martinoli, R.M. Goodman, Distributed odor source localization. IEEE Sens. J. **2**(3), 260–271 (2002)
347. R. Hegselmann, U. Krause, Opinion dynamics and bounded confidence models, analysis, and simulation. J. Artif. Soc. Soc. Simul. **5**(3), 1–24 (2002)

348. M.K. Heinrich, M. Wahby, M.D. Soorati, D.N. Hofstadler, P. Zahadat, P. Ayres, K. Støy, H. Hamann, Self-organized construction with continuous building material: higher flexibility based on braided structures, in *Proceedings of the 1st International Workshop on Self-Organising Construction (SOCO)* (IEEE, 2016), pp. 154–159. https://doi.org/10.1109/FAS-W.2016.43
349. D. Helbing, Traffic and related self-driven many-particle systems. Rev. Mod. Phys. **73**(4), 1067–1141 (2001). https://doi.org/10.1103/RevModPhys.73.1067
350. D. Helbing, P. Mukerji, Crowd disasters as systemic failures: analysis of the love parade disaster. EPJ Data Sci. **1** (2012)
351. D. Helbing, J. Keltsch, P. Molnár, Modelling the evolution of human trail systems. Nature **388**, 47–50 (1997)
352. D. Helbing, F. Schweitzer, J. Keltsch, P. Molnár, Active walker model for the formation of human and animal trail systems. Phys. Rev. E **56**(3), 2527–2539 (1997)
353. D. Helbing, P. Molnár, I.J. Farkas, K. Bolay, Self-organizing pedestrian movement. Environ. Plann. B. **28**(3), 361–383 (2001)
354. D.P. Helmbold, C.E. McDowell, Modelling speedup (n) greater than n. IEEE Trans. Parallel Distrib. Syst. **1**(2), 250–256 (1990)
355. J.M. Hereford, A distributed particle swarm optimization algorithm for swarm robotic applications, in *IEEE International Conference on Evolutionary Computation* (IEEE, 2006), pp. 1678–1685
356. J.M. Hereford, Analysis of BEECLUST swarm algorithm, in *Proceedings of the IEEE Symposium on Swarm Intelligence (SIS 2011)* (IEEE, 2011), pp. 192–198
357. V.-L. Heuthe, E. Panizon, H. Gu, C. Bechinger, Counterfactual rewards promote collective transport using individually controlled swarm microrobots. Sci. Robot. **9**(97), eado5888 (2024)
358. F. Higgins, A. Tomlinson, K.M. Martin, Threats to the swarm: security considerations for swarm robotics. Int. J. Adv. Secur. **2**(2&3) (2009)
359. R.W. Hockney, C.R. Jesshope, *Parallel Computers 2: Architecture, Programming and Algorithms* (CRC Press, 2019)
360. T. Hogg, Coordinating microscopic robots in viscous fluids. Auton. Agent. Multi-Agent Syst. **14**(3), 271–305 (2006)
361. J.H. Holland, *Adaptation in Natural and Artificial Systems* (University of Michigan Press, Ann Arbor, MI, 1975)
362. J.H. Holland, *Emergence—From Chaos to Order* (Oxford University Press, New York, 1998)
363. O. Holland, C. Melhuish, S. Hoddell, Chorusing and controlled clustering for minimal mobile agents, in *Proceeding of European Conference on Artificial Life (ECAL)* (1997)
364. J. Hu, A.E. Turgut, T. Krajník, B. Lennox, F. Arvin, Occlusion-based coordination protocol design for autonomous robotic shepherding tasks. IEEE Trans. Cogn. Dev. Syst. **14**(1), 126–135 (2020)
365. Y. Hu, H. Chen, J.-G. Lou, J. Li, Distributed density estimation using non-parametric statistics, in *27th International Conference on Distributed Computing Systems (ICDCS'07)* (IEEE, 2007), p. 28
366. S.P. Hubbell, L.K. Johnson, E. Stanislav, B. Wilson, H. Fowler, Foraging by bucket-brigade in leaf-cutter ants. Biotropica 210–213 (1980)
367. C. Huepe, G. Zschaler, A.-L. Do, T. Gross, Adaptive-network models of swarm dynamics. New J. Phys. **13**, 073022 (2011)
368. J. Humann, K.A. Pollard, Human factors in the scalability of multirobot operation: a review and simulation, in *IEEE International Conference on Systems, Man and Cybernetics (SMC)* (IEEE, 2019), pp. 700–707
369. K. Humpert, Trails, Tracks and Traces (Edition Esefeld & Traub, 2020)
370. W. Hunt, T. Godfrey, M.D. Soorati, Conversational language models for human-in-the-loop multi-robot coordination, in *Proceedings of the 23rd International Conference on Autonomous Agents and Multiagent Systems* (International Foundation for Autonomous Agents and Multiagent Systems (AAMAS), Richland, SC, 2024), pp. 2809–2811. ISBN 9798400704864

371. M. Hüttenrauch, A. Šošić, G. Neumann, Deep reinforcement learning for swarm systems. J. Mach. Learn. Res. **20**(54), 1–31 (2019)
372. S. Ichihashi, S. Kuroki, M. Nishimura, K. Kasaura, T. Hiraki, K. Tanaka, S. Yoshida, Swarm body: embodied swarm robots, in *Proceedings of the CHI Conference on Human Factors in Computing Systems, CHI '24* (Association for Computing Machinery, New York, NY, USA, 2024). ISBN 9798400703300. https://doi.org/10.1145/3613904.3642870
373. A.J. Ijspeert, A. Martinoli, A. Billard, L.M. Gambardella, Collaboration through the exploitation of local interactions in autonomous collective robotics: the stick pulling experiment. Auton. Robots **11**, 149–171 (2001). ISSN 0929-5593. https://doi.org/10.1023/A:1011227210047
374. A. Ilgün, K. Angelov, M. Stefanec, S. Schönwetter-Fuchs, V. Stokanic, J. Vollmann, D.N. Hofstadler, M.H. Kärcher, H. Mellmann, V. Taliaronak, et al., Bio-hybrid systems for ecosystem level effects, in *Artificial Life Conference Proceedings 33* (MIT Press, Cambridge, MA, 2021), pp. 41–50. https://doi.org/10.1162/isal_a_00396
375. A.G. Ingham, G. Levinger, J. Graves, V. Peckham, The Ringelmann effect: Studies of group size and group performance. J. Exp. Soc. Psychol. **10**(4), 371–384 (1974). ISSN 0022-1031. https://doi.org/10.1016/0022-1031(74)90033-X, http://www.sciencedirect.com/science/article/pii/002210317490033X
376. D.E. Jackson, M. Holcombe, F.L.W. Ratnieks, Trail geometry gives polarity to ant foraging networks. Nature **432**, 907–909 (2004)
377. A. Jadbabaie, J. Lin, A.S. Morse, Coordination of groups of mobile autonomous agents using nearest neighbor rules. IEEE Trans. Autom. Control **48**(6), 988–1001 (2003)
378. V. Jadhav, R. Pasqua, C. Zanon, M. Roy, G. Tredan, R. Bon, V. Guttal, G. Theraulaz, Collective responses of flocking sheep (Ovis aries) to a herding dog (border collie). Commun. Biol. **7**(1), 1543 (2024)
379. J. Jaeger, S. Surkova, M. Blagov, H. Janssens, D. Kosman, K.N. Kozlov, C.M. Manu, C.E. Vanario-Alonso, M. Samsonova, D.H. Sharp, J. Reinitz, Dynamic control of positional information in the early *Drosophila* embryo. Nature **430**, 368–371 (2004). https://doi.org/10.1038/nature02678
380. N. Jakobi, P. Husbands, I. Harvey, Noise and the reality gap: the use of simulation in evolutionary robotics, in *Proceedings of the Third European Conference on Advances in Artificial Life*, Lecture Notes in Computer Science, vol. 929 (Springer, Berlin, 1995), pp.704–720
381. C.H. Janson, Spatial movement strategies: theory, evidence, and challenges, in *On the Move: How and Why Animals Travel in Groups*. ed. by S. Boinski, P.A. Garber (University of Chicago Press, Chicago, 2000), pp.165–203
382. F. Jansson, M. Hartley, M. Hinsch, I. Slavkov, N. Carranza, T.S.G. Olsson, R.M. Dries, J.H. Grönqvist, A.F.M. Marée, J. Sharpe, et al., Kilombo: a kilobot simulator to enable effective research in swarm robotics (2015). arXiv:1511.04285
383. Jasmine, Swarm robot—project website (2025). http://www.swarmrobot.org/
384. R.L. Jeanne, E.V. Nordheim, Productivity in a social wasp: per capita output increases with swarm size. Behav. Ecol. **7**(1), 43–48 (1996)
385. R. Jeanson, C. Rivault, J.-L. Deneubourg, S. Blanco, R. Fournier, C. Jost, G. Theraulaz, Self-organized aggregation in cockroaches. Anim. Behav. **69**, 169–180 (2005)
386. K.H. Jensen, W. Kim, N.M. Holbrook, J.W.M. Bush, Optimal concentrations in transport systems. J. R. Soc. Interface **10**(83), 20130138 (2013)
387. A. Jeradi, M.M. Racissi, A. Farinelli, N. Brooks, P. Scerri, et al., Focused exploration for cooperative robotic watercraft, in *Proceedings of the International Workshop on Artificial Intelligence and Robotics* (2015), pp. 83–93
388. W. Ji, H. Chen, M. Chen, G. Zhu, L. Xu, R. Groß, R. Zhou, M. Cao, S. Zhao, GenSwarm: scalable multi-robot code-policy generation and deployment via language models (2025). arXiv:2503.23875
389. C.-J. Jin, K.-D. Shi, R. Jiang, D. Li, S. Fang, Simulation of bi-directional pedestrian flow under high densities using a modified social force model. Chaos, Solitons Fractals **172**, 113559 (2023). ISSN 0960-0779. https://doi.org/10.1016/j.chaos.2023.113559, https://www.sciencedirect.com/science/article/pii/S0960077923004605

390. S. Johnson, *Emergence: The Connected Lives of Ants, Brains, Cities, and Software* (Scribner, New York, 2001)
391. J. Jones, The emergence and dynamical evolution of complex transport networks from simple low-level behaviours. Int. J. Unconventional Comput. **6**(2), 125–144 (2010)
392. J. Jones, Characteristics of pattern formation and evolution in approximations of physarum transport networks. Artif. Life **16**(2), 127–153 (2010)
393. J.L. Jones, D. Roth, *Robot Programming: A Practical Guide to Behavior-Based Robotics* (McGraw Hill, 2004)
394. S. Jones, M. Studley, S. Hauert, A.F.T. Winfield, A two teraflop swarm. Front. Robot. AI **5**, 11 (2018)
395. M. Kac, Random walk and the theory of Brownian motion. Am. Math. Monthly **54**, 369 (1947)
396. J. Kaduk, M. Cavdan, K. Drewing, A. Vatakis, H. Hamann, Effects of human-swarm interaction on subjective time perception: Swarm size and speed, in *Proceedings of the 2023 ACM/IEEE International Conference on Human-Robot Interaction (HRI)* (2023), pp. 456–465
397. J. Kaduk, M. Cavdan, K. Drewing, H. Hamann, From one to many: How active robot swarm sizes influence human cognitive processes, in *33rd IEEE International Conference on Robot and Human Interactive Communication (RO-MAN)* (2024a)
398. J. Kaduk, F. Weilbeer, H. Hamann, Emotional tandem robots: How different robot behaviors affect human perception while controlling a mobile robot, in *IEEE/RSJ International Conference on Intelligent Robots and Systems (IROS)* (2024b), pp. 2465–2470. https://doi.org/10.1109/IROS58592.2024.10801974
399. T.K. Kaiser, H. Hamann, Innate motivation for robot swarms by minimizing surprise: from simple simulations to real-world experiments. IEEE Trans. Robot. **38**(6), 3582–3601 (2022)
400. T.K. Kaiser, M.J. Begemann, T. Plattenteich, L. Schilling, G. Schildbach, H. Hamann, ROS2SWARM—a rOS 2 package for swarm robot behaviors, in *International Conference on Robotics and Automation (ICRA)* (IEEE, 2022), pp. 6875–6881
401. Z. Kakish, K. Elamvazhuthi, S. Berman, Using reinforcement learning to herd a robotic swarm to a target distribution, in *International Symposium Distributed Autonomous Robotic Systems* (Springer, Berlin, 2021), pp. 401–414
402. K. Kalthoff, Pattern formation in early insect embryogenesis—data calling for modification of a recent model. J. Cell Sci. **29**(1), 1–15 (1978)
403. A.P. Kanakia, *Response Threshold Based Task Allocation in Multi-Agent Systems Performing Concurrent Benefit Tasks with Limited Information*. Ph.D. thesis, University of Colorado Boulder, 2015
404. G. Kapellmann-Zafra, N. Salomons, A. Kolling, R. Groß, Human-robot swarm interaction with limited situational awareness, in *International Conference on Swarm Intelligence (ANTS 2016)*. ed. by M. Dorigo. Lecture Notes in Computer Science. (Springer, Berlin, 2016a), pp. 125–136
405. G. Kapellmann-Zafra, N. Salomons, A. Kolling, R. Groß, Human-robot swarm interaction with limited situational awareness, in *Swarm Intelligence: 10th International Conference (ANTS)* (Springer, Berlin, 2016b), pp. 125–136
406. I. Karsai, T. Schmickl, Regulation of task partitioning by a "common stomach": a model of nest construction in social wasps. Behav. Ecol. **22**, 819–830 (2011). https://doi.org/10.1093/beheco/arr060
407. N. Kashiri, A. Abate, S.J. Abram, A. Albu-Schaffer, P.J. Clary, M. Daley, S. Faraji, R. Furnemont, M. Garabini, H. Geyer et al., An overview on principles for energy efficient robot locomotion. Front. Robot. AI **5**, 129 (2018)
408. D. Katabi, M. Handley, C. Rohrs, Congestion control for high bandwidth-delay product networks, in *Proceedings of the Conference on Applications, Technologies, Architectures, and Protocols for Computer Communications* (2002), pp. 89–102
409. E. Kaufmann, L. Bauersfeld, A. Loquercio, M. Müller, V. Koltun, D. Scaramuzza, Champion-level drone racing using deep reinforcement learning. Nature **620**(7976), 982–987 (2023)

410. R.L. Kay, An application of the Fuoss-Onsager conductance theory to the alkali halides in several solvents. J. Am. Chem. Soc. **82**(9), 2099–2105 (1960)
411. P. Kedia, M. Rao, GenGrid: a generalised distributed experimental environmental grid for swarm robotics, in *IEEE International Conference on Robotics and Automation (ICRA)* (IEEE, 2021), pp. 1910–1917
412. P. Kedia, C. Apolinsky, H. Hamann, Developing SailSwarm: small uncrewed sailing vessels for maritime environments (2024). https://doi.org/10.5281/zenodo.15470231
413. M. Kegeleirs, M. Birattari, Towards applied swarm robotics: current limitations and enablers. Front. Robot. AI **12**, 1607978 (2025)
414. M. Kegeleirs, D. Garzón Ramos, M. Birattari, Random walk exploration for swarm mapping, in *Annual Conference Towards Autonomous Robotic Systems* (Springer, Berlin, 2019), pp. 211–222
415. M. Kegeleirs, G. Grisetti, M. Birattari, Swarm SLAM: challenges and perspectives. Front. Robot. AI **8**, 618268 (2021)
416. N. Kemsaram, A. Delibasi, J. Hardwick, B. Gautam, D.M. Plasencia, S. Subramanian, A cooperative contactless object transport with acoustic robots (2025). arXiv:2506.13957
417. D. Kengyel, H. Hamann, P. Zahadat, G. Radspieler, F. Wotawa, T. Schmickl, Potential of heterogeneity in collective behaviors: a case study on heterogeneous swarms, in *PRIMA 2015: Principles and Practice of Multi-agent Systems*, ed. by Q. Chen, P. Torroni, S. Villata, J. Hsu, A. Omicini. Lecture Notes in Computer Science, vol. 9387 (Springer, Berlin, 2015), pp. 201–217
418. J. Kennedy, R.C. Eberhart, Particle swarm optimization, in *IEEE International Conference on Neural Networks* (IEEE Press, Los Alamitos, CA, 1995)
419. J. Kennedy, R.C. Eberhart, *Swarm Intelligence* (Morgan Kaufmann, 2001)
420. W.O. Kermack, A.G. McKendrick, A contribution to the mathematical theory of epidemics. Proc. R. Soc. Lond. Ser. A, Containing Pap. Math. Phys. Character **115**(772), 700–721 (1927)
421. A.-M. Kermarrec, M. van Steen, Gossiping in distributed systems. ACM SIGOPS Oper. Syst. Rev. **41**(5), 2–7 (2007)
422. S. Kernbach, R. Thenius, O. Kernbach, T. Schmickl, Re-embodiment of honeybee aggregation behavior in an artificial micro-robotic swarm. Adapt. Behav. **17**, 237–259 (2009)
423. S. Kernbach, D. Häbe, O. Kernbach, R. Thenius, G. Radspieler, T. Kimura, T. Schmickl, Adaptive collective decision-making in limited robot swarms without communication. Int. J. Robot. Res. **32**(1), 35–55 (2013)
424. M.A. Kessler, B.T. Werner, Self-organization of sorted patterned ground. Science **299**, 380–383 (2003)
425. A. Kettler, M. Szymanski, H. Wörn, The wanda robot and its development system for swarm algorithms, in *Advances in Autonomous Mini Robots* (2012), pp. 133–146
426. L. Keysberg, N. Wakamiya, Towards flexible swarms: comparison of flocking models with varying complexity. Artif. Life Robot. (2025)
427. Y. Khaluf, M. Birattari, F. Rammig, Probabilistic analysis of long-term swarm performance under spatial interferences, in *Proceeding of Theory and Practice of Natural Computing*. ed. by A.-H. Dediu, C. Martín-Vide, B. Truthe, M.A. Vega-Rodríguez (Springer, Berlin, 2013), pp. 121–132. ISBN 978-3-642-45008-2, https://doi.org/10.1007/978-3-642-45008-2_10
428. Y. Khaluf, M. Birattari, H. Hamann, A swarm robotics approach to task allocation under soft deadlines and negligible switching costs, in *Simulation of Adaptive Behavior (SAB 2014)*, ed. by A.P. del Pobil, E. Chinellato, E. Martinez-Martin, J. Hallam, E. Cervera, A. Morales. Lecture Notes in Computer Science, vol. 8575 (Springer, Berlin, 2014), pp. 270–279
429. Y. Khaluf, C. Pinciroli, G. Valentini, H. Hamann, The impact of agent density on scalability in collective systems: noise-induced versus majority-based bistability. Swarm Intell. **11**(2), 155–179 (2017). ISSN 1935-3820. https://doi.org/10.1007/s11721-017-0137-6
430. Y. Khaluf, P. Simoens, H. Hamann, The neglected pieces of designing collective decision-making processes. Front. Robot. AI **6**, 16 (2019)
431. T. Kida, Y. Sueoka, H. Shigeyoshi, Y. Tsunoda, Y. Sugimoto, K. Osuka, Verification of acoustic-wave-oriented simple state estimation and application to swarm navigation. J. Robot. Mechatron. **33**(1), 119–128 (2021)

432. R. Killick, P. Fearnhead, I.A. Eckley, Optimal detection of changepoints with a linear computational cost. J. Am. Stat. Assoc. **107**(500), 1590–1598 (2012)
433. L.H. Kim, S. Follmer, UbiSwarm: ubiquitous robotic interfaces and investigation of abstract motion as a display. Proc. ACM Interact. Mob. Wearable Ubiquitous Technol. **1**(3), 66:1–66:20 (2017). ISSN 2474-9567. https://doi.org/10.1145/3130931
434. L.H. Kim, S. Follmer, Swarmhaptics: haptic display with swarm robots, in *Proceedings of the 2019 CHI Conference on Human Factors in Computing Systems, CHI '19* (Association for Computing Machinery, New York, NY, USA, 2019), pp. 1–13. ISBN 9781450359702. https://doi.org/10.1145/3290605.3300918
435. L.H. Kim, D.S. Drew, V. Domova, S. Follmer, User-defined swarm robot control, in *Proceedings of the CHI Conference on Human Factors in Computing Systems, CHI '20* (Association for Computing Machinery, New York, NY, USA, 2020), pp. 1–13. ISBN 9781450367080. https://doi.org/10.1145/3313831.3376814
436. A.J. King, S.J. Portugal, D. Strömbom, R.P. Mann, J.A. Carrillo, D. Kalise, G. de Croon, H. Barnett, P. Scerri, R. Groß, D.R. Chadwick, M. Papadopoulou, Biologically inspired herding of animal groups by robots. Methods Ecol. Evol. **14**(2), 478–486 (2023)
437. J. Klein, Continuous 3D agent-based simulations in the breve simulation environment, in *Proceedings of NAACSOS Conference (North American Association for Computational, Social, and Organizational Sciences)* (2003)
438. K. Klein, B. Sommer, H.T. Nim, A. Flack, K. Safi, M. Nagy, S.P. Feyer, Y. Zhang, K. Rehberg, A. Gluschkow, et al., Fly with the flock: immersive solutions for animal movement visualization and analytics. J. R. Soc. Interface **16**(153), 20180794 (2019)
439. N. Koenig, A. Howard, Design and use paradigms for Gazebo, an open-source multi-robot simulator, in *IEEE/RSJ International Conference on Intelligent Robots and Systems (IROS)*, vol. 3 (IEEE, 2004), pp. 2149–2154
440. A. Kolling, P. Walker, N. Chakraborty, K. Sycara, M. Lewis, Human interaction with robot swarms: a survey. IEEE Trans. Hum. Mach. Syst. **46**(1), 9–26 (2016). ISSN 2168-2291. https://doi.org/10.1109/THMS.2015.2480801
441. L. König, S. Mostaghim, H. Schmeck, Decentralized evolution of robotic behavior using finite state machines. Int. J. Intell. Comput. Cybern. **2**(4), 695–723 (2009)
442. S. Koos, J.-B. Mouret, S. Doncieux, The transferability approach: crossing the reality gap in evolutionary robotics. IEEE Trans. Evol. Comput. **17**(1), 122–145 (2012)
443. J.R. Koza, *Genetic Programming: On the Programming of Computers by Means of Natural Selection* (MIT Press, 1992). ISBN 0-262-11170-5
444. L.P. Krapivsky, S. Redner, E. Ben-Naim *A Kinetic View of Statistical Physics* (Cambridge University Press, 2010)
445. J. Krause, G.D. Ruxton, *Living in Groups* (Oxford University Press, 2002)
446. C.P. Kruskal, L. Rudolph, M. Snir, Efficient parallel algorithms for graph problems. Algorithmica **5**, 43–64 (1990). https://doi.org/10.1007/BF01840376
447. C.R. Kube, E. Bonabeau, Cooperative transport by ants and robots. Robot. Auton. Syst. **30**, 85–101 (2000)
448. C.R. Kube, H. Zhang, Collective robotic intelligence, in *From Animals to Animats 2: Proceedings of the Second International Conference on Simulation of Adaptive Behavior* (MIT Press, 1993). ISBN 9780262287159. https://doi.org/10.7551/mitpress/3116.003.0062
449. A. Kubík, On emergence in evolutionary multiagent systems, in *Proceedings of the 6th European Conference on Artificial Life* (2001), pp. 326–337
450. A. Kubík, Toward a formalization of emergence. Artif. Life **9**, 41–65 (2003)
451. J. Kuckling, R. Luckey, V. Avrutin, A. Vardy, A. Reina, H. Hamann, Do we run large-scale multi-robot systems on the edge? more evidence for two-phase performance in system size scaling, in *IEEE International Conference on Robotics and Automation (ICRA)* (2024), pp. 4562–4568. https://doi.org/10.1109/ICRA57147.2024.10610771
452. G. Kumar, N. Dukkipati, K. Jang, H.M.G. Wassel, X. Wu, B. Montazeri, Y. Wang, K. Springborn, C. Alfeld, M. Ryan, et al., Swift: delay is simple and effective for congestion control in the datacenter, in *Proceedings of the Annual conference of the ACM Special Interest Group*

on Data Communication on the Applications, Technologies, Architectures, and Protocols for Computer Communication (2020), pp. 514–528
453. D. Kunertova, Drones have boots: learning from Russia's war in Ukraine. Contemp. Secur. Policy **44**(4), 576–591 (2023)
454. D. Kunertova, The war in Ukraine shows the game-changing effect of drones depends on the game. Bull. Atomic Sci. **79**(2), 95–102 (2023)
455. Y. Kuramoto, *Chemical Oscillations, Waves, and Turbulence* (Springer, Berlin, 1984)
456. H.L. Kwa, J. Philippot, R. Bouffanais, Effect of swarm density on collective tracking performance. Swarm Intell. **17**(3), 253–281 (2023a)
457. H.L. Kwa, J. Philippot, R. Bouffanais, The impact of agent density and environmental factors on target tracking swarms, in *Artificial Life Conference Proceedings*, vol. 35 (MIT Press, 2023b)
458. M. Kwiatkowska, G. Norman, D. Parker, PRISM 4.0: Verification of probabilistic real-time systems, in *Proceedings of 23rd International Conference on Computer Aided Verification (CAV'11)*, ed. by G. Gopalakrishnan, S. Qadeer. Lecture Notes in Computer Science, vol. 6806 (Springer, Berlin, 2011), pp. 585–591
459. P.-Y. Lajoie, G. Beltrame, Swarm-SLAM: sparse decentralized collaborative simultaneous localization and mapping framework for multi-robot systems. IEEE Robot. Autom. Lett. **9**(1), 475–482 (2023)
460. A. Lama, M. di Bernardo, Shepherding and herdability in complex multiagent systems. Phys. Rev. Res. **6**(3), L032012 (2024)
461. L. Lamport, R. Shostak, M. Pease, The Byzantine generals problem. ACM Trans. Program. Lang. Syst. **4**(3), 382–401 (1982). ISSN 0164-0925. https://doi.org/10.1145/357172.357176
462. T. Landgraf, H. Nguyen, S. Forgo, J. Schneider, J. Schröer, C. Krüger, H. Matzke, R.O. Clément, J. Krause, R. Rojas, Interactive robotic fish for the analysis of swarm behavior, in *Advances in Swarm Intelligence: 4th International Conference* (Springer, Berlin, 2013), pp. 1–10
463. H.K. Lau, I. Bate, P. Cairns, J. Timmis, Adaptive data-driven error detection in swarm robotics with statistical classifiers. Robot. Auton. Syst. **59**(12), 1021–1035 (2011)
464. D. Lazer, A. Friedman, The network structure of exploration and exploitation. Adm. Sci. Q. **52**, 667–694 (2007)
465. M.L. Goc, L.H. Kim, A. Parsaei, J.-D. Fekete, P. Dragicevic, S. Follmer, Zooids: building blocks for swarm user interfaces, in *Proceedings of the 29th Annual Symposium on User Interface Software and Technology* (ACM, 2016), pp. 97–109
466. S. Lee, S. Hauert, A data-driven method to identify fault mitigation strategies in robot swarms, in *International Conference on Swarm Intelligence (ANTS)* (Springer, Berlin, 2024), pp. 16–28
467. S. Lee, E. Milner, S. Hauert, A data-driven method for metric extraction to detect faults in robot swarms. IEEE Robot. Autom. Lett. **7**(4), 10746–10753 (2022)
468. R.E. Lee, Jr., Aggregation of lady beetles on the shores of lakes (Coleoptera: Coccinellidae). Am. Midl. Nat. **104**(2), 295–304 (1980)
469. J. Lehman, K.O. Stanley, Exploiting open-endedness to solve problems through the search for novelty, in *Artificial Life XI: Proceedings of the Eleventh International Conference on the Simulation and Synthesis of Living Systems*. ed. by S. Bullock, J. Noble, R. Watson, M.A. Bedau (MIT Press, 2008), pp. 329–336
470. J. Lehman, K.O. Stanley, Improving evolvability through novelty search and self-adaptation, in *Proceedings of the 2011 IEEE Congress on Evolutionary Computation (CEC'11)* (IEEE, 2011), pp. 2693–2700
471. X. Lei, Y. Xiang, M. Duan, X. Peng, Exploring the criticality hypothesis using programmable swarm robots with Vicsek-like interactions. J. R. Soc. Interface **20**(204), 20230176 (2023)
472. R. Leidenfrost, W. Elmenreich, Firefly clock synchronization in an 802.15.4 wireless network. EURASIP J. Embed. Syst. 186406 (2009)
473. A. Lein, R.T. Vaughan, Adaptive multi-robot bucket brigade foraging. Artif. Life **11**, 337 (2008)

474. T. Lemaire, R. Alami, S. Lacroix, A distributed tasks allocation scheme in multi-UAV context, in *Proceedings of the IEEE International Conference on Robotics and Automation (ICRA)*, vol. 4 (IEEE Press, 2004), pp. 3622–3627. https://doi.org/10.1109/ROBOT.2004.1308816
475. S.C. Lenaghan, Y. Wang, N. Xi, T. Fukuda, T. Tarn, W.R. Hamel, M. Zhang, Grand challenges in bioengineered nanorobotics for cancer therapy. IEEE Trans. Biomed. Eng. **60**(3), 667–673 (2013)
476. N.E. Leonard, S.A. Levin, Collective intelligence as a public good. Collective Intell. **1**(1), 26339137221083293 (2022)
477. I. Leoncini, C. Rivault, Could species segregation be a consequence of aggregation processes? Example of *Periplaneta americana* (L.) and *P. fuliginosa* (serville). Ethology **111**(5), 527–540 (2005)
478. K. Lerman, A model of adaptation in collaborative multi-agent systems. Adapt. Behav. **12**(3–4), 187–198 (2004)
479. K. Lerman, A. Galstyan, Mathematical model of foraging in a group of robots: effect of interference. Auton. Robot. **13**, 127–141 (2002)
480. K. Lerman, A. Galstyan, A. Martinoli, A. Ijspeert, A macroscopic analytical model of collaboration in distributed robotic systems. Artif. Life **7**, 375–393 (2001)
481. K. Lerman, A. Martinoli, A. Galstyan, A review of probabilistic macroscopic models for swarm robotic systems, in *Swarm Robotics—SAB 2004 International Workshop*. ed. by E. Şahin, W.M. Spears, Lecture Notes in Computer Science, vol. 3342 (Springer, Berlin, Germany, 2005), pp. 143–152
482. K. Lerman, C. Jones, A. Galstyan, M.J. Matarić, Analysis of dynamic task allocation in multi-robot systems. Int. J. Robot. Res. **25**(3), 225–241 (2006)
483. O. Levenspiel, Chemical reaction engineering. Ind. Eng. Chem. Res. **38**(11), 4140–4143 (1999)
484. P. Levi, S. Kernbach (eds.), *Symbiotic Multi-robot Organisms: Reliability, Adaptability, Evolution* (Springer, Berlin, 2010)
485. M. Lewis, J. Wang, P. Scerri, Teamwork coordination for realistically complex multi robot systems, in *NATO Symposium on Human Factors of Uninhabited Military Vehicles as Force Multipliers* (2006)
486. M. Lewis, H. Wang, S.Y. Chien, P. Velagapudi, P. Scerri, K. Sycara, Choosing autonomy modes for multirobot search. Hum. Factors **52**(2), 225–233 (2010)
487. G. Li, D. St-Onge, C. Pinciroli, A. Gasparri, E. Garone, G. Beltrame, Decentralized progressive shape formation with robot swarms. Auton. Robot. **43**, 1505–1521 (2019)
488. M.Z.F. Li, A generic characterization of equilibrium speed-flow curves. Transp. Sci. **42**(2), 220–235 (2008)
489. N. Li, J.R. Marden, Designing games for distributed optimization. IEEE J. Sel. Top. Signal Process. **7**(2), 230–242 (2013)
490. S. Li, R. Kong, Y. Guo, Cooperative distributed source seeking by multiple robots: algorithms and experiments. IEEE/ASME Trans. Mechatron. **19**(6), 1810–1820 (2014)
491. Y. Li, R. Miao, H.H. Liu, Y. Zhuang, F. Feng, L. Tang, Z. Cao, M. Zhang, F. Kelly, M. Alizadeh, et al., HPCC: high precision congestion control, in *Proceedings of the ACM Special Interest Group on Data Communication* (ACM, 2019b), pp. 44–58
492. M.J. Lighthill, G.B. Whitham, On kinematic waves. II. A theory of traffic flow on long crowded roads. Proc. R. Soc. Lond. **A229**(1178), 317–345 (1955)
493. J. Lind, Issues in agent-oriented software engineering, in *Agent-Oriented Software Engineering*. ed. by M. Wooldridge (Springer, Berlin, Heidelberg, New York, 2000)
494. H-S. Liu, S. Kuroki, T. Kozuno, W.-F. Sun, C.-Y. Lee, Language-guided pattern formation for swarm robotics with multi-agent reinforcement learning, in *IEEE/RSJ International Conference on Intelligent Robots and Systems (IROS)* (IEEE, 2024), pp. 8998–9005
495. M. Liu, L. Deferme, T. Van Eyck, F. Yang, S. Michiels, A. Abadie, S. Alvarado-Marin, F. Maksimovic, G. Miyauchi, J. Jayakumar, et al., CapBot: Enabling battery-free swarm robotics, in *IEEE International Conference on Robotics and Automation (ICRA)* (2025)

496. Z.T. Liu, Y. Shi, Y. Zhao, H. Chaté, X.-G. Shi, T.H. Zhang, Activity waves and freestanding vortices in populations of subcritical quincke rollers. Proc. Natl. Acad. Sci. **118**(40), e2104724118 (2021)
497. S. Lloyd, Least squares quantization in PCM. IEEE Trans. Inf. Theory **28**(2), 129–137 (1982). https://doi.org/10.1109/TIT.1982.1056489
498. T. Lochmatter, X. Raemy, L. Matthey, S. Indra, A. Martinoli, A comparison of casting and spiraling algorithms for odor source localization in laminar flow, in *IEEE International Conference on Robotics and Automation (ICRA)* (IEEE, 2008), pp. 1138–1143
499. N.K. Long, K. Sammut, D. Sgarioto, M. Garratt, H.A. Abbass, A comprehensive review of shepherding as a bio-inspired swarm-robotics guidance approach. IEEE Trans. Emerg. Top. Comput. Intell. **4**(4), 523–537 (2020)
500. R. Lowe, Y.I. Wu, A. Tamar, J. Harb, P. Abbeel, I. Mordatch, Multi-agent actor-critic for mixed cooperative-competitive environments. Adv. Neural Inf. Process. Syst. (NeurIPS) **30** (2017)
501. L. Lü, Z.-K. Zhang, T. Zhou, Zipf's law leads to Heaps' law: analyzing their relation in finite-size systems. PLoS ONE **5**(12), e14139 (2010)
502. L. Ludwig, M. Gini, Robotic swarm dispersion using wireless intensity signals, in *Distributed Autonomous Robotic Systems*, vol. 7 (Springer, 2006), pp. 135–144
503. S. Luke, C. Cioffi-Revilla, L. Panait, K. Sullivan, Mason: a new multi-agent simulation toolkit, in *Proceedings of the 2004 Swarmfest Workshop*, vol. 8 (2004), pp. 44
504. Q. Luo, H. Wang, Y. Zheng, J. He, Research on path planning of mobile robot based on improved ant colony algorithm. Neural Comput. Appl. **32**(6), 1555–1566 (2020)
505. S. Macenski, T. Foote, B. Gerkey, C. Lalancette, W. Woodall, Robot operating system 2: Design, architecture, and uses in the wild. Sci. Robot. **7**(66), eabm6074 (2022)
506. C.A. Mack, Fifty years of moore's law. IEEE Trans. Semicond. Manuf. **24**(2), 202–207 (2011)
507. R. Madhavan, K. Fregene, L.E. Parker, Terrain aided distributed heterogeneous multirobot localization and mapping. Auton. Robot. **17**, 23–39 (2004)
508. H. Mahmoud, *Pólya urn Models* (Chapman and Hall/CRC, Boca Raton, FL, USA, 2008)
509. N. Majcherczyk, C. Pinciroli, SwarmMesh: a distributed data structure for cooperative multi-robot applications, in *IEEE International Conference on Robotics and Automation (ICRA)* (2020), pp. 4059–4065. https://doi.org/10.1109/ICRA40945.2020.9197403
510. E.B. Mallon, N.R. Franks, Ants estimate area using Buffon's needle. Proc. R. Soc. Lond. B **267**(1445), 765–770 (2000)
511. B.B. Mandelbrot, R. Pignoni, *The Fractal Geometry of Nature*, vol. 173 (WH freeman New York, 1983)
512. P.D. Manrique, M. Klein, Y.S. Li, C. Xu, P.M. Hui, N.F. Johnson, Getting closer to the goal by being less capable. Sci. Adv. **5**(2), eaau5902 (2019)
513. M.C. Marchetti, J.-F. Joanny, S. Ramaswamy, T.B. Liverpool, J. Prost, M. Rao, R.A. Simha, Hydrodynamics of soft active matter. Rev. Modern Phys. **85**(3), 1143–1189 (2013)
514. L.S. Marcolino, L. Chaimowicz, Traffic control for a swarm of robots: avoiding target congestion, in *IEEE/RSJ International Conference on Intelligent Robots and Systems (IROS)* (IEEE, 2009), pp. 1955–1961
515. T.J.S. Maria, F. Griffin, J.A. Deja, Set the stage: enabling storytelling with multiple robots through roleplaying metaphors (2025). arXiv:2508.01736
516. P. Mariano, Z. Salem, R. Mills, P. Zahadat, L. Correia, T. Schmickl, Design choices for adapting bio-hybrid systems with evolutionary computation, in *Proceedings of the Genetic and Evolutionary Computation Conference Companion, GECCO '17* (ACM, New York, NY, USA, 2017), pp. 211–212. ISBN 978-1-4503-4939-0. https://doi.org/10.1145/3067695.3076044
517. M. Marsili, Y.-C. Zhang, Interacting individuals leading to Zipf's law. Phys. Rev. Lett. **80**(12), 2741 (1998)
518. A. Martinoli, *Swarm Intelligence in Autonomous Collective Robotics: From Tools to the Analysis and Synthesis of Distributed Control Strategies*. Ph.D. thesis, Ecole Polytechnique Fédérale de Lausanne, 1999

519. A. Martinoli, K. Easton, W. Agassounon, Modeling swarm robotic systems: a case study in collaborative distributed manipulation. Int. J. Robot. Res. **23**(4), 415–436 (2004)
520. N. Masuda, M.A. Porter, R. Lambiotte, Random walks and diffusion on networks. Phys. Rep. **716–717**, 1–58 (2017). ISSN 0370-1573. https://doi.org/10.1016/j.physrep.2017.07.007. https://www.sciencedirect.com/science/article/pii/S0370157317302946
521. M.J. Matarić, Integration of representation into goal-driven behavior-based robots. IEEE Trans. Robot. Autom. **8**(3), 304–312 (1992)
522. M.J. Matarić, Minimizing complexity in controlling a mobile robot population, in *IEEE International Conference on Robotics and Automation* (IEEE, 1992b), pp. 830–835
523. M.J. Matarić, Designing emergent behaviors: from local interactions to collective intelligence. *Proceedings of the Second International Conference on From Animals to Animats 2: Simulation of Adaptive Behavior* (1993), pp. 432–441
524. M.J. Matarić, Designing and understanding adaptive group behavior. Adapt. Behav. **4**, 51–80 (1995)
525. M.J. Matarić, Reinforcement learning in the multi-robot domain. Auton. Robot. **4**(1), 73–83 (1997)
526. D. Mateo, Y.K. Kuan, R. Bouffanais, Effect of correlations in swarms on collective response. Sci. Rep. **7**, 10388 (2017). https://doi.org/10.1038/s41598-017-09830-w
527. N. Mathews, A.L. Christensen, E. Ferrante, R. O'Grady, M. Dorigo, Establishing spatially targeted communication in a heterogeneous robot swarm, in *Proceedings of the 9th International Conference on Autonomous Agents and Multiagent Systems (AAMAS)* (International Foundation for Autonomous Agents and Multiagent Systems, 2010), pp. 939–946
528. K. Matsuka, A.O. Feldman, E.S. Lupu, S.-J. Chung, F.Y. Hadaegh, Decentralized formation pose estimation for spacecraft swarms. Adv. Space Res. **67**(11), 3527–3545 (2021)
529. P. Matzinger, The danger model: a renewed sense of self. Science **296**(5566), 301–305 (2002)
530. R. Mayet, J. Roberz, T. Schmickl, K. Crailsheim, Antbots: a feasible visual emulation of pheromone trails for swarm robots, in *Swarm Intelligence: 7th International Conference, ANTS 2010*, ed. by M. Dorigo, M. Birattari, G.A. Di Caro, R. Doursat, A.P. Engelbrecht, D. Floreano, L.M. Gambardella, R. Groß, E. Şahin, H. Sayama, T. Stützle. Lecture Notes in Computer Science, vol. 6234 (Springer, Berlin, 2010), pp. 84–94. ISBN 978-3-642-15460-7. https://doi.org/10.1007/978-3-642-15461-4
531. S. Mayya, G. Notomista, D. Shell, S. Hutchinson, M. Egerstedt, Non-uniform robot densities in vibration driven swarms using phase separation theory, in *IEEE/RSJ International Conference on Intelligent Robots and Systems (IROS)* (IEEE, 2019a), pp. 4106–4112
532. S. Mayya, P. Pierpaoli, M. Egerstedt, Voluntary retreat for decentralized interference reduction in robot swarms, in *International Conference on Robotics and Automation (ICRA)* (IEEE, 2019b), pp. 9667–9673
533. M.A. McEvoy, N. Correll, Materials that couple sensing, actuation, computation, and communication. Science **347**(6228) (2015). ISSN 0036-8075. https://doi.org/10.1126/science.1261689, http://science.sciencemag.org/content/347/6228/1261689
534. H. McGloin, M. Studley, R. Mawle, A. Winfield, Introducing the concept of repurposing robots; to increase their useful life, reduce waste, and improve sustainability in the robotics industry, in *Sustainable Ethics and Values in the Design and Regulation of Robotics and AI*, ed. by R.W. de Bruin, M.O. Tokhi, L. Belder, M.I.A. Ferreira, N.S. Govindarajulu, M.F. Silva (CLAWAR Association, 2023), pp. 16–27. ISBN 978-1-7396142-1-8, https://clawar.org/icres2023/
535. J. McLurkin, J. Smith, Distributed algorithms for dispersion in indoor environments using a swarm of autonomous mobile robots, in *Distributed Autonomous Robotic Systems Conference* (2004)
536. J. McLurkin, J. Smith, J. Frankel, D. Sotkowitz, D. Blau, B. Schmidt, Speaking swarmish: human-robot interface design for large swarms of autonomous mobile robots, in *AAAI Spring Symposium: To Boldly Go Where No Human-Robot Team has Gone Before* (2006), pp. 72–75
537. J. McLurkin, A.J. Lynch, S. Rixner, T.W. Barr, A. Chou, K. Foster, S. Bilstein, A low-cost multi-robot system for research, teaching, and outreach, in *Distributed Autonomous Robotic Systems* (Springer, 2013), pp. 597–609

538. H. Meinhardt, *Models of Biological Pattern Formation* (Academic, New York, 1982)
539. H. Meinhardt, *The Algorithmic Beauty of Sea Shells* (Springer, 2003)
540. H. Meinhardt, A. Gierer, Applications of a theory of biological pattern formation based on lateral inhibition. J. Cell Sci. **15**(2), 321–346 (1974). ISSN 0021-9533, http://view.ncbi.nlm.nih.gov/pubmed/4859215
541. H. Meinhardt, A. Gierer, Pattern formation by local self-activation and lateral inhibition. BioEssays **22**, 753–760 (2000)
542. H. Meinhardt, M. Klingler, A model for pattern formation on the shells of molluscs. J. Theor. Biol. **126**, 63–69 (1987)
543. T. Meister, R. Thenius, D. Kengyel, T. Schmickl, Cooperation of two different swarms controlled by BEECLUST algorithm, in *Mathematical Models for the Living Systems and Life Sciences (ECAL)* (2013), pp. 1124–1125
544. C. Melhuish, O. Holland, S. Hoddell, Convoying: using chorusing to form travelling groups of minimal agents, Robot. Auton. Syst. **28**(2–3), 207–216 (1999)
545. C. Melhuish, M. Wilson, A. Sendova-Franks, Patch sorting: multi-object clustering using minimalist robots, in *Advances in Artificial Life: 6th European Conference, ECAL*. ed. by J. Kelemen, P. Sosík (Springer, 2001), pp. 543–552. ISBN 978-3-540-44811-2, https://doi.org/10.1007/3-540-44811-X_62
546. C. Melhuish, O. Holland, S. Hoddell, Collective sorting and segregation in robots with minimal sensing, in *5th International Conference on the Simulation of Adaptive Behaviour (SAB)* (MIT Press, 1998)
547. D. Mellinger, M. Shomin, N. Michael, V. Kumar, Cooperative grasping and transport using multiple quadrotors, in *Distributed Autonomous Robotic Systems* (Springer, Berlin, 2013), pp. 545–558
548. D. Merkle, M. Middendorf, A. Scheidler, Swarm controlled emergence-designing an anti-clustering ant system, in *IEEE Swarm Intelligence Symposium* (IEEE, 2007), pp. 242–249
549. D. Merkle, M. Middendorf, A. Scheidler, Organic computing and swarm intelligence, in *Swarm Intelligence: Introduction and Applications*. ed. by C. Blum, D. Merkle (Springer, Berlin, 2008)
550. E. Merlo, C. Pinciroli, J. Panerati, M. Famelis, G. Beltrame, Automated extraction and checking of property models from source code for robot swarms, in *Proceedings of the 4th International Workshop on Robotics Software Engineering* (2022), pp. 47–54
551. B. Meyer, K. Ehlers, C. Isokeit, E. Maehle, The development of the modular hard-and software architecture of the autonomous underwater vehicle MONSUN, in *41st International Symposium on Robotics* (VDE, 2014)
552. B. Meyer, C. Renner, E. Maehle, Versatile sensor and communication expansion set for the autonomous underwater vehicle MONSUN, in *Advances in Cooperative Robotics: Proceedings of the 19th International Conference on CLAWAR 2016* (World Scientific, 2016), pp. 250–257
553. B. Meyer, M. Beekman, A. Dussutour, Noise-induced adaptive decision-making in ant-foraging, in *Simulation of Adaptive Behavior (SAB)*, Lecture Notes in Computer Science, vol. 5040 (Springer, Berlin, 2008), pp. 415–425
554. D. Mezey, D. Deffner, R.H.J.M. Kurvers, P. Romanczuk, Visual social information use in collective foraging. PLOS Comput. Biol. **20**(5), 1–24 (2024). https://doi.org/10.1371/journal.pcbi.1012087
555. D. Mezey, R. Bastien, Y. Zheng, N. McKee, D. Stoll, H. Hamann, P. Romanczuk, Purely vision-based collective movement of robots. NPJ Robot. **3**(1), 11 (2025)
556. John Stuart Mill, *A System of Logic: Ratiocinative and Inductive* (John W. Parker and Son, London, 1843)
557. A. Millard, *Exogenous fault detection in swarm robotic systems*. Ph.D. thesis, University of York, 2016
558. D. Milutinovic, P. Lima, Modeling and optimal centralized control of a large-size robotic population. IEEE Trans. Robot. **22**(6), 1280–1285 (2006)

559. D. Milutinovic, P. Lima, *Cells and Robots: Modeling and Control of Large-Size Agent Populations* (Springer, Berlin, Germany, 2007)
560. R.E. Mirollo, S.H. Strogatz, Synchronization of pulse-coupled biological oscillators. SIAM J. Appl. Math. **50**(6), 1645–1662 (1990)
561. C.W. Misner, K.S. Thorne, J.A. Wheeler, *Gravitation* (Princeton University Press, 2017)
562. E. Miya, Suggestion on superlinear speed up terminology. Netw. News Posting (1988)
563. G. Miyauchi, Y.K. Lopes, R. Groß, Sharing the control of robot swarms among multiple human operators: a user study, in *2023 IEEE/RSJ International Conference on Intelligent Robots and Systems (IROS)* (2023), pp. 8847–8853. https://doi.org/10.1109/IROS55552.2023.10342457
564. S. Mkhatshwa, G. Nitschke, Body and brain quality-diversity in robot swarms. ACM Trans. Evol. Learn. Optim. **5**(1) (2025). https://doi.org/10.1145/3664656
565. V. Mnih, K. Kavukcuoglu, D. Silver, A. Graves, I. Antonoglou, D. Wierstra, M. Riedmiller, Playing Atari with deep reinforcement learning (2013). arXiv:1312.5602
566. V. Mnih, K. Kavukcuoglu, D. Silver, A.A. Rusu, J. Veness, M.G. Bellemare, A. Graves, M. Riedmiller, A.K. Fidjeland, G. Ostrovski et al., Human-level control through deep reinforcement learning. Nature **518**(7540), 529–533 (2015)
567. M.E. Möbius, B.E. Lauderdale, S.R. Nagel, H.M. Jaeger, Brazil-nut effect: Size separation of granular particles. Nature **414**(6861), 270–270 (2001)
568. C. Moeslinger, T. Schmickl, K. Crailsheim, Emergent flocking with low-end swarm robots, in *Swarm Intelligence*, ed. by M. Dorigo, M. Birattari, G. Di Caro, R. Doursat, A. Engelbrecht, D. Floreano, L. Gambardella, R. Groß, E. Sahin, H. Sayama, T. Stützle. Lecture Notes in Computer Science, vol. 6234 (Springer, Berlin, 2010), pp. 424–431
569. C. Moeslinger, T. Schmickl, K. Crailsheim, A minimalist flocking algorithm for swarm robots, in *Advances in Artificial Life: Darwin Meets von Neumann (ECAL'09*, Lecture Notes in Computer Science, vol. 5778 (Springer, Berlin, 2011), pp. 357–382
570. R. Moioli, P.A. Vargas, P. Husbands, Exploring the kuramoto model of coupled oscillators in minimally cognitive evolutionary robotics tasks, in *WCCI 2010 IEEE World Congress on Computational Intelligence—CEC IEEE* (2010), pp. 2483–2490
571. M. Mokhtari, P.H.A. Mohamadi, M. Aernouts, R.K. Singh, B. Vanderborght, M. Weyn, J. Famaey, Energy-aware multi-robot task scheduling using meta-heuristic optimization methods for ambiently-powered robot swarms. Robot. Auton. Syst. **186**, 104898 (2025)
572. V.M. Monajjemi, J. Wawerla, R. Vaughan, G. Mori, HRI in the sky: creating and commanding teams of UAVs with a vision-mediated gestural interface, in *IEEE/RSJ International Conference on Intelligent Robots and Systems (IROS)* (2013), pp. 617–623. https://doi.org/10.1109/IROS.2013.6696415
573. F. Mondada, E. Franzi, P. Ienne, Mobile robot miniaturisation: a tool for investigation in control algorithms, in *Proceedings of the 3rd International Symposium on Experimental Robotics* (Springer, Berlin, 1994), pp. 501–513
574. F. Mondada, G.C. Pettinaro, A. Guignard, I. Kwee, D. Floreano, J.-L. Deneubourg, S. Nolfi, L.M. Gambardella, M. Dorigo, SWARM-BOT: a New Distributed Robotic Concept. Auton. Robots special Issue Swarm Robot. **17**(2–3), 193–221 (2004)
575. F. Mondada, M. Bonani, A. Guignard, S. Magnenat, C. Studer, D. Floreano, Superlinear physical performances in a SWARM-BOT, in *Proceedings of the 8th European Conference on Artificial Life (ECAL)*, ed. by M.S. Capcarrere. Lecture Notes in Computer Science, vol. 3630 (Springer, Berlin, 2005), pp. 282–291. ISBN 978-3-540-28848-0
576. F. Mondada, M. Bonani, X. Raemy, J. Pugh, C. Cianci, A. Klaptocz, S. Magnenat, J.-C. Zufferey, D. Floreano, A. Martinoli, The e-puck, a robot designed for education in engineering, in*Proceedings of the 9th Conference on Autonomous Robot Systems and Competitions*, vol. 1 (2009), pp. 59–65
577. L. Mondada, M.E. Karim, F. Mondada, Electroencephalography as implicit communication channel for proximal interaction between humans and robot swarms. Swarm Intell. **10**(4), 247–265 (2016). ISSN 1935-3820. https://doi.org/10.1007/s11721-016-0127-0
578. M.M. de Oca, E. Ferrante, A. Scheidler, C. Pinciroli, M. Birattari, M. Dorigo, Majority-rule opinion dynamics with differential latency: a mechanism for self-organized collective

decision-making. Swarm Intell. **5**, 305–327 (2011). ISSN 1935-3812. https://doi.org/10.1007/s11721-011-0062-z

579. J.-B. Mouret, J. Clune, Illuminating search spaces by mapping elites (2015). arXiv:1504.04909
580. M. Moussaid, S. Garnier, G. Theraulaz, D. Helbing, Collective information processing and pattern formation in swarms, flocks, and crowds. Top. Cogn. Sci. **1**(3), 469–497 (2009)
581. M. Moussaïd, D. Helbing, G. Theraulaz, How simple rules determine pedestrian behavior and crowd disasters. Proc. Natl. Acad. Sci. **108**(17), 6884–6888 (2011)
582. M. Moussaïd, M. Kapadia, T. Thrash, R.W. Sumner, M. Gross, D. Helbing, C. Hölscher, Crowd behaviour during high-stress evacuations in an immersive virtual environment. J. R. Soc. Interface **13**(122), 20160414 (2016)
583. J.D. Murray, A prepattern formation mechanism for animal coat markings. J. Theor. Biol. **88**, 161–199 (1981)
584. J.D. Murray, On the mechanochemical theory of biological pattern formation with application to vasculogenesis. C.R. Biol. **326**(2), 239–252 (2003)
585. V.C. Müller, Ethics of artificial intelligence and robotics, in *The Stanford Encyclopedia of Philosophy*. ed. by E.N. Zalta, U. Nodelman (Metaphysics Research Lab, Stanford University, Fall, 2025)
586. S. Na, Y. Qiu, A.E. Turgut, J. Ulrich, T. Krajník, S. Yue, B. Lennox, F. Arvin, Bio-inspired artificial pheromone system for swarm robotics applications. Adapt. Behav. **29**(4), 395–415 (2021). https://doi.org/10.1177/1059712320918936
587. S. Na, T. Rouček, J. Ulrich, J. Pikman, T. Krajník, B. Lennox, F. Arvin, Federated reinforcement learning for collective navigation of robotic swarms. IEEE Trans. Cogn. Dev. Syst. **15**(4), 2122–2131 (2023)
588. S. Nagavalli, L. Luo, N. Chakraborty, K. Sycara, Neglect benevolence in human control of robotic swarms, in *IEEE International Conference on Robotics and Automation (ICRA)* (2014), pp. 6047–6053. https://doi.org/10.1109/ICRA.2014.6907750
589. J. Nagi, A. Giusti, L.M. Gambardella, G.A. Di Caro, Human-swarm interaction using spatial gestures, in *2014 IEEE/RSJ International Conference on Intelligent Robots and Systems* (2014), pp. 3834–3841. https://doi.org/10.1109/IROS.2014.6943101
590. R. Nair, T. Ito, M. Tambe, S. Marsella, Task allocation in the RoboCup rescue simulation domain: a short note, in *RoboCup 2001: Robot Soccer World Cup V*, vol. 2377, ed. by A. Birk, S. Coradeschi, S. Tadokoro (Springer, Berlin, 2002), pp. 1–22. https://doi.org/10.1007/3-540-45603-1_129
591. I. Napolitano, A. Lama, F. De Lellis, M. di Bernardo, mergent cooperative strategies for multi-agent shepherding via reinforcement learning (2024). arXiv:2411.05454
592. R. De Nardi, O.E. Holland, Ultraswarm: a further step towards a flock of miniature helicopters, in *Swarm Robotics—Second SAB 2006 International Workshop*, ed. by E. Şahin, W.M. Spears, A.F.T. Winfield. Lecture Notes in Computer Science, vol. 4433 (Springer, Berlin, 2007), pp. 116–128
593. J. Nauta, S. Van Havermaet, P. Simoens, Y. Khaluf, Enhanced foraging in robot swarms using collective Lévy walks, *24th European Conference on Artificial Intelligence (ECAI)* (IOS Press, 2020), pp. 171–178
594. Z. Néda, E. Ravasz, T. Vicsek, Y. Brechet, A.L. Barabási, Physics of the rhythmic applause. Phys. Rev. E **61**, 6987–6992 (2000)
595. V. Niculescu, T. Polonelli, M. Magno, L. Benini, NanoSLAM: Enabling fully onboard slam for tiny robots. IEEE Int. Things J. **11**(8), 13584–13607 (2023)
596. G. Nitschke, S. Didi, Evolutionary policy transfer and search methods for boosting behavior quality: RoboCup keep-away case study. Front. Robot. AI **4**, 62 (2017)
597. S. Nolfi, D. Floreano, *Evolutionary Robotics: The Biology, Intelligence, and Technology of Self-Organizing Machines* (MIT Press, 2000)
598. S. Nolfi, J. Bongard, P. Husbands, D. Floreano, Evolutionary robotics, in *Springer Handbook of Robotics* (Springer, 2016), pp. 2035–2068
599. S. Nouyan, A. Campo, M. Dorigo, Path formation in a robot swarm: self-organized strategies to find your way home. Swarm Intell. **2**(1), 1–23 (2008)

600. S. Nouyan, R. Groß, M. Bonani, F. Mondada, M. Dorigo, Teamwork in self-organized robot colonies. IEEE Trans. Evol. Comput. **13**(4), 695–711 (2009)
601. R. O'Grady, R. Groß, F. Mondada, M. Bonani, M. Dorigo, Self-assembly on demand in a group of physical autonomous mobile robots navigating rough terrain, in *Advances in Artificial Life, 8th European Conference (ECAL)* (Springer, Berlin, 2005), pp. 272–281
602. J. O'Keeffe, Anticipating degradation: a predictive approach to fault tolerance in robot swarms (2025). arXiv:2504.01594
603. R. Olfati-Saber, R.M. Murray, Consensus problems in networks of agents with switching topology and time-delays. IEEE Trans. Autom. Control **49**(9), 1520–1533 (2004)
604. R. Olfati-Saber, A. Fax, R.M. Murray, Consensus and cooperation in networked multi-agent systems. Proc. IEEE **95**(1), 215–233 (2007)
605. D.R. Olsen Jr., S.B. Wood, Fan-out: Measuring human control of multiple robots, in *Proceedings of the SIGCHI Conference on Human Factors in Computing Systems* (2004), pp. 231–238
606. L. Onsager, R.M. Fuoss, Irreversible processes in electrolytes diffusion, conductance and viscous flow in arbitrary mixtures of strong electrolytes. J. Phys. Chem. **36**(11), 2689–2778 (1932)
607. E.H. Østergaard, G.S. Sukhatme, M.J. Matarić, Emergent bucket brigading: a simple mechanisms for improving performance in multi-robot constrained-space foraging tasks, in *Proceedings of the fifth international conference on Autonomous agents (AGENTS'01)*. ed. by E. André, S. Sen, C. Frasson, J.P. Müller (ACM, New York, NY, USA, 2001), pp. 29–35. ISBN 1-58113-326-X, https://doi.org/10.1145/375735.375825
608. C. Osterloh, T. Pionteck, E. Maehle, MONSUN II: a small and inexpensive AUV for underwater swarms, in *ROBOTIK 2012—7th German Conference on Robotics* (2012)
609. W. Ostwald, Studien über die Bildung und Umwandlung fester Körper. Z. Phys. Chem. **22**(1), 289–330 (1897)
610. A. Özdemir, M. Gauci, R. Groß, Shepherding with robots that do not compute, in *Artificial Life Conference Proceedings* (MIT Press One Rogers Street, Cambridge, MA). ISBN 02142-1209, 2017, pp. 332–339
611. A. Özdemir, M. Gauci, S. Bonnet, R. Groß, Finding consensus without computation. IEEE Robot. Autom. Lett. **3**(3), 1346–1353 (2018). https://doi.org/10.1109/LRA.2018.2795640
612. K.P. O'Keeffe, H. Hong, S.H. Strogatz, Oscillators that sync and swarm. Nat. Commun. **8**(1), 1504 (2017)
613. A. Paas, E.B.J. Coffey, G. Beltrame, D. St-Onge, Towards evaluating the impact of swarm robotic control strategy on operators' cognitive load, in *31st IEEE International Conference on Robot and Human Interactive Communication (RO-MAN)* (IEEE, 2022), pp. 217–223
614. A. Pacheco, U. Denis, R. Zakir, V. Strobel, A. Reina, M. Dorigo, Toychain: a simple Blockchain for research in swarm robotics (2024). arXiv:2407.06630
615. T.M. Paine, M.R. Benjamin, A model for multi-agent autonomy that uses opinion dynamics and multi-objective behavior optimization, in *2024 IEEE International Conference on Robotics and Automation (ICRA)* (IEEE, 2024), pp. 8305–8311
616. B. Pang, Y. Song, C. Zhang, R. Yang, Effect of random walk methods on searching efficiency in swarm robots for area exploration. Appl. Intell. **51**(7), 5189–5199 (2021)
617. M. Papadopoulou, M. Ball, P. Bartashevich, A.L.J. Burns, V. Chiara, M.A. Clark, B.R. Costelloe, M. Fele, F. French, S. Hauert, et al., Active interactions between animals and technology: biohybrid approaches for animal behaviour research. Animal Behav. 123160 (2025)
618. J.K. Parrish, L. Edelstein-Keshet, Complexity, pattern, and evolutionary trade-offs in animal aggregation. Science **284**(5411), 99–101 (1999.) ISSN 0036-8075. https://doi.org/10.1126/science.284.5411.99
619. H.V.D. Parunak, S.A. Brueckner, Engineering swarming systems, in *Methodologies and Software Engineering for Agent Systems* (Springer, Berlin, 2004), pp. 341–376
620. J.M. Pastor, A. Garcimartín, P.A. Gago, J.P. Peralta, C. Martín-Gómez, L.M. Ferrer, D. Maza, D.R. Parisi, L.A. Pugnaloni, I. Zuriguel, Experimental proof of faster-is-slower in systems of frictional particles flowing through constrictions. Phys. Rev. E **92**, 062817 (2015)

621. J. Patel, C. Pinciroli, Improving human performance using mixed granularity of control in multi-human multi-robot interaction, in *29th IEEE International Conference on Robot and Human Interactive Communication (RO-MAN)* (IEEE, 2020), pp. 1135–1142
622. J. Patel, P. Sonar, C. Pinciroli, On multi-human multi-robot remote interaction: a study of transparency, inter-human communication, and information loss in remote interaction. Swarm Intell. **16**(2), 107–142 (2022)
623. D. Payton, M. Daily, R. Estowski, M. Howard, C. Lee, Pheromone robotics. Auton. Robot. **11**(3), 319–324 (2001)
624. H. Pedersen, K. Singh, Organic labyrinths and mazes, in *Proceedings of the 4th International Symposium on Non-photorealistic Animation and Rendering* (2006), pp. 79–86
625. F.T. Peires, Tensile tests for cotton yarns. J. Textile Inst. 355–368 (1926)
626. B. Pendleton, M. Goodrich, Scalable human interaction with robotic swarms, in *AIAA Infotech Aerospace Conference* (2013), p. 4731
627. K. Petersen, R. Nagpal, J. Werfel, Termes: an autonomous robotic system for three-dimensional collective construction, in *Proceedings of Robotics: Science and Systems VII* (2011), pp. 257–264
628. K. Pfister, H. Hamann, Collective decision-making and change detection with Bayesian robots in dynamic environments, in *IEEE/RSJ International Conference on Intelligent Robots and Systems* (IEEE, 2023), pp. 8814–8819
629. L. Pichierri, A. Testa, G. Notarstefano, Crazychoir: flying swarms of crazyflie quadrotors in ROS 2. IEEE Robot. Autom. Lett. **8**(8), 4713–4720 (2023). https://doi.org/10.1109/LRA.2023.3286814
630. D. Pickem, M. Lee, M. Egerstedt, The GRITSBot in its natural habitat: a multi-robot testbed, in *IEEE International Conference on Robotics and Automation (ICRA)* (IEEE, 2015), pp. 4062–4067
631. M. Pilch, J. Altmann, D. Suter, Survey of the status of small armed and unarmed uninhabited aircraft. Technical Report Report No. 1, Chair Experimental Physics III, TU Dortmund University, February 2021. Preventive Arms Control for Small and Very Small Armed Aircraft and Missiles
632. C. Pinciroli, G. Beltrame, Buzz: an extensible programming language for heterogeneous swarm robotics, in *IEEE/RSJ International Conference on Intelligent Robots and Systems (IROS)* (IEEE, 2016), pp. 3794–3800
633. C. Pinciroli, V. Trianni, R. O'Grady, G. Pini, A. Brutschy, M. Brambilla, N. Mathews, E. Ferrante, G. Di Caro, F. Ducatelle, et al., ARGoS: a modular, multi-engine simulator for heterogeneous swarm robotics, in *2011 IEEE/RSJ International Conference on Intelligent Robots and Systems* (IEEE, 2011), pp. 5027–5034
634. C. Pinciroli, V. Trianni, R. O'Grady, G. Pini, A. Brutschy, M. Brambilla, N. Mathews, E. Ferrante, G. Caro, F. Ducatelle, M. Birattari, L.M. Gambardella, M. Dorigo, ARGoS: a modular, parallel, multi-engine simulator for multi-robot systems. Swarm Intell. **6**(4), 271–295 (2012). ISSN 1935-3812. https://doi.org/10.1007/s11721-012-0072-5
635. C. Pinciroli, M.S. Talamali, A. Reina, J.A.R. Marshall, V. Trianni, Simulating Kilobots within ARGoS: models and experimental validation, in *International Conference on Swarm Intelligence (ANTS)* (Springer, Berlin, 2018), pp. 176–187
636. G. Pini, A. Brutschy, M. Birattari, M. Dorigo, Interference reduction through task partitioning in a robotic swarm, in *Sixth International Conference on Informatics in Control, Automation and Robotics–ICINCO* (2009), pp. 52–59
637. G. Pini, A. Brutschy, G. Francesca, M. Dorigo, M. Birattari, Multi-armed bandit formulation of the task partitioning problem in swarm robotics, in *8th International Conference on Swarm Intelligence (ANTS)* (Springer, Berlin, 2012), pp. 109–120
638. R. Pinsler, M. Maag, O. Arenz, G. Neumann, Inverse reinforcement learning of bird flocking behavior, in *ICRA Swarms Workshop* (2018)
639. B. Piranda, J. Bourgeois, Designing a quasi-spherical module for a huge modular robot to create programmable matter. Auton. Robot. **42**(8), 1619–1633 (2018)

640. A. Pirrone, A. Reina, T. Stafford, J.A.R. Marshall, F. Gobet, Magnitude-sensitivity: rethinking decision-making. Trends Cogn. Sci. **26**(1), 66–80 (2022)
641. M. Planck, Über einen Satz der statistischen Dynamik und seine Erweiterung in der Quantentheorie. Sitzungsberichte der Preußischen Akademie der Wissenschaften **24**(324–341) (1917)
642. G. Podevijn, R. O'Grady, N Mathews, A. Gilles, C. Fantini-Hauwel, M. Dorigo, Investigating the effect of increasing robot group sizes on the human psychophysiological state in the context of human–swarm interaction. Swarm Intell. 1–18 (2016). ISSN 1935-3820. https://doi.org/10.1007/s11721-016-0124-3
643. L.-A. Poissonnier, S. Motsch, J. Gautrais, C. Buhl, A. Dussutour, Experimental investigation of ant traffic under crowded conditions. eLife **8**, e48945 (2019). ISSN 2050-084X. https://doi.org/10.7554/eLife.48945
644. G. Pólya, F. Eggenberger, Über die Statistik verketteter Vorgänge. Z. Angew. Math. Mech. **3**(4), 279–289 (1923)
645. G. Popkin, The physics of life. Nature **529**, 16–18 (2016). https://doi.org/10.1038/529016a
646. M.A. Potter, L.A. Meeden, A.C. Schultz, Heterogeneity in the coevolved behaviors of mobile robots: the emergence of specialists, in *International Joint Conference on Artificial Intelligence (IJCAI)* (Morgan Kaufmann, 2001), pp. 1337–1343
647. S.h Pourmehr, V.M. Monajjemi, R. Vaughan, G. Mori, you two! Take off!: creating, modifying and commanding groups of robots using face engagement and indirect speech in voice commands, in *IEEE/RSJ International Conference on Intelligent Robots and Systems (IROS)* (2013), pp. 137–142. https://doi.org/10.1109/IROS.2013.6696344
648. J. Prasetyo, G. De Masi, P. Ranjan, E. Ferrante, The best-of-n problem with dynamic site qualities: achieving adaptability with stubborn individuals, in *Swarm Intelligence: 11th International Conference, (ANTS)* (Springer, Berlin, 2018), pp. 239–251
649. W.H. Press, S.A. Teukolsky, W.T. Vetterling, B.P. Flannery, *Numerical Recipes in C++* (Cambridge University Press, 2002)
650. I. Prigogine, *The End of Certainty: Time, Chaos, and The New Laws of Nature* (Free Press, 1997)
651. A. Prorok, N. Correll, A. Martinoli, Multi-level spatial models for swarm-robotic systems. Int. J. Robot. Res. **30**(5), 574–589 (2011)
652. A. Prorok, M.A. Hsieh, V. Kumar, Fast redistribution of a swarm of heterogeneous robots, in *International Conference on Bio-inspired Information and Communications Technologies (BICT)* (2015)
653. A. Prorok, M.A. Hsieh, V. Kumar, Formalizing the impact of diversity on performance in a heterogeneous swarm of robots, in *IEEE International Conference on Robotics and Automation (ICRA)* (2016), pp. 5364–5371. https://doi.org/10.1109/ICRA.2016.7487748
654. A. Prorok, M. Malencia, L. Carlone, G.S. Sukhatme, B.M. Sadler, V. Kumar, Beyond robustness: a taxonomy of approaches towards resilient multi-robot systems (2021). arXiv:2109.12343
655. J. Pugh, A. Martinoli, Multi-robot learning with particle swarm optimization, in *Proceedings of the Fifth International Joint Conference on Autonomous Agents and Multiagent Systems* (2006), pp. 441–448
656. F. Qian, A. Gerber, Z.M. Mao, S. Sen, O. Spatscheck, W. Willinger, Tcp revisited: a fresh look at tcp in the wild, in *Proceedings of the 9th ACM SIGCOMM Conference on Internet Measurement* (2009), pp. 76–89
657. M. Quigley, K. Conley, B. Gerkey, J. Faust, T. Foote, J. Leibs, R. Wheeler, A.Y. Ng et al., ROS: an open-source robot operating system, in *ICRA Workshop on Open Source Software* (Kobe, Japan, 2009)
658. F. Raischel, F. Kun, H.J. Herrmann, Fiber bundle models for composite materials, in *Conference on Damage in Composite Materials* (2006)
659. J.F. Ramaley, Buffon's noodle problem. Am. Math. Mon. **76**(8), 916–918 (1969). http://www.jstor.org/stable/2317945
660. Jonatan Pena Ramirez and Henk Nijmeijer, The secret of the synchronized pendulums. Phys. World **33**(1), 36 (2020)

661. A. Rankin, M. Maimone, J. Biesiadecki, N. Patel, D. Levine, O. Toupet, Mars curiosity rover mobility trends during the first 7 years. J. Field Robot. **38**(5), 759–800 (2021)
662. M. Raoufi, P. Romanczuk, H. Hamann, Estimation of continuous environments by robot swarms: Correlated networks and decision-making, in *2023 IEEE International Conference on Robotics and Automation (ICRA)* (IEEE, 2023a), pp. 5486–5492
663. M. Raoufi, P. Romanczuk, H. Hamann, Individuality in swarm robots with the case study of Kilobots: Noise, bug, or feature? in *Artificial Life Conference Proceedings*, vol. 35 (MIT Press, 2023b)
664. M. Raoufi, P. Romanczuk, H. Hamann, LARS: Light augmented reality system for swarm, in *International Conference on Swarm Intelligence (ANTS)*, vol. 14987 (Springer, Berlin, 2024)
665. S. Rasouli, K. Dautenhahn, C.L. Nehaniv, Simulation of a bio-inspired flocking-based aggregation behaviour in swarm robotics. Biomimetics **9**(11) (2024). ISSN 2313-7673. https://doi.org/10.3390/biomimetics9110668, https://www.mdpi.com/2313-7673/9/11/668
666. F.L.W. Ratnieks, C. Anderson, Task partitioning in insect societies. Insectes Sociaux **46**(2), 95–108 (1999). https://doi.org/10.1007/s000400050119
667. I. Rechenberg, *Evolutionsstrategie* (Optimierung technischer Systeme nach Prinzipien der biologischen Evolution, Frommann Holzboog, 1973). ISBN: 3-7728-0373-3
668. A. Reina, M. Dorigo, V. Trianni, Towards a cognitive design pattern for collective decision-making, in *Swarm Intelligence*, ed. by M. Dorigo, M. Birattari, S. Garnier, H. Hamann, M.M. de Oca, C. Solnon, T. Stützle. Lecture Notes in Computer Science, vol. 8667 (Springer International Publishing, 2014), pp. 194–205. ISBN 978-3-319-09951-4, https://doi.org/10.1007/978-3-319-09952-1_17
669. A. Reina, G. Valentini, C. Fernández-Oto, M. Dorigo, V. Trianni, A design pattern for decentralised decision making. PLOS ONE **10**(10), 1–18 (2015). https://doi.org/10.1371/journal.pone.0140950
670. A. Reina, A.J. Cope, E. Nikolaidis, J.A.R. Marshall, C. Sabo, ARK: augmented reality for Kilobots. IEEE Robot. Autom. Lett. **2**(3), 1755–1761 (2017)
671. A. Reina, R. Zakir, G. De Masi, E. Ferrante, Cross-inhibition leads to group consensus despite the presence of strongly opinionated minorities and asocial behaviour. Commun. Phys. **6**(1), 236 (2023)
672. M. Resnick, *Turtles, Termites, and Traffic Jams* (MIT Press, 1994)
673. A.M. Reynolds, C.J. Rhodes, The Lévy flight paradigm: random search patterns and mechanisms. Ecology **90**(4), 877–887 (2009)
674. C.W. Reynolds, Flocks, herds, and schools. Comput. Graph. **21**(4), 25–34 (1987)
675. F. Riedo, M. Chevalier, S. Magnenat, F. Mondada, Thymio II, a robot that grows wiser with children, in *IEEE Workshop on Advanced Robotics and Its Social Impacts (ARSO 2013)* (IEEE, 2013), pp. 187–193
676. M. Ringelmann, Recherches sur les moteurs animés: Travail de l'homme. Annales de l'Insitut National Agronomique **12**(1) (1913)
677. H. Risken, *The Fokker-Planck Equation* (Springer, Berlin, Germany, 1984)
678. S. Riso, D. Adăscăliţei, Human–robot interaction: what changes in the workplace? (2024). https://www.eurofound.europa.eu/en/publications/2024/human-robot-interaction-what-changes-workplace. Accessed 10 Jan 2025
679. E. Rohmer, S.P.N. Singh, M. Freese, V-REP: a versatile and scalable robot simulation framework, in *2013 IEEE/RSJ International Conference on Intelligent Robots and Systems* (IEEE, 2013), pp. 1321–1326
680. D. Romano, E. Donati, G. Benelli, C. Stefanini, A review on animal-robot interaction: from bio-hybrid organisms to mixed societies. Biol. Cybern. **113**, 201–225 (2019)
681. A. Rosenblueth, N. Wiener, J. Bigelow, Behavior, purpose and teleology. Philos. Sci. **10**(1), 8–24 (1943). ISSN 00318248, 1539767X, http://www.jstor.org/stable/184878
682. A. Rosenfeld, G.A. Kaminka, S. Kraus, A study of scalability properties in robotic teams, in *Coordination of Large-Scale Multiagent Systems*. ed. by P. Scerri, R. Vincent, R. Mailler (Springer, US, Boston, MA, 2006), pp. 27–51. ISBN 978-0-387-27972-5, https://doi.org/10.1007/0-387-27972-5_2

683. S. Ross, G. Gordon, D. Bagnell, A reduction of imitation learning and structured prediction to no-regret online learning, in *Proceedings of the Fourteenth International Conference on Artificial Intelligence and Statistics. JMLR Workshop and Conference Proceedings* (2011), pp. 627–635
684. M. Rubenstein, W.-M. Shen, Scalable self-assembly and self-repair in a collective of robots, in *Proceedings of the IEEE/RSJ International Conference on Intelligent Robots and Systems (IROS)* (St. Louis, Missouri, USA, 2009)
685. M. Rubenstein, C. Ahler, R. Nagpal, Kilobot: a low cost scalable robot system for collective behaviors, in *IEEE International Conference on Robotics and Automation (ICRA)* (2012), pp. 3293–3298. https://doi.org/10.1109/ICRA.2012.6224638
686. M. Rubenstein, A. Cornejo, R. Nagpal, Programmable self-assembly in a thousand-robot swarm. Science **345**(6198), 795–799 (2014)
687. D.E. Rumelhart, G.E. Hinton, R.J. Williams, Learning representations by back-propagating errors. Nature **323**, 533–536 (1986)
688. B. Russell, Vagueness. Australas. J. Psychol. Philos. **1**(2), 84–92 (1923)
689. R.A. Russell, Heat trails as short-lived navigational markers for mobile robots. Proc. IEEE Int. Conf. Robot. Autom. **4**, 3534–3539 (1997)
690. S.J. Russell, P. Norvig *Artificial Intelligence: A Modern Approach* (Prentice Hall, 1995)
691. A.S. Amjadi, M. Raoufi, A.E. Turgut, A self-adaptive landmark-based aggregation method for robot swarms. Adapt. Behav. **30**(3), 223–236 (2022)
692. A. Saha, J.A.R. Marshall, A. Reina, Memory and communication efficient algorithm for decentralized counting of nodes in networks. PloS one **16**(11), e0259736 (2021)
693. M. Salahshour, I.D. Couzin, Allocentric flocking. bioRxiv (2025)
694. E. Salvato, G. Fenu, E. Medvet, F.A. Pellegrino, Crossing the reality gap: A survey on sim-to-real transferability of robot controllers in reinforcement learning. IEEE Access **9**, 153171–153187 (2021)
695. M. Santos, M. Egerstedt, From motions to emotions: can the fundamental emotions be expressed in a robot swarm? Int. J. Soc. Robot. **13**(4), 751–764 (2021)
696. G. Sartoretti, M.-O. Hongler, Interacting Brownian swarms: some analytical results. Entropy **18**(1), 27 (2016)
697. G. Sartoretti, Y. Wu, W. Paivine, T.K. Satish Kumar, S. Koenig, H. Choset, Distributed reinforcement learning for multi-robot decentralized collective construction, in *Distributed Autonomous Robotic Systems: The 14th International Symposium* (Springer, Berlin, 2019), pp. 35–49
698. H. Sato, C.W. Berry, B.E. Casey, G. Lavella, Y. Yao, J.M. VandenBrooks, M.M. Maharbiz, A cyborg beetle: insect flight control through an implantable, tetherless microsystem, in *2008 IEEE 21st International Conference on Micro Electro Mechanical Systems* (2008), pp. 164–167. https://doi.org/10.1109/MEMSYS.2008.4443618
699. A.V. Savkin, Coordinated collective motion of groups of autonomous mobile robots: analysis of Vicsek's model. IEEE Trans. Autom. Control **49**(6), 981–982 (2004)
700. H. Sayama, Swarm chemistry. Artif. Life **15**(1), 105–114 (2009). https://doi.org/10.1162/artl.2009.15.1.15107
701. H. Sayama, Robust morphogenesis of robotic swarms. IEEE Comput. Intell. Mag. **5**(3), 43–49 (2010)
702. S. Sayin, E. Couzin-Fuchs, I. Petelski, Y. Günzel, M. Salahshour, C.-Y. Lee, J.M. Graving, L. Li, O. Deussen, G.A. Sword, I.D. Couzin, The behavioral mechanisms governing collective motion in swarming locusts. Science **387**(6737), 995–1000 (2025)
703. A. Scheidler, Dynamics of majority rule with differential latencies. Phys. Rev. E **83**(3), 031116 (2011)
704. A. Scheidler, D. Merkle, M. Middendorf, Swarm controlled emergence for ant clustering. Int. J. Intell. Comput. Cybern. **6**(1), 62–82 (2013)
705. F. Schilling, J. Lecoeur, F. Schiano, D. Floreano, Learning vision-based flight in drone swarms by imitation. IEEE Robot. Autom. Lett. **4**(4), 4523–4530 (2019)
706. J.W.P. Schmelzer, *Nucleation Theory and Applications* (Wiley, 2006)

707. T. Schmickl, R. Thenius, C. Moslinger, J. Timmis, A. Tyrrell, M. Read, J. Hilder, J. Halloy, A. Campo, C. Stefanini, L. Manfredi, S. Orofino, S. Kernbach, T. Dipper, and D. Sutantyo. Cocoro – the self-aware underwater swarm, in *2011 Fifth IEEE Conference on Self-Adaptive and Self-Organizing Systems Workshops* (2011a), pp. 120–126. https://doi.org/10.1109/SASOW.2011.11
708. T. Schmickl, K. Crailsheim, Costs of environmental fluctuations and benefits of dynamic decentralized foraging decisions in honey bees. Adapt. Behav. Animals, Animats, Softw. Agents Robots Adapt. Syst. **12**, 263–277 (2004)
709. T. Schmickl, K. Crailsheim, Trophallaxis among swarm-robots: a biologically inspired strategy for swarm robotics, in *First IEEE/RAS-EMBS International Conference on Biomedical Robotics and Biomechatronics* (IEEE, 2006), pp. 377–382
710. T. Schmickl, H. Hamann, BEECLUST: a swarm algorithm derived from honeybees, in *Bio-inspired Computing and Communication Networks*. ed. by Y. Xiao (CRC Press, Boca Raton, FL, USA, 2011), pp. 95–137
711. T. Schmickl, C. Möslinger, K. Crailsheim, Collective perception in a robot swarm, in *Swarm Robotics—Second SAB 2006 International Workshop*, ed. by E. Şahin, W.M. Spears, A.F.T. Winfield. Lecture Notes in Computer Science, vol. 4433 (Springer, Berlin, 2007)
712. T. Schmickl, R. Thenius, C. Möslinger, G. Radspieler, S. Kernbach, K. Crailsheim, Get in touch: Cooperative decision making based on robot-to-robot collisions. Auton. Agent. Multi-Agent Syst. **18**(1), 133–155 (2008)
713. T. Schmickl, H. Hamann, H. Wörn, K. Crailsheim, Two different approaches to a macroscopic model of a bio-inspired robotic swarm. Robot. Auton. Syst. **57**(9), 913–921 (2009). https://doi.org/10.1016/j.robot.2009.06.002
714. T. Schmickl, H. Hamann, K. Crailsheim, Modelling a hormone-inspired controller for individual- and multi-modular robotic systems. Math. Comput. Model. Dyn. Syst. **17**(3), 221–242 (2011)
715. T. Schmickl, J. Stradner, H. Hamann, L. Winkler, K. Crailsheim, Major feedback loops supporting artificial evolution in multi-modular robotics, in *New Horizons in Evolutionary Robotics*, ed. by S. Doncieux, N. Bredèche, J.-B. Mouret. Studies in Computational Intelligence, vol. 341 (Springer, Berlin, 2011c), pp. 195–209. ISBN 978-3-642-18271-6, https://doi.org/10.1007/978-3-642-18272-3
716. T.C. Schneirla, A unique case of circular milling in ants, considered in relation to trail following and the general problem of orientation. Am. Mus. Novit. **1253**, 1–25 (1944)
717. T.C. Schneirla, Army Ants: A Study in Social Organization, in ed. by H.R. Topoff (W.H. Freeman and Company, San Francisco, CA, 1971)
718. M. Schranz, M. Umlauft, M. Sende, W. Elmenreich, Swarm robotic behaviors and current applications. Front. Robot. AI **7**, 36 (2020)
719. M. Schranz, W. Elmenreich, F. Arvin, *Engineering Swarms of Cyber-Physical Systems* (CRC Press, 2025)
720. A.C. Schultz, J.J. Grefenstette, W. Adams, Robo-shepherd: Learning complex robotic behaviors, in *Proceedings of the International Symposium on Robotics and Automation (ICRA)*, vol. 6, ed. by M. Jamshidi, F. Pin, P. Dauchez (ASME Press, 1996), pp. 763–768
721. R. Schumacher, Book review: Achim Stephan: Emergenz. Von der Unvorhersagbarkeit zur Selbstorganisation. Dresden/München: Dresden University Press, 1999. Eur. J. Philos. **10**(3), 415–419 (2002)
722. F. Schweitzer, Brownian agent models for swarm and chemotactic interaction, in *Fifth German Workshop on Artificial Life. Abstracting and Synthesizing the Principles of Living Systems*. ed. by D. Polani, J. Kim, T. Martinetz (Akademische Verlagsgesellschaft Aka, 2002), pp. 181–190
723. F. Schweitzer, *Brownian Agents and Active Particles. On the Emergence of Complex Behavior in the Natural and Social Sciences* (Springer, Berlin, Germany, 2003)
724. F. Schweitzer, L. Schimansky-Geier, Clustering of active walkers in a two-component system. Phys. A **206**, 359–379 (1994)

725. F. Schweitzer, K. Lao, F. Family, Active random walkers simulate trunk trail formation by ants. Biosystems **41**, 153–166 (1997)
726. T.D. Seeley, P.K. Visscher, T. Schlegel, P.M. Hogan, N.R. Franks, J.A.R. Marshall, Stop signals provide cross inhibition in collective decision-making by honeybee swarms. Science **335**(6064), 108–111 (2012). https://doi.org/10.1126/science.1210361
727. G. Sempo, S. Depickère, J.-M. Amé, C. Detrain, J. Halloy, J.-L. Deneubourg, Integration of an autonomous artificial agent in an insect society: Experimental validation, in *From Animals to Animats 9: 9th International Conference on Simulation of Adaptive Behavior, SAB 2006, Rome, Italy, September 25-29, 2006. Proceedings*. ed. by S. Nolfi, G. Baldassarre, R. Calabretta, J.C.T. Hallam, D. Marocco, J.-A. Meyer, O. Miglino, D. Parisi (Springer, Berlin, 2006), pp. 703–712. ISBN 978-3-540-38615-5, https://doi.org/10.1007/11840541_58
728. J. Seyfried, M. Szymanski, N. Bender, R. Estaña, M. Thiel, H. Wörn, The I-SWARM project: Intelligent small world autonomous robots for micro-manipulation, in *Swarm Robotics Workshop: State-of-the-Art Survey*. ed. by E. Şahin, W.M. Spears (Springer, Berlin, 2005), pp. 70–83
729. A.J.C. Sharkey, Swarm robotics and minimalism. Connection Sci. **19**(3), 245–260 (2007)
730. Q. Shi, H. Ishii, S. Kinoshita, A. Takanishi, S. Okabayashi, N. Iida, H. Kimura, S. Shibata, Modulation of rat behaviour by using a rat-like robot. Bioinspiration Biomimetics **8**(4), 046002 (2013)
731. Z. Shi, J. Zhang, Z. Shang, M. Fan, W. Song, The effect of obstacle layouts on regulating luggage-laden pedestrian flow through bottlenecks. Physica A: Stat. Mech. Appl. **608**, 128255 (2022). ISSN 0378-4371, https://doi.org/10.1016/j.physa.2022.128255, https://www.sciencedirect.com/science/article/pii/S0378437122008135
732. A.N. Shiryaev, *Optimal Stopping Rules*, vol. 8 (Springer Science & Business Media, 2007)
733. R.M. Sibly, Optimal group size is unstable. Animal Behav. (1983). https://doi.org/10.1016/S0003-3472(83)80250-4
734. A. Sion, A. Reina, M. Birattari, E. Tuci, Controlling robot swarm aggregation through a minority of informed robots, in *International Conference on Swarm Intelligence* (Springer, Berlin, 2022), pp. 91–103
735. F. Sivrikaya, B. Yener, Time synchronization in sensor networks: a survey. IEEE Netw. **18**(4), 45–50 (2004)
736. I. Slavkov, D. Carrillo-Zapata, N. Carranza, X. Diego, F. Jansson, J. Kaandorp, S. Hauert, J. Sharpe, Morphogenesis in robot swarms. Sci. Robot. **3**(25), eaau9178 (2018)
737. R.G. Smith, The contract net protocol: high-level communication and control in a distributed problem solver. IEEE Trans. Comput. **29**(12), 1104–1113 (1980)
738. V. Sood, S. Redner, Voter model on heterogeneous graphs. Phys. Rev. Lett. **94**(17), 178701 (2005)
739. L. Soriano Marcolino, Y. Tavares dos Passos, A.A. Fonseca de Souza, A. dos Santos Rodrigues, L. Chaimowicz, Avoiding target congestion on the navigation of robotic swarms. Auton. Robots **41**(6), 1297–1320 (2017)
740. D. Sorin, M. Hill, D. Wood, *A Primer on Memory Consistency and Cache Coherence* (Morgan & Claypool Publishers, 2011)
741. O. Soysal, E. Şahin, A macroscopic model for self-organized aggregation in swarm robotic systems, in *Swarm Robotics—Second SAB 2006 International Workshop*, ed. by E. Şahin, W.M. Spears, A.F.T. Winfield. Lecture Notes in Computer Science, vol. 4433 (Springer, Berlin, 2007), pp. 27–42
742. V. Springel, S.D.M. White, A. Jenkins, C.S. Frenk, N. Yoshida, L. Gao, J. Navarro, R. Thacker, D. Croton, J. Helly, J.A. Peacock, S. Cole, P. Thomas, H. Couchman, A. Evrard, J. Colberg, F. Pearce, Simulations of the formation, evolution and clustering of galaxies and quasars. Nature **435**, 629–636 (2005). http://www.nature.com/nature/journal/v435/n7042/full/nature03597.html
743. V.H. Sridhar, L. Li, D. Gorbonos, M. Nagy, B.R. Schell, T. Sorochkin, N.S. Gov, I.D. Couzin, The geometry of decision-making in individuals and collectives. Proc. Natl. Acad. Sci. **118**(50), e2102157118 (2021)

744. K.O. Stanley, R. Miikkulainen, Evolving neural networks through augmenting topologies. Evol. Comput. **10**, 99–127 (2002)
745. P.M. Stella, R.C. Ewell, J.J. Hoskin, Design and performance of the MER (Mars Exploration Rovers) solar arrays, in *Conference Record of the Thirty-first IEEE Photovoltaic Specialists Conference* (IEEE, 2005), pp. 626–630
746. A. Stephan, *Emergenz: Von der Unvorhersagbarkeit zur Selbstorganisation* (Dresden University Press, Dresden, Munich, 1999)
747. P.A. Stephens, W.J. Sutherland, R.P. Freckleton, What is the Allee effect? Oikos 185–190 (1999)
748. S. Stepney, F. Polack, H. Turner, Engineering emergence, in *CEC 2006: 11th IEEE International Conference on Engineering of Complex Computer Systems* (IEEE Press, Los Alamitos, CA, 2006)
749. R. Stern, N. Sturtevant, A. Felner, S. Koenig, H. Ma, T. Walker, J. Li, D. Atzmon, L. Cohen, T.K. Kumar, et al., Multi-agent pathfinding: definitions, variants, and benchmarks, in *Proceedings of the International Symposium on Combinatorial Search*, vol. 10 (2019), pp. 151–158
750. P. Stone, R.S. Sutton, S. Singh, Reinforcement learning for 3 versus 2 keepaway, in *RoboCup 2000: Robot Soccer World Cup IV*. ed. by P. Stone, T. Balch, G. Kraetzschmar (Springer, Berlin, 2001), pp. 249–258. ISBN 978-3-540-45324-6
751. K. Støy, R. Nagpal, Self-repair through scale independent self-reconfiguration, in *Proceedings. 2004 IEEE/RSJ International Conference on Intelligent Robots and Systems, 2004. (IROS 2004)*, vol. 2 (IEEE, 2004), pp. 2062–2067
752. V. Strobel, E. Castelló Ferrer, M. Dorigo, Managing Byzantine robots via Blockchain technology in a swarm robotics collective decision-making scenario, in *Proceedings of the 17th International Conference on Autonomous Agents and Multiagent Systems (AAMAS)* (International Foundation for Autonomous Agents and Multiagent Systems (IFAAMAS), 2018), pp. 541–549
753. S.H. Strogatz, Spontaneous synchronization in nature. In *Proceedings of International Frequency Control Symposium* (IEEE, 1997), pp. 2–4
754. S.H. Strogatz, Exploring complex networks. Nature **410**(6825), 268–276 (2001). http://www.nature.com/nature/journal/v410/n6825/abs/410268a0.html
755. S.H. Strogatz, D.M. Abrams, A. McRobie, B. Eckhardt, E. Ott, Theoretical mechanics: crowd synchrony on the millennium bridge. Nature **438**(7064), 43–44 (2005)
756. D. Strömbom, R.P. Mann, A.M. Wilson, S. Hailes, A. Jennifer Morton, D.J.T. Sumpter, A.J. King, Solving the shepherding problem: heuristics for herding autonomous, interacting agents. J. R. Soc. Interface **11**(100), 20140719 (2014)
757. K. Sugawara, M. Sno, Cooperative acceleration of task performance: foraging behavior of interacting multi-robots system. Physica D **100**, 343–354 (1997)
758. K. Sugawara, T. Kazama, T. Watanabe, Foraging behavior of interacting robots with virtual pheromone, in *Proceedings of 2004 IEEE/RSJ International Conference on Intelligent Robots and Systems* (IEEE Press, Los Alamitos, CA, 2004), pp. 3074–3079
759. G. Sun, R. Zhou, Z. Ma, Y. Li, R. Groß, Z. Chen, S. Zhao, Mean-shift exploration in shape assembly of robot swarms. Nat. Commun. **14**(1), 3476 (2023)
760. Q. Sun, W. Qi, H. Liu, X. Ji, H. Qian, Toward long-term sailing robots: state of the art from energy perspectives. Front. Robot. AI **8**, 787253 (2022)
761. Z. Sun, A. Feng, J. Yu, W. Zhao, Y. Huang, Development of autonomous sailboat sails and future perspectives: a review. Renew. Sustain. Energy Rev. **207**, 114918 (2025). ISSN 1364-0321, https://doi.org/10.1016/j.rser.2024.114918, https://www.sciencedirect.com/science/article/pii/S1364032124006440
762. D.K. Sutantyo, S. Kernbach, P. Levi, V.A. Nepomnyashchikh, Multi-robot searching algorithm using Lévy flight and artificial potential field, in *2010 IEEE Safety Security and Rescue Robotics* (IEEE, 2010), pp. 1–6
763. R.S. Sutton, A.G. Barto, *Reinforcement Learning: An Introduction* (MIT Press, Cambridge, MA, USA, 1998)

764. R. Suzuki, A. Karim, T. Xia, H. Hedayati, N. Marquardt, Augmented reality and robotics: A survey and taxonomy for ar-enhanced human-robot interaction and robotic interfaces, in *Proceedings of the CHI Conference on Human Factors in Computing Systems* (2022), pp. 1–33
765. K. Sznajd-Weron, J. Sznajd, Opinion evolution in closed community. Int. J. Mod. Phys. C **11**(06), 1157–1165 (2000)
766. M. Szopek, T. Schmickl, R. Thenius, G. Radspieler, K. Crailsheim, Dynamics of collective decision making of honeybees in complex temperature fields. PLoS ONE **8**(10), e76250 (2013). https://doi.org/10.1371/journal.pone.0076250
767. J. Szpirer, D. Garzón Ramos, M. Birattari, Automatic design of robot swarms that perform composite missions: an approach based on inverse reinforcement learning, in *IEEE/RSJ International Conference on Intelligent Robots and Systems (IROS)* (IEEE, 2024), pp. 5791–5798
768. K. Szwaykowska, L. Mier-y-Teran Romero, I.B. Schwartz, Collective motions of heterogeneous swarms. IEEE Trans. Autom. Sci. Eng. **12**(3), 810–818 (2015)
769. J. Tabony, D. Job, Gravitational symmetry breaking in microtubular dissipative structures. Proc. Natl. Acad. Sci. **89**(15), 6948–6952 (1992)
770. M.S. Talamali, A. Saha, J.A.R. Marshall, A. Reina, When less is more: robot swarms adapt better to changes with constrained communication. Sci. Robot. **6**(56), eabf1416 (2021)
771. S. Tao, F. Xiang, A. Shukla, Y. Qin, X. Hinrichsen, X. Yuan, C. Bao, X. Lin, Y. Liu, T.-K. Chan, et al., Maniskill3: GPU parallelized robotics simulation and rendering for generalizable embodied AI (2024). arXiv:2410.00425
772. D. Tarapore, P.U. Lima, J. Carneiro, A. Lyhne Christensen, To err is robotic, to tolerate immunological: fault detection in multirobot systems. Bioinspiration Biomimetics **10**(1), 016014 (2015)
773. D. Tarapore, A. Lyhne Christensen, J. Timmis, Generic, scalable and decentralized fault detection for robot swarms. PLOS ONE **12**(8), 1–29 (2017). https://doi.org/10.1371/journal.pone.0182058
774. D. Tarapore, R. Groß, K.-P. Zauner, Sparse robot swarms: moving swarms to real-world applications. Front. Robot. AI **7**, 83 (2020)
775. C. Taylor, C. Nowzari, The impact of catastrophic collisions and collision avoidance on a swarming behavior. Robot. Auton. Syst. **140**, 103754 (2021). ISSN 0921-8890, https://doi.org/10.1016/j.robot.2021.103754, https://www.sciencedirect.com/science/article/pii/S0921889021000397
776. K. Teknomo, Application of microscopic pedestrian simulation model. Transp. Res. F: Traffic Psychol. Behav. **9**(1), 15–27 (2006)
777. E.W. Tekwa, M.A. Whalen, P.T. Martone, M.I. O'Connor, Theory and application of an improved species richness estimator. Philos. Trans. R. Soc. B **378**(1881), 20220187 (2023)
778. P. Thalamy, B. Piranda, J. Bourgeois, A survey of autonomous self-reconfiguration methods for robot-based programmable matter. Robot. Auton. Syst. **120**, 103242 (2019)
779. R. Nagpal et al., The Kilobot Project, Self-organizing systems research group. website (2013). https://ssr.seas.harvard.edu/kilobots
780. R. Thenius, D. Moser, J. Cherian Varughese, S. Kernbach, I. Kuksin, O. Kernbach, E. Kuksina, N. Mišković, S. Bogdan, T. Petrović, et al., SubCULTron-cultural development as a tool in underwater robotics, in *Artificial Life and Intelligent Agents Symposium* (Springer, Berlin, 2016), pp. 27–41
781. G. Theraulaz, E. Bonabeau, Coordination in distributed building. Science **269**, 686–688 (1995a)
782. G. Theraulaz, E. Bonabeau, Modelling the collective building of complex architectures in social insects with lattice swarms. J. Theor. Biol. **177**, 381–400 (1995b)
783. G. Theraulaz, E. Bonabeau, S.C. Nicolis, R.V. Solé, V. Fourcassié, S. Blanco, R. Fournier, J.-L. Joly, P. Fernández, A. Grimal, P. Dalle, J.-L. Deneubourg, Spatial patterns in ant colonies. Proc. Natl. Acad. Sci. **99**(15), 9645–9649 (2002)
784. D.W. Thompson, *On Growth and Form: The Complete Revised Edition* (Cambridge University Press, 1917)

785. S. Thrun, W. Burgard, D. Fox, *Probabilistic Robotics* (MIT Press, 2005)
786. J. Toner, Y. Tu, Flocks, herds, and schools: a quantitative theory of flocking. Phys. Rev. E **58**(4), 4828–4858 (1998)
787. V. Trianni, *Evolutionary Swarm Robotics—Evolving Self-Organising Behaviours in Groups of Autonomous Robots, Studies in Computational Intelligence*, vol. 108 (Springer, Berlin, Germany, 2008)
788. V. Trianni, R. Groß, T.H. Labella, E. Şahin, M. Dorigo, Evolving aggregation behaviors in a swarm of robots, in *Advances in Artificial Life (ECAL 2003)*, ed. by W. Banzhaf, J. Ziegler, T. Christaller, P. Dittrich, J.T. Kim. Lecture Notes in Artificial Intelligence, vol. 2801 (Springer, Berlin, 2003), pp. 865–874
789. V. Trianni, T.H. Labella, M. Dorigo, Evolution of direct communication for a swarm-bot performing hole avoidance, in *Ant Colony Optimization and Swarm Intelligence (ANTS 2004)*, ed. by M. Dorigo, M. Birattari, C. Blum, L. Maria Gambardella, F. Mondada, T. Stützle. Lecture Notes in Computer Science, vol. 3172 (Springer, Berlin, 2004), pp. 130–141
790. V. Trianni, S. Nolfi, M. Dorigo, Cooperative hole avoidance in a swarm bot. Robot. Auton. Syst. **54**(2), 97–103 (2006). ISSN 0921-8890, https://doi.org/10.1016/j.robot.2005.09.018, https://www.sciencedirect.com/science/article/pii/S0921889005001478. Intelligent Autonomous Systems
791. V. Trianni, J. IJsselmuiden, R. Haken, et al., The SAGA concept: Swarm robotics for agricultural applications. Technical report, National Research Council (CNR), 2016. Available at https://laral.istc.cnr.it/saga/
792. E. Tuci, R. Groß, V. Trianni, F. Mondada, M. Bonani, M. Dorigo, Cooperation through self-assembly in multi-robot systems. ACM Trans. Auton. Adapt. Syst. (TAAS) **1**(2), 115–150 (2006)
793. E. Tuci, M.H.M. Alkilabi, O. Akanyeti, Cooperative object transport in multi-robot systems: a review of the state-of-the-art. Front. Robot. AI **5**, 59 (2018)
794. A. Turgut, H. Çelikkanat, F. Gökçe, E. Şahin, Self-organized flocking in mobile robot swarms. Swarm Intell. **2**(2), 97–120 (2008)
795. A.M. Turing, The chemical basis of morphogenesis. Philos. Trans. R. Soc. Lond. Ser. B, Biol. Sci. **B237**(641), 37–72 (1952)
796. C.J. Turner, R. Ma, J. Chen, J. Oyekan, Human in the loop: Industry 4.0 technologies and scenarios for worker mediation of automated manufacturing. IEEE Access **9**, 103950–103966 (2021). https://doi.org/10.1109/ACCESS.2021.3099311
797. P. Twu, Y. Mostofi, M. Egerstedt, A measure of heterogeneity in multi-agent systems, in *American Control Conference* (2014), pp. 3972–3977
798. A. Tyrrell, G. Auer, C. Bettstetter, Fireflies as role models for synchronization in ad hoc networks, in *Proceedings of the 1st International Conference on Bio-Inspired Models of Network, Information and Computing Systems* (ACM, 2006)
799. G.E. Uhlenbeck, L.S. Ornstein, On the theory of the Brownian motion. Phys. Rev. **36**, 823–841 (1930)
800. G. Valentini, *Achieving Consensus in Robot Swarms: Design and Analysis of Strategies for the best-of-n Problem* (Springer, 2017). ISBN 978-3-319-53608-8, https://doi.org/10.1007/978-3-319-53609-5
801. G. Valentini, H. Hamann, M. Dorigo, Self-organized collective decision making: the weighted voter model, in *Proceedings of the 13th International Conference on Autonomous Agents and Multiagent Systems (AAMAS 2014)*. ed. by A. Lomuscio, P. Scerri, A. Bazzan, M. Huhns (IFAAMAS, 2014), pp. 45–52
802. G. Valentini, H. Hamann, M. Dorigo, Self-organized collective decisions in a robot swarm, in *AAAI-15 Video Proceedings* (AAAI Press, 2015a). https://www.youtube.com/watch?v=5lz_HnOLBW4
803. G. Valentini, H. Hamann, M. Dorigo, Efficient decision-making in a self-organizing robot swarm: on the speed versus accuracy trade-off, in *Proceedings of the 14th International Conference on Autonomous Agents and Multiagent Systems (AAMAS 2015)*. ed. by R. Bordini, E. Elkind, G. Weiss, P. Yolum (IFAAMAS, 2015b), pp. 1305–1314. http://dl.acm.org/citation.cfm?id=2773319

804. G. Valentini, D. Brambilla, H. Hamann, M. Dorigo, Collective perception of environmental features in a robot swarm, in *10th International Conference on Swarm Intelligence, ANTS 2016*, Lecture Notes in Computer Science, vol. 9882 (Springer, Berlin, 2016a), pp. 65–76
805. G. Valentini, E. Ferrante, H. Hamann, M. Dorigo, Collective decision with 100 Kilobots: speed versus accuracy in binary discrimination problems. J. Auton. Agents Multi-Agent Syst. **30**(3), 553–580 (2016b). https://doi.org/10.1007/s10458-015-9323-3
806. G. Valentini, E. Ferrante, M. Dorigo, The best-of-n problem in robot swarms: Formalization, state of the art, and novel perspectives. Front. Robot. AI **4**, 9 (2017). ISSN 2296-9144, https://doi.org/10.3389/frobt.2017.00009, http://journal.frontiersin.org/article/10.3389/frobt.2017.00009
807. G. Valentini, A. Antoun, M. Trabattoni, B. Wiandt, Y. Tamura, E. Hocquard, V. Trianni, M. Dorigo, Kilogrid: a novel experimental environment for the kilobot robot. Swarm Intell. **12**(3), 245–266 (2018)
808. J. van den Berg, S.J. Guy, M. Lin, D. Manocha, Reciprocal n-body collision avoidance, in *Robotics Research*. ed. by C. Pradalier, R. Siegwart, G. Hirzinger (Springer, Berlin, 2011), pp. 3–19. ISBN 978-3-642-19457-3
809. A. van den Oord, O. Vinyals, K. Kavukcuoglu, Neural discrete representation learning, in *Proceedings of the 31st International Conference on Neural Information Processing Systems, NIPS'17* (Curran Associates Inc., Red Hook, NY, USA, 2017), pp. 6309–6318. ISBN 9781510860964
810. S. Van Havermaet, P. Simoens, T. Landgraf, Y. Khaluf, Steering herds away from dangers in dynamic environments. R. Soc. Open Sci. **10**(5), 230015 (2023)
811. S. Van Havermaet, Y. Khaluf, P. Simoens, Reactive shepherding along a dynamic path. Sci. Rep. **14**(1), 14915 (2024)
812. N.G. van Kampen, *Stochastic Processes in Physics and Chemistry* (North-Holland, Amsterdam, 1981)
813. F. van Wageningen-Kessels, H. Van Lint, K. Vuik, S. Hoogendoorn, Genealogy of traffic flow models. EURO J. Transp. Logistics **4**(4), 445–473 (2015)
814. A. van Wynsberghe, J. Donhauser, The dawning of the ethics of environmental robots. Sci. Eng. Ethics **24**, 1777–1800 (2018)
815. D. Vanderelst, A. Winfield, The dark side of ethical robots, in *Proceedings of the 2018 AAAI/ACM Conference on AI, Ethics, and Society* (2018), pp. 317–322
816. A. Vardy, Robot distancing: planar construction with lanes, in *International Conference on Swarm Intelligence (ANTS)* (Springer, 2020), pp. 229–242
817. A. Vardy, The lasso method for multi-robot foraging, in *19th Conference on Robots and Vision (CRV)* (2022), pp. 106–113. https://doi.org/10.1109/CRV55824.2022.00022
818. A. Vardy, The swarm within the labyrinth: planar construction by a robot swarm. Artif. Life Robot. **28**(1), 117–126 (2023)
819. J. Cherian Varughese, H. Hornischer, P. Zahadat, R. Thenius, F. Wotawa, T. Schmickl, A swarm design paradigm unifying swarm behaviors using minimalistic communication. Bioinspiration Biomimetics **15**(3), 036005 (2020). https://doi.org/10.1088/1748-3190/ab6ed9
820. R. Vaughan, Massively multi-robot simulation in stage. Swarm Intell. **2**, 189–208 (2008)
821. M. Vergassola, E. Villermaux, B.I. Shraiman, 'infotaxis' as a strategy for searching without gradients. Nature **445**(7126), 406–409 (2007)
822. P. Vestartas, M. Katherine Heinrich, M. Zwierzycki, D. Andres Leon, A. Cheheltan, R. La Magna, P. Ayres, Design tools and workflows for braided structures, in *Humanizing Digital Reality: Design Modelling Symposium Paris 2017*. ed. by K. De Rycke, C. Gengnagel, O. Baverel, J. Burry, C. Mueller, M. Man Nguyen, P. Rahm, M. Ramsgaard Thomsen (Springer, Singapore, 2018), pp. 671–681. ISBN 978-981-10-6611-5, https://doi.org/10.1007/978-981-10-6611-5_55
823. T. Vicsek, A. Zafeiris, Collective motion. Phys. Rep. **517**(3–4), 71–140 (2012)
824. T. Vicsek, A. Czirók, E. Ben-Jacob, I. Cohen, O. Shochet, Novel type of phase transition in a system of self-driven particles. Phys. Rev. Lett. **75**(6), 1226–1229 (1995)

825. M. Vigelius, B. Meyer, G. Pascoe, Multiscale modelling and analysis of collective decision making in swarm robotics. PLOS ONE **9**(11), 1–19 (2014). https://doi.org/10.1371/journal.pone.0111542
826. V. Vijay, K.A. Pant, M. Cho, Y. Guo, J.M. Goppert, I. Hwang, Range-based multi-robot integrity monitoring for cyberattacks and faults: an anchor-free approach. IEEE Robot. Autom. Lett. **10**(3), 2630–2637 (2025). https://doi.org/10.1109/LRA.2025.3534068
827. V. Villani, B. Capelli, C. Secchi, C. Fantuzzi, L. Sabattini, Humans interacting with multi-robot systems: a natural affect-based approach. Auton. Robot. **44**, 601–616 (2020)
828. D. Vinković, A. Kirman, A physical analogue of the Schelling model. Proc. Natl. Acad. Sci. **103**(51), 19261–19265 (2006)
829. M. Viroli, J. Beal, F. Damiani, G. Audrito, R. Casadei, D. Pianini, From distributed coordination to field calculus and aggregate computing. J. Logical Algebraic Methods Program. **109**, 100486 (2019)
830. M.L. Visinsky, J.R. Cavallaro, I.D. Walker, Robotic fault detection and fault tolerance: a survey. Reliab. Eng. Syst. Saf. **46**(2), 139–158 (1994)
831. G.M. Viswanathan, V. Afanasyev, S.V. Buldyrev, E.J. Murphy, P.A. Prince, H. Eugene Stanley, Lévy flight search patterns of wandering albatrosses. Nature **381**(6581), 413–415 (1996)
832. G.M. Viswanathan, M.G.E. Da Luz, E.P. Raposo, H. Eugene Stanley, *The Physics of Foraging: An Introduction to Random Searches and Biological Encounters* (Cambridge University Press, 2011)
833. K. von Frisch, *Animal Architecture* (Harcourt, 1974)
834. J. von Neumann, *The Theory of Self-Reproducing Automata*, in ed. by A. Burks (University of Illinois Press, Champaign, IL, 1966)
835. M. von Smoluchowski, Zur kinetischen Theorie der Brownschen Molekularbewegung und der Suspensionen. Ann. Phys. **21**, 756–780 (1906)
836. P.W. Voorhees, The theory of Ostwald ripening. J. Stat. Phys. **38**(1), 231–252 (1985)
837. A. Šošić, W.R. KhudaBukhsh, A.M. Zoubir, H. Koeppl, Inverse reinforcement learning in swarm systems, in *Proceedings of the 16th Conference on Autonomous Agents and Multi-Agent Systems* (International Foundation for Autonomous Agents and Multiagent Systems (AAMAS), 2017), pp. 1413–1421
838. M. Wahby, A. Weinhold, H. Hamann, Revisiting BEECLUST: aggregation of swarm robots with adaptiveness to different light settings, in *Proceedings of the 9th EAI International Conference on Bio-inspired Information and Communications Technologies (BICT 2015)* (ICST, 2016), pp. 272–279. https://doi.org/10.4108/eai.3-12-2015.2262877
839. M. Wahby, M. Katherine Heinrich, D. Nicolas Hofstadler, E. Neufeld, I. Kuksin, P.Zahadat, T. Schmickl, P. Ayres, H. Hamann, Autonomously shaping natural climbing plants: a bio-hybrid approach. R. Soc. Open Sci. **5**, 180296 (2018). https://doi.org/10.1098/rsos.180296
840. M. Wahby, J. Petzold, C. Eschke, T. Schmickl, H. Hamann, Collective change detection: Adaptivity to dynamic swarm densities and light conditions in robot swarms, in *Artificial Life Conference Proceedings* (MIT Press, 2019), pp. 642–649
841. M. Waibel, B. Keays, F. Augugliaro, *Drone shows: Creative Potential and Best Practices*. Technical report, ETH Zurich, 2017
842. D. Wälchli, P. Weber, M. Chatzimanolakis, R. Katzschmann, P. Koumoutsakos, Inverse reinforcement learning for objective discovery in collective behavior of artificial swimmers. Phys. Rev. Fluids **10**(6), 064901 (2025)
843. M. Walker, T. Phung, T. Chakraborti, T. Williams, D. Szafir, Virtual, augmented, and mixed reality for human-robot interaction: a survey and virtual design element taxonomy. ACM Trans. Human-Robot Interaction **12**(4), 1–39 (2023)
844. V. Walter, V. Spurnỳ, M. Petrlík, T. Báca, D. Zaitlík, L.I. Demkiv, M. Saska, Extinguishing real fires by fully autonomous multirotor UAVs in the MBZIRC 2020 competition. Field Robot. **2**(1), 406–436 (2022)
845. D. Wang, K. Ohnishi, X. Weiliang, Multimodal haptic display for virtual reality: a survey. IEEE Trans. Industr. Electron. **67**(1), 610–623 (2020). https://doi.org/10.1109/TIE.2019.2920602

846. H. Wang, M. Rubenstein, A fast, accurate, and scalable probabilistic sample-based approach for counting swarm size, in *IEEE International Conference on Robotics and Automation (ICRA)* (IEEE, 2020), pp. 7180–7185
847. K. Wardega, M. von Hippel, R. Tron, C. Nita-Rotaru, W. Li, Hola robots: Mitigating plan-deviation attacks in multi-robot systems with co-observations and horizon-limiting announcements, in *Proceedings of the 2023 International Conference on Autonomous Agents and Multiagent Systems* (International Foundation for Autonomous Agents and Multiagent Systems, AAMAS, Richland, SC, 2023), pp. 2553–2555. ISBN 9781450394321
848. C.J.C.H. Watkins, P. Dayan, Q-learning. Mach. Learn. **8**(3), 279–292 (1992)
849. R.A. Watson, S.G. Ficici, J.B. Pollack, Embodied evolution: Distributing an evolutionary algorithm in a population of robots. Robot. Auton. Syst. **39**(1), 1–18 (2002). ISSN 0921-8890
850. D.J. Watts, S.H. Strogatz, Collective dynamics of 'small-world' networks. Nature **393**(6684), 440–442 (1998)
851. B. Webb, Can robots make good models of biological behaviour? Behav. Brain Sci. **24**, 1033–1050 (2001)
852. B. Webb, Robots in invertebrate neuroscience. Nature **417**, 359–363 (2002)
853. B. Webb, T. Scutt, A simple latency-dependent spiking-neuron model of cricket phonotaxis. Biol. Cybern. **82**, 247–269 (2000)
854. S. Weinberg, Reductionism redux. N. Y. Rev. Books **42**(15), 5 (1995)
855. C.B. Weinstock, J.B. Goodenough, On system scalability. Technical Note CMU/SEI-2006-TN-012, Carnegie Mellon University, Software Engineering Institute, 2006
856. H. Wells, P.H. Wells, P. Cook, The importance of overwinter aggregation for reproductive success of monarch butterflies (*Danaus plexippus L.*). J. Theor. Biol. **147**(1), 115–131 (1990). ISSN 0022-5193. https://doi.org/10.1016/S0022-5193(05)80255-3, http://www.sciencedirect.com/science/article/pii/S0022519305802553
857. J. Werfel, K. Petersen, R. Nagpal, Designing collective behavior in a termite-inspired robot construction team. Science **343**(6172), 754–758 (2014)
858. S. Whiteson, N. Kohl, R. Miikkulainen, P. Stone, Evolving keepaway soccer players through task decomposition, in *Genetic and Evolutionary Computation-GECCO 2003* (Springer, 2003), pp. 201–212
859. S. Whiteson, P. Stone, Evolutionary function approximation for reinforcement learning. J. Mach. Learn. Res. **7**, 877–917 (2006)
860. R.J. Williams, Simple statistical gradient-following algorithms for connectionist reinforcement learning. Mach. Learn. **8**(3), 229–256 (1992)
861. J. Wilson, G. Chance, P. Winter, S. Lee, E. Milner, D. Abeywickrama, S. Windsor, J. Downer, K. Eder, J. Ives, et al., Trustworthy swarms, in *Proceedings of the First International Symposium on Trustworthy Autonomous Systems* (2023)
862. M. Wilson, C. Melhuish, A.B. Sendova-Franks, S. Scholes, Algorithms for building annular structures with minimalist robots inspired by brood sorting in ant colonies. Auton. Robot. **17**, 115–136 (2004)
863. R.P. Wilson, F. Quintana, V.J. Hobson, Construction of energy landscapes can clarify the movement and distribution of foraging animals. Proc. R. Soc. B: Biol. Sci. **279**(1730), 975–980 (2012)
864. S. Wilson, T.P. Pavlic, G.P. Kumar, A. Buffin, S.C. Pratt, S. Berman, Design of ant-inspired stochastic control policies for collective transport by robotic swarms. Swarm Intell. **8**(4), 303–327 (2014)
865. S. Wilson, R. Gameros, M. Sheely, M. Lin, K. Dover, R. Gevorkyan, M. Haberland, A. Bertozzi, S. Berman, Pheeno, a versatile swarm robotic research and education platform. IEEE Robot. Autom. Lett. **1**(2), 884–891 (2016)
866. A.F.T. Winfield, J. Nembrini, Safety in numbers: Fault tolerance in robot swarms. Int. J. Model. Ident. Control **1**(1), 30–37 (2006)
867. A.F.T. Winfield, C.J. Harper, J. Nembrini, Towards dependable swarms and a new discipline of swarm engineering, in *International Workshop on Swarm Robotics* (Springer, Berlin, 2004), pp. 126–142

868. A.F.T. Winfield, J. Sa, M.-C. Fernández-Gago, C. Dixon, M. Fisher, On formal specification of emergent behaviours in swarm robotic systems. Int. J. Adv. Robot. Syst. **2**(4), 363–370 (2005). https://doi.org/10.5772/5769
869. A.T. Winfree, Biological rhythms and the behavior of populations of coupled oscillators. J. Theor. Biol. **16**(1), 15–42 (1967)
870. T.A. Witten Jr., L.M. Sander, Diffusion-limited aggregation, a kinetic critical phenomenon. Phys. Rev. Lett. **47**(19), 1400–1403 (1981). https://doi.org/10.1103/PhysRevLett.47.1400
871. M. Wittlinger, R. Wehner, H. Wolf, The ant odometer: stepping on stilts and stumps. Science **312**(5782), 1965–1967 (2006)
872. H. Wolf, Odometry and insect navigation. J. Exp. Biol. **214**(10), 1629–1641 (2011). ISSN 0022-0949, https://doi.org/10.1242/jeb.038570, http://jeb.biologists.org/content/214/10/1629
873. T. De Wolf, T. Holvoet, Emergence versus self-organisation: different concepts but promising when combined, in *Proceedings of the workshop on Engineerings Self Organising Applications*, ed. by S. Brueckner, G. Di Marzo Serugendo, A. Karageorgos, R. Nagpal. Lecture Notes in Computer Science, vol. 3464 (Springer, Berlin, 2005), pp. 1–15
874. L. Wolpert, One hundred years of positional information. Trends Genet. **12**(9), 359–364 (1996). ISSN 0168-9525, http://view.ncbi.nlm.nih.gov/pubmed/8855666
875. M. Robson Wright, *An Introduction to Aqueous Electrolyte Solutions* (Wiley, New York, 2007)
876. X. Xing Chen, J. Huang, Odor source localization algorithms on mobile robots: a review and future outlook. Robot. Auton. Syst. **112**, 123–136 (2019). ISSN 0921-8890, https://doi.org/10.1016/j.robot.2018.11.014, https://www.sciencedirect.com/science/article/pii/S0921889018303014
877. H. Xu, L. Wang, Y. Zhang, K. Qiu, S. Shen, Decentralized visual-inertial-uwb fusion for relative state estimation of aerial swarm, in *IEEE International Conference on Robotics and Automation (ICRA)* (IEEE, 2020a), pp. 8776–8782
878. X. Hao, Y. Zhang, B. Zhou, L. Wang, X. Yao, G. Meng, S. Shen, Omni-swarm: a decentralized omnidirectional visual-inertial-UWB state estimation system for aerial swarms. IEEE Trans. Robot. **38**(6), 3374–3394 (2022)
879. M. Xu, W. Dai, C. Liu, X. Gao, W. Lin, G.-J. Qi, H. Xiong, Spatial-temporal transformer networks for traffic flow forecasting (2020). arXiv:2001.02908
880. H. Yamaguchi, T. Arai, G. Beni, A distributed control scheme for multiple robotic vehicles to make group formations. Robot. Auton. Syst. **36**(4), 125–147 (2001)
881. D. Yamins, Towards a theory of "local to global" in distributed multi-agent systems, in *Proceedings of the Fourth International Joint Conference on Autonomous Agents and Multiagent Systems (AAMAS'05)* (2005), pp. 183–190
882. D. Yamins, *A Theory of Local-to-Global Algorithms for One-Dimensional Spatial Multi-Agent Systems*. Ph.D. thesis, Harvard University, November 2007
883. D. Yamins, R. Nagpal, Automated global-to-local programming in 1-D spatial multi-agent systems, in *Proceedings of 7th International Conference on Autonomous Agents and Multiagent Systems (AAMAS 2008)*. ed. by L. Padgham, D.C. Parkes, J.P. Müller, S. Parsons (Estoril, Portugal, 2008)
884. C.N. Yang, The spontaneous magnetization of a two-dimensional Ising model. Phys. Rev. **85**(5), 808–816 (1952). https://doi.org/10.1103/PhysRev.85.808
885. T. Yasuda, K. Ohkura, reinforcement learning technique with an adaptive action generator for a multi-robot system, in *The tenth International Conference on Simulation of Adaptive Behavior (SAB'08)*, Lecture Notes in Artificial Intelligence, vol. 5040 (Springer, Berlin, 2008), pp. 250–259
886. C.A. Yates, R. Erban, C. Escudero, I.D. Couzin, J. Buhl, I.G. Kevrekidis, P.K. Maini, D.J.T. Sumpter, Inherent noise can facilitate coherence in collective swarm motion. Proc. Natl. Acad. Sci. **106**(14), 5464–5469 (2009). https://doi.org/10.1073/pnas.0811195106, http://www.pnas.org/content/106/14/5464.abstract

887. B. Yu, H. Kasaei, M. Cao, Co-NavGPT: Multi-robot cooperative visual semantic navigation using large language models (2023). arXiv:2310.07937
888. P. Zahadat, T. Schmickl, Division of labor in a swarm of autonomous underwater robots by improved partitioning social inhibition. Adapt. Behav. **24**(2), 87–101 (2016). https://doi.org/10.1177/1059712316633028
889. P.Zahadat, D.J. Christensen, S.D. Katebi, K. Støy, Sensor-coupled fractal gene regulatory networks for locomotion control of a modular snake robot, in *Proceedings of the 10th International Symposium on Distributed Autonomous Robotic Systems (DARS)* (2010), pp. 517–530
890. P. Zahadat, S. Hahshold, R. Thenius, K. Crailsheim, T. Schmickl, From honeybees to robots and back: division of labor based on partitioning social inhibition. Bioinspiration Biomimetics **10**(6), 066005 (2015). https://doi.org/10.1088/1748-3190/10/6/066005
891. P. Zahadat, D. Nicolas Hofstadler, T. Schmickl, Vascular morphogenesis controller: a generative model for developing morphology of artificial structures, in *Proceedings of the Genetic and Evolutionary Computation Conference, GECCO'17* (ACM, New York, NY, USA, 2017), pp. 163–170. ISBN 978-1-4503-4920-8, https://doi.org/10.1145/3071178.3071247
892. A. Zakiev, T. Tsoy, E. Magid, Swarm robotics: remarks on terminology and classification, in *International Conference on Interactive Collaborative Robotics* (Springer, Berlin, 2018), pp. 291–300
893. R. Zakir, M. Dorigo, A. Reina, Miscommunication between robots can improve group accuracy in best-of-n decision-making, in *IEEE/RSJ International Conference on Intelligent Robots and Systems (IROS)* (IEEE, 2024a), pp. 9014–9021
894. R. Zakir, M. Salahshour, M. Dorigo, A. Reina, Heterogeneity can enhance the adaptivity of robot swarms to dynamic environments, in *International Conference on Swarm Intelligence (ANTS)* (Springer, Berlin, 2024b), pp. 112–126
895. J. Zhang, R. Alert, J. Yan, N.S. Wingreen, S. Granick, Active phase separation by turning towards regions of higher density. Nat. Phys. **17**(8), 961–967 (2021)
896. H. Zhao, T. Thrash, M. Kapadia, K. Wolff, C. Hölscher, D. Helbing, V.R. Schinazi, Assessing crowd management strategies for the 2010 Love Parade disaster using computer simulations and virtual reality. J. R. Soc. Interface **17**(167), 20200116 (2020)
897. H. Zheng, J. Panerati, G. Beltrame, A. Prorok, An adversarial approach to private flocking in mobile robot teams. IEEE Robot. Autom. Lett. **5**(2), 1009–1016 (2020)
898. G. Zhou, T. He, S. Krishnamurthy, J.A. Stankovic, Impact of radio irregularity on wireless sensor networks, in *Proceedings of the 2nd International Conference on Mobile Systems, applications, and Services* (ACM, 2004), pp. 125–138
899. S. Zhou, M.J. Phielipp, J.A. Sefair, S.I. Walker, H. Ben Amor, Clone swarms: Learning to predict and control multi-robot systems by imitation, in *IEEE/RSJ International Conference on Intelligent Robots and Systems (IROS)* (2019), pp. 4092–4099. https://doi.org/10.1109/IROS40897.2019.8967824
900. F. Zhu, Y. Ren, F. Kong, H. Wu, S. Liang, N. Chen, W. Xu, F. Zhang, Swarm-LIO: Decentralized swarm LiDAR-inertial odometry, in *IEEE International Conference on Robotics and Automation (ICRA)* (IEEE, 2023), pp. 3254–3260
901. F. Zhu, Y. Ren, L. Yin, F. Kong, Q. Liu, R. Xue, W. Liu, Y. Cai, G. Lu, H. Li, et al., Swarm-LIO2: Decentralized, efficient LiDAR-inertial odometry for UAV swarms. IEEE Trans. Robot. (2024a)
902. P. Zhu, Z. Zeng, W. Yao, W. Dai, H. Lu, Z. Zhou, DVRP-MHSI: Dynamic visualization research platform for multimodal human-swarm interaction (2024b). arXiv:2408.10861
903. R.K.P. Zia, B. Schmittmann, Watching a drunkard for 10 nights: a study of distributions of variances. Am. J. Phys. **71**(9), 859–865 (2003)
904. B.M. Zoss, D. Mateo, Y. Kong Kuan, G. Tokić, M. Chamanbaz, L. Goh, F. Vallegra, R. Bouffanais, D.K.P. Yue, Distributed system of autonomous buoys for scalable deployment and monitoring of large waterbodies. Auton. Robots **42**(8), 1669–1689 (2018)
905. V. Zykov, E. Mytilinaios, M. Desnoyer, H. Lipson, Evolved and designed self-reproducing modular robotics. IEEE Trans. Robot. **23**(2), 308–319 (2007)

Index

H. Hamann, *Swarm Robotics*,
https://doi.org/10.1007/978-3-032-10584-4

X

Z